化工设备英汉图解手册

宋天民 主 编

赵 岩　禹晓伟　袁庆斌
史维良　尹成江　副主编

中国石化出版社

内容提要 Abstract

本书由中国石化、中国石油、中国海油所属石油化工企业、研究院、工程公司和辽宁石油化工大学联合编写，它是已出版发行的《炼油设备英汉图解手册》的姊妹篇；以图示形式讲解各种石油化工设备的结构和主要零部件，同时给出相应的中英文名称，以便读者学习时对照；内容涵盖了石油化工装置中的主要设备，包括反应设备、塔设备、加热炉、换热设备、容器与储罐、泵、压缩机、专用阀门和其它石油化工设备等。作为一本工具书，体现出专、全、精、准、新的特点。

本书可供石油化工企业设备工程技术人员和管理人员使用，也可以供从事石油化工工程设计、贸易的技术人员，以及高等院校相关专业的师生参考。

图书在版编目(CIP)数据

化工设备英汉图解手册/宋天民主编．—北京：
中国石化出版社，2017.1
ISBN 978-7-5114-4370-0

Ⅰ．①化… Ⅱ．①宋… Ⅲ．①石油化工设备-图解
Ⅳ．①TE65-64

中国版本图书馆CIP数据核字(2016)第312240号

未经本社书面授权，本书任何部分不得被复制、抄袭，或者以任何形式或任何方式传播。版权所有，侵权必究。

中国石化出版社出版发行
地址：北京市朝阳区吉市口路9号
邮编：100020 电话：(010)59964500
发行部电话：(010)59964526
http://www.sinopec-press.com
E-mail:press@sinopec.com
北京科信印刷有限公司印刷
全国各地新华书店经销

*

787×1092毫米16开本54.75印张1340千字
2017年1月第1版 2017年1月第1次印刷
定价：198.00元

《化工设备英汉图解手册》编委会

主　　任：宋天民

副 主 任：赵　岩　禹晓伟　袁庆斌　史维良　王子康

编　　委：王百森　尹成江　朱　晔　张国信　张国福　虞永清
　　　　　远战红　魏志刚　褚卫彬　韩庆平　刘晓亮　白　桦

参编人员：（按姓氏笔画排列）

万　龙	万颖胜	尹成江	王　为	王长新	王世栋	王　乐
王　平	王　光	王百森	王达夫	王茂廷	王道立	车晓刚
史维良	任立伟	任名晨	刘开连	刘方臣	刘达彬	刘丽喆
刘春旺	刘　洋	刘盈盈	刘晓亮	刘墨文	孙少杰	孙华山
孙　旭	孙　锐	巩　波	庄海洋	朱立辉	朱志祥	朱冠龙
朱　晔	闫桂艳	阳利锋	何睿丰	冷松原	吴立军	吴茶平
宋天民	宋尔明	宋爱华	张世成	张国信	张国辉	张国福
张联合	张瑞十	李春树	李家奇	李海根	李　涛	李漠盆
杜开宇	杨良卫	杨佩佩	杨胜长	远战红	邵雪梅	闵　涛
陈　怡	周世文	尚　斌	林　海	范根方	金　鑫	段　勇
禹晓伟	胡小青	赵卫东	赵心灵	赵成发	赵志强	赵　佳
赵　岩	徐鑫金	聂　鑫	袁庆斌	郭丽娜	郭维隆	钱银先
高忠原	高　磊	常　亮	章颖顾	黄卫东	黄法武	黄梓友
黄　鑫	彭遂平	舒　文	董晓明	韩　光	韩庆平	韩建荒
韩洪义	鲁　敏	蒲云洲	虞永清	褚卫彬	雷翔光	管建军
魏　冬	魏志刚					

参编单位：（排名不分先后）

辽宁石油化工大学
中国海油惠州炼化分公司
中国石化天津石化分公司
中国石化安庆石化分公司
中国石化巴陵石化分公司
中国石化金陵石化分公司
中国石化齐鲁石化分公司
中国石化燕山石化分公司
中国石化扬子石化公司
中国石化洛阳石化工程公司
中国石油锦州石化分公司
中国石油抚顺石化分公司
中国石油大庆石化分公司
中国石油辽阳石化分公司

前　言

《化工设备英汉图解手册》(以下简称《手册》)是一部全面介绍石油化工设备结构的工具书,是《炼油设备英汉图解手册》的姊妹篇。《手册》以图示形式介绍了石油化工设备的结构和主要零部件。《手册》以石油化工为主,不包含精细化工和无机化工。

《手册》编写的目的在于为石化企业工程技术人员和管理人员快捷掌握石油化工设备的结构提供一部参考书,以中英文对照的形式给出各种设备的结构和主要零部件的名称,是为了提升石油化工企业设备的管理水平和提高设备工程技术人员的专业英语水平。

《手册》在编写过程中,着力收集国内外先进石油化工设备资料,力求使《手册》成为石油化工设备大全。《手册》以图示形式介绍石油化工设备的结构和主要零部件,一目了然,具有实用性;《手册》以石油化工产品为章,将石油化工产品归纳成乙烯、合成树脂、合成橡胶等章节,实属首例,具有创新性;《手册》中有些设备的结构图是编者设计的,具有新颖性;设备结构采用英汉对照标注,便于学习专业英语,具有鲜明的特色。

《手册》共分9章:第1章 乙烯;第2章 合成树脂;第3章 合成橡胶;第4章 合成纤维;第5章 合成原料;第6章 芳烃;第7章 合成氨与尿素;第8章 煤化工;第9章 其它化工设备。

《手册》由辽宁石油化工大学牵头组织编写,中国石化、中国石油、中国海油所属石油化工企业和相关的研究院、工程公司等参加了编写。辽宁石油化工大学刘晓亮、赵成发、聂鑫、邵雪梅、王为和中国石化抚顺石油化工研究院王旭负责英文翻译,中国石油大学魏耀东教授和辽宁石油化工大学赵成发教授负责英文翻译的审定。

《手册》在编写过程中,得到了中国石油化工集团公司王子康、黄志华、李兆斌、周书俭、王光、周培荣、贾鹏林、何承厚、孙新文,中国石油天然气集团公司周敏等专家的具体指导,中国石化出版社对《手册》的编写和出版给予了大力支持和帮助,在此一并表示衷心感谢。

由于编者水平有限,《手册》中会有错误和不妥之处,敬请读者批评指正,以便改进,不胜感谢。

目 录

1 乙烯 Ethylene ……………………………………………………………… (1)
1.1 裂解炉 Cracking furnace ……………………………………………… (1)
1.1.1 斯通-韦伯斯特(S&W)公司的 USC 型裂解炉 USC cracking furnace of Stone & Webster (S&W) ………………………………………… (1)
1.1.2 鲁姆斯(Lummus)公司的 SRT 型裂解炉 SRT cracking furnace of Lummus …………………………………………………………… (4)
1.1.3 TECHNIP(KTI 荷兰)公司的 GK 型裂解炉 GK cracking furnace of TECHNIP KTI ……………………………………………………… (7)
1.1.4 毫秒裂解炉 Kellogg milli-second furnace ………………………… (8)
1.1.5 Pyrocrack 型裂解炉 Pyrocrack cracking furnace ………………… (9)
1.1.6 布朗路特公司的 HSLR 型裂解炉 HSLR cracking furnace of Kellogg Brown & Root ……………………………………………………… (9)
1.1.7 辐射段炉管 Furnace tube in radiation section ……………………… (10)
1.1.8 裂解炉附属设备 Cracking furnace accessories …………………… (11)
1.2 废热锅炉 Waste heat boiler ……………………………………………… (13)
1.2.1 立式废热锅炉 Vertical type waste heat boiler ……………………… (14)
1.2.2 卧式废热锅炉 Horizontal waste heat boiler ………………………… (15)
1.2.3 椭圆形管板废热锅炉 Oval tube plate waste heat boiler …………… (15)
1.2.4 碟形管板废热锅炉 Dish tube plate waste heat boiler ……………… (16)
1.2.5 薄管板废热锅炉 Thin tube plate waste heat boiler ………………… (16)
1.3 压缩机 Compressor ……………………………………………………… (16)
1.3.1 裂解气压缩机 Cracking gas compressor …………………………… (17)
1.3.2 裂解气汽轮机 Cracking gas turbine ………………………………… (19)
1.3.3 丙烯制冷压缩机 Propylene refrigeration compressor ……………… (24)
1.3.4 乙烯制冷压缩机 Ethylene refrigeration compressor ……………… (24)
1.3.5 甲烷膨胀压缩机 Methane expansion compressor …………………… (25)
1.3.6 废碱氧化空压机 Waste alkaline oxidation air compressor ………… (26)
1.3.7 液环压缩机 Liquid ring compressor ………………………………… (27)
1.4 反应设备 Reaction device ……………………………………………… (27)
1.4.1 加氢反应器 Hydrogenation reactor …………………………………… (27)
1.4.2 脱砷反应器 Dearsenication reactor …………………………………… (30)
1.4.3 甲烷化反应器 Methanation reactor …………………………………… (32)

1.4.4 废碱氧化反应器 Waste alkali oxidation reactor ………………………… (33)
1.5 塔设备 Column equipment ……………………………………………………… (34)
 1.5.1 急冷塔 Quench column …………………………………………………… (34)
 1.5.2 汽提塔 Stripping column ………………………………………………… (36)
 1.5.3 精馏塔 Distillation column ……………………………………………… (40)
 1.5.4 洗涤塔 Washing column ………………………………………………… (50)
1.6 干燥设备 Drying equipment …………………………………………………… (52)
 1.6.1 裂解气干燥器 Drier of cracking gas ……………………………………… (52)
 1.6.2 液体干燥器 Liquid drier …………………………………………………… (53)
 1.6.3 氢气干燥器 Hydrogen drying tower …………………………………… (54)
1.7 换热设备 Heat exchange equipment …………………………………………… (55)
 1.7.1 加热器 Heater ……………………………………………………………… (55)
 1.7.2 换热器 Heat exchanger …………………………………………………… (57)
 1.7.3 冷却器及冷凝器 Coolers and condenser ……………………………… (60)
 1.7.4 再沸器 Reboiler …………………………………………………………… (62)
 1.7.5 汽化器 Vaporiser ………………………………………………………… (64)
 1.7.6 冷箱 Cold box ……………………………………………………………… (65)
1.8 罐 Tank ……………………………………………………………………………… (66)
 1.8.1 分离罐 Separation tank …………………………………………………… (66)
 1.8.2 缓冲罐 Buffer tank ………………………………………………………… (68)
 1.8.3 闪蒸罐 Flash tank ………………………………………………………… (70)
 1.8.4 回流罐 Reflux tank ………………………………………………………… (71)
 1.8.5 储罐 Storage tank ………………………………………………………… (72)
1.9 泵 Pump …………………………………………………………………………… (74)
 1.9.1 油泵 Oil pump ……………………………………………………………… (74)
 1.9.2 水泵 Water pump ………………………………………………………… (75)
 1.9.3 烯烃泵 Olefin pump ……………………………………………………… (78)
 1.9.4 碱泵 Alkali pump ………………………………………………………… (78)
 1.9.5 润滑油泵 Lubrication pump ……………………………………………… (79)
 1.9.6 注剂泵 Agent injection pump …………………………………………… (82)
 1.9.7 蒸汽喷射泵 Steam injection pump ……………………………………… (83)
 1.9.8 其它泵 Other pumps ……………………………………………………… (84)
1.10 特种阀门 Special valves ……………………………………………………… (85)
 1.10.1 蝶阀 Butterfly valve ……………………………………………………… (85)
 1.10.2 闸阀 Gate valve …………………………………………………………… (88)
 1.10.3 截止阀 Globe valve ……………………………………………………… (90)
 1.10.4 止回阀 Check valve ……………………………………………………… (92)
 1.10.5 旋塞阀 Plug valve ………………………………………………………… (94)

1.10.6	安全阀 Safety valve	(94)
1.10.7	疏水阀 Traps	(95)
1.10.8	裂解气阀 Cracking gas valve	(96)
1.10.9	防喘振阀 Anti-surge valve	(97)
1.10.10	无冲击止回阀 Non impact check valve	(97)

1.11 其他设备 Other equipments ......(98)
    1.11.1 聚结器 Coalescer ......(98)
    1.11.2 除氧器 Deaerator ......(98)
    1.11.3 膨胀机 Expander ......(99)
    1.11.4 密封 Seal ......(99)

2 合成树脂 Synthetic resin ......(101)

2.1 聚乙烯 Polyethylene ......(101)
    2.1.1 反应设备 Reaction equipment ......(101)
    2.1.2 塔设备 Column equipment ......(108)
    2.1.3 泵 Pump ......(109)
    2.1.4 压缩机 Compressor ......(116)
    2.1.5 挤压造粒机 Extrusion granulator ......(135)
    2.1.6 加料器 Feeder ......(151)
    2.1.7 振动筛 Vibrating screen ......(153)
    2.1.8 催化剂配制槽搅拌器 Catalyst preparation slot stirrer ......(154)
    2.1.9 料仓、干燥器及风机 Feed bin、dryer and blower ......(154)
    2.1.10 其它设备 Other equipments ......(157)

2.2 聚丙烯 Polypropylene ......(160)
    2.2.1 反应设备 Reaction equipment ......(160)
    2.2.2 塔设备 Column equipment ......(163)
    2.2.3 泵 Pump ......(166)
    2.2.4 压缩机 Compressor ......(170)
    2.2.5 挤压造粒机 Extrusion granulator ......(171)
    2.2.6 搅拌混合器 Stirring mixer ......(176)
    2.2.7 其它设备 Other equipments ......(177)

2.3 聚氯乙烯 Polyvinyl chloride ......(179)
    2.3.1 聚合釜 Polymerizers ......(179)
    2.3.2 流化床 Fluid bed ......(180)
    2.3.3 螺旋板换热器 Spiral plate heat exchanger ......(181)
    2.3.4 分离器 Separator ......(181)
    2.3.5 罐 Tank ......(182)
    2.3.6 离心机 Centrifuge ......(185)

2.3.7　振动筛 Vibrating screen …… (187)
2.3.8　振荡加料器 Oscillating feeder …… (188)
2.3.9　特种阀门 Special valves …… (188)
2.4　聚苯乙烯 Polystyrene …… (189)
2.4.1　反应设备 Reaction equipment …… (189)
2.4.2　脱挥器 Devolatilizer …… (191)
2.4.3　加热炉 Heating furnace …… (192)
2.4.4　切胶机 Rubber cutting machine …… (194)
2.4.5　换热设备 Heat exchanger …… (195)
2.4.6　罐 Tank …… (199)
2.4.7　泵 Pump …… (206)
2.4.8　造粒设备 Granulation equipment …… (209)
2.4.9　干燥设备 Drying equipment …… (210)
2.4.10　输送设备 Conveying equipment …… (214)
2.5　丙烯腈-丁二烯-苯乙烯(ABS) Acrylonitrile-butadiene-styrene …… (216)
2.5.1　聚合釜 Polymerizer …… (216)
2.5.2　凝聚罐 Cohesion tank …… (217)
2.5.3　离心脱水机 Centrifugal hydroextractor …… (218)
2.5.4　流化床干燥器 Fluidized-bed dryer …… (218)
2.5.5　挤压造粒机 Extrusion granulator unit …… (219)
2.5.6　计量秤 Measurement scale …… (222)

3　合成橡胶 Synthetic rubber …… (223)
3.1　顺丁橡胶 Butadiene rubber …… (223)
3.1.1　聚合釜 Polymerization kettle …… (223)
3.1.2　塔设备 Column equipment …… (227)
3.1.3　塔顶冷凝器 Butadiene distillation overhead condenser …… (229)
3.1.4　泵 Pump …… (229)
3.1.5　分离设备 Separation equipment …… (232)
3.1.6　脱水干燥设备 Dehydration drying equipment …… (234)
3.1.7　成形包装设备 Forming packaging equipment …… (235)
3.1.8　输送设备 Conveying equipment …… (239)
3.2　丁苯橡胶 Styrene butadiene rubber …… (241)
3.2.1　聚合反应器 Polymerization reactor …… (241)
3.2.2　塔设备 Column …… (242)
3.2.3　换热设备 Heat-exchange equipment …… (244)
3.2.4　泵 Pump …… (245)
3.2.5　氨制冷压缩机 Ammonia refrigerating compressor …… (248)

    3.2.6 其它设备 Other equipment ········· (249)
  3.3 丁基橡胶 Butyl rubber ············· (252)
    3.3.1 反应设备 Reaction equipment ········· (252)
    3.3.2 塔设备 Column equipment ··········· (258)
    3.3.3 换热设备 Heat-exchange equipment ······· (263)
    3.3.4 泵 Pump ··················· (264)
    3.3.5 压缩机 Compressor ············· (266)
    3.3.6 干燥设备 Drying equipment ··········· (269)
    3.3.7 包装设备 Packaging equipment ········· (274)
  3.4 丁腈橡胶 Nitrile rubber ············· (278)
    3.4.1 聚合釜 Polymeric kettle ············· (279)
    3.4.2 丙烯腈汽提塔 Acrylonitrile stripper ······· (282)
    3.4.4 机泵 Blower and pump ············· (284)
    3.4.5 分离设备 Parting device ············· (286)
    3.4.6 脱水挤压机 Dehydration extrusion machine ····· (288)
    3.4.7 包装设备 Forming packaging equipment ····· (290)
  3.5 SBS橡胶 SBS(Styrene-butadiene-styrene)rubber ····· (292)
    3.5.1 聚合釜 Polymerizing pot ············· (293)
    3.5.2 闪蒸罐 Flash drum ··············· (293)
    3.5.3 凝聚釜 Condensation kettle ··········· (294)
    3.5.4 洗涤水罐/洗胶罐 Washing tank/ Washing glue tank ··· (294)
    3.5.5 泵 Pump ··················· (295)
    3.5.6 换热设备 Heat-exchange equipment ······· (298)
    3.5.7 振动输送设备 Vibrating conveyer equipment ····· (299)
    3.5.8 干燥设备 Dehydration drying equipment ····· (300)

4 合成纤维 Synthetic fiber ··············· (301)
  4.1 聚酯 Polyester ··················· (301)
    4.1.1 反应设备 Reactor equipment ··········· (301)
    4.1.2 塔设备 Column equipment ··········· (307)
    4.1.3 料仓 Hopper ··················· (310)
    4.1.4 换热设备 Heat exchange equipment ······· (313)
    4.1.5 切粒设备 Grain-sized dicing equipment ····· (315)
    4.1.6 泵 Pump ··················· (321)
    4.1.7 其它设备 Other equipments ··········· (327)
  4.2 涤纶 Polyester ··················· (333)
    4.2.1 搅拌釜式反应器 Stirred tank reactor ······· (333)

4.2.2 卧式预缩聚釜 Horizontal pre-polycondensation kettle ……（334）
4.2.3 卧式后缩聚釜 Horizontal post-polycondensation kettle ……（335）
4.2.4 纺前设备 Pre-spinning equipment ……（336）
4.2.5 纺丝设备 Spinning equipment ……（339）
4.2.6 纺后设备 Post-spinning equipment ……（352）
4.3 腈纶 Acrylic fibers ……（374）
4.3.1 纺丝原液制备设备 Spinning solution preparation equipment ……（374）
4.3.2 纺丝设备 Spinning-equipment ……（385）
4.3.3 后加工设备 Post-processing equipment ……（394）
4.3.4 溶剂回收设备 Solvent-recovery equipment ……（405）
4.4 锦纶 Nylon ……（406）
4.4.1 熔融聚合设备 Melt polymerization equipment ……（406）
4.4.2 切片设备 Slice equipment ……（411）
4.4.3 纺丝设备 Spinning equipment ……（416）
4.4.4 牵伸加捻设备 Draft twisting device ……（429）
4.4.5 后加工设备 Post-processing equipment ……（433）

5 合成原料 Synthetic materials ……（444）

5.1 精对苯二甲酸（PTA）Purified terephthalic acid（PTA）……（444）
5.1.1 反应设备 Reaction equipment ……（444）
5.1.2 塔设备 Comlumn equipment ……（449）
5.1.3 换热设备 Heat-exchanger equipment ……（454）
5.1.4 加热炉 Heating furnace ……（461）
5.1.5 泵 Pump ……（463）
5.1.6 压缩机 Compressor ……（466）
5.1.7 其它设备 Other equipments ……（470）
5.2 己内酰胺 Caprolactam ……（478）
5.2.1 反应设备 Conversion equipment ……（478）
5.2.2 塔设备 Column equipment ……（481）
5.2.3 离子交换器 Ion exchanger ……（483）
5.2.4 换热设备 Heat-exchange equipment ……（483）
5.2.5 泵 Pump ……（484）
5.2.6 离心机 Centrifugal machine ……（485）
5.3 环己酮 Cyclohexanone ……（486）
5.3.1 反应设备 Reaction equipment ……（486）
5.3.2 塔设备 Column equipment ……（491）
5.3.3 泵 Pump ……（494）
5.3.4 压缩机 Compressor ……（495）

目 录

- 5.4 乙二醇 Ethylene glycol …… (496)
  - 5.4.1 反应设备 Reaction equipment …… (496)
  - 5.4.2 塔设备 Column equipment …… (497)
  - 5.4.3 换热设备 Heat-exchange equipment …… (500)
  - 5.4.4 容器与储罐 Container and tank …… (502)
  - 5.4.5 泵 Pump …… (504)
  - 5.4.6 压缩机 Compressor …… (513)
- 5.5 丙烯腈 Acrylonitrile …… (515)
  - 5.5.1 丙烯腈反应器 Acrylonitnle reactor …… (515)
  - 5.5.2 塔设备 Column equipment …… (516)
  - 5.5.3 容器与储罐 Container and tank …… (520)
  - 5.5.4 换热设备 Heat exchanger …… (526)
  - 5.5.5 工业炉 Industrial furnace …… (530)
  - 5.5.6 泵 Pump …… (532)
  - 5.5.7 压缩机 Compressor …… (534)
- 5.6 聚酰胺 Polyamide …… (537)
  - 5.6.1 反应设备 Reaction equipment …… (537)
  - 5.6.2 塔设备 Column equipment …… (540)
  - 5.6.3 换热设备 Heat exchange equipment …… (543)
  - 5.6.4 泵 Pumps …… (544)
  - 5.6.5 氮气压缩机 Nitrogen compressor …… (548)
  - 5.6.6 预萃取水罐 Preliminary extraction tank …… (550)
  - 5.6.7 其它设备 Other equipments …… (551)
- 5.7 羟胺肟化 Hydroxylamine oximation …… (554)
  - 5.7.1 反应设备 Reaction equipment …… (554)
  - 5.7.2 塔设备 Column equipment …… (559)
  - 5.7.3 换热设备 Heat exchange equipment …… (562)
  - 5.7.4 泵 Pump …… (564)
  - 5.7.5 容器与储罐 Container and tank …… (566)
  - 5.7.6 其它设备 Other equipments …… (568)

6 芳烃 Arene …… (572)

- 6.1 二甲苯分馏 Xylene fractionation …… (574)
  - 6.1.1 白土反应器 Clay reactor …… (574)
  - 6.1.2 塔设备 Column equipment …… (575)
  - 6.1.3 加热炉 Heating furnace …… (583)
  - 6.1.4 换热设备 Heat exchange equipment …… (595)
  - 6.1.5 泵 Pump …… (597)

6.2 吸附分离 Adsorption separation ……………………………………………………… (608)
　　6.2.1 吸附塔 Adsorption column …………………………………………………… (608)
　　6.2.2 分馏塔 Fractionator ……………………………………………………………… (616)
　　6.2.3 回流罐 Reflux tank ……………………………………………………………… (619)
　　6.2.4 换热设备 Heat exchange equipment ……………………………………… (621)
　　6.2.5 泵 Pump ……………………………………………………………………………… (624)
　　6.2.6 特种阀门 Special valves ……………………………………………………… (626)
6.3 歧化-烷基转移 Disproportionation-transalkylation ………………………… (633)
　　6.3.1 歧化反应器 Disproportionation reactor ………………………………… (633)
　　6.3.2 塔设备 Column device ………………………………………………………… (634)
　　6.3.3 歧化循环氢压缩机 Disproportionation recycle hydrogen compressor ……… (635)
　　6.3.4 歧化进料罐 Disproportionation feed tank ……………………………… (641)
　　6.3.5 歧化加热炉 Disproportionation furnace ………………………………… (642)
　　6.3.6 换热设备 Heat exchanger equipment …………………………………… (643)
　　6.3.7 泵 Pump ……………………………………………………………………………… (646)
6.4 异构化 Isomerization …………………………………………………………………… (648)
　　6.4.1 异构化反应器 Isomerization reactor ……………………………………… (648)
　　6.4.2 塔设备 Column equipment …………………………………………………… (649)
　　6.4.3 异构化循环氢压缩机 Isomerization recycle hydrogen compressor ……… (651)
　　6.4.4 异构化分液罐 Isomerization sub tank …………………………………… (652)
　　6.4.5 塔顶换热器 Overhead heat exchanger …………………………………… (653)
　　6.4.6 异构化进料泵 Isomerization feed pump ………………………………… (653)
6.5 芳烃抽提 Aromatics extraction ……………………………………………………… (654)
　　6.5.1 塔设备 Column equipment …………………………………………………… (654)
　　6.5.2 非芳烃蒸馏塔再沸器 Non-aromatic hydrocarbon distillation
　　　　　column reboiler ………………………………………………………………… (656)
　　6.5.3 泵 Pump ……………………………………………………………………………… (657)

7 合成氨与尿素 Synthetic ammonia and Urea ……………………………………… (659)
7.1 合成氨 Synthetic ammonia …………………………………………………………… (659)
　　7.1.1 反应设备 Reaction equipment ……………………………………………… (659)
　　7.1.2 塔设备 Column equipment …………………………………………………… (667)
　　7.1.3 换热设备 Heat exchanger equipment …………………………………… (669)
　　7.1.4 泵 Pump ……………………………………………………………………………… (680)
　　7.1.5 压缩机 Compressor ……………………………………………………………… (686)
　　7.1.6 液氨罐 Ammonia tank ………………………………………………………… (704)
7.2 尿素 Urea …………………………………………………………………………………… (705)
　　7.2.1 反应设备 Reaction equipment ……………………………………………… (705)

目 录

    7.2.2　塔设备 Column equipment ……………………………………………………（707）
    7.2.3　换热设备 Heat-exchange equipment …………………………………………（717）
    7.2.4　泵 Pump ……………………………………………………………………………（719）
    7.2.5　压缩机 Compressor ………………………………………………………………（722）
    7.2.6　其它设备 Other equipments ……………………………………………………（731）

8　煤化工 Coal chemical industry ………………………………………………………（733）
8.1　粉煤化工 Powdered coal chemical industry
    8.1.1　反应设备 Reaction equipment …………………………………………………（733）
    8.1.2　塔设备 Tower equipment ………………………………………………………（736）
    8.1.3　换热设备 Heat-exchange equipment …………………………………………（739）
    8.1.4　泵 Pump ……………………………………………………………………………（743）
    8.1.5　压缩机 Compressor ………………………………………………………………（746）
    8.1.6　其它设备 Other equipments ……………………………………………………（753）
8.2　水煤浆化工 Coal water slurry chemical industry ………………………………（755）
    8.2.1　反应设备 Reaction equipment …………………………………………………（755）
    8.2.2　塔设备 Column equipment ……………………………………………………（760）
    8.2.3　换热设备 Heat-exchange equipment …………………………………………（763）
    8.2.4　容器与罐 Storage tank and vessel ……………………………………………（765）
    8.2.5　特种阀门 Special valves …………………………………………………………（767）
    8.2.6　泵 Pump ……………………………………………………………………………（772）
    8.2.7　磨煤机 Coal pulverizer …………………………………………………………（775）
    8.2.8　真空过滤机 Vacuum filter ………………………………………………………（776）
    8.2.9　煤浆搅拌器 Coal slurry blender ………………………………………………（777）
    8.2.10　捞渣机 Slag conveyor …………………………………………………………（777）
8.3　MTO化工 MTO chemical industry ………………………………………………（778）
    8.3.1　反应与再生设备 Reactor-regenerator system ………………………………（778）
    8.3.2　塔设备 Column equipment ……………………………………………………（796）

9　其它化工设备 Other chemical equipment ………………………………………（798）
9.1　空分 Air separation ……………………………………………………………………（798）
    9.1.1　塔设备 Column ……………………………………………………………………（798）
    9.1.2　净化设备 Purification equipment ………………………………………………（802）
    9.1.3　空气过滤器 Air filter ……………………………………………………………（804）
    9.1.4　加热器 The heater ………………………………………………………………（805）
    9.1.5　汽化器 Vaporizer …………………………………………………………………（807）
    9.1.6　储罐 Storage tanks ………………………………………………………………（809）
    9.1.7　低温泵 Cryogenic pump …………………………………………………………（812）

9.1.8 压缩机 Compressor ………………………………………………………… (815)
9.1.9 冷水机组 Cold water unit ……………………………………………… (819)
9.1.10 膨胀机 Expansion machine ………………………………………… (820)
9.2 双氧水 Hydrogen peroxide …………………………………………………… (825)
9.2.1 塔设备 Column equipment …………………………………………… (825)
9.2.2 过滤设备 Filtration equipment ……………………………………… (830)
9.2.3 换热设备 Heat-exchange equipment ……………………………… (831)
9.2.4 泵 Pump ……………………………………………………………… (832)
9.2.5 压缩机 Compressor …………………………………………………… (836)
9.2.6 尾气吸附设备 Oxidized exhaust gas adsorption equipment …… (837)
9.3 火炬 Flare …………………………………………………………………… (838)
9.3.1 火炬 Flare …………………………………………………………… (839)
9.3.2 火炬气分液罐 Flare gas knock-out drum ………………………… (841)
9.3.3 火炬气水封罐 Flare gas water sealed drum ……………………… (841)
9.4 装车设备 Truck-loading facility …………………………………………… (842)
9.4.1 火车装车设施 Train loading facility ………………………………… (843)
9.4.2 火车装车鹤管 Train loading crane tube …………………………… (844)
9.4.3 汽车装车设施 Auto loading facility ………………………………… (845)
9.4.4 汽车装车鹤管 Auto loading crane tube …………………………… (846)
9.4.5 油气回收装置 Vapor recovery equipment ………………………… (848)
9.5 循环水 Circulating water …………………………………………………… (850)
9.5.1 冷却塔 Cooling column ……………………………………………… (850)
9.5.2 循环水泵 Circulating water pump …………………………………… (850)
9.5.3 多功能水泵控制阀 Multifunctional water pump control valve …… (851)
9.5.4 监测换热器 Monitoring heat exchanger …………………………… (851)
9.5.5 纤维球过滤器 Fiber ball filter ……………………………………… (852)
9.5.6 纤维束过滤器 Fiber bundle filter …………………………………… (853)
9.6 污水处理 Sewage treatment ………………………………………………… (853)
9.6.1 臭氧发生器 Ozonizer ………………………………………………… (853)
9.6.2 气浮机 Flotation machine …………………………………………… (854)
9.6.3 污泥脱水机 Sludge centrifugal dewatering machine …………… (854)
9.6.4 刮泥机 Mud scraper ………………………………………………… (855)
9.6.5 潜水搅拌器 Submersible mixer ……………………………………… (855)
9.6.6 滗水器 Water decanter ……………………………………………… (856)
9.6.7 螺旋输送机 Shaftless screw conveyor ……………………………… (857)

1 乙烯 Ethylene

乙烯是石油化工的基础原料,乙烯的产量、规模和技术水平,标志着一个国家石油化学工业的发展水平。乙烯装置的设备主要由裂解炉、塔设备、反应设备、换热设备、冷箱和成套机组等组成。

Ethylene is the basic raw material of the petrochemical industry. The production capacity, scale and technical level of ethylene mark the level of petroleum chemical industry of a country. Ethylene unit is mainly composed of cracking furnace, column equipment, reaction equipment, heat exchange equipment, cold box, and complete sets of units, etc.

1.1 裂解炉 Cracking furnace

乙烯裂解炉是乙烯生产装置的核心设备,主要作用是把液化石油气、炼厂气、石脑油或加氢尾油等各类原料裂解成乙烯、丙烯、丁二烯和各种副产品。裂解炉由炉体、辐射段、对流段、燃烧器和引风机等组成,其他设备还有急冷换热器、汽包、消音器和清焦罐等。

Ethylene cracking furnace is the key device of the ethylene production unit, whose main function is to process liquefied petroleum gas, refinery gas, naphtha, hydrogenation tail oil and other kinds of raw materials into cracking gas, and eventually into ethylene, propylene, butadiene and various kinds of by-products. The cracking furnace is composed of furnace body, radiation section, convection section, burner and induced draft fan, etc. Other devices include quenching heat exchanger, steam drum, quencher, muffler and coke cans, etc.

1.1.1 斯通－韦伯斯特(S&W)公司的 USC 型裂解炉 USC cracking furnace of Stone & Webster (S&W)

USC 型裂解炉是美国斯通－韦伯斯特(Stone & Webster)公司开发的一种炉型。USC 型裂解炉为单排双辐射立管式裂解炉,辐射段炉管为 M 型、W 型和 U 型。

USC cracking furnace is a kind of furnace developed by the U. S. Stone & Webster, which is a kind of single row, double radiation tube cracking furnace, with M type, W type and U type radiation coil.

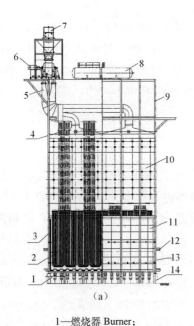
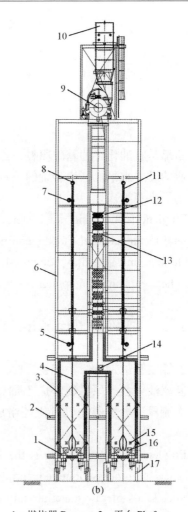

(a)

1—燃烧器 Burner；
2—辐射段炉管 Furnace tube in radiation section；
3—耐火衬里 Refractory liner；
4—换热器 Heat exchanger；
5—烟道 Flue；6—引风机 Induced draft fan；
7—烟囱 Chimney；8—汽包 Steam drum；
9—炉体 Furnace body；
10—对流段 Convection section；
11—辐射段 Radiation section；
12—平台 Platform；13—看火孔 Kiln eye；
14—基础 Base

(b)

1—燃烧器 Burner；2—平台 Platform；
3—耐火衬里 Refractory liner；
4—辐射段炉管 Furnace tube in radiation section；
5—水入口 Water inlet；
6—炉体 Furnace body；
7—水出口 Water outlet；
8—裂解气出口 Cracking gas outlet；
9—引风机 Induced draft fan；
10—烟囱 Chimney；
11—急冷换热器 Quenching heat exchanger；
12—对流室 Convection chamber；
13—对流段炉管 Furnace tube in convection section；
14—横跨段人孔 Manhole in crossover section；
15—看火孔 Kiln eye；16—检修门 Access door；
17—基础 Base

1 乙烯 Ethylene

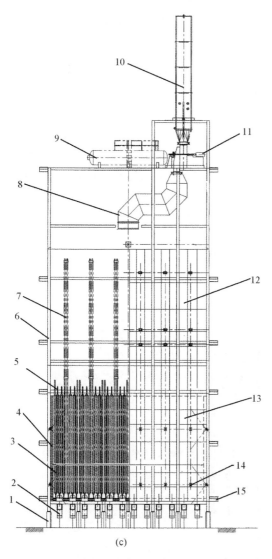

(c)

1—基础 Base；2—燃烧器 Burner；
3—辐射段炉管 Furnace tube in radiation section；
4—耐火衬里 Refractory liner；
5—定位管 Locating pipe；6—炉体 Furnace body；
7—急冷换热器 Quenching heat exchanger；
8—烟道 Flue；9—汽包 Steam drum；
10—烟囱 Chimney；11—引风机 Induced draft fan；
12—对流段 Convection section；
13—辐射段 Radiation section；
14—看火孔 Kiln eye；15—平台 Platform

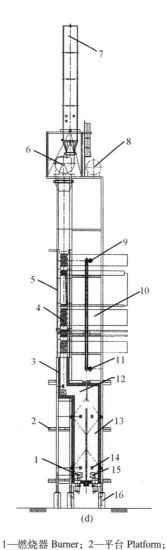

(d)

1—燃烧器 Burner；2—平台 Platform；
3—耐火衬里 Refractory liner；
4—对流段炉管 Furnace tube in convection section；
5—炉体 Furnace body；6—引风机 Induced draft fan；
7—烟囱 Chimney；8—汽包 Steam drum；
9—水出口 Water outlet；
10—对流段 Convection section；
11—水入口 Water inlet；
12—横跨段 Crossover section；
13—辐射段 Radiation section；
14—看火孔 Kiln eye；15—检修门 Access door；
16—基础 Base

图 1.1.1　USC 型裂解炉结构图
Structure drawing of USC cracking furnace

1.1.2 鲁姆斯(Lummus)公司的 SRT 型裂解炉 SRT cracking furnace of Lummus

SRT 型裂解炉是美国鲁姆斯(Lummus)公司开发的一种炉型，是目前世界上大型乙烯装置中应用最多的炉型。SRT 型裂解炉通常采用一个对流段配置一个辐射段，对流段布置在辐射段上部的一侧，对流段顶部设置烟囱和引风机。

SRT cracking furnace is a kind of furnace developed by the U. S. Lummus Corporation, which is the most widely used furnace of large-scale ethylene unit. SRT cracking furnace usually configures a radiation section through a convertion section, with the convection section on the upper side of the radiation section, chimney and induced draft fan on the top of the convection section.

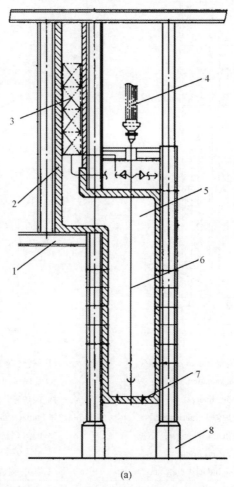

(a)

1—钢结构 Steel structure；2—耐火衬里 Refractory liner；3—对流段 Convection section；
4—急冷换热器 Quenching heat exchanger；5—横跨段 Crossover section；
6—辐射段炉管 Furnace tube in radiation section；7—燃烧器 Burner；8—基础 Base

1 乙烯 Ethylene

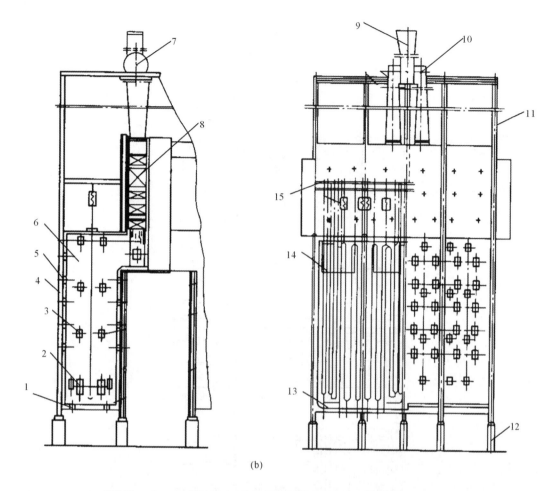

(b)

1—燃烧器 Burner；2—检修门 Access door；3—看火孔 Kiln eye；4—炉墙 Furnace wall；5—侧壁烧嘴 Side wall burner；6—辐射段 Radiation section；7—引风机 Induced draft fan；8—对流段 Convection section；9—烟囱 Chimney；10—烟道 Flue；11—炉体 Furnace body；12—基础 Base；13—导向管 Guide tube；14—辐射段炉管 Furnace tube in radiation section；15—弹簧吊架 Spring hanger

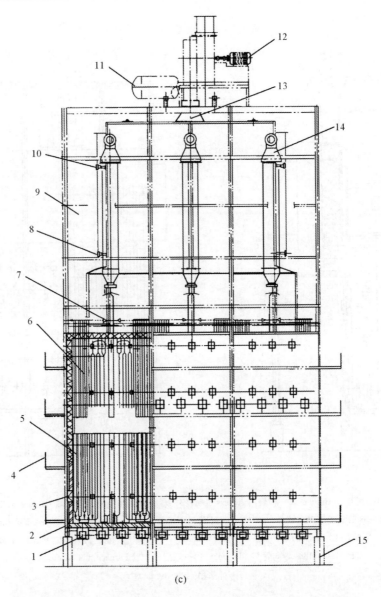

(c)

1—燃烧器 Burner；2—炉体 Furnace body；3—看火孔 Kiln eye；4—平台 Platform；
5—辐射段炉管 Furnace tube in radiation section；6—辐射段 Radiation section；
7—集合管 Concentrated pipe；8—水入口 Water inlet；9—对流段 Convection section；
10—水出口 Water outlet；11—汽包 Steam drum；12—引风机 Induced draft fan；
13—烟道 Flue；14—急冷换热器 Quenching heat exchanger；15—基础 Base

1 乙烯 Ethylene

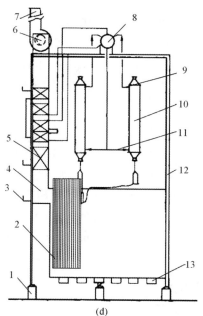

(d)

1—基础 Base；2—辐射段炉管 Furnace tube in radiation section；3—平台 Platform；4—横跨段 Crossover section；5—对流段 Convection section；6—引风机 Induced draft fan；7—烟囱 Chimney；8—汽包 Steam drum；9—水汽出口 Vapor outlet；10—急冷换热器 Quenching heat exchanger；11—水汽入口 Vapor inlet；12—炉体 Furnace body；13—燃烧器 Burner

图 1.1.2　SRT 型裂解炉结构图 Structure drawing of SRT cracking furnace

1.1.3　TECHNIP(KTI 荷兰)公司的 GK 型裂解炉 GK cracking furnace of TECHNIP KTI

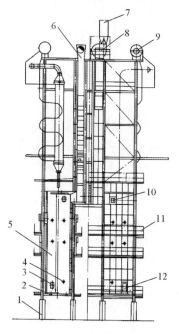

1—基础 Base；2—燃烧器 Burner；3—人孔 Manhole；4—看火孔 Kiln eye；5—辐射段 Radiation section；6—烟道 Flue；7—烟囱 Chimney；8—引风机 Induced draft fan；9—汽包 Drum；10—防爆门 Explosion-proof door；11—平台 Platform；12—检修门 Access door

图 1.1.3　GK 型裂解炉结构图
Structure drawing of GK cracking furnace

· 7 ·

1.1.4 毫秒裂解炉 Kellogg milli-second furnace

毫秒裂解炉是美国凯洛格（Kellogg）公司研究开发的一种炉型。该炉在高裂解温度下，使物料在炉管内的停留时间缩短到 0.05～0.1s(50～100ms)，因此被称为毫秒裂解炉。KBR 裂解炉是凯洛格公司的毫秒裂解炉，简称 SC 炉。目前，SC 炉是停留时间最短、裂解温度最高和选择性最好的超短停留时间毫秒管式裂解炉。

Milli-second furnace is a kind of furnace developed by the U. S. Kellogg Company, which shortens the residence time of the material in the furnace to 0.05 to 0.1 seconds (50～100ms) at the high cracking temperature, thus comes the name Kellogg milli-second furnace.

KBR cracking furnace is the milli-second furnace of Kellogg Company, hereinafter referred to as SC furnace. Currently, SC furnace is the pipeline milli-second cracking furnace with the shortest residence time, the highest pyrolysis temperature and the greatest selectivity.

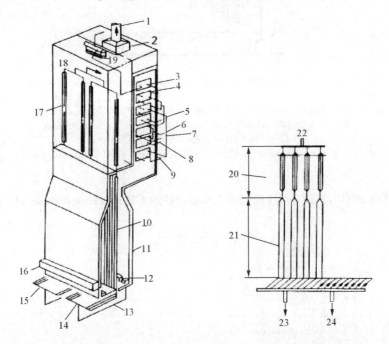

图 1.1.4 毫秒裂解炉结构图 Structure drawing of Kellogg milli-second furnace

1、8—高压蒸汽 High pressure steam; 2—引风机 Induced draft fan; 3、6—锅炉给水 Boiler feed water;
4—原料油 Raw oil; 5—稀释蒸汽 Dilution steam; 7—过热高压蒸汽 Overheat high pressure steam;
9、13—横跨管 Crossover tube; 10—辐射炉管 Radiation tube; 11—炉体 Furnace body;
12—燃烧器 Burner; 14、15—猪尾管 Pigtail tube; 16—助燃空气管道 Combustion air pipe;
17—第一废热锅炉 Primary waste heat boiler; 18—裂解气 Cracked gas; 19—汽包 Steam drum;
20—第一急冷锅炉 Primary quenching boiler; 21—反应管 Reaction tube;
22—至第二废热锅炉 To secondary waste heat boiler;
23、24—物料出口 Material outlet

1.1.5 Pyrocrack 型裂解炉 Pyrocrack cracking furnace

林德公司开发的 Pyrocrack 型裂解炉,通常为双辐射段、单对流段结构。

Pyrocrack cracking furnace is usually double radiation setion and single convection section structure, which is developed by Linde Group.

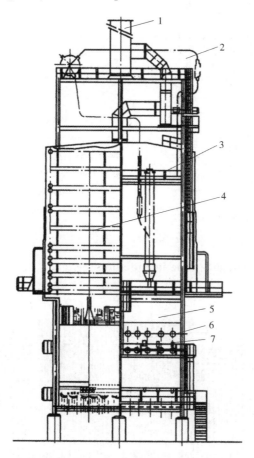

图 1.1.5 Pyrocrack 型裂解炉结构图 Structure drawing of pyrocrack cracking furnace
1—烟囱 Chimney;2—汽包 Steam drum;3—急冷换热器 Quenching heat exchanger;
4—对流段 Convection section;5—辐射段 Radiation section;6—侧壁燃烧器 Side wall burner;
7—看火孔 Kiln eye

1.1.6 布朗路特公司的 HSLR 型裂解炉 HSLR cracking furnace of Kellogg Brown & Root

布朗路特公司开发出的 HSLR 型裂解炉,通常采用一个辐射段配置一个对流段结构。

HSLR(High Selectivity Long Runlength) cracking furnace developed by Kellogg Brown & Root is usually a single furnace structure, one radiation section configuring one convection section.

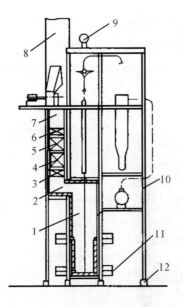

图 1.1.6 HSLR 型裂解炉结构图
Structure drawing of HLSR cracking furnace

1—辐射段 Radiation section；2—横跨段 Crossover section；3—蒸汽过热器 Steam superheater；4—混合预热器 Mixing preheater；5—锅炉给水预热器 Boiler feed water preheater；6—原料预热器；7—对流段 Convection section；8—烟囱 Chimney；9—汽包 Steam drum；10—炉体 Furnace body；11—燃烧器 Burner；12—基础 Base

1.1.7 辐射段炉管 Furnace tube in radiation section

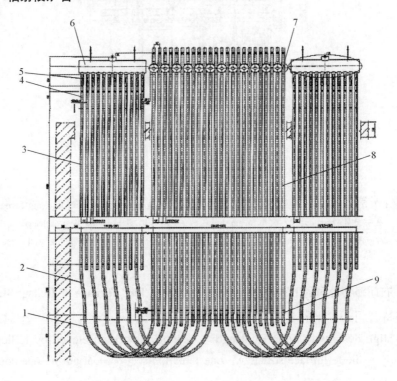

图 1.1.7 辐射段炉管结构图 Structure drawing of furnace tube in radiation section
1—U 形炉管 U type coils；2—S 型炉管 S type coils；3—一程管 First process coils；4—辐射段炉管 Furnace tube in radiation section；5—文丘里管 Venture coils；6—集合管 Collection coils；7—热电偶接口 Thermocouple interface；8—辐射段出口管 Radiation section outlet coils；9—Y 形炉管 Y type coils

1.1.8 裂解炉附属设备 Cracking furnace accessories

1.1.8.1 汽包 Steam drum

每台裂解炉都有一个汽包,用以产生高质量的超高压蒸汽。

Each cracking furnace has a steam drum to produce ultrahigh pressure steam.

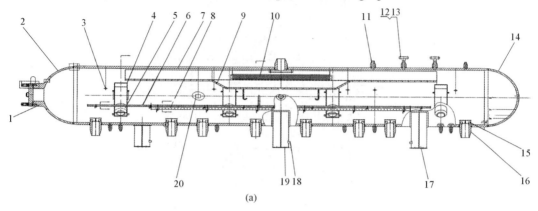

(a)

1—人孔 Manhole;2—左封头 Left head;3、5、11、12—接管 Connecting pipe;4—上升管 Riser;
6、8—支承件 Support;7—筒体 Cylinder;9—汽水分离器 Steam separator;10—丝网除沫器
Wire mesh demister;13—法兰 Flange;14—右封头 Right head;15—防涡器
Vortex breaker;16—下降管 Downcomer;17—滑动支座 Sliding support;18—接地板
Grounding plate;19—固定支座 Fixed support;20—铭牌 Nameplate

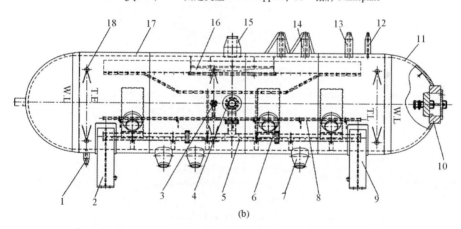

(b)

1—间断排污口 Intermittent drainage outlet;2—固定支座 Fixed support;
3—连续排污口 Continuous drainage outlet;4—给水管接口 Feed water connection;
5—给水分配管 Feed water distribution pipe;6—上升管接口 Rising pipe connection;
7—下降管接口 Down pipe connection;8—排污分配管 Blow down distribution pipe;
9—滑动支座 Sliding support;10—人孔 Manhole;11—右封头 Right head;
12—压力表接口 Pressure gauge connection;13—放空口 Vent;
14—安全阀接口 Safety valve connection;15—蒸汽出口 Steam outlet;
16—丝网除沫器 Wire mesh demister;17—筒体 Cylinder;
18—液位计接口 Level gauge connection

图 1.1.8 汽包结构图
Structure drawing of steam drum

1.1.8.2 急冷换热器 Quenching heat exchanger

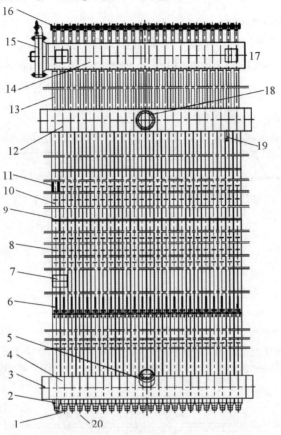

图 1.1.9 急冷换热器结构图
Structure drawing of quenching heat exchanger
1—入口带叉锥体 Inlet cone with a fork;
2—排污管 Blow down pipe;
3—联箱端盖 Connection box end cover;
4—下联箱 Down connection box;
5—给水入口 Feed water inlet;
6—支座 Support; 7—铭牌 Nameplate;
8—保温托板 Heat insulation plate;
9—固定夹 Stationary clamp;
10—外管 Outer pipe; 11—内管 Inner pipe;
12—上联箱 Up connection box;
13—裂解气外接管 Cracking gas outer connection pipe;
14—裂解气联箱 Cracking gas connection box;
15—裂解气联箱法兰 Cracking gas connection box flange;
16—清焦接头 Decoking joint;
17—裂解气出口 Cracking gas outlet;
18—蒸汽出口 Steam outlet;
19—排污管 Blow down pipe;
20—裂解气入口 Cracking gas inlet

1.1.8.3 混合器 Commingler

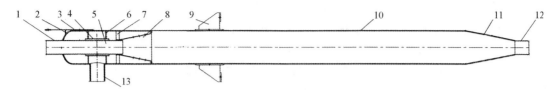

图 1.1.10　混合器结构图 Structure drawing of commingler

1—入口管 Inlet pipe；2—封头 Head；3—吊耳 Lug；4—支承板 Support plate；5—内部接管 Inside connection；
6—环板 Annular plate；7—筋板 Reinforcing plate；8—扩散异径管 Diffusion reducer；9—支座 Support；
10—筒体 Cylinder；11—锥体 Cone；12—出口管 Outlet pipe；13—蒸汽入口管 Steam inlet pipe

1.1.8.4 清焦罐 Decoking tank

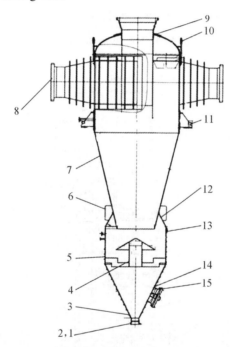

图 1.1.11　清焦罐结构图 Structure drawing of decoking tank

1、15—法兰 Flange；2、14—接管 Connecting pipe；3—伴热盘管 Heating coil；
4—内管 inner pipe；5、12—下筒体 Lower cylinder；6—筋板 Reinforcing plate；
7—上筒体 Upper cylinder；8—入口管 Inlet pipe；9—出口管 Outlet pipe；
10—吊耳 Lug；11—支座 Support；13—铭牌 Nameplate

1.2　废热锅炉 Waste heat boiler

废热锅炉是回收高温裂解气余热的一种换热设备，在结构上有列管式、盘管式、插入式、双套管式和 U 形管式等多种。

Waste heat boiler is a heat exchanger, which can be used to recover the waste heat of high temperature cracking gas There are many types, tube, coil, double-pipe, U pipe.

列管式废热锅炉有普通列管式和新型管板列管式两种。

There are two types of shell-tube waste heat boiler: common column tube type waste heat boiler and new tube plate type waste heat boiler.

普通列管式废热锅炉,相当于普通列管式固定管板换热器。

Common column tube type waste heat boiler is equivalent to ordinary column pipe fixed tube sheet heat exchanger.

新型管板列管式废热锅炉的新型管板是指椭圆形、蝶形和薄板等管板,其特点是管板较薄,管板的挠性变形比较容易,因此,也被称为挠性管板废热锅炉。

The new tube plates refer to oval tube plate, butterfly tube plates, thin plates and so on, which are characterized by thin tube plates and more flexible deformation of tube sheets, hence comes the name flexible pipe type waste heat boiler.

1.2.1 立式废热锅炉 Vertical type waste heat boiler

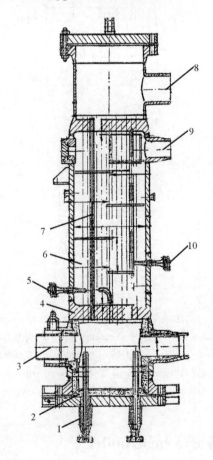

图 1.2.1 立式废热锅炉结构图 Structure drawing of vertical type waste heat boiler
1—支脚 Support foot; 2—耐热衬里 Heat-resistant lining; 3—工艺气入口 Process gas inlet;
4—管板 Tube sheet; 5—排污口 Sewage draining exit; 6—换热管 Heat exchange tube; 7—拉杆 Tie rod;
8—工艺气出口 Process gas outlet; 9—蒸汽出口 Steam outlet; 10—工艺水入口 Process water inlet

1.2.2 卧式废热锅炉 Horizontal waste heat boiler

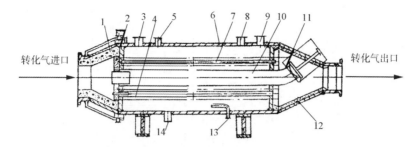

图 1.2.2 卧式废热锅炉结构图 Structure drawing of horizontal waste heat boiler
1—耐热衬里 Heat-resistant lining；2—管板 Tube sheet；3、5、8、9—汽水混合物出口 Outlet of steam and water；
4—换热管 Heat exchange tube；6—筒体 Cylinder；7—拉杆 Tie rod；10—中心管 Center pipe；11—弯管 Bend pipe；
12—锥体 Cone；13—排污口 Sewage outlet；14—排放口 Discharge outlet

1.2.3 椭圆形管板废热锅炉 Oval tube plate waste heat boiler

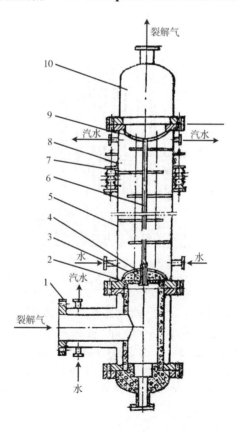

图 1.2.3 椭圆形管板废热锅炉结构图 Structure drawing of oval tube plate waste heat boiler
1—下三通 Lower three links；2—保护板 Protection board；3—保护套管 Protective casing；
4—下管板 Lower tube plate；5—筒体 Cylinder；6—换热管 Heat exchange tube；7—折流板 Baffle；
8—膨胀节 Expansion joint；9—上管板 Upper tube plate；10—上三通 Upper three links

1.2.4　碟形管板废热锅炉 Dish tube plate waste heat boiler

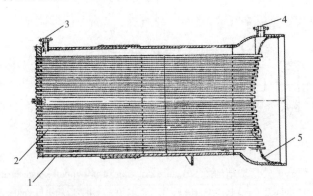

图 1.2.4　碟形管板废热锅炉结构图 Structure drawing of dish tube plate waste heat boiler
1—筒体 Casing；2—换热管 Heat exchange tube；3—给水入口 Feed-water inlet；
4—排水出口 Drainage outlet；5—碟形管板 Dish tube plate

1.2.5　薄管板废热锅炉 Thin tube plate waste heat boiler

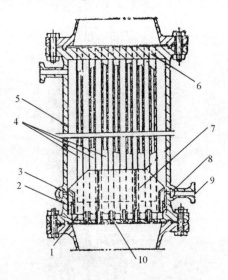

图 1.2.5　薄管板废热锅炉结构图
Structure drawing of thin tube plate waste heat boiler
1、6—管板 Tube sheet；2—支承环 Support ring；3—导流室 Diversion chamber；
4—换热管 Heat exchange tube；5—筒体 Cylinder；7—隔板 Partition；
8—环形通道 Annular channel；9—给水管 Feed pipe；10—拉杆 Tie rod

1.3　压缩机 Compressor

压缩机为工艺气体提供压力能，其中包括裂解气压缩机、丙烯制冷压缩机和乙烯制冷压缩机，有的装置制冷系统采用二元制冷压缩机组，有的装置采用三元制冷压缩机组，此外,还有甲烷膨胀压缩机、废碱氧化空压机、火炬气回收压缩机和液环压缩机等。

Compressors, which include the cracking gas compressor, propylene refrigeration compressor and ethylene refrigeration compressor, provide pressure and kinetic energy to the process gas. Binary or ternary refrigeration compressors are used by some refrigeration systems. Others, such as methane expansion compressor, the waste alkaline oxidation air compressor, flare gas recovery compressor and liquid ring compressor, are also applied.

1.3.1 裂解气压缩机 Cracking gas compressor

裂解气压缩机是多级离心式压缩机,其机组由主机和辅助系统组成。主机包括压缩机和汽轮机,辅助系统包括凝液系统、油系统和密封系统。

Cracking gas compressor is a multistage centrifugal compressor, which is composed by the main engine and auxiliary systems. The main engine includes turbine and compressor. The auxiliary systems involve condensate system, oil system and sealing system.

1.3.1.1 裂解气压缩机结构 Cracked gas compressor structure

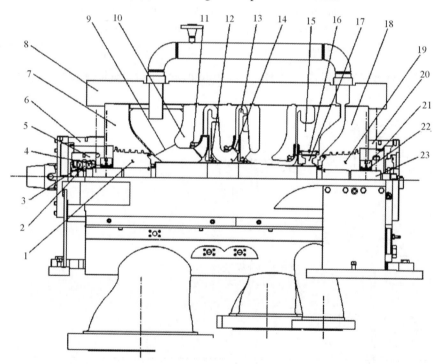

图 1.3.1 裂解气压缩机结构图 Structure drawing of cracked gas compressor
1—壳体 Shell；2—推力盘 Thrust disc；3—止推轴承 Thrust bearing；4、22—可倾瓦轴承 Tilting-pad bearing；
5、21—轴承护圈 Bearing guard ring；6、20—轴承箱 Bearing box；7、18—端盖 End cover；
8、19—干气密封 Dry gas seal；9—入口导向叶片 Inlet guiding impeller；10—入口隔板 Inlet baffle；
11、12—级间迷宫密封 Labyrinth seal；13—隔板 Baffle；14—叶轮 Wheel；15—出口隔板 Outlet baffle；
16—平衡活塞密封 Balance piston seal；17—平衡活塞 Balance piston；23—轴 Shaft

1.3.1.2 裂解气压缩机低压缸 Low pressure cylinder of cracking gas compressor

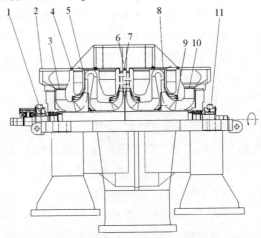

图 1.3.2 裂解气压缩机低压缸结构图
Structure drawing of low pressure cylinder of cracking gas compressor

1、11—轴承箱 Bearing box；2—壳体 Shell；3、10—入口 Inlet；4—螺栓 Bolt；
5、8—O 形密封圈 O sealing ring；6—扩散器 Diffuser；7—固定销 Fixed pin；9—锁紧螺母 Blocking nut

1.3.1.3 裂解气压缩机高压缸 High pressure cylinder of cracking gas compressor

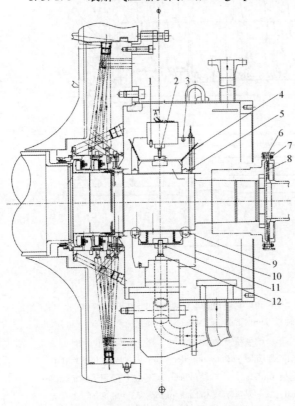

图 1.3.3 裂解气压缩机高压缸结构图
Structure drawing of high pressure cylinder of cracking gas compressor

1、2、5、6、7、12—螺栓 Bolt；3—手孔 Handhole；4、10—壳体 Casing；8—锥体 Cone；9—锁紧螺母 Blocking nut；11—O 形密封圈 O sealing ring

1.3.2 裂解气汽轮机 Cracking gas turbine

1.3.2.1 裂解气汽轮机结构 Cracking gas turbine structure

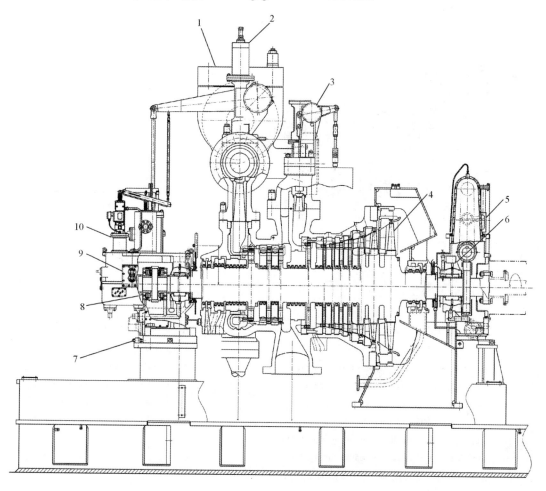

图 1.3.4 裂解气汽轮机结构图 Structure drawing of cracking gas turbine

1—进气阀 Inlet valve

2、10—控制系统 Control system;

3—抽气控制 Gas extraction control;

4—转子 Rotor;

5—盘车装置 Barring equipment;

6、8—轴承 Bearing;

7—限位器 Stopper;

9—紧急制动系统 Braking system

1.3.2.2 裂解气汽轮机主要零部件 Main components of cracked gas turbine

1. 气缸 Cylinder
2. 导叶持环 Guide vane retaining ring

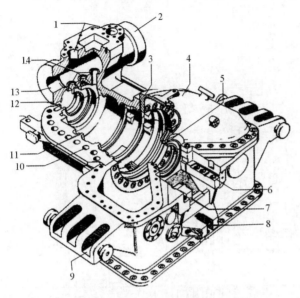

图 1.3.5 气缸结构图 Structure drawing of cylinder

1—调节汽阀阀杆装配孔
Assemble hole of regulating steam valve stem;
2—速关阀阀壳 Quick closing valve shell;
3—导叶持环支承 Guide blade supporting ring lap;
4—排汽缸 Exhaust cylinder;
5—后汽封装配凸环 Rear shaft assembling convex ring;
6—后轴承座安装面 Rear bearing seat mounting surface;
7—后轴承座导向键 Rear bearing seat guiding key;
8—水平调整键 Key for horizontal adjustment;
9—尾部猫爪 Tear cat pad;
10—中分面螺栓孔 Horizontal bolt hole;
11—导向持环装配凸环
Guide blade supporting ring assembling convex ring;
12—前汽封装配凸环 Front shaft assembling convex ring;
13—进汽室 Suction chamber;
14—调节汽阀阀座装配孔
Assembling hole of regulating steam valve stem;

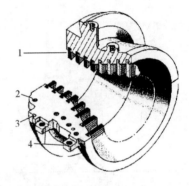

图 1.3.6 导叶持环结构图
Structure drawing of guide vane retaining ring

1—导叶片 Guide vane;
2—中分面螺栓孔 Horizontal bolt hole;
3—中分面凹槽 Horizontal groove;
4—水平调整键 Adjusting key

3. 转子 Rotor

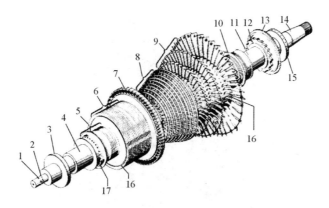

图 1.3.7 转子结构图 Structure drawing of rotor

1—危急保安器 Crisis protector；
2—轴位移凸肩 Convex shoulder of the shaft displacement；3—推力盘 Thrust disc；
4—前轴承轴颈 Front bearing journal；5—前汽封 Front steam seal；
6—平衡活塞汽封 Balance piston shaft section；7—调节段 Control stage；
8—压力段 Pressure stage；9—低压段 Low pressure stage；
10—后汽封 Rear steam seal；11—后轴承轴颈 Rear bearing journal；
12—盘车棘轮 Turning gear；13—盘车齿轮 Oil worm gear turning rotating gear；
14—联轴器轴段 Coupling shaft section；15—后端平衡面 Rear balance surface；
16—主平衡面 Main balance surface；17—前端平衡面 Front balance surface

4. 内缸 Inner cylinder

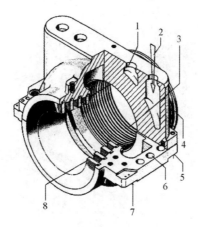

图 1.3.8 内缸结构图 Structure drawing of inner cylinder

1—角形密封环装配孔 Angular seal ring assembling hole；2—蒸汽入口 Steam inlet；
3—平衡活塞汽封 Balance piston shaft section；4—定位槽 Positioning groove；
5—前支承面 Front supporting surface；6—中分面螺栓孔 Horizontal bolt hole；
7—后支承面 Rear supporting surface；8—导叶 Guide blade

5. 推力轴承 Thrust bearing

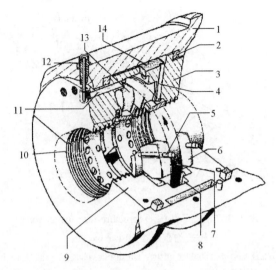

图 1.3.9　推力轴承结构图 Structure drawing of thrust bearing

1—前轴承座 Front bearing seat；2—环形垫片 Annular gasket；3、9—推力轴承体 Thrust bearing；
4—进油孔 Fuel feed hole；5—圆柱销 Cylindrical pin；6、13—推力块 Thrust block；
7—推力盘 Thrust plate；8—内油槽 Inner oil sump；10—油封齿 Oil seal teeth；
11—排油孔 Oil drain hole；12—温度计接口 Thermometer interface；14—外油槽 External oil sump

6. 调节汽阀 Regulating valve

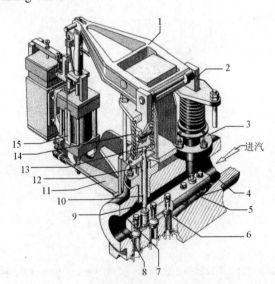

图 1.3.10　调节汽阀结构图 Structure drawing of regulating valve

1—杠杆 Lever；2—连接板 Connection plate；3—阀盖 Valve cover；4—气缸进汽室 Cylinder steam chamber；
5—阀梁 Valve stem；6—阀碟 Valve disc；7—衬套 Bush；8—阀座 Valve seat；9—阀杆 Valve pole；
10—下导向套筒 Lower guide sleeve；11—托架 Bracket；12—上导向套筒 Upper guide sleeve；
13—支架 Support；14—弹簧 Spring；15—油动机 Oil motivator

7. 油动机 Oil motivator

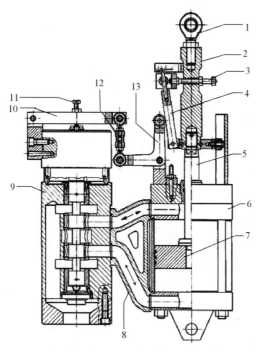

图 1.3.11 油动机结构图 Structure drawing of oil motivator

1—万向连杆 Universal connecting rod；2—拉杆 Pulling rod；3、11—调节螺栓 Adjustment bolt；
4—反馈板 Feedback block；5—活塞杆 Piston rod；6—油缸 Oil cylinder；7—活塞 Piston；
8—油路 Oil circuit；9—错油门油缸 Pilot oil valve cylinder；10—反馈杠杆 Feedback lever；
12—弯角杠杆 Bend angel lever；13—肘形杆 Elbow pole

8. 速关阀 Quick closing valve

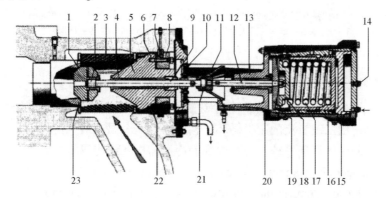

图 1.3.12 速关阀结构图 Structure drawing of quick closing valve

1—蝶阀 Butterfly valve；2—卸载阀 Unloading valve；3—蒸汽滤网 Steam strainer；4—导向套筒 Guiding sleeve；
5—阀盖 Valve cover；6—螺栓 Bolt；7—弓形环 Bow ring；8—压环 Pressure ring；9—阀杆 Valve stem；
10—隔热板 Thermal baffle；11—油杯 Lubricator；12—活塞杆 Piston rod；13—支座 Support；
14—压力表接口 Pressure gauge interface；15—试验活塞 Testing piston；16—活塞 Piston；
17—弹簧 Spring；18—弹簧座 Spring seat；19—活塞盘 Piston plate；20—油缸 Oil cylinder；
21—联轴节 Coupling；22—密封环 Seal ring；23—阀座 Valve seat

1.3.3 丙烯制冷压缩机 Propylene refrigeration compressor

丙烯制冷压缩机是利用丙烯作介质，通过多级压缩提高其压力，使其在较高的温度下冷凝，然后进行节流膨胀，在较低的温度下汽化，获得深冷分离所需的冷量。

Propylene is used as a medium in propylene refrigeration compressor to increase its pressure through a multi-stage compression, so as to be condensed at a higher temperature. Then it is throttled in expansion and gasified under low temperature to get the required cooling capacity for cryogenic separation.

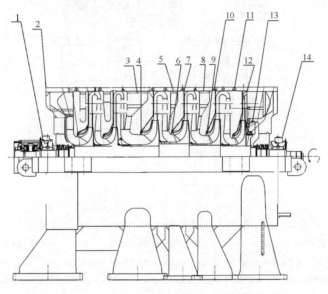

图 1.3.13　丙烯制冷压缩机结构图
Structure drawing of propylene refrigeration compressor

1、14—轴承 Bearings and related components；2—机壳 Casing；3—二段入口挡板 The second inlet baffle；
4、6、10、11、13—螺栓 Bolt；5—三段入口挡板 The third inlet baffle；7、9—O 形密封圈 O sealing ring；
8—四段入口挡板 The fourth inlet baffle；12—放空挡板 Vent baffle

1.3.4 乙烯制冷压缩机 Ethylene refrigeration compressor

乙烯制冷压缩机提供裂解气低温分离装置所需 -102～-40℃各温度级的冷量。多数乙烯制冷压缩机采用三级节流的制冷循环，相应提供三个温度级别的冷剂。该系统为多段压缩、多级节流的封闭循环系统，并与丙烯制冷压缩机系统构成重叠制冷。

Ethylene refrigeration compressor provides cooling capacity ranging from -102℃ to -40℃ required by low temperature pyrolysis gas separation device. Most ethylene refrigeration compressors adopt three-level throttling refrigeration cycle and offer corresponding refrigerant temperature to each level. It is a multi-stage compression and multi-level throttling closed loop system and constitutes a cascade refrigeration with propylene refrigeration compressor system.

1.3.4.1 乙烯制冷压缩机低压缸 Low pressure cylinder of ethylene refrigeration compressor

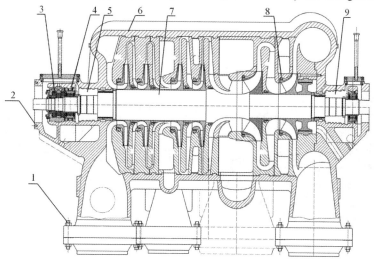

图 1.3.14　乙烯制冷压缩机低压缸结构图
Structure drawing of low pressure cylinder of ethylene refrigeration compressor
1—螺栓 Bolt；2—轴承箱 Bearing box；3—推力轴承 Thrust bearing；4—径向轴承 Radial bearing；
5、9—迷宫密封 Labyrinth seal；6—机壳 Casing；7—主轴 Shaft；8—锁紧螺母 Blocking nut

1.3.4.2 乙烯制冷压缩机高压缸 High pressure cylinder of ethylene refrigeration compressor

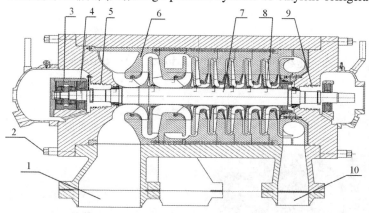

图 1.3.15　乙烯制冷压缩机高压缸结构图
Structure drawing of high pressure cylinder of ethylene refrigeration compressor
1—入口管 Inlet；2—定位螺栓 Jackbolt；3—推力轴承 Thrust bearing；4—径向轴承 Radial bearing；
5、9—迷宫密封 Labyrinth seal；6—机壳 Case；7—主轴 Shaft；8—叶轮 Wheel；10—出口管 Outlet pipe

1.3.5 甲烷膨胀压缩机 Methane expansion compressor

甲烷膨胀压缩机分活塞式和透平式两种。透平式膨胀压缩机具有效率高、流量大、无摩擦件、运转平稳和工作可靠等优点，因此，现在广泛使用的是透平式膨胀压缩机。

There are two kinds of methane expansion compressors: piston type and turbine type. Turbine expansion compressor is used with the advantage of high efficiency, high flow rate, no friction, smooth and reliable operation, etc. Therefore, turbine expansion compressor is widely used.

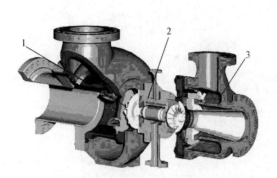

图 1.3.16 甲烷膨胀压缩机外形图
Outside drawing of methane expansion compressor
1—压缩部分 Compressor part;
2—总成部分 Assembly part;
3—膨胀部分 Expansion part

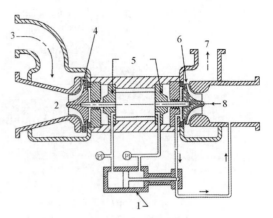

图 1.3.17 甲烷膨胀压缩机结构图
Structure drawing of methane expansion compressor
1—自动推力均衡器 Automatic thrust equalizer installation;
2—膨胀机出口 Expander discharge;
3—膨胀机入口 Expander suction;
4—膨胀机叶轮 Expander wheel;
5—轴承箱 Bearing box; 6—压缩机叶轮 Compressor wheel;
7—压缩机出口 Compressor discharge;
8—压缩机入口 Compressor suction

1.3.6 废碱氧化空压机 Waste alkaline oxidation air compressor

1—排空阀 Vent valve; 2—第 3 级压缩机机头 Level 3 compressor head; 3—齿轮箱 Gearing box; 4—油泵 Oil Pump; 5—定位器 Locator; 6—进气导向叶片 Inlet guide vane; 7—第 1 级压缩机机头 Level 1 compressor head; 8—第 2 中间冷却器 Second intercooler; 9、11、20—水调节阀 Water valve; 10—恒温旁通阀 Thermostat by-pass valve; 12—第 1 中间冷却器 First intercooler; 13—油冷却器 Oil cooler; 14—辅助油泵 Auxiliaryoil pump; 15—辅助油泵电机 Auxiliary motor; 16—水分离器 Water separator; 17—冷却水进口 Cooling water inlet; 18—冷却水出口 Cooling water outlet; 19—后冷却器 After-cooler; 21—压缩空气出口 Compressed air outlet

图 1.3.18 废碱氧化空压机结构图
Structure drawing of waste alkaline oxidation air compressor

1.3.7 液环压缩机 Liquid ring compressor

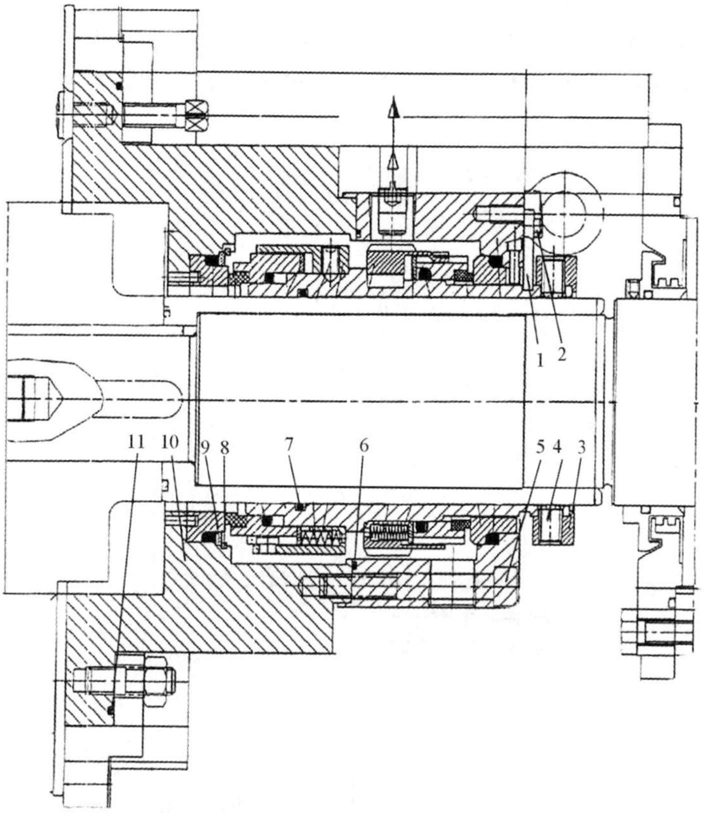

图 1.3.19 液环压缩机结构图 Structure drawing of liquid ring compressor
1—装配架 Mounting bracket；2—六角螺栓 Hexagon bolt；3、8—定位环 Locating ring；
4—定位螺栓 Set bolt；5—螺钉 Screw；6、7、9、11—O形密封环 O sealing ring；
10—转接头 Connector

1.4 反应设备 Reaction device

乙烯装置中的反应器主要有碳二加氢反应器、碳三加氢反应器和甲烷化反应器，此外，还有脱砷反应器和废碱氧化反应器等。

C_2 hydrogenation reactor, C_3 hydrogenation reactor and methanation reactor are mainly adopted in ethylene plant. In addition, dearsenication reactor and waste alkali oxidation reactor are also used.

1.4.1 加氢反应器 Hydrogenation reactor

1.4.1.1 碳二加氢反应器 C_2 hydrogenation reactor

碳二加氢反应器的作用是脱除碳二组分中的乙炔，以满足乙烯装置目标产品聚合级乙烯对乙炔含量的要求。碳二加氢反应器为单层绝热固定床反应器，由壳体、催化剂支承件、筛网和惰性球等组成，反应类型为气固非均相连续式反应。

C_2 hydrogenation reactor can be used to remove C_2 components from acetylene so as to meet the requirements of acetylene content in polymer grade ethylene, the target product. C_2 hydrogenation reactor is a single adiabatic fixed-bed reactor, which is composed by casing, catalyst supports, screen cloth and inert ball, etc. Continuous gas-solid heterogeneous reaction is produced.

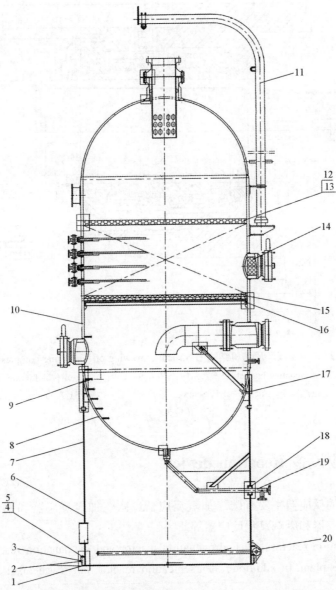

图1.4.1 碳二加氢反应器结构图 Structure drawing of C_2 hydrogenation reactor

1—基础环板 Base plate; 2—筋板 Reinforcing plate; 3—接地板 Grounding plate; 4—盖板 Cover plate; 5—垫板 Plate; 6—铭牌 Nameplate; 7—裙座 Skirt; 8—球形封头 Spherical head; 9—梯子 Ladder; 10—筒体 Cylinder; 11—吊柱 Davit; 12—支承架 Support frame; 13—不锈钢丝网 Stainless steel wire mesh; 14—人孔 Manhole; 15—瓷球 Porcelain ball; 16—填料支承 Packing support; 17—隔气圈 Stop gas ring; 18—补强板 Support plate; 19—接管 Connecting pipe; 20—尾部吊耳 Rear lug

1.4.1.2 碳三加氢反应器 C₃ hydrogenation reactor

碳三加氢反应器的作用是脱除碳三馏分中的甲基乙炔和丙二烯。碳三加氢反应与碳二加氢反应一样都是非均相反应，一般采用气固反应或液固反应。碳三加氢反应器由壳体、进口分配器、带升气管的筛板塔盘、出口收集器和催化剂排放口等组成。

C_3 hydrogenation reactor can be used to remove methyl acetylene and propadiene from C_3-fraction. And like C_2 hydrogenation reaction, C_3 hydrogenation reaction is also a heterogeneous reaction, which generally produces gas-solid reactions or liquid-solid reactions. C_3 hydrogenation reactor is composed by casing, inlet distributor, sieve tray with chimney, outlet collector and catalyst outfall.

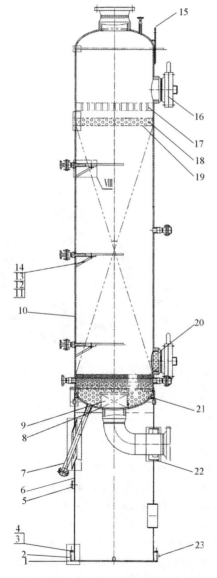

1—基础环板 Foundation ring plate；2、14—筋板 Reinforcing plate；3—盖板 Cover plate；4—垫板 Plate；5—铭牌 Nameplate；6—裙座 Skirt；7—接管 Connecting pipe；8—封头 Head；9—收集器 Collector；10—筒体 Cylinder；11—连接板 Connecting plate；12—U 形螺栓 U bolt；13—螺母 Nut；15—吊耳 Lug；16、20—人孔 Manhole；17—分布器 Distributor；18—瓷球 Porcelain ball；19—不锈钢丝网 Stainless steel wire mesh；21—隔气圈 Stop gas ring；22—引出管 Outlet pipe；23—接地板 Grounding plate

图 1.4.2 碳三加氢反应器结构图
Structure drawing of C_3 hydrogenation reactor

1.4.2 脱砷反应器 Dearsenication reactor
1.4.2.1 碳二加氢脱砷反应器 C_2 dearsenication reactor

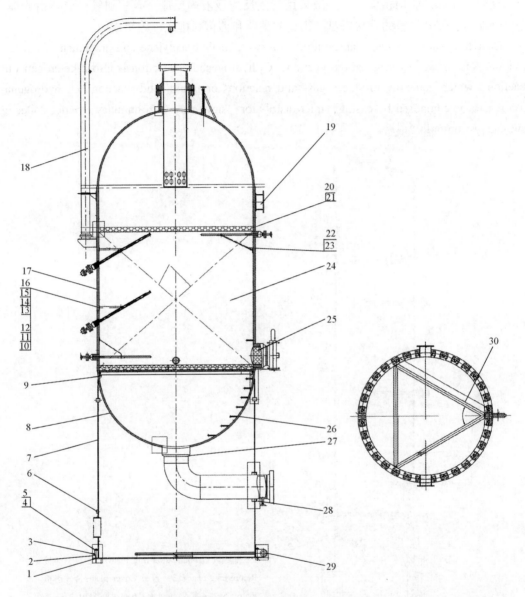

图 1.4.3 碳二加氢脱砷反应器结构图 Structure drawing of C_2 dearsenication reactor

1—基础环板 Foundation ring plate；2、14—筋板 Reinforcing plate；3—接地板 Grounding plate；4—盖板 Cover plate；
5、10、13、23—垫板 Plate；6—铭牌 Nameplate；7—裙座 Skirt；8—封头 Head；
9—支承件 Supporting parts；11—支承圆钢 Supporting round steel；12—半圆管 Semicircle pipe；15—U 形螺栓 U bolt；
16—螺母 Nut；17—筒体 Cylinder；18—塔顶吊柱 Top davit；19、27—接管 Connecting pipe；
20—瓷球 Porcelain ball；21—不锈钢丝网 Stainless steel wire mesh；22—支承角钢 Supporting angle steel；
24—催化剂 Catalyst；25—人孔 Manhole；26—梯子 Ladder；28—引出管 Outlet pipe；
29—尾部吊耳 Rear lug；30—支承管 Supporting pipe

1.4.2.2 碳三加氢脱砷反应器 C₃ dearsenication reactor

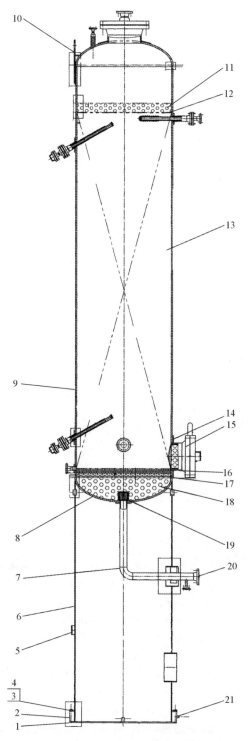

图 1.4.4 碳三加氢脱砷反应器结构图
Structure drawing of C₃ dearsenication reactor

1—基础环板 Foundation ring plate；2—筋板 Reinforcing plate；3—盖板 Cover plate；4—垫板 Plate；5—铭牌 Nameplate；6—裙座 Skirt；7—接管 Connecting pipe；8—封头 Head；9—筒体 Cylinder；10—吊耳 Lug；11、16、17、18—瓷球 Porcelain ball；12—不锈钢丝网 Stainless steel wire mesh；13—催化剂 Catalyst；14—补强圈 Reinforcing ring；15—人孔 Manhole；19—出口捕集器 Outlet collector；20—引出管 Outlet pipe；21—接地板 Grounding plate

1.4.3 甲烷化反应器 Methanation reactor

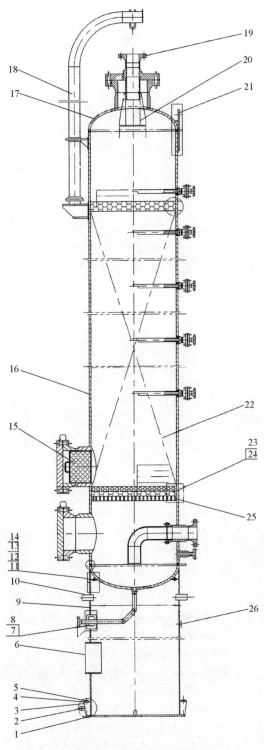

图1.4.5 甲烷化反应器结构图
Structure drawing of methanation reactor

1—基础环板 Foundation ring plate；2—接地板 Grounding plate；3—筋板 Reinforcing plate；4—盖板 Cover plate；5、14—垫板 Plate；6—检修口 Access hole；7—引出管 Outlet pipe；8—支承板 Supporting plate；9—裙座 Skirt；10—排气管 Vent pipe；11—螺栓 Bolt；12—螺母 Nut；13—弧形板 Curved plate；15—人孔 Manhole；16—筒体 Cylinder；17—封头 Head；18—塔顶吊柱 Top davit；19—接管 Connecting pipe；20—缓冲板 Buffering plate；21—吊耳 Lug；22—催化剂 Catalyst；23—瓷球 Porcelain ball；24—不锈钢丝网 Wire mesh；25—填料支承件 Packing support；26—铭牌 Nameplate

1.4.4 废碱氧化反应器 Waste alkali oxidation reactor

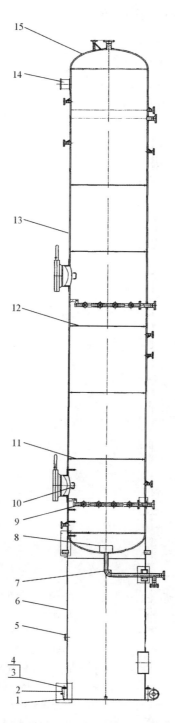

图 1.4.6 废碱氧化反应器结构图 Structure drawing of waste alkali oxidation reactor
1—基础环板 Foundation ring plate; 2—筋板 Reinforcing plate; 3—盖板 Cover plate; 4—垫板 Plate; 5—铭牌 Nameplate;
6—裙座 Skirt; 7—引出管 Outlet pipe; 8—防涡器 Vortex breaker; 9—空气分布器 Air distributor;
10—人孔 Manhole; 11、12—塔盘 Tray; 13—筒体 Cylinder; 14—吊耳 Lug; 15—封头 Head

1.5 塔设备 Column equipment

1.5.1 急冷塔 Quench column

1.5.1.1 急冷油塔 Quenching oil column

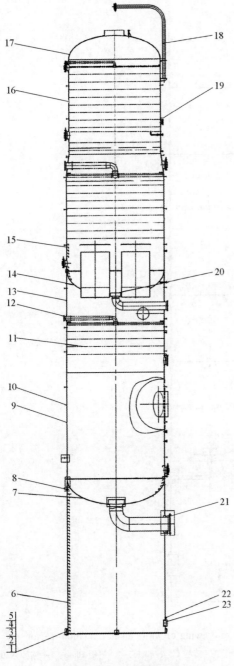

图1.5.1 急冷油塔结构图
Structure drawing of quenching oil column

1—基础环板 Foundation ring plate；2—筋板 Reinforcing plate；3—盖板 Cover plate；4—垫板 Plate；5—接地板 Grounding plate；6—裙座 Skirt；7、17—封头 Head；8—排气管 Exhaust pipe；9—保温支承圈 Insulation supporting ring；10、13、16—筒体 Cylinder；11—塔盘 Tray；12、19—接管 Connecting pipe；14—集油器 Oil collector；15—梯子 Ladder；18—塔顶吊柱 Top davit；20—防涡器 Vortex bleaker；21—引出管 Outlet pipe；22—铭牌 Nameplate；23—检修口 Access hole

1.5.1.2 急冷水塔 Quenching water column

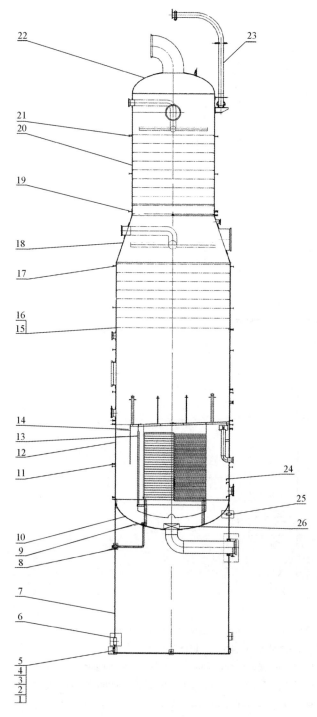

图 1.5.2 急冷水塔结构图
Structure drawing of quenching water column

1—基础环板 Foundation ring plate；2—筋板 Reinforcing plate；3—盖板 Cover plate；4—垫板 Plate；5—接地板 Grounding plate；6—检修口 Access hole；7—裙座 Skirt；8—引出管 Outlet pipe；9、26—防涡器 Vortex breaker；10、22—封头 Head；11—接管 Connecting pipe；12、20—筒体 Cylinder；13—油水分离器 Oil water separator；14—挡板 Upper baffle plate；15—塔盘支承 Tray support；16—塔盘 Tray；17、19、21—保温支承圈 Insulation supporting ring；18—锥体 Cone；23—塔顶吊柱 Top davit；24—梯子 Ladder；25—排气管 Exhaust pipe

1.5.2 汽提塔 Stripping column

1.5.2.1 重燃料油汽提塔 Heavy fuel oil stripper column

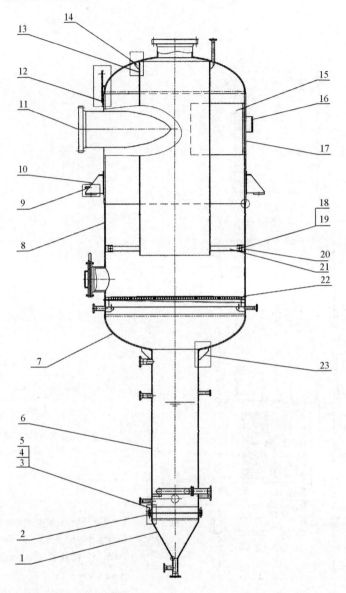

图 1.5.3 重燃料油汽提塔结构图 Structure drawing of heavy fuel oil stripper column

1—锥体 Cone；2—法兰 Flange；3、18—螺柱 Stud；4、19—螺母 Nut；5—垫片 Gasket；
6、8、17—筒体 Cylinder；7—封头 Head；9—接地板 Grounding plate；10—支座 Support；
11—进料口 Feed inlet；12—吊耳 Lug；13—导流筒 Draft tube；14、20、23—筋板 Reinforcing plate；
15—防冲挡板 Impingement baffle；16—铭牌 Nameplate；21、22—连接板 Connecting plate

1.5.2.2 轻燃料油汽提塔 Light fuel oil stripper column

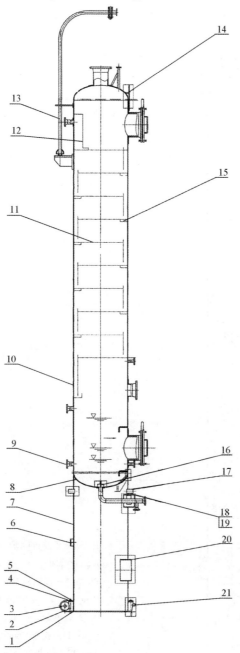

图1.5.4 轻燃料油汽提塔结构图 Structure drawing of light fuel oil stripper column

1—基础环板 Foundation ring plate；2—筋板 Reinforcing plate；3—吊耳 Lug；4—盖板 Cover plate；
5—垫板 Plate；6—铭牌 Nameplate；7—裙座 Skirt；8—封头 Head；9—接管 Connecting pipe；10—筒体 Cylinder；
11—塔盘 Tray；12—防冲挡板 Impingement baffle；13—塔顶吊柱 Top davit；14—吊耳 Lug；
15—塔盘支承 Tray support；16—防涡器 Vortex breaker；17—排气口 Exhaust port；
18—引出管 Outlet pipe；19—补强板 Support plate；
20—检修口 Access hole；21—接地板 Grounding plate

1.5.2.3 汽油汽提塔 Gas stripper column

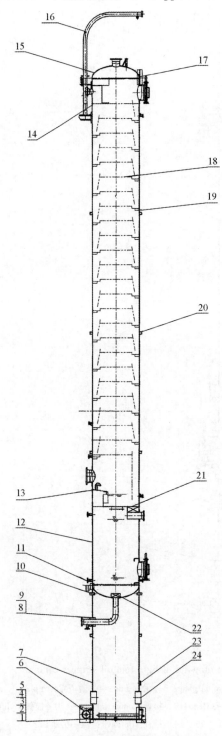

图 1.5.5 汽油汽提塔结构图
Structure drawing of light fuel oil stripper column

1—基础环板 Foundation ring plate；2—筋板 Reinforcing plate；3—盖板 Cover plate；4—垫板 Plate；5—接地板 Grounding plate；6—尾部吊耳 Rear lug；7—裙座 Skirt；8—引出管 Outlet pipe；9—补强板 Support plate；10—排气管 Exhaust pipe；11—接管 Connecting pipe；12—筒体 Cylinder；13—挡板 Baffle；14—防冲挡板 Impingement baffle；15—封头 Head；16—塔顶吊柱 Top davit；17—板式吊耳 Plate lug；18—塔盘 Tray；19—塔盘支承 Tray support；20—保温支承圈 Insulation supporting ring；21—受液盘 Liquid receiving plate；22—防涡器 Vortex breaker；23—铭牌 Nameplate；24—检修口 Access hole

1.5.2.4 工艺水汽提塔 Process water stripper column

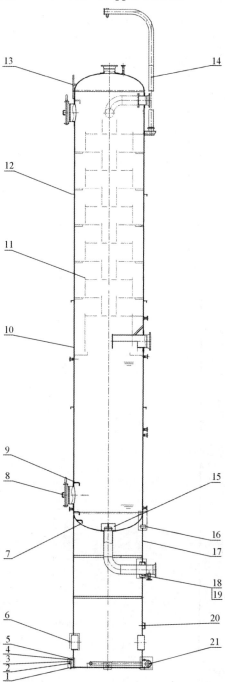

图 1.5.6 工艺水汽提塔结构图 Structure drawing of process water stripper column

1—基础环板 Foundation ring plate；2—接地板 Grounding plate；3—筋板 Reinforcing plate；4—盖板 Cover plate；
5—垫板 Plate；6—检修口 Access hole；7—封头 Head；8—接管 Connecting pipe；9—梯子 Ladder；
10—筒体 Cylinder；11—塔盘 Tray；12—保温支承圈 Insulation supporting ring；13—吊耳 Lug；14—塔顶吊柱 Top davit；
15—防涡器 Vortex breaker；16—排气口 Exhaust port；17—裙座 Skirt；18—引出管 Outlet pipe；
19—补强板 Supporting plate；20—铭牌 Nameplate；21—尾部吊耳 Rear lug

1.5.3 精馏塔 Distillation column

1.5.3.1 高压脱丙烷塔 High pressure depropanizing column

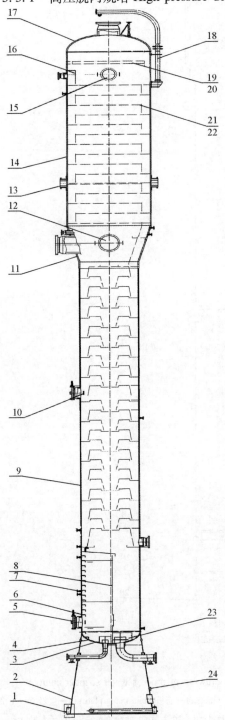

图 1.5.7 高压脱丙烷塔结构图
Structure drawing of high pressure depropanizing column

1—接地板 Grounding plate；2—裙座 Skirt；3、23—防涡器 Vortex breaker；4—下封头 Lower head；5、12、15—人孔 Manhole；6—梯子 Ladder；7—接管 Connecting pipe；8—塔釜隔板 Column reactor partition；9、14—筒体 Cylinder；10—把手 Hand grip；11—锥体 Cone；13—吊耳 Lug；16—防冲挡板 Impingement baffle；17—上封头 Upper head；18—塔顶吊柱 Top davit；19—丝网除沫器 Wire mesh demister；20—丝网除沫器支承 Wire mesh demister support；21—塔盘 Tray；22—塔盘支承 Tray support；24—铭牌 Nameplate

1.5.3.2 低压脱丙烷塔 Low pressure depropanizing column

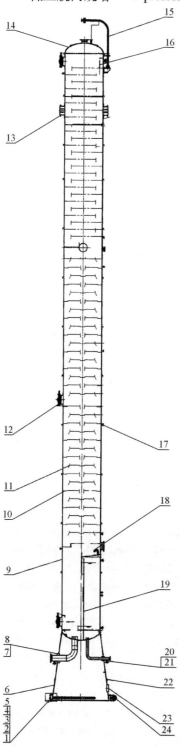

图1.5.8 低压脱丙烷塔结构图
Structure drawing of Low pressure depropanizer

1—基础环板 Foundation ring plate；2—接地板 Grounding plate；3—筋板 Reinforcing plate；4—环形盖板 Circular plate；5—垫板 Plate；6—裙座 Skirt；7、20—引出管 Outlet pipe；8、21—补强板 Supporting plate；9—筒体 Cylinder；10—塔盘 Tray；11—塔盘支承 Support；12—接管 Connecting pipe；13—吊耳 Lug；14—封头 Head；15—塔顶吊柱 Top davit；16—防冲挡板 Impingement baffle；17—保温支承圈 Insulation supporting ring；18—挡板 Baffle；19—塔釜隔板 Column reactor partition；22—铭牌 Nameplate；23—检修口 Access hole；24—尾部吊耳 Rear lug

1.5.3.3 脱甲烷塔 Demethanizing column

1. 脱甲烷塔 Demethanizing column

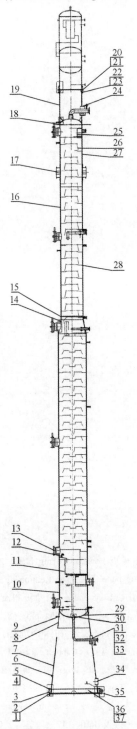

图 1.5.9 脱甲烷塔结构图
Structure drawing of demethanizer

1、20—基础环板 Foundation ring plate；2、21—筋板 Reinforcing plate；3—接地板 Grounding plate；4、22—盖板 Cover plate；5、23、36—垫板 Plate；6—铭牌 Nameplate；7—裙座 Skirt；8、19—裙座筒体 Skirt cylinder；9—排气管 Exhaust pipe；10、15、16、27—筒体 Cylinder；11—受液盘 Liquid receiving plate；12、25—防冲挡板 Impingement baffle；13—接管 Connecting pipe；14—变径段 Reducers；17—吊耳 Lug；18、29—封头 Head；24、31—引出管 Outlet pipe；26—塔盘 Tray；28—塔盘支承 Internal parts support；30—防涡器 Vortex breaker；32—垫块 Plate；33—螺钉 Screw；34—检修口 Access hole；35—尾部吊耳 Rear lug；37—底座 Base

2. 预脱甲烷塔 Pre-demethanizing column

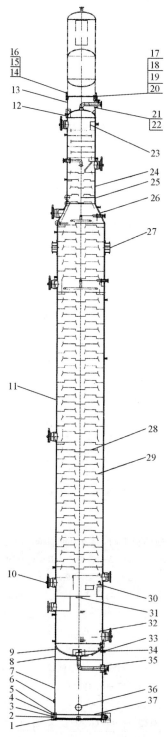

图 1.5.10 预脱甲烷塔结构图
Structure drawing of pre-demethanizing column

1、17—基础环板 Foundation ring plate；2—接地板 Grounding plate；3、18—筋板 Reinforcing plate；4、19—盖板 Cover plate；5、20—垫板 Plate；6—铭牌 Nameplate；7、8—裙座 Skirt；9、12—封头 Head；10—接管 Connecting pipe；11、13、24—筒体 Cylinder；14—螺柱 Stud；15—螺母 Nut；16—垫圈 Casket；21、35—引出管 Outlet pipe；22—补强板 Supporting plate；23、30—防冲挡板 Impingement baffle；25、28—塔盘 Tray；26—变径段 Reducers；27—吊耳 Lug；29—塔盘支承 Tray support；31—受液盘 Liquid receiving plate；32—梯子 Ladder；33—排气管 Exhaust pipe；34—防涡器 Vortex breaker；36—检修口 Access hole；37—底座 Base

1.5.3.4 脱乙烷塔 Deethanizing column

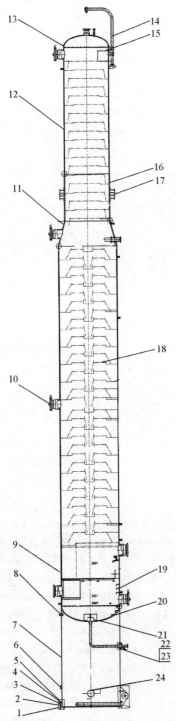

图 1.5.11 脱乙烷塔结构图
Structure drawing of deethanizing column

1—基础环板 Foundation ring plate；2—接地板 Grounding plate；3—筋板 Reinforcing plate；4—盖板 Cover plate；5—垫板 Plate；6—铭牌 Nameplate；7—裙座 Skirt；8—排气口 Exhaust port；9、12—筒体 Cylinder；10—接管 Connecting pipe；11—变径段 Reducers；13—上封头 Upper head；14—塔顶吊柱 Top davit；15—防冲挡板 Impingement baffle；16—加强筒节 Strengthening tube section；17—吊耳 Lug；18—塔盘 Tray；19—梯子 Ladder；20—下封头 Lower head cap；21—防涡器 Vortex breaker；22—引出管 Outlet pipe；23—补强板 Supporting plate；24—检修口 Access hole

1.5.3.5 脱丁烷塔 Debutanizing column

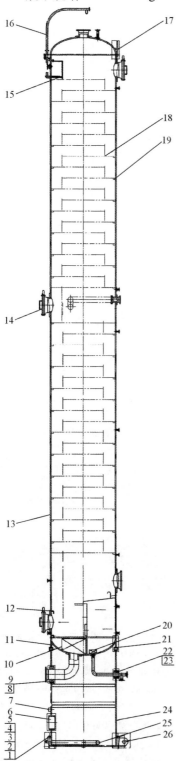

1—基础环板 Foundation ring plate；2—接地板 Grounding plate；3—筋板 Reinforcing plate；4—盖板 Cover plate；5—垫板 Plate；6—检修口 Access hole；7—铭牌 Nameplate；8、22—引出管 Outlet pipe；9、23—补强板 Support plate；10、20—防涡器 Vortex breaker；11—封头 Head；12—梯子 Ladder；13—筒体 Cylinder；14—人孔 Manhole；15—防冲挡板 Impingement baffle；16—塔顶吊柱 Top davit；17—板式吊耳 Plate lug；18—塔盘 Tray；19—保温支承圈 Insulation supporting ring；21—排气管 Exhaust pipe；24—裙座 Skirt；25—支承管 Supporting tube；26—尾部吊耳 Rear lug

图 1.5.12 脱丁烷塔结构图
Structure drawing of debutanizing column

1.5.3.6 脱辛烷塔 De-octane column

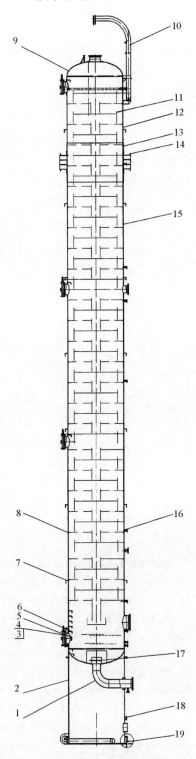

图 1.5.13 脱辛烷塔结构图
Structure drawing of de-octane column

1—引出管 Outlet pipe；2—裙座 Skirt；3—人孔 Manhole；4—补强圈 Reinforcing ring；5—把手 Hand grip；6—梯子 Ladder；7—保温支承圈 Insulation supporting ring；8—塔盘支承 Tray support；9—封头 Head；10—塔顶吊柱 Top davit；11—塔盘 Tray；12、13、15—筒体 Cylinder；14—吊耳 Lug；16—接管 Connecting pipe；17—防涡器 Vortex breaker；18—铭牌 Nameplate；19—接地板 Grounding plate

1.5.3.7 脱戊烷塔 Depentanizing column

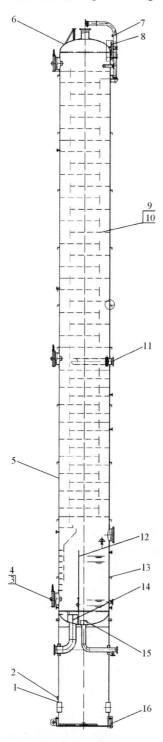

图 1.5.14 脱戊烷塔结构图
Structure drawing of depentanizing column

1—裙座 Skirt；2—铭牌 Nameplate；3—人孔 Manhole；4—补强圈 Reinforcing ring；5—筒体 Cylinder；6—封头 Head；7—塔顶吊柱 Top davit；8—吊耳 Lug；9—塔盘 Tray；10—塔盘支承 Tray support；11—接管 Connecting pipe；12—塔釜隔板 Column kettle baffle；13—保温支承圈 Insulation supporting ring；14、15—防涡器 Vortex breaker；16—接地板 Grounding plate

1.5.3.8 乙烯塔 Ethylene column

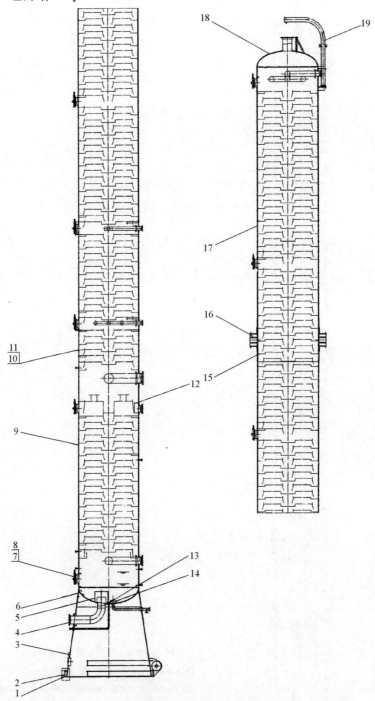

图 1.5.15 乙烯塔结构图 Structure drawing of ethylene column

1—基础环板 Foundation ring plate；2—接地板 Grounding plate；3—裙座 Skirt；4—引出管 Outlet pipe；
5、13、14—防涡器 Vortex breaker；6、18—封头 Head；7—人孔 Manhole；
8—补强圈 Reinforcing ring；9、15、17—筒体 Cylinder；10—塔盘 Tray；11—塔盘支承 Tray support；
12—防冲挡板 Impingement baffle；16—吊耳 Lug；19—塔顶吊柱 Top davit

1.5.3.9 丙烯塔 Propylene column

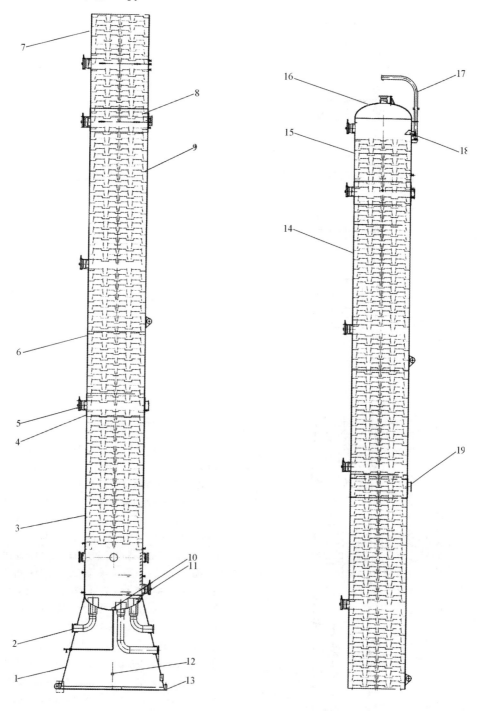

图 1.5.16 丙烯塔结构图 Structure drawing of propylene column

1—裙座 Skirt；2—引出管 Outlet pipe；3、4、6、7、14、15—筒体 Cylinder；5—人孔 Manhole；
8—塔盘 Tray；9—塔盘支承 Tray support；10、11—防涡器 Vortex breaker；12—铭牌 Nameplate；
13—接地板 Grounding plate；16—封头 Head；17—塔顶吊柱 Top davit；18、19—接管 Connecting pipe

1.5.4 洗涤塔 Washing column

1.5.4.1 碱洗塔 Alkaline column

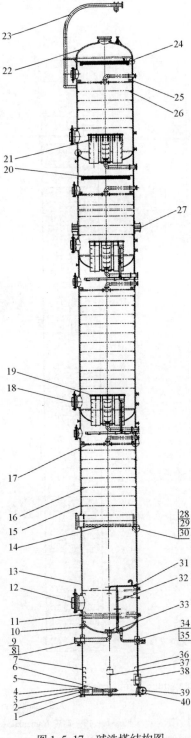

1—基础环板 Foundation ring plate；2—接地板 Grounding plate；3、40—筋板 Reinforcing plate；4—盖板 Cover plate；5—垫板 Plate；6—三角支承管 Triangle support；7—裙座 Skirt；8、34—引出管 Outlet pipe；9、29、35—补强板 Supporting plate；10—排气口 Exhaust port；11—接管 Connecting pipe；12、18—人孔 Manhole；13—筒体 Cylinder；14—进气管 Intake pipe；15、26—塔盘支承 Tray support；16—塔盘 Tray；17—物料分布器 Material distributor；19、21—烟囱 Chimney；20—丝网除沫器 Wire mesh demister；22—封头 Head；23—塔顶吊柱 Top davit；24—丝网除沫器支承 Wire mesh demister support；25—分布器支承 Distributor support；27—吊耳 Lug；28—进气管 Inlet pipe；30—支承垫板 Supporting plate；31—罩板 Shield plate；32—隔板 Baffle；33—防涡器 Vortex breaker；36—梯子 Ladder；37—铭牌 Nameplate；38—检修口 Access hole；39—尾部吊耳 Rear lug

图 1.5.17 碱洗塔结构图
Structure drawing of alkaline column

1.5.4.2 废碱氧化水洗塔 Waste alkaline oxidation washing column

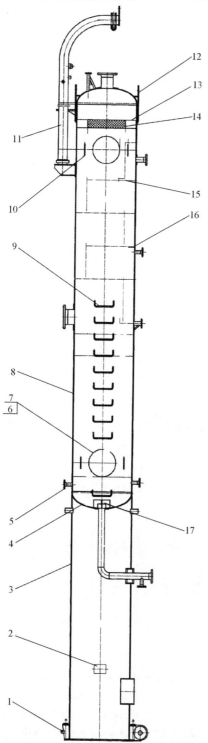

图 1.5.18 废碱氧化水洗塔结构图
Structure drawing of waste alkaline oxidation washing column

1—接地板 Grounding plate；2—铭牌 Nameplate；3—裙座 Skirt；4—封头 Head；5—接管 Connecting pipe；6—人孔 Manhole；7—补强圈 Reinforcing ring；8—筒体 Cylinder；9—梯子 Ladder；10—把手 Handle；11—塔顶吊柱 Top davit；12—吊耳 Lug；13—丝网除沫器支承 Wire mesh demister support；14—丝网除沫器 Wire mesh demister；15—塔盘 Tray；16—塔盘支承 Tray support；17—防涡器 Vortex breaker

1.6 干燥设备 Drying equipment

干燥设备包括裂解气干燥器、液体干燥器和氢气干燥器等。
Drying apparatus include drier of cracking gas, liquid drier tower and hydrogen drier, etc.

1.6.1 裂解气干燥器 Drier of cracking gas

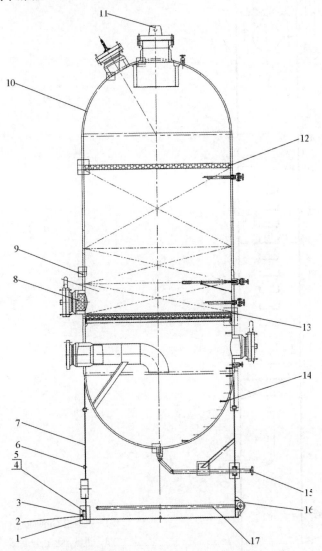

图 1.6.1 裂解气干燥器结构图 Structure drawing of drier of cracking gas
1—基础环板 Foundation ring plate；2—筋板 Reinforcing plate；3—接地板 Grounding plate；4—盖板 Cover plate；
5—垫板 Plate；6—铭牌 Nameplate；7—裙座 Skirt；8—人孔 Manhole；9—筒体 Cylinder；
10—球形封头 Spherical head；11—吊耳 Lug；12—不锈钢丝网 Stainless steel wire mesh；
13—干燥剂填料 Drier packing；14—梯子 Ladder；15—引出管 Outlet pipe；16—尾部吊耳 Rear lug；
17—支承管 Supporting steel pipe

1.6.2 液体干燥器 Liquid drier

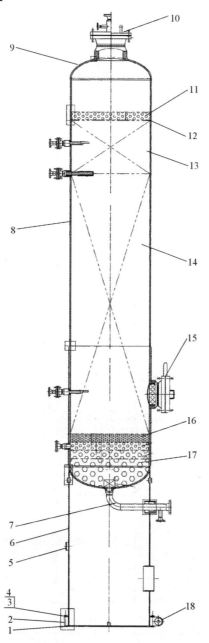

图 1.6.2 液体干燥器结构图 Structure drawing of liquid drier

1—基础环板 Foundation ring plate；2—筋板 Reinforcing plate；3—盖板 Cover plate；4—垫板 Plate；
5—铭牌 Nameplate；6—裙座 Skirt；7—引出管 Outlet pipe；8—筒体 Cylinder；
9—封头 Head；10—吊耳 Lug；11、16、17—瓷球 Porcelain ball；
12—不锈钢丝网 Stainless steel wire mesh；13—干燥剂填料(防护层) Desiccant filler(protective layer)；
14—干燥剂填料(主填料层) Desiccant filler (main filler layer)；15—人孔 Manhole；
18—尾部吊耳 Rear lug

1.6.3 氢气干燥器 Hydrogen drying tower

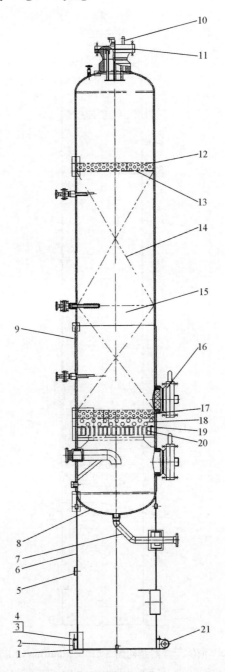

图 1.6.3 氢气干燥器结构图 Structure drawing of hydrogen drier
1—基础环板 Foundation ring plate；2—筋板 Reinforcing plate；3—盖板 Cover plate；4—垫板 Plate；
5—铭牌 Nameplate；6—裙座 Skirt；7—引出管 Outlet pipe；8—封头 Head；9—筒体 Cylinder；
10—吊耳 Lug；11—垫片 Gasket；12、17、18—瓷球 Porcelain ball；13、19—不锈钢丝网 Stainless steel wire mesh；
14、15—干燥剂填料 Drier packing；16—人孔 Manhole；20—填料支承 Packing support；
21—尾部吊耳 Rear lug

1.7 换热设备 Heat exchange equipment

1.7.1 加热器 Heater

1.7.1.1 浮头式加热器 Floating head heaters

1. 加热器结构 Structure of heater

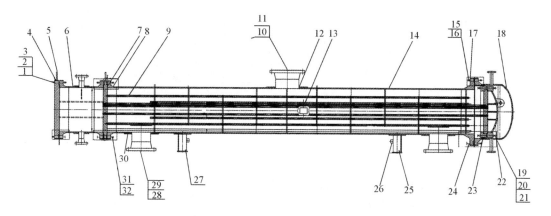

图 1.7.1 加热器结构图 Structure drawing of heater

1、19—螺母 Nut；2、15、20、31—螺栓 Stud；3—垫片 Gasket；4—平盖 Flat cover；5—吊耳 Lug；
6—管箱 Channel box；7—筒体垫片 Cylinder gasket；8—筒体法兰 Shell hange；
9—管束 Tube bundle；10、28—法兰 Flange；11、29—接管 Connecting pipe；12、30—补强圈 Reinforcing ring；
13—铭牌 Nameplate；14—筒体 Cylinder；16—外头盖垫片 Outer head gasket；
17—外头盖法兰 Outer head flange；18—外头盖 Outer head；
21—浮头垫片 Floating head gasket；22—浮头 Floating head；23—钩圈 Hook ring；
24—顶丝 Jackscrew；25、27—支座 Support；26—接地板 Grounding plate；
32—带肩螺柱 Shoulder stud

2. 加热器管束 Heater tube bundle

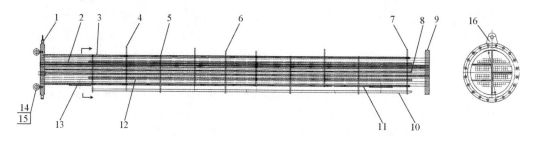

图 1.7.2 加热器管束结构图 Structure drawing of heater tube bundle

1—固定管板 Fixed tube sheet；2—换热管 Heat exchange tube；3、11—拉杆 Tie rod；4、5、6—折流板 Baffle plate；
7—支承板 Supporting plate；8—旁路挡板 Bypass baffle；9—浮动管板 Floating tub esheet；
10—滑杆 Sliding rod；12—定距管 Spacer tube；13—防冲挡板 Impingement baffle；
14—环首螺钉 Eye bolt；15—丝堵 Plug；16—吊耳 Lug

1.7.1.2 固定管板式过热器 Fixed tube plate superheater

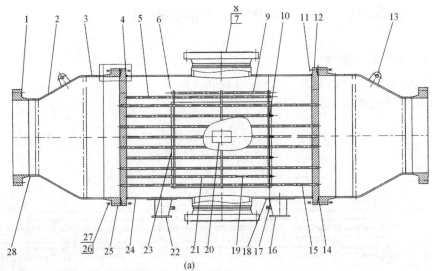

(a)

1、7、25—法兰 Flange；2—锥形封头 Conical head；3、28—短节 Swage nipple；4—左管板 Left tubesheet；
5—定距管 Spacer tube；6、23—挡板 Baffle；8—接管 Connecting pipe；9—防冲挡管 Bumping tube；
10、27—螺母 Nut；11—顶丝 Jackscrew；12—右管板 Right tubesheet；13—吊耳 Lug；14—垫片 Gasket；
15—换热管 Heat exchange tube；16—滑动支座 Sliding support；17—接地板 Grounding plate；
18—补强板 Supporting plate；19、21—拉杆 Tie rod；20—铭牌 Nameplate；22—固定支座 Fixed support；
24—筒体 Cylinder；26—双头螺柱 Double end stud

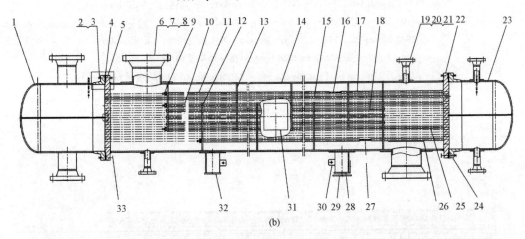

(b)

1—左管箱 Left channel box；2—双头螺栓 Double end stud；3—螺母 Nut；4、24—缠绕垫 Wound gasket；
5—左管板 Left tubesheet；6、19—法兰 Flange；7、20—接管 Connecting pipe；8、21—补强圈 Reinforcing ring；
9—螺母 Nut；10、11、27—定距管 Spacer tube；12、13—折流板 Baffle plate；
14—筒体 Cylinder；15、25—换热管 Heat exchange tube；16、17—拉杆 Tie rod；
18—挡管 Dummy tube；22—右管板 Right tubesheet；23—右管箱 Right channel box；
26—防冲挡板 Impingement baffle；28—滑动支座 Sliding saddle；29—滑板 Base slide plate；
30—接地板 Grounding plate；31—铭牌 Nameplate；32—固定支座 Fixed support；33—顶丝 Jackscrew

图 1.7.3 固定管板式过热器结构图 Structure drawing of fixed tube plate superheater

1.7.1.3 U形管式加热器 U-tube heater

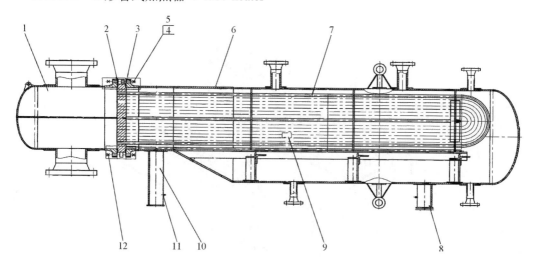

图 1.7.4 U形管式加热器结构图 Structure drawing of U-tube heater
1—管箱 Channel box；2、3—垫片 Gasket；4—带肩螺柱 Shoulder stud；5—螺母 Nut；
6—筒体 Cylinder；7—U形管束 U tube bundle；8、10—支座 Support；9—铭牌 Nameplate；
11—接地板 Grounding plate；12—顶丝 Jackscrew

1.7.2 换热器 Heat exchanger

1.7.2.1 固定管板式换热器 Fixed tube plate heat exchanger

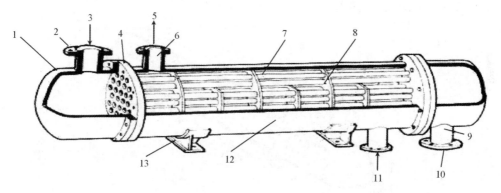

图 1.7.5 固定管板式换热器结构图 Structure drawing of fixed tube plate heat exchanger
1—管箱 Channel；2、6、9—接管 Connection pipe；3—管程入口 Tube inlet；
4—管板 Tube sheet；5—壳程出口 Shell outlet；7—换热管 Heat exchange tube；
8—折流板 Baffle；10—管程出口 Tube outlet；11—壳程入口 Shell inlet；
12—筒体 Cylinder；13—支座 Support

1.7.2.2 浮头式换热器 Floating head heat exchanger

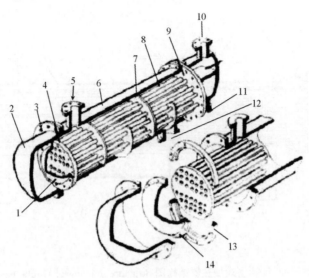

图 1.7.6 浮头式换热器结构图 Structure drawing of floating head heat exchanger
1—浮头 Floating head；2—管箱 Channel；3—法兰 Flange；4—浮头管板 Floating tubesheet；
5—壳程入口 Shell inlet；6—筒体 Shell；7—折流板 Baffle；8—换热管 Heat exchange tube；
9—固定管板 Stationary tubesheet；10—管程入口 Tube inlet；11—管程出口 Tube outlet；
12—壳程出口 Shell outlet；13—钩圈 Hooking ring；14—浮头盖 Floating head cover

1.7.2.3 U形管式换热器 U-tube heat exchanger

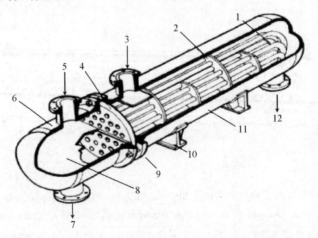

图 1.7.7 U形管式换热器结构图 Structure drawing of U-tube heat exchanger
1—U形管 U tube；2—折流板 Baffle；3—壳程入口 Shell inlet；4—管板 Tube sheet；
5—管程入口 Tube inlet；6—管箱 Channel；7—管程出口 Tube outlet；
8—隔板 Pass partition；9—法兰 Flange；10—支座 Support；11—筒体 Shell；
12—壳程出口 Shell outlet

1 乙烯 Ethylene

1.7.2.4 釜式换热器 Kettle-type heat exchanger

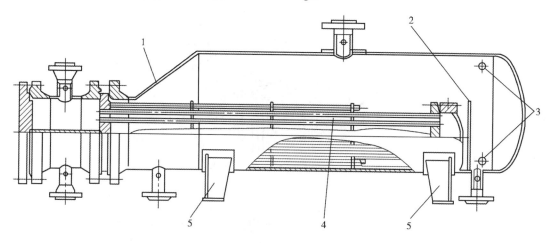

图 1.7.8 釜式换热器结构图 Structure drawing of kettle-type heat exchanger

1—锥体 Shell；2—堰板 Weir plate；3—液位计接口 Level gauge connection；
4—管束 Tube；5—支座 Support

1.7.2.5 套管式换热器 Tube heat exchanger

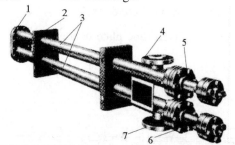

图 1.7.9 套管式换热器结构图 Structure drawing of tube heat exchanger

1—管箱法兰 Channelflange；2—支座 Support；3—外管 Outsidetube；
4、6、7—接管法兰 Connectionflange；5—内管 Insidetube；

1.7.2.6 板式换热器 Plate heat exchanger

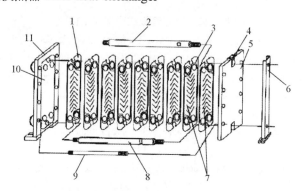

图 1.7.10 板式换热器结构图 Structure drawing of plate heat exchanger

1、7—换热板 Plate pack；2—上导杆 Carrying bar(top)；3—密封胶垫 Seal rubber gasket；4—滚轮 Scroll wheel；
5—活动夹紧板 Moveable cover；6—支柱 Support column；8—下导杆 Carrying bar(bottom)；9—夹紧螺栓 Frame nuts；
10—橡胶板 Rubber sheet；11—固定夹紧板 Stationary frame plate

1.7.2.7　板翅式换热器 Plate-fin heat exchanger

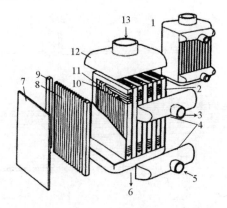

图 1.7.11　板翅式换热器结构图 Structure drawing of plate-fin heat exchanger

1—外形 Outline drawing；2—板束 Buddle；3—壳程出口 Shell outlet；
4、12—封头 Head；5—壳程入口 Shell inlet；6—管程出口 Tube outlet；
7—侧板 Cap sheet；8—换热翅片 Heat transfer fin；9—封条 Side bar；
10—导流片 Deflector；11—分配段 Distributor fin；13—管程入口 Tube inlet

1.7.3　冷却器及冷凝器 Coolers and condenser

1.7.3.1　固定管板式冷却器 Fixed tube plate condenser

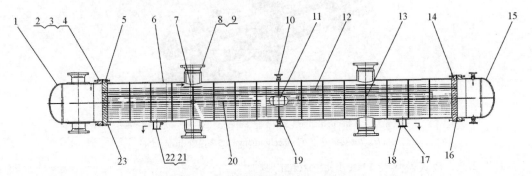

图 1.7.12　固定管板式冷却器结构图
Structure drawing of fixed tube plate condenser

1—左管箱 Left channel box；2、16—双头螺栓 Stud；3—螺母 Nut；4 垫片 Gasket；
5—左管板 Left tube plate；6—筒体 Cylinder；7—防冲挡板 Impingement baffle；
8—法兰 Flange；9、10—接管 Connecting pipe；11—铭牌 Nameplate；
12—换热管 Heat exchange tube；13、19—支承板 Support plate；
14—右管板 Right tube plate；15—右管箱 Right channel box；
17、21—支座 Support；18—滑板 Sliding plate；
20—纵向隔板 Longitudinal baffle；22—接地板 Grounding plate；
23—顶丝 Jackscrew

1.7.3.2 浮头式冷却器 Floating head cooler

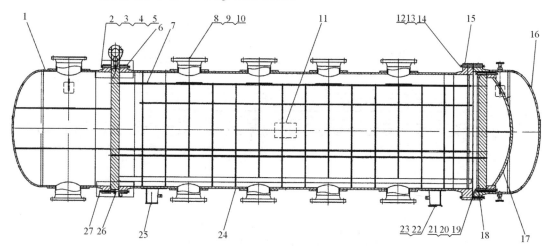

图 1.7.13 浮头式冷却器结构图 Structure drawing of floating head cooler

1—管箱 Channel box；2、12、19—双头螺栓 Stud；3、13、20—螺母 Nut；4、5、14、21—垫片 Gasket；
6、9、15—法兰 Flange；7—管束 Tube bundle；8—接管 Connecting pipe；10—补强圈 Reinforcing ring；
11—铭牌 Nameplate；16—外头盖 Outer head cover；17—浮头盖 Floating head planting；
18—钩圈 Hook ring；22—滑动支座 Sliding support；23—接地板 Grounding plate；24—筒体 Cylinder；
25—固定支座 Fixed support；26—带肩双头螺柱 Double end shoulder stud；27—顶丝 Jackscrew

1.7.3.3 U形管式过冷器 U-shaped tube subcooler

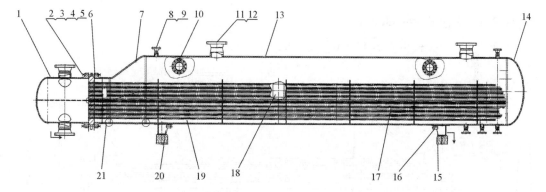

图 1.7.14 U形管式过冷器结构图 Structure drawing of U-shaped tube subcooler

1—管箱 Channel box；2—双头螺栓 Stud；3—螺母 Nut；4—带肩螺柱 Shoulder stud；
5—管箱垫片 Channel box gasket；6、11—法兰 Flange；7—锥体 Cone；
8—接管 Connecting pipe；9、12—补强圈 Reinforcing ring；
10—防冲挡板 Impingement baffle；13—筒体 Cylinder；14—封头 Head；
15、20—支座 Support；16—接地板 Grounding plate；17—管束 Tube bundle；
18—铭牌 Nameplate；19—导轨 Guide rail；21—短节 Swage nipple

1.7.3.4 固定管板式冷凝器 Fixed tube plate condenser

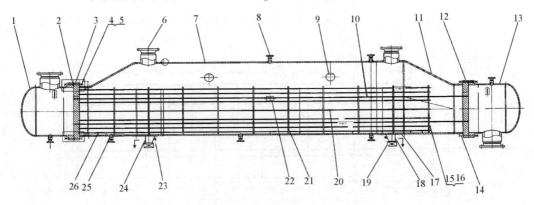

图 1.7.15 固定管板式冷凝器结构图
Structure drawing of fixed tube plate condenser

1—左管箱 Left channel box；2—垫片 Gasket；3—左管板 Left tube plate；
4—双头螺柱 Double end stud；5、16—螺母 Nut；6、8、9、25—接管 Connecting pipe；
7、17—筒体 Cylinder；10—换热管 Heat exchange tube；
11—锥体 Cone；12—右管板 Right tube plate；13—右管箱 Right channel box；
14—顶丝 Jackscrew；15—拉杆 Tie rod；18—滑动支座 Sliding support；
19—绝热层 Heat insulating layer；20、26—定距管 Spacer tube；
21—支承板 Support plate；22—铭牌 Nameplate；
23—接地板 Grounding plate；24—固定支座 Fixed support

1.7.4 再沸器 Reboiler

1.7.4.1 浮头式再沸器 Floating reboiler

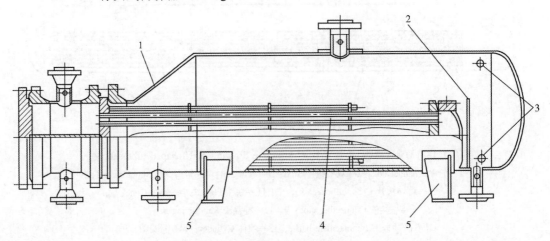

图 1.7.16 浮头式再沸器结构图 Structure drawing of floating reboiler
1—管程接管 Tube side connection；2—锥体 Cone section；3—管束 Bundle；4—壳程接管 Shell side connection；
5—小浮头 Floating head；6—支座 Support；7—管箱 Channel box

1.7.4.2 固定管板式再沸器 Fixed tube plate reboiler

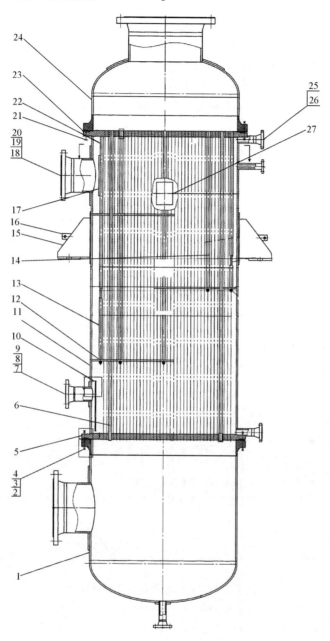

图 1.7.17 固定管板式再沸器结构图
Structure drawing of fixed tube plate reboiler

1—下管箱 Lower channel box；2—双头螺栓 Double end stud；3、23—螺母 Nut；4—垫片 Gasket；
5—下管板 Lower tube plate；6—换热管 Heat exchange tube；7、18、25—法兰 Flange；
8、19、26—接管 Connecting pipe；9、20—补强圈 Reinforcing ring；10、17—防冲挡板 Impingement baffle；
11—筒体 Cylinder；12—折流板 Baffle plate；13、22—拉杆 Tie rod；14—定距管 Spacer tube；
15—支座 Support；16—接地板 Grounding plate；21—顶丝 Jackscrew；24—上管箱 Upper channel box；
27—铭牌 Nameplate

1.7.4.3　U形管式再沸器 U-shaped tube reboiler

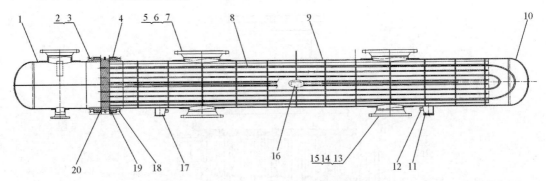

图 1.7.18　U形管式再沸器结构图
Structure drawing of U-shaped tube reboiler

1—管箱 Channel box；2—双头螺栓 Double-threaded screw；3—螺母 Nut；
4、5、13—法兰 Flange；6、14—接管 Connecting pipe；
7、15—补强圈 Reinforcing ring；8—管束 Tube bundle；9—筒体 Cylinder；
10—封头 Head；11、17—支座 Support；12—接地板 Grounding plate；
16—铭牌 Nameplate；18—顶丝 Jackscrew；19、20—垫片 Gasket

1.7.5　汽化器 Vaporiser

1.7.5.1　冷火炬排放汽化器 Cold flare emission vaporiser

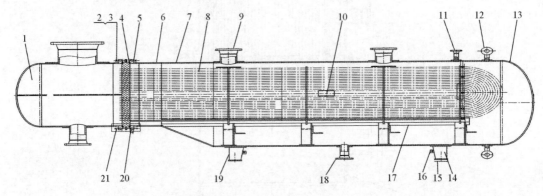

图 1.7.19　冷火炬排放汽化器结构图
Structure drawing of cold flare emission vaporiser

1—管箱 Channel box；2—双头螺栓 Double end stud；3—螺母 Nut；4—垫片 Gasket；
5—法兰 Flange；6—短节 Strengthening segment；7—筒体 Cylinder；
8—管束 Tube bundle；9、11、12、18—接管 Connecting pipe；10—铭牌 Nameplate；
13—封头 Head；14、19—支座 Support；15—滑板 Sliding plate；
16—接地板 Grounding plate；17—导轨 Guide rail；
20—挡块 Stopper；21—顶丝 Jackscrew

1 乙烯 Ethylene

1.7.5.2 循环丙烷蒸发器 Recycling propane evaporator

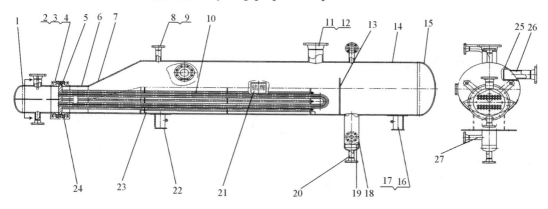

图 1.7.20 循环丙烷蒸发器结构图 Structure drawing of recycling propane evaporator
1—管箱 Channel box；2—双头螺栓 Double end stud；3—螺母 Nut；4—管箱垫片 Channel box gasket；
5—筒体垫片 Cylinder gasket；6、14—筒体 Cylinder；7—锥体 Cone；8、12、18、20、26、27—接管 Connecting pipe；
9、11、24—法兰 Flange；10—管束 Tube bundle；13—挡板 Baffle；15—封头 Head；
16—接地板 Grounding plate；17—滑动支座 Sliding support；19—管帽 Pipe cap；21—铭牌 Nameplate；
22—固定支座 Fixed support；23—导轨 Guide rail；25—防冲挡板 Impingement baffle

1.7.6 冷箱 Cold box

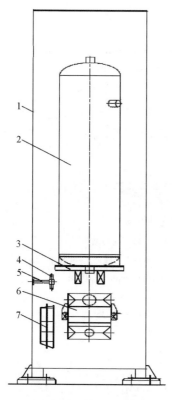

图 1.7.21 冷箱结构图
Structure drawing of cold box

1—冷箱 Cold box；2—闪蒸罐 Flash tank；3—支架 Support；4—内管 Inner pipe；5—内管架 Inner pipe rack；6—蒸发器 Evaporator；7—梯子 Ladder

· 65 ·

1.8 罐 Tank

1.8.1 分离罐 Separation tank

1.8.1.1 气-液分离罐 Gas-liquid separation tank

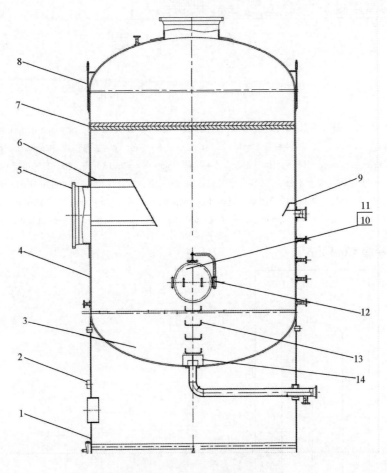

图 1.8.1 气-液分离罐结构图
Structure drawing of gas - liquid separation tank

1—裙座 Skirt；2—铭牌 Nameplate；3—封头 Head；4—筒体 Cylinder；
5—接管 Connecting pipe；6、9—防冲挡板 Impingement baffle；
7—丝网除沫器 Wire mesh demister；8—吊耳 Lug；
10—人孔 Manhole；11—补强圈 Reinforcing ring；
12—把手 Hand grip；13—梯子 Ladder；
14—防涡器 Vortex breaker

1.8.1.2 液-液分离罐 Liquid-liquid separation tank

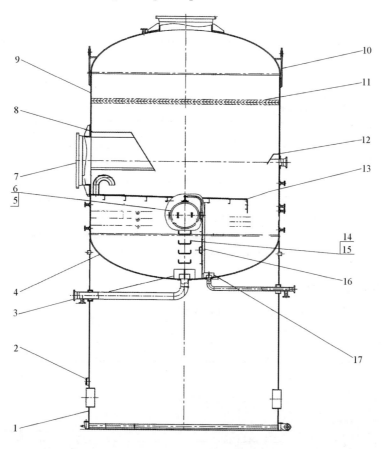

图 1.8.2 液-液分离罐结构图
Structure drawing of liquid - liquid separation tank
1—裙座 Skirt；2—铭牌 Nameplate；3、17—防涡器 Vortex breaker
4—封头 Head；5—人孔 Manhole；6—补强圈 Reinforcing ring；
7—接管 Connecting pipe；8、12—防冲挡板 Impingement baffle；
9—筒体 Cylinder；10—吊耳 Lug；11—丝网除沫器 Wire mesh demister；
13、16—隔板 Partition；14—梯子 Ladder；15—把手 Handle

1.8.1.3 热火炬分离罐 Hot flare separation tank

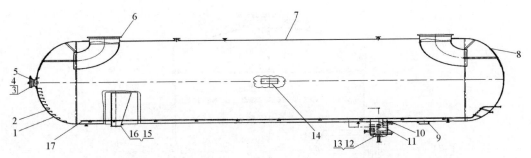

图 1.8.3 热火炬分离罐结构图 Structure drawing of hot flare separation tank
1、8、12—封头 Head；2—梯子 Ladder；3—人孔 Manhole；4、10—补强圈 Reinforcing ring；
5—把手 Hand grip；6—接管 Connecting pipe；7、11—筒体 Cylinder；9—固定支座 Fixed support；
13—防涡器 Vortex breaker；14—铭牌 Nameplate；15—滑动支座 Sliding support；
16—接地板 Grounding plate；17—直管 Straight pipe

1.8.2 缓冲罐 Buffer tank

1.8.2.1 卧式缓冲罐 Horizontal buffer tank

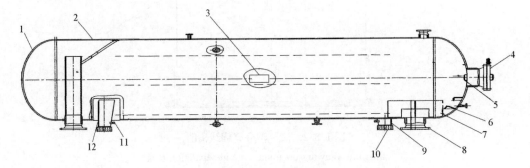

图 1.8.4 卧式缓冲罐结构图 Structure drawing of horizontal buffer tank
1—封头 Head；2—筒体 Cylinder；3—铭牌 Nameplate；4—人孔 Manhole；
5—把手 Hand grip；6—梯子 Ladder；7—防涡器 Vortex breaker；
8—接管 Connecting pipe；9—固定支座 Fixed support；
10—保冷垫木 Cold skid；11—接地板 Grounding plate；
12—滑动支座 Sliding support

1.8.2.2 立式缓冲罐 Vertical buffer tank

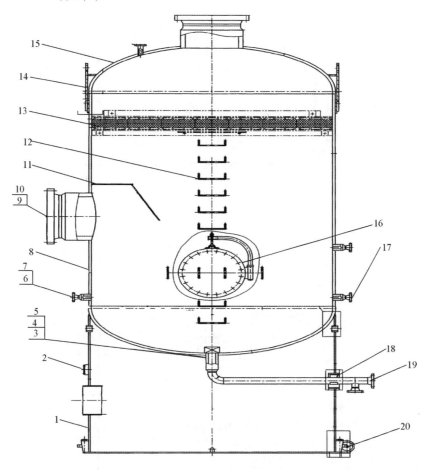

图 1.8.5 立式缓冲罐结构图

Structure drawing of vertical buffer tank

1—裙座 Skirt；2—铭牌 Nameplate；3—弯头 Elbow；
4、7、10、17、19—接管 Connecting pipe；
5—防涡器 Vortex breaker；6、9—法兰 Flange；
8—筒体 Cylinder；11—防冲挡板 Impingement baffle；
12—梯子 Ladder；13—丝网除沫器 Wire mesh demister；
14—吊耳 Lug；15—封头 Head；
16—人孔 Manhole；18—补强板 Support plate；20—尾部吊耳 Rear lug

1.8.3 闪蒸罐 Flash tank

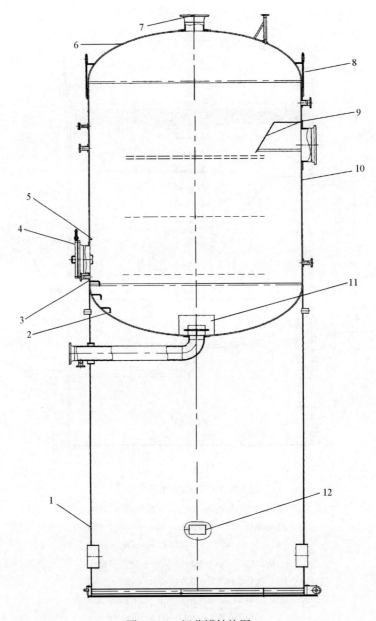

图 1.8.6 闪蒸罐结构图
Structure drawing of flash tank
1—裙座 Skirt；2—梯子 Ladder；3—补强圈 Reinforcing ring；
4—人孔 Manhole；5—把手 Hand grip；
6—封头 Head；7—接管 Connecting pipe；8—吊耳 Lug；
9—防冲挡板 Impingement baffle；10—筒体 Cylinder；
11—防涡器 Vortex breaker；12—铭牌 Nameplate

1.8.4 回流罐 Reflux tank
1.8.4.1 立式回流罐 Vertical reflux tank

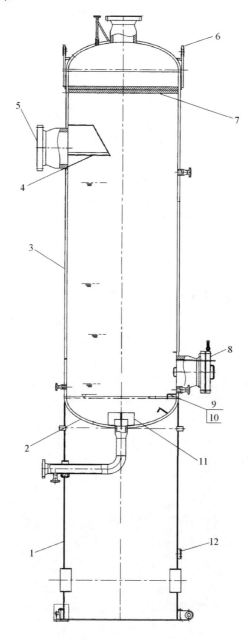

图 1.8.7 立式回流罐结构图 Structure drawing of vertical reflux tank
1—裙座 Skirt；2—封头 Head；3—筒体 Cylinder；
4—防冲挡板 Impingement baffle；5—接管 Connecting pipe；
6—吊耳 Lug；7—丝网除沫器 Wire mesh demister；8—人孔 Manhole；
9—梯子 Ladder；10—把手 Hand grip；11—防涡器 Vortex breaker；
12—铭牌 Nameplate

1.8.4.2 卧式回流罐 Horizontal reflux tank

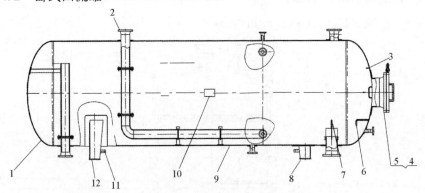

图1.8.8 卧式回流罐结构图 Structure drawing of horizontal reflux tank

1—封头 Head；2—接管 Connecting pipe；3—把手 Hand grip；4—人孔 Manhole；
5—补强圈 Reinforcing ring；6—梯子 Ladder；7—防涡器 Vortex breaker；
8、12—支座 Support；9—筒体 Cylinder；10—铭牌 Nameplate；11—接地板 Grounding plate

1.8.5 储罐 Storage tank

1.8.5.1 常压储罐 Atmospheric storage tank

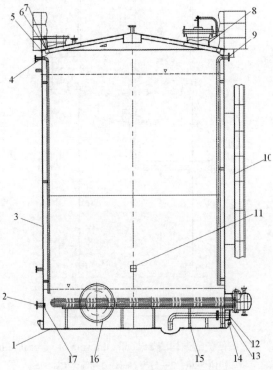

图1.8.9 常压储罐结构图 Structure drawing of atmospheric storage tank

1—底板 Foundation plate；2、4、9、12—接管 Connecting pipe；3—筒体 Cylinder；5—顶部角钢 Top angle steel；
6—罐顶支承 Tank roof support；7—顶板 Roof；8、16—人孔 Manhole；10—梯子 Ladder；
11—铭牌 Nameplate；13—地脚螺栓 Anchor bolt；14—接地板 Grounding plate；
15—换热器支架 Heat exchanger stent；17—防涡器 Vortex breaker

1.8.5.2 球罐 Spherical tank

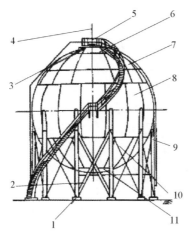

图 1.8.10 球罐结构图 Structure drawing of spherical tank

1—底板 Foundation plate; 2—拉杆 Tie rod; 3—平台 Plateform; 4—避雷针 Lighting rod;
5—安全阀 Relief valve; 6—上极板 Top crown; 7—北温带板 Upper temperature zone plate;
8—赤道带板 Equator zone plate; 9—支柱 Support; 10—南温带板 Lower temperature zone plate;
11—下极板 Bottom crown

1.8.5.3 内浮顶罐 Inner floating roof tank

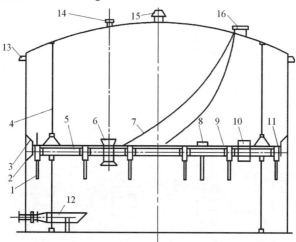

图 1.8.11 内浮顶罐结构图 Structure drawing of inner floating roof tank

1—支柱 Support; 2—滑动件 Sliding parts; 3—舌形密封 Tongue shape seal;
4—防旋转装置 Anti spin device; 5—浮顶 Floating roof; 6,14—量油孔 Hole for oil measurement;
7—静电导出装置 Statichaul-off gear; 8—真空阀 Vacuum valve; 9—铺板 Floor plate;
10—人孔 Manhole; 11—消泡沫挡板 Firecontrol foam baffle; 12—油品入口扩散管 Oil inlet diffuser;
13—罐壁通气孔 Tank wall vent hole; 15—罐顶通气孔 Tank top vent hole;
16—罐顶人孔 Tank top manhole

1.9 泵 Pump

1.9.1 油泵 Oil pump

1.9.1.1 重质油泵 Heavy oil pump

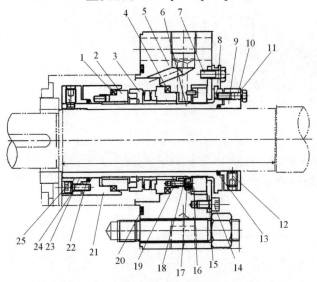

1—防转销 Stop pin；2、21—静环 Static ring；3—波纹管 Bellows；4、13、22—O 形密封圈 O sealing ring；5—卡环 Clamp ring；6—轴套 Sleeve；7、10—螺栓 Bolt；8—定位块 Stock locater block；9—驱动块 Drive block；11、18、24—弹簧垫圈 Spring washer；12—顶丝 Jackscrew；14、19、25—圆柱头螺钉 Cylindrical head screws；15—压板 Plate pinch；16—节流衬套 Throttling bush；17—弹簧 Spring；20—垫片 Gasket；23—压紧环 Clamping ring

图 1.9.1 重质油泵结构图 Structure drawing of heavy oil pump

1.9.1.2 轻质油泵 Light oil pump

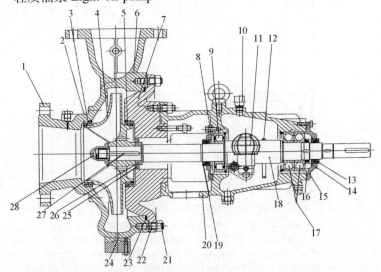

图 1.9.2 轻质油泵结构图 Structure drawing of light oil pump

1—吸入口 Suction；2、5—壳体密封环 Casing sealing ring；3—叶轮密封环 Impeller labyrinth seal ring；4—叶轮 Wheel；6—排出口 Discharge；7—泵盖 Pump cover；8、15—轴承压盖 Bearing gland；9、17—轴承 Bearing；10—放空口 Vent；11—油杯 Constant Level oiler；12—甩油环 Oil flinger；13、19—迷宫密封 Labyrinth seal；14、21、28—螺母 Nut；16—O 形密封圈 O sealing ring；18—轴 Shaft；20—轴承架 Bearing bracket；22—螺栓 Nut；23、24、26—垫片 Gasket；25—轴套 Sleeve；27—键 Key

1.9.2 水泵 Water pump

1.9.2.1 锅炉给水泵 Boiler feed water pump

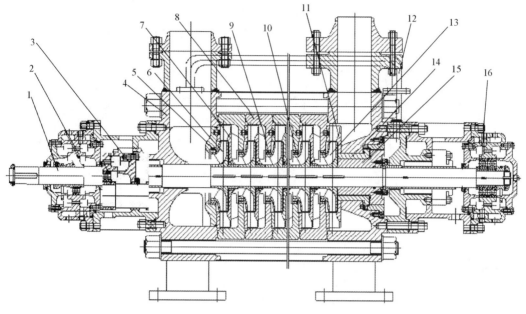

（a）节段式锅炉给水泵

1—轴 Shaft；2—轴承 Bearing；3—机械密封 Mechanical seal；4—吸入段 Suction casing；
5—首级叶轮密封环 First stage impeller seal ring；6—首级叶轮 First stage impeller；7—导叶 Diffuser；
8—隔板 Baffle；9—叶轮密封环 Impeller labyrinth seal ring；10—壳体密封环 Casing sealing ring；
11—末级导叶 End-stage guide vane；12—平衡管 Balance pipe；13—末级叶轮 End-stage Impeller；
14—平衡座 Balancing Block；15—平衡盘 Balance disk；16—推力轴承 Thrust bearing

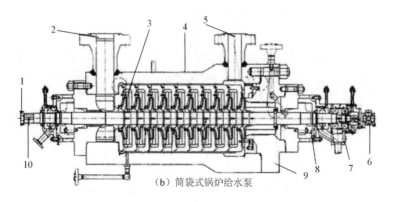

（b）筒袋式锅炉给水泵

1—联轴器 Coupling；2—吸入口 Suction；3—叶轮 Wheel；4、9—壳体 Casing；
5—排出口 Discharge；6—轴头油泵 Main oil；7—轴承 Bearing；8—机械密封 Mechanical seal；10—键 Key

图 1.9.3 锅炉给水泵结构图 Structure drawing of boiler feed water pump

1.9.2.2 急冷水泵 Quench water pump

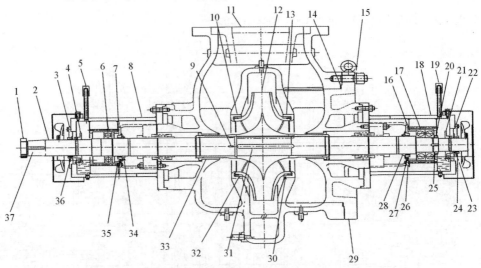

图 1.9.4 急冷水泵结构图 Structure drawing of quench water pump

1—联轴器 Coupling；2—轴 Shaft；3、7、16、22—压盖 Gland；4、21—甩油环 Oil slinger；5、19—放空口 Vent；
6、17—轴承 Bearing；8、18—轴承箱 Bearing box；9、31、37—键 Key；10、13—壳体密封环 Casing sealing ring；
11—排出口 Dichorge；12—叶轮 Wheel；14、15、24、25、26—垫圈 Gasket；20—螺母 Nut；
23、27、35、36—轴承压盖 Bearing gland；28、34—导流板 Deflector；29—泵壳 Shell；
30、32—叶轮密封环 Impeller labyrinth seal ring；33—轴套 Sleeve

1.9.2.3 自吸泵 Self-priming pump

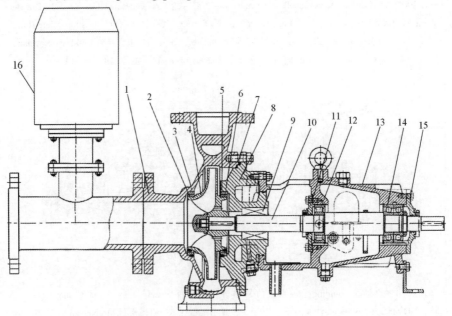

图 1.9.5 自吸泵结构图 Structure drawing of self-priming pump

1—泵壳 Pump casing；2、6—叶轮密封 Impeller seal；3、5—壳体密封 Casing sealing ring；
4—叶轮 Wheel；7—螺栓 Gasket；8—泵盖 Pump cover；9—机械密封 Mechanical seal；10—泵轴 Pump shaft；
11、15—轴承压盖 Bearing gland；12、14—轴承 Bearing；13—轴承箱 Bearing box；16—抽真空罐 Vacuum tank

1.9.2.4 液下泵 Submerged pump

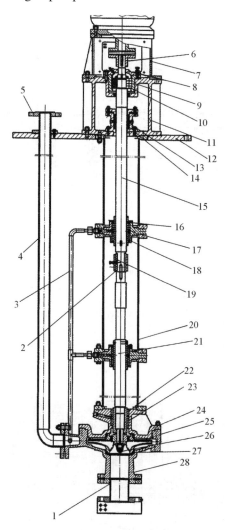

图 1.9.6 液下泵结构图
Structure drawing of submerged pump

1—吸入管 Suction pipe；2—联轴键 Spindle key；3—自冲洗管 Self flushing pipe；
4—出液管 Outlet pipe；5—出液口 Liquid outlet；6、19—联轴器 Coupling；
7—电机支架 Motor bracket；8—轴承压盖 Bearing cover；9—轴承箱 Bearing box；
10—轴承 Bearing；11—轴承架 Bearing bracket；12—底板 Base plate；
13—水封环 Water seal cage；14—填料函 Stuffing box；15—上轴 Upper shaft；
16—中间轴套 Middle shaft sleeve；17—滑动轴承箱 Sliding bearing box；
18—中间轴衬 Intermediate bush；20—支承管 Supporting tube；21—下轴 Lower shaft；
22—下轴衬 Lower shaft bushing；23—下轴套 Lower shaft sleeve；
24—泵盖 Pump cover；25—叶轮 Wheel；26—叶轮螺母 Impeller nut；
27—叶轮密封环 Impeller labyrinth seal ring；28—泵壳 Pump casing

1.9.3 烯烃泵 Olefin pump

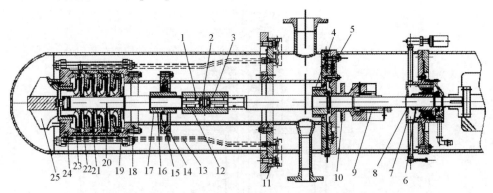

图 1.9.7 烯烃泵结构图 Structure drawing of olefin pump

1、20—泵轴 Pump shaft；2—定位环 Locating ring；3—驱动轴 Drive shaft；
4—平衡鼓套 Balance drum sets；5—平衡鼓 Balance drum；6—轴承套 Bearing bush；
7—轴承箱 Bearing box；8—压盖 Gland；9—机械密封 Mechanical seal；
10—密封箱体 Sealed box；11—进出口段 Inlet and outlet section；12—连接箱 Connecting box；
13—软管 Hose；14、18—圆柱管 Cylindrical tube；15—支承板 Support plat；16—下轴套 Lower shaft sleeve；
17—滑动轴承 Sliding Bearing；19—出口端盖 Outlet cover；21—导叶 Diffuser；22—中段壳体 Intermediate housing；
23—叶轮 Wheel；24—入口端盖 Inlet cap；25—扩散管 Diffusing pipe

1.9.4 碱泵 Alkali pump

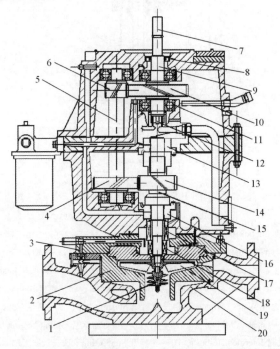

图 1.9.8 碱泵结构图 Structure drawing of alkali pump

1—诱导轮 Inducer；2—扩压器 Diffuser；3—分离器 Extractor；
4、6—传动齿轮 Transmission gear；5—惰性轴 Inertia shaft；7—连接轴 Connecting shaft；
8—轴封 Shaft seal；9—驱动轴 Low-speed shaft；10—驱动齿轮 Driving gear；11、16—轴承 Bearing；
12—内部润滑泵 Internal lubrication pump；13—轴颈轴承 Journal bearing；14—从动齿轮 Driven gear；
15—从动轴 High-speed shaft；17、18—机械密封 Mechanical seal；19—叶轮 Wheel；
20—泵体 Pump body

1.9.5 润滑油泵 Lubrication pump

1.9.5.1 单螺杆泵 Single-screw pump

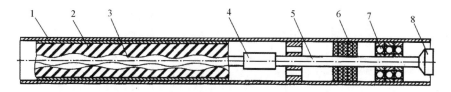

图 1.9.9 单螺杆泵结构图 Structure drawing of single-screw pump

1—泵壳 Pump casing；2—衬套 Bushing；3—螺杆 Screw；4—联轴节 Shaft coupling；
5—中间传动轴 Intermediate driving shaft；6—密封装置 Sealing device；
7—径向止推轴承 Radial thrust bearing；8—联轴器 Coupling

1.9.5.2 三螺杆泵 Three-screw pump

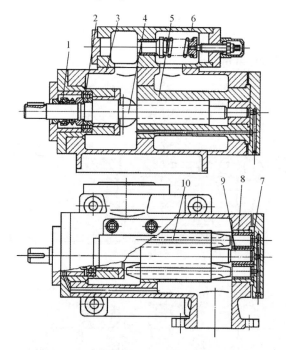

图 1.9.10 三螺杆泵结构图 Structure drawing of three-screw pump

1—机械密封 Mechanical seal；2—滚动轴承 Rolling bearing；
3—泵体 Pump body；4—主动螺杆 Driving spindle；
5—泵套 Pump sets；6—安全阀组 Valve group；7—止推垫 Gasket；
8—从动螺杆衬套 Driven screw bushing；
9—主动螺杆衬套 Driving screw bushing；
10—从动螺杆 Driven screw

1.9.5.3 双螺杆泵 Two-screw pump

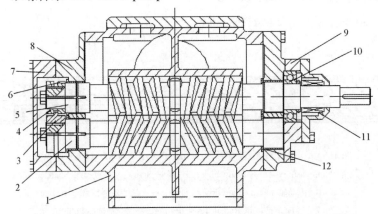

图 1.9.11 双螺杆泵结构图 Structure drawing of two-screw pump

1—泵体 Pump body；2—从动螺杆 Driven screw；3—从动轮 Driven gear；4—主动轮 Drive gear；
5—主动螺杆 Driving screw；6—衬套 Bushing；7—后端盖 Rear end cover；8—后盖 Rear cover；
9—轴承座 Bearing seat；10—轴承 Bearing；11—机械密封 Mechanical seal；12—轴套 Sleeve

1.9.5.4 齿轮泵 Gear pump

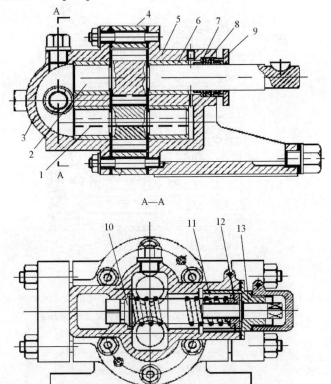

图 1.9.12 齿轮泵（外啮合）结构图 Structure drawing of gear pump（outer gearing）

1—从动齿轮 Driven gear；2—主动齿轮 Driving gear；3—阀体 Valve body；4—泵体 Pump body；
5—轴承座 Bearing seat；6—轴套 Sleeve；7—密封 Seal；8—挡圈 Retainer ring；
9—压紧盖 Tightening cap；10—阀芯 Spool valve；11—弹簧 Spring；
12—弹簧座 Spring seat；13—弹簧轴盖 Spring shaftcover

1 乙烯 Ethylene

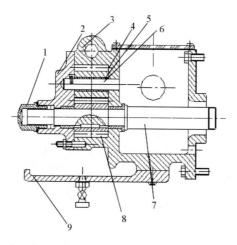

图 1.9.13 齿轮泵(内啮合)结构图 Structure drawing of gear pump (internal gearing)
1—盖型螺母 Cap nut；2—泵盖 Pump cover；3—泵体 Pump body；
4—从动齿轮 Driven gear；5—轴套 Sleeve；6—小轴 Small shaft；
7—主轴 Shaft；8—主动齿轮 Driving gear；9—集油托盘 Oil collecting tray

1.9.6 注剂泵 Agent injection pump

1.9.6.1 隔膜式注剂泵 Diaphragm agent injection pump

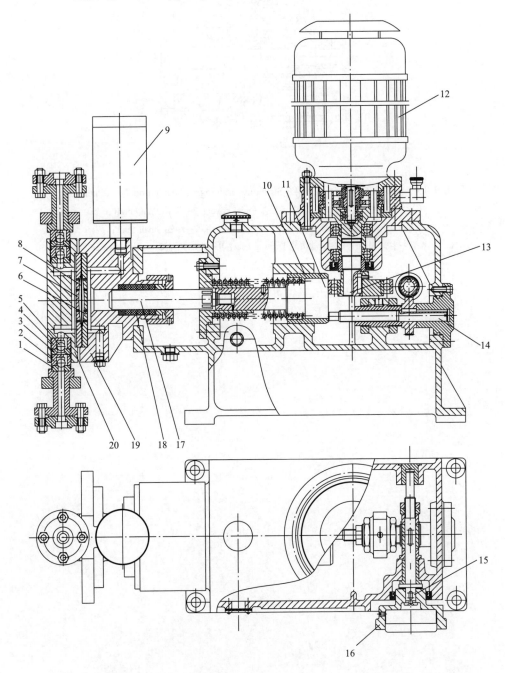

图 1.9.14　隔膜式注剂泵结构图 Structure drawing of diaphragm agent injection pump
1—下阀座 Lower valve seat；2—下导向套 Lower guide sleeve；3—上阀座 Upper valve seat；
4—上导向套 Upper guide sleeve；5—阀球 Valve ball；6—前限制板 Front limit plate；7—后限制板 Rear limit plate；
8—隔膜 Diaphragm；9—补油阀 Filling valve；10—导向套 Guide sleeve；11—十字头 Crosshead；
12—电机 Motor；13—输出轴 Output shaft；14—调节杆 Regulating bar；15—调节轴 Regulating shaft；
16—调节转盘 Knob；17—柱塞 Plunger；18—密封环 Sealing ring；19—前缸头 Front cylinder head；
20—后缸头 Rear cylinder head

1.9.6.2 柱塞式注剂泵 Piston agent injection pump

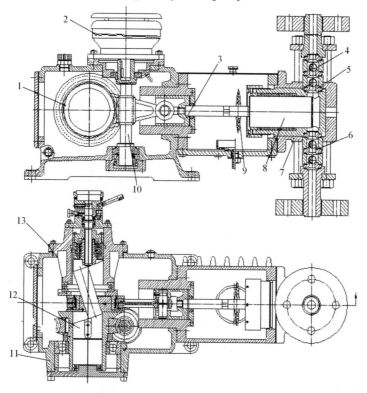

图1.9.15 柱塞式注剂泵结构图 Structure drawing of piston agent injection pump
1—蜗轮 Worm wheel；2—电机 Motor；3—十字头 Crosshead；4—阀球 Valve ball；
5—排阀座 Row valve seat；6—吸阀座 Suction valve seat；7—液缸体 Liquid cylinder；8—柱塞 Plunger；
9—挡液环 Liquid retaining ring；10—蜗杆 Worm rod；11—传动箱 Transmission block body；
12—蜗轮轴 Worm shaft；13—上套筒 Upper sleeve

1.9.7 蒸汽喷射泵 Steam injection pump

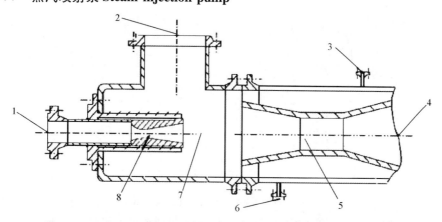

图1.9.16 蒸汽喷射泵结构图 Structure drawing of steam injection pump
1—喷射蒸汽进口 Jet steam inlet；2—吸入口 Suction；3—保温蒸汽进口 Thermal steam inlet；
4—排气口 Exhaust port；5—扩压器 Diffuser；6—保温蒸汽出口 Thermal steam outlet；
7—喷嘴室 Nozzle box；8—喷嘴 Jet nozzle

1.9.8 其它泵 Other pumps

1.9.8.1 屏蔽泵 Shielding pump

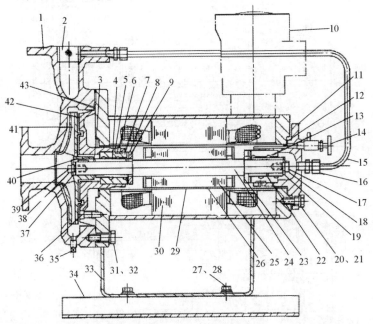

图 1.9.17 屏蔽泵结构图 Structure drawing of shielding pump

1—泵体 Pump body；2—过滤器 Filter；3—调整垫片 Adjustment gasket；4—轴套 Sleeve；
5—垫片 Gasket；6—紧定螺钉 Cock screw；7、13—轴承 Bearing；8、23—推力环 Thrust ring；
9—销 Pin；10 接线盒 Terminal box；11、42—端盖 Cover；12、43—密封垫片 Sealing washer；
14—排气阀 Vent valve；15—循环管 Circulation pipe；16—活接头 Union；17、21、28、31、36、39—螺钉 Screw；
18、38—止推螺母 Thrust nut；19、20、27、32、37—垫片 Gasket；22—轴套 Sleeve；
24—轴 Shaft；25—转子 Rotor；26—转子屏蔽套 Rotor can；29—定子屏蔽套 Stator can；30—定子 Stator；
33—机架 Pump support；34—底座 Foundation plate；35—丝堵 Plug；40—键 Key；41—叶轮 Wheel

1.9.8.2 磁力泵 Magnetic pump

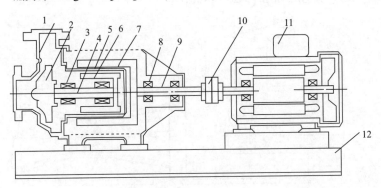

图 1.9.18 磁力泵结构图 Structure drawing of magnetic pump

1—泵体 Pump body；2—叶轮 Wheel；3—滑动轴承 Journal bearing；4—泵内轴 Pump inner shaft；
5—隔离套 Shroud；6—内磁钢 Inner magnetic body；7—外磁钢 Outer magnetic body；
8—止推轴承 Thrust bearing；9—驱动轴 Driving shaft；10—联轴器 Coupling；
11—电机 Motor；12—底座 Baseplate

1.9.8.3 计量泵 Metering pump

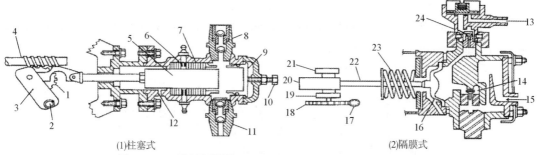

(1)柱塞式   (2)隔膜式

图 1.9.19 计量泵结构图 Structure drawing of metering pump

1—齿轮 Gear；2—滑动枢轴 Sliding pivot；3—摇摆臂 Yoke；4—电机轴 Motor shaft；5—柱塞 Plunger；
6—灯笼环 Lantern ring；7—填料 Packing；8—出口阀座 Discharge valve seat；9—填料压盖 Stuffing gland；
10—填料压紧螺栓 Packing back-up bolt；11—吸入阀座 Suction valve seat；12—填料液排出孔 Packing drain；
13—出口 Outlet；14—吸入阀 Suction valve；15—入口 Inlet；16—隔膜 Diaphragm；
17—带蜗杆的电机轴 Motor shaft with worm；18—驱动齿轮 Drive gear；19、21—轴承 Bearing；
20—偏心轮 Eccentric；22—推杆 Push rod；23—复位弹簧 Return spring；24—排出阀 Discharge valve

1.10 特种阀门 Special valves

1.10.1 蝶阀 Butterfly valve

1.10.1.1 蝶阀 Butterfly valve

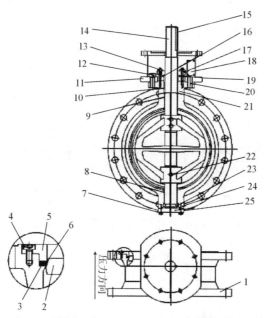

图 1.10.1 蝶阀结构图 Structure drawing of butterfly valve

1—阀体 Valve body；2—碟板 Disc；3—阀座 Valve seat；4、7、12—螺栓 Bolt；5、17—压板 Flange；6—阀座垫片 Seat gasket；8—衬套 Lower sleeve；9—上套筒 Upper sleeve；10—填料 Packing；11—支架 Support；13—螺母 Nut；14—阀杆 Stem；15—平键 Flat key；16—螺柱 Gland bolt；18—铭牌 Nameplate；19—压套 Gland；20、22—销 Pin；21—填料函 Packing washing；23—止推轴承 Block bearing；24—垫片 Gasket；25—底盖 Bottom cap

1.10.1.2 偏心蝶阀 Eccentric butterfly valve

偏心蝶阀分为单偏心蝶阀、双偏心蝶阀和三偏心蝶阀。

Eccentric butterfly valve includes single eccentric butterfly valve, double-eccentric butterfly valve and three-eccentric center butterfly valve.

1. 单偏心蝶阀 Single eccentric butterfly valve

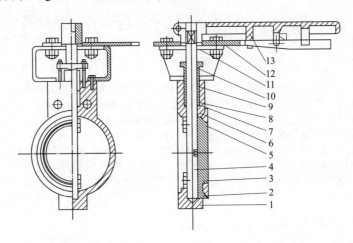

图1.10.2 单偏心蝶阀结构图 Structure drawing of single eccentric butterfly valve

1—阀体 Valve body；2—阀套 Valve sleeve；3—碟板 Dish plate；4—阀杆 Valve stem；5—密封 Seal；
6—轴套 Sleeve；7—填料函 Stuffing box；8—填料 Stuffing；9—填料压盖 Stuffing gland；10—压盖螺栓 Gland bolt；
11—支架 Support；12—指示板 Indicator plate；13—手柄 Handle

2. 双偏心蝶阀 Double-eccentric butterfly valve

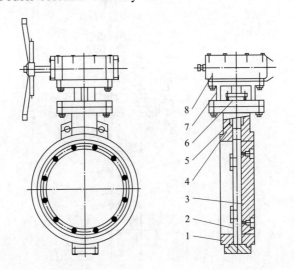

图1.10.3 双偏心蝶阀结构图 Structure drawing of double-eccentric butterfly valve

1—阀体 Valve body；2—阀套 Valve sleeve；3—碟板 Dish plate；
4—填料函 Stuffing box；5—填料 Stuffing；6—填料压盖 Stuffing gland；
7—支架 Support；8—蜗轮蜗杆 Worm gearing and worm shaft

3. 三偏心蝶阀 Three-eccentric center butterfly valve

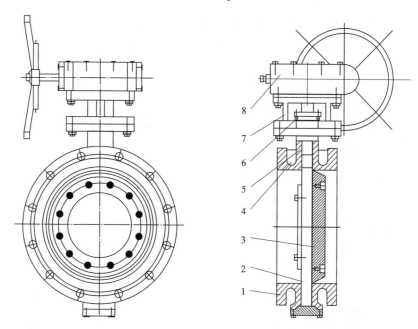

图 1.10.4　三偏心蝶阀结构图 Structure drawing of three-eccentric center butterfly valve
1—阀体 Valve body；2—阀套 Valve sleeve；3—碟板 Dish plate；
4—填料函 Stuffing box；5—填料 Stuffing；6—填料压盖 Stuffing gland；
7—支架 Support；8—蜗轮蜗杆 Worm gearing and worm shaft

1.10.2 闸阀 Gate valve

1.10.2.1 低温楔式单闸板闸阀 Low temperature wedge single gate valve
1.10.2.2 明杆楔式单闸板闸阀 Stem single wedge gate valve

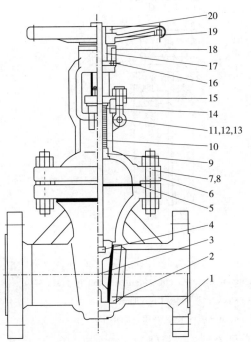

图 1.10.5 低温楔式单闸板闸阀结构图
Structure drawing of low temperature wedge single gate valve

1—阀体 Valve body；2—阀座 Valve seat；
3—闸板 Gate disc；4—阀杆 Valve rod；
5—垫片 Gasket；6—阀盖 Valve cover；
7—螺栓 Bolt；8、13—螺母 Nut；
9—上密封座 Upper sealing seat；
10—填料 Packing；11—销 Pin；
12—活节螺栓 Gland bolt；
14—填料压盖 Stuffing gland；
15—填料压板 Packing gland；
16—油杯 Lubricator；
17—阀杆螺母 Valve rod nut；
18—轴承压盖 Bearing gland；
19—手轮 Handwheel；
20—手轮螺母 Handwheel nut

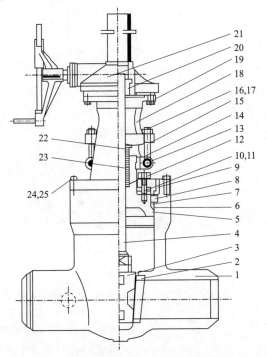

图 1.10.6 明杆楔式单闸板闸阀结构图
Structure drawing of stem single wedge gate valve

1—阀体 Valve body；2—阀座 Valve seat；
3—闸板 Gate disc；4—阀杆 Valve rod；
5—阀盖 Valve cover；6—密封环 Seal ring；
7—垫片 Gasket；
8—四开环 Segment ring；
9—牵制环 Supporting plate；
10、19、24—螺栓 Bolt；
11、17、25—螺母 Bonnet nut；
12—填料函 Stuffing box；
13—销 Pin；14—卡箍 Spit ring；
15—填料压板 Packing gland；
16—活节螺栓 Gland bolt；18—支架 Support；
20—阀杆螺母 Stem nut；21—齿轮箱 Gear box；
22—填料压盖 Stuffing gland；23—填料 Packing

1 乙烯 Ethylene

1.10.2.3 双闸板平板闸阀 Double disc flat gate valve

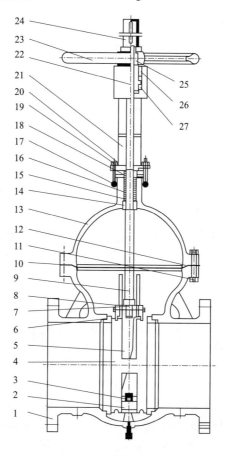

图 1.10.7 双闸板平板闸阀结构图
Structure drawing of double disc flat gate valve

1—阀体 Valve body；2—副闸板架 Support；3—弹簧 Spring；
4—闸板 Gate disc；5—主闸板架 Support；
6—阀座 Valve seat；7—键 Key；8—定位块 Gauge block；
9—阀杆 Stem；10、25—螺栓 Bolt；11—螺母 Nut；
12—垫片 Gasket；13—阀盖 Valve cover；
14—上密封座 Top seal seat；15—填料 Packing；
16—隔环 Baffle plate；17—销 Pin；
18—填料压盖 Stuffing gland；19—填料压板 Packing gland；
20—活节螺栓 Gland bolt；21—支架 Support；
22—油杯 Lubricator；23—手轮 Handwheel；
24—阀杆罩 Stem enclosure；
26—阀杆螺母 Stem nut；27—轴承 Bearing

1.10.3 截止阀 Globe valve

1.10.3.1 直通式截止阀 Straight-through globe valve
1.10.3.2 直流式截止阀 Oblique stop valve

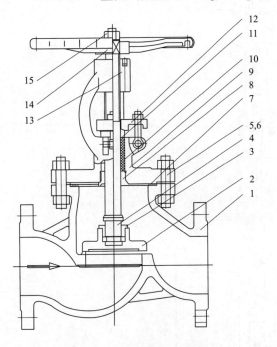

图 1.10.8 直通式截止阀结构图
Structure drawing of straight-through globe valve
1—阀体 Valve body；2—阀瓣 Valve disc；
3—压盖 Disc cover；4—阀杆 Stem；5—螺栓 Bolt；
6、15—螺母 Nut；7—垫片 Gasket；8—密封座 Sealing seat；
9—阀盖 Valve cover；10—填料 Packing；
11—填料压盖 Stuffing gland；12—填料压板 Gland flange；
13—阀杆螺母 Stem nut；14—手轮 Handwheel

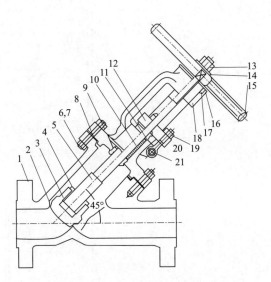

图 1.10.9 直流式截止阀结构图
Structure drawing of oblique stop valve
1—阀体 Valve body；2—阀座 Valve seat；
3—阀瓣 Valve disc；4—压盖 Disc cover；
5—阀杆 Stem；6、13、19—螺母 Nut；
7、17、20—螺栓 Bolt；8、14—垫片 Gasket；
9—密封座 Sealing seat；10—填料 Packing；
11—填料压盖 Stuffing gland；12—填料压板 Gland flange；
15—手轮 Handwheel；16—阀杆螺母 Stem nut；
18—阀盖 Valve cover；21—销 Pin

1.10.3.3 角式截止阀 Angle stop valve

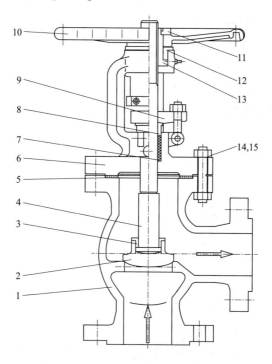

图 1.10.10 角式截止阀结构图 Structure drawing of angle stop valve
1—阀体 Valve body；2—阀瓣 Valve bisc；3、12—压盖 Disc cover；4—阀杆 Stem；
5—垫片 Gasket；6—阀盖 Valve cover；7—填料 Stuffing；8—填料函 Stuffing box；
9—填料压盖 Stuffing gland；10—手轮 Handwheel；11、14—螺母 Nut；
13—阀杆螺母 Stem nut；15—螺栓 Bolt

1.10.3.4 低温截止阀 Low temperature globe valve

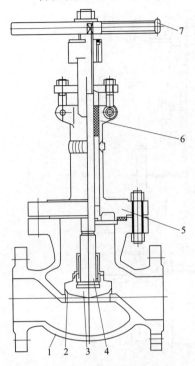

图 1.10.11 低温截止阀结构图
Structure drawing of low temperature globe valve

1—阀体 Valve body；2—阀座 Valve seat；3—阀瓣 Valve disc；4—阀杆 Valve rod；5—阀盖 Valve cover；6—支架 Support；7—手轮 Handwheel

1.10.4 止回阀 Check valve

1.10.4.1 旋启式止回阀 Swing check valve

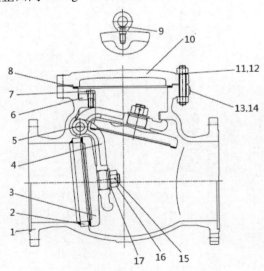

图 1.10.12 旋启式止回阀结构图 Structure drawing of swing check valves

1—阀体 Valve body；2—阀座 Valve seat；3—阀瓣 Valve disc；4—摇杆 Hinge；5—销轴 Hinge pin；6—支架 Support；7、11—螺栓 Bolt；8—垫片 Gasket；9—吊环 Lifting eye bolt；10—阀盖 Valve cover；12、17—螺母 Nut；13—铆钉 Rive；14—铭牌 Nameplate；15—防转销 Pin；16—阀瓣螺母 Nut

1.10.4.2 升降式止回阀 Lift check valve

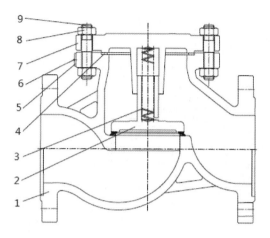

图 1.10.13 升降式止回阀结构图 Structure drawing of lift check valve
1—阀体 Valve body；2—阀瓣 Valve disc；3—弹簧 Spring；4—垫片 Gasket；5—铆钉 Rive；
6—铭牌 Nameplate；7—阀盖 Valve cover；8—螺栓 Bolt；9—螺母 Nut

1.10.4.3 轴流式止回阀 Aaxial flow check valve

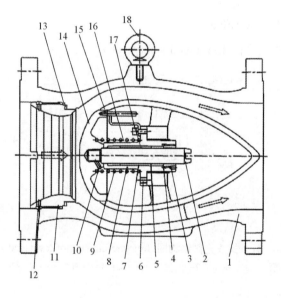

图 1.10.14 轴流式止回阀结构图 Structure drawing of aaxial flow check valve
1—阀体 Valve body；2—阀杆 Shaft；3、6、10、12、15—螺栓 Bolt；4—压盖 Gland；
5—轴承 Bearing；7—隔环 Retainer；8、17—导向套 Guiding bushing；
9、16—弹簧 Spring；11—阀座 Valve seat；13—垫片 Gasket；14—阀瓣 Valve disc；
18—吊环 Lifting eye bolt

1.10.5 旋塞阀 Plug valve

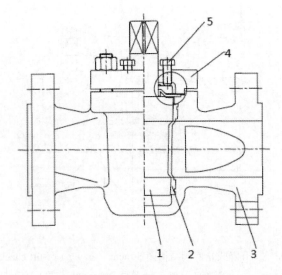

图 1.10.15 旋塞阀结构图 Structure drawing of plug valve
1—阀座 Valve seat；2—衬套 Bushing；
3—阀体 Valve Body；4—阀盖 Valve cover；5—调整螺栓 Setting bolt

1.10.6 安全阀 Safety valve

1.10.6.1 杠杆式安全阀 Leveraged safety valve

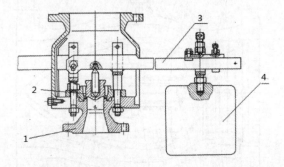

图 1.10.16 杠杆式安全阀结构图 Structure drawing of leveraged safety valve
1—阀体 Valve Body；2—阀瓣 Valve disc；3—杠杆 Lever；4—重锤 Tup

1.10.6.2 弹簧式安全阀 Spring-loaded safety valve

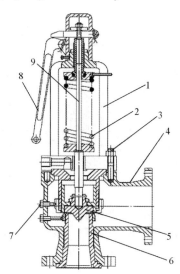

图 1.10.17 弹簧式安全阀结构图 Structure drawing of spring-loaded safety valves
1—阀盖 Valve cover；2—弹簧 Spring；3—螺栓 Stud；4—阀体 Valve body；5—阀盘 Valve disc；
6—阀座 Valve seat；7—定位销 Ring pin；8—手柄 Handle；9—阀杆 Stem

1.10.7 疏水阀 Traps

1.10.7.1 机械型疏水阀 Mechanical trap

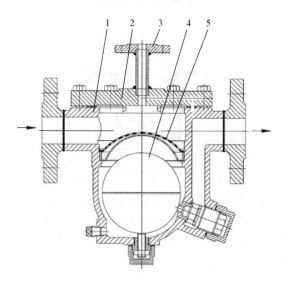

图 1.10.18 机械型疏水阀结构图 Structure drawing of mechanical trap
1—阀体 Valve body；2—阀盖 Valve cover；3—法兰 Flange；
4—浮球 Floating ball；5—过滤器 Filter

1.10.7.2 浮球式空气疏水阀 Floating ball air trap

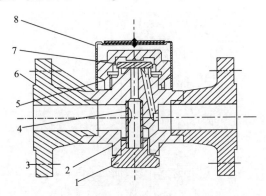

图 1.10.19　浮球式空气疏水阀结构图 Structure drawing of floating ball air trap
1—螺塞 Plug；2—阀体 Valve Body；3—螺栓、螺母 Bolt，nut；4—过滤器 Filter；
5—阀盖 Valve cover；6—法兰 Flange；7—圆盘芯 Disc cone；8—阀帽 Valve cap

1.10.8　裂解气阀 Cracking gas valve

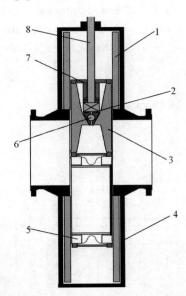

图 1.10.20　裂解气阀结构图 Structure drawing of cracking gas valve
1—导板 Guiding plate；2—楔形块 Woven shape block；3—闸板 Gate；4—阀体 Valve Body；
5—膨胀环 Inflation ring；6—球体 Ball；7—载体 Carrier；8—阀杆 Stem

1.10.9 防喘振阀 Anti-surge valve

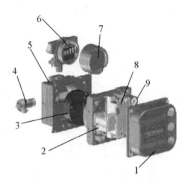

图 1.10.21　防喘振阀结构图
Structure drawing of anti-surge valve

1—防护盖 Protecting cover；2—I/P 转换器 I/P converter；3—电路板 Circuit board；
4—行程传感器 Distance sensor；5—壳体 Casing；6—接线盒 Junction box；
7—接线盒盖 Junction box cover；8—气动放大器 Pneumatic amplifier；
9—压力表 Pressure gauge

1.10.10 无冲击止回阀 Non impact check valve

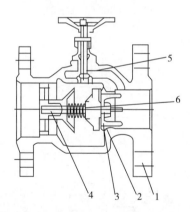

图 1.10.22　无冲击止回阀结构图
Structure drawing of non impact check valve

1—阀体 Valve body；2、5—阀座 Valve seat；3—阀瓣 Valve disk；
4—阀杆 Valve stem；6—弹簧 Spring

1.11 其他设备 Other equipments

1.11.1 聚结器 Coalescer

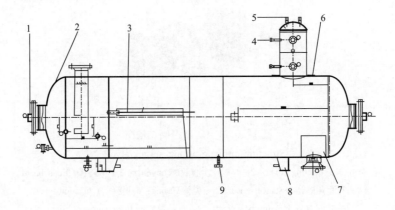

图 1.11.1　聚结器结构图 Structure drawing of coalescer

1—人孔 Manhole；2—封头 Head；3—拉杆 Tie rod；
4、5、9—接管 Connecting pipe；6—补强圈 Reinforcing ring；
7—挡板 Baffle；8—支座 Support

1.11.2 除氧器 Deaerator

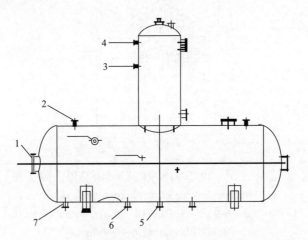

图 1.11.2　除氧器结构图 Structure drawing of deaerator

1—人孔 Manhole；2—汽平衡管接口 Steam balance nozzle；
3—高加疏水器入口 High hydrophobic device inlet；
4—凝结水出口 Condensation outlet；
5—强化换热接口 Strengthening heat transfer interface；
6—放水管 Outlet pipe；7—水平衡管接口 Water balance nozzle

1.11.3 膨胀机 Expander

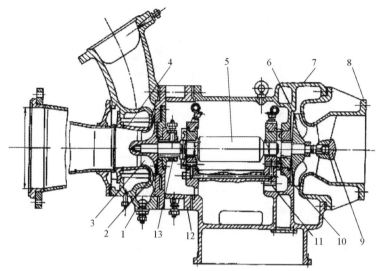

图 1.11.3 膨胀机结构图 Structure drawing of expander
1—蜗壳 Volute casing；2—喷嘴 Nozzle；3—工作轮 Working impeller；4—扩压器 Diffuser；5—主轴 Main shaft；
6—膨胀机叶轮 Expander impeller；7—膨胀机蜗壳 Expander volute；8—膨胀机端盖 Expander cover；9—转速计 Tachometer；
10—轴承座 Bearing seat；11—机体 Compressor body；12—中间体 Midbody；13—轴封 Shaft seal

1.11.4 密封 Seal

1.11.4.1 填料密封 Packing seal

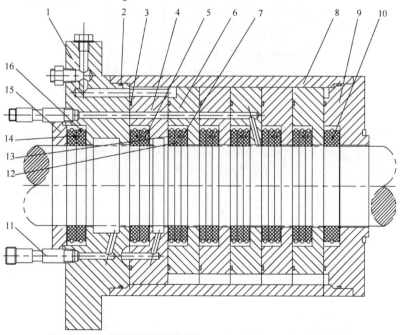

图 1.11.4 填料密封结构图 Structure drawing of packing seal
1—填料法兰 Packing flange；2、3—O 形密封圈 O sealing ring；4、6、9—填料盒 Packing box；
5、12—切向环 Tangential ring；7—径向环 Radial ring；8—水套 Water sleeve；
10—节流环 Restrictor ring；11—接头 Joint；13—阻流环 Restrictive ring；
14—圆柱销 Cylindrical pin；15—端盖 End cover；16—弹簧 Spring

1.11.4.2 机械密封 Mechanical seal

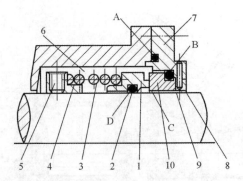

图 1.11.5 机械密封结构图 Structure drawing of mechanical seals

1—动环 Rotating ring；2—动环辅助密封圈 Rotating auxiliary seal ring；3—弹簧 Spring；4—弹簧座 Spring seat；5—紧定螺钉 Tightening screw；6—密封腔 Seal chamber；7—密封压盖 Seal gland；8—防转销 anti-rotating pin；9—静环辅助密封圈 Static auxiliary seal ring；10—静环 static ring；A、B、C、D—泄漏点 Leak point；

1.11.4.3 干气密封 Dry gas seal

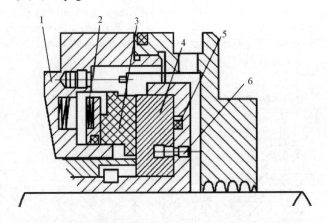

图 1.11.6 干气密封结构图 Structure drawing of dry gas seal

1—弹簧座 Spring seat；2—弹簧 Spring；3—静环 Static ring；
4—动环 Rotating ring；5—密封圈 Seal ring；6—销轴 Pin

2　合成树脂 Synthetic resin

2.1　聚乙烯 Polyethylene

聚乙烯是以乙烯为原料经催化剂催化聚合而得到的一种化合物，聚乙烯产品按照生产方式和分子结构的不同，分为低密度聚乙烯(LDPE)、线性低密度聚乙烯(LLDPE)和高密度聚乙烯(HDPE)三类。工艺流程主要有高压管式法、气相法、淤浆法和釜式法等。本章主要对前两种生产方法的设备进行介绍。

Polyethylene is a kind of compound produced by ethylene through the catalysis and polymerized by catalyst. According to different production methods and molecular structure, polyethylene products are divided into three categories: low-density polyethylene (LDPE), linear low density polyethylene(LLDPE) and high-density polyethylene(HDPE). Polyethylene producing methods mainly include high pressure tubular process, gas phase method, slurry packing method and kettle type method, etc. This chapter focuses on the major equipment of the first two production methods.

2.1.1　反应设备 Reaction equipment

石油化工生产过程中所有完成化学反应的设备统称为反应设备。聚乙烯反应器分流化床反应器、固定床反应器和管式反应器等。

Reaction equipment refers to all devices used to complete chemical reaction in the process of petrochemical production. Polyethylene reactors include fluidized bed reactors, fixed-bed reactors and tubular reactors, etc.

2.1.1.1　聚乙烯流化床反应器 Polyethylene fluidized bed reactor

1. 高密度聚乙烯流化床反应器 Fluidized bed reactor of high-density polyethylene

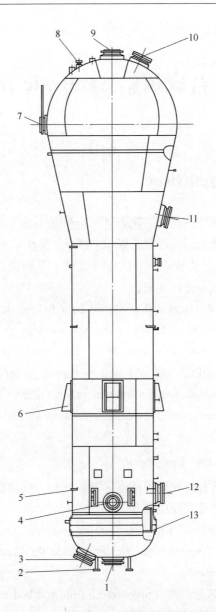

图 2.1.1 高密度聚乙烯流化床反应器结构图
Structure drawing of fluidized bed reactor of high-density polyethylene

1—乙烯入口 Ethylene inlet;

2—放空口 Vent;

3、10、11、12—人孔 Manhole;

4—聚乙烯出口 Polyethylene outlet;

5—保温支承圈 Insulation supporting ring;

6—支座 Support;

7—吊耳 Lug;

8—安全阀接口 Safety valve interface;

9—乙烯出口 Ethylene outlet;

13—分布板 Distribution plate

2. 线性低密度聚乙烯流化床反应器 Fluidized bed reactor of linear low density polyethylene

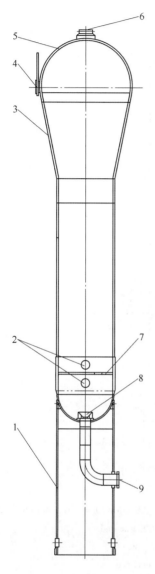

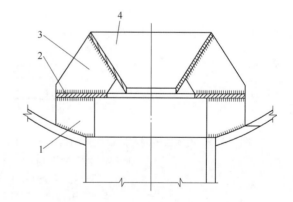

图 2.1.2 线性低密度聚乙烯流化床反应器结构图
Structure drawing of fluidized bed reactor of linear low density polyethylene
1—裙座 Skirt；2—人孔 Manhole；
3—锥形扩大段 Conical expanded section；
4—吊耳 Lug；5—球形封头 Spherical head；
6—乙烯出口 Ethylene outlet；
7—分布板 Distribution plate；
8—导流器 Guide device；
9—乙烯入口 Ethylene inlet

图 2.1.3 导流器 Guide device
1、3—支承板 Supporting plate；
2—环板 Ring plate；
4—锥形板 Conical plate

2.1.1.2 乙烯加氢固定床反应器 Ethylene hydrogenation fixed bed reactor

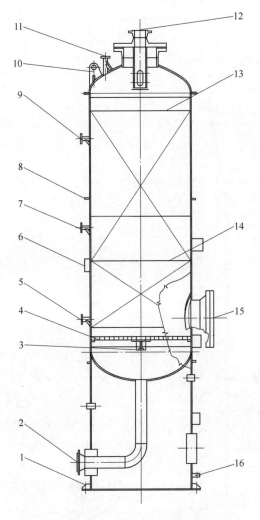

图 2.1.4 乙烯加氢固定床反应器结构图 Structure drawing of Reactor for hydrogenation of ethylene

1—地脚螺栓 Anchor bolt；2—乙烯出口 Ethylene outlet；
3—支架 Support；4—格栅板 Grid plate；
5、7、9—温度计接口 Thermometer interface；
6—铭牌 Nameplate；8—保温支承圈 Insulation support ring；
10—吊耳 Lug；11—安全阀接口 Safety valve interface；
12—乙烯入口 Ethylene inlet；13、14—浮动筛网 Floating mesh；
15—人孔 Manhole；16—接地板 Grounding plate

2.1.1.3 高压聚乙烯环管反应器 High pressure polyethylene loop reactor

1. 高压聚乙烯环管反应器结构 High pressure polyethylene loop reactor structure

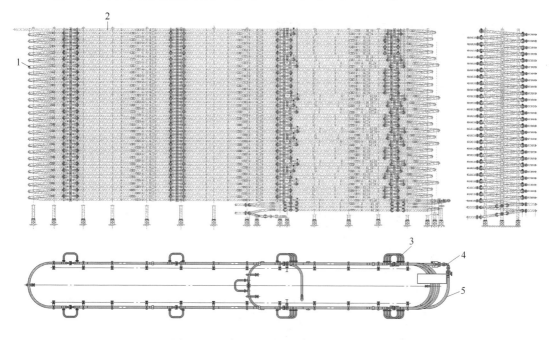

图 2.1.5　高压聚乙烯环管反应器结构图
Structure drawing of high pressure polyethylene loop reactor
1—弯管 Bend tube；2—直管 Straight tube；3—夹套管 Jacket tube；4—出料阀 Outlet valve；5—冷却器 Cooler

2. 高压聚乙烯环管反应器主要零部件 Main components of high pressure polyethylene loop reactor

1) 连接部件 Connecting parts

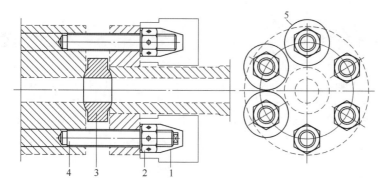

图 2.1.6　连接部件 Connecting parts
1—螺栓保护盖 Nut cap；2—螺母 Nut；
3—透镜垫 Lens gasket；4—螺栓 Bolt；5—拉伸器 Stretcher

2) 直管 Straight tube

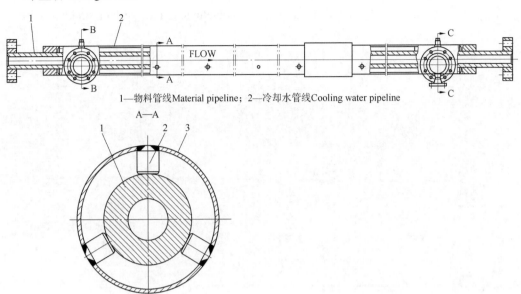

1—物料管线 Material pipeline；2—冷却水管线 Cooling water pipeline

1—物料管线 Material pipeline；2—固定管堵 Fixed tube plug；3—冷却水管线 Cooling water pipeline

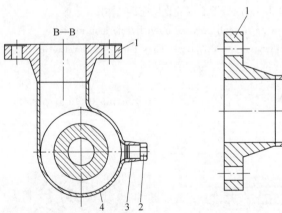

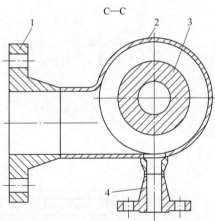

1—法兰 Flange；2—管堵 Tube plug；
3—排放口 Discharge outlet；
4—冷却水管线 Cooling water pipeline

1、4—法兰 Flange；
2—冷却水管线 Cooling water pipeline；
3—物料管线 Material pipeline

图 2.1.7　直管 Straight tube

2 合成树脂 Synthetic resin

3) 弯管 Bend tube

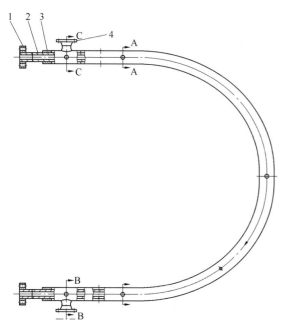
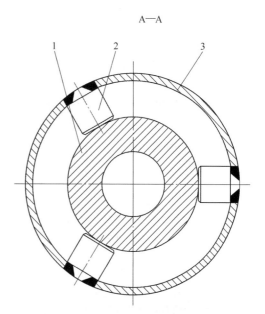

1、4—法兰 Flange；
2—物料管线 Material pipeline；
3—冷却水管线 Cooling water pipeline

1—物料管线 Material pipeline；
2—固定管堵 Fixed tube plug；
3—冷却水管线 Cooling water pipeline

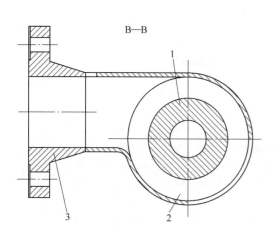
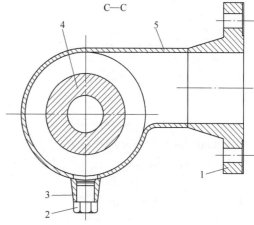

1—物料管线 Material pipeline；
2—冷却水管线 Cooling water pipeline；
3—法兰 Flange

1—法兰 Flange；2—管堵 Tube plug；
3—排放口 Discharge outlet；
4—物料管线 Material pipeline；
5—冷却水管线 Cooling water pipeline

图 2.1.8　弯管 Bend tube

2.1.2 塔设备 Column equipment

2.1.2.1 丙烯轻组分汽提塔 Propylene light component stripper
2.1.2.2 丁烯轻组分汽提塔 Butene light component stripper

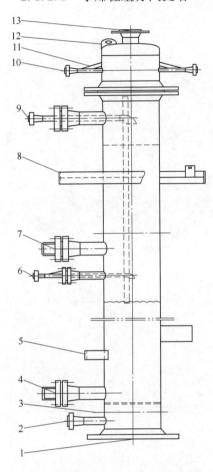
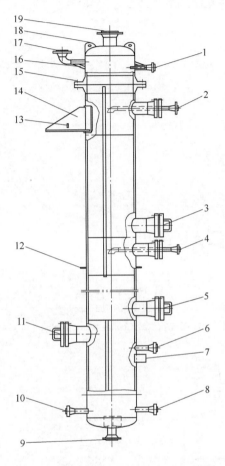

图2.1.9 丙烯轻组分汽提塔结构图
Structure drawing of propylene light component stripper

1—丙烯气入口 Propylene gas inlet；
2—液态丙烯入口 Liquid propylene inlet；
3—保温支承圈 Insulation supporting ring；
4、7—手孔 Hand hole；
5—铭牌 Nameplate；
6、9—温度计接口 Thermometer interface；
8—支座 Support；
10—冷却水入口 Cooling water inlet；
11—筋板 Reinforcing plate；
12—吊耳 Lug；
13—轻组分出口 Light component outlet

图2.1.10 丁烯轻组分汽提塔结构图
Structure drawing of butene light component stripper

1—冷却水入口 Cooling water outlet；
2、4—温度计接口 Thermometer interface；
3、5、11—手孔 Hand hole；
6、8—压力计接口 Pressure pipe interface；
7—铭牌 Nameplate；
9—丁烯气入口 Butene gas inlet；
10—液态丁烯入口 Liquid butene inlet；
12—保温支承圈 Insulation supporting ring；
13—接地板 Grounding plate；
14—支座 Support；15—紧固螺栓 Fastening bolt；
16—筋板 Reinforcing plate；17—冷却水出口 Cooling water outlet；
18—吊耳 Lug；19—轻组分出口 Light component outlet

2 合成树脂 Synthetic resin

2.1.2.3 脱气塔 Degassing column

2.1.3 泵 Pump

2.1.3.1 丁烯加料柱塞泵 Butene feed piston pump

1. 丁烯加料柱塞泵外形 Butene feed plunger pump outline

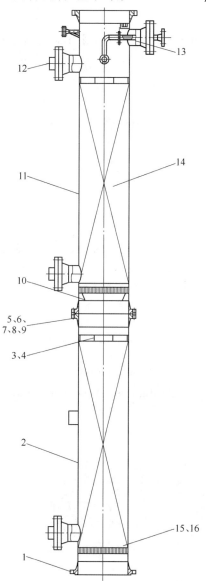

图 2.1.11 脱气塔结构图
Structure drawing of degassing column
1、5—法兰 Flange；2、11—筒体 Cylinder；
3—支承圈 Support ring；4—床层限位器 Bed restrictor；
6—螺柱 Stud；7—螺母 Nut；8—垫片 Gasket；
9—顶丝 Jackscrew；10—集液锥 Liquid collecting cone；
12—入孔 Manhole；13—喷淋装置 Spray device；
14—填料 Packing；15—格栅 Grille；16—格栅支承 Grille support

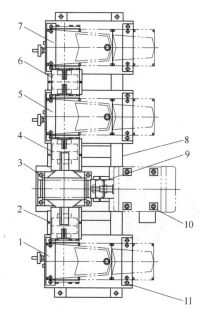

图 2.1.12 丁烯加料柱塞泵外形图
Outline drawing of butene feed plunger pump
1、5、7—曲轴箱 Crankcase；
2、4、6、9—蛇簧联轴器 Grid coupling；
3—蜗轮蜗杆减速器 Worm reduction box；
8—机座 Motor seat；10—电机 Motor；
11—紧固螺栓 Fastening bolt

2. 丁烯加料柱塞泵结构 Butene feed piston pump structure

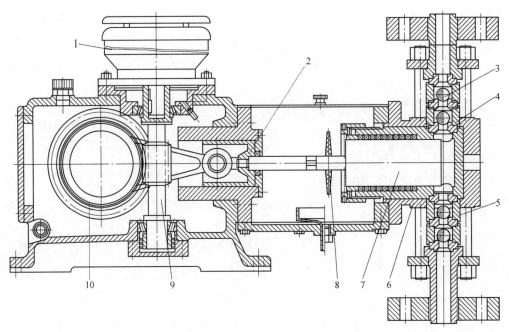

图 2.1.13 丁烯加料柱塞泵结构图 Structure drawing of butene feed piston pump
1—电机 Motor；2—十字头 Crosshead；3—阀球 Valve ball；4—排出阀座 Discharge valve seat；
5—吸入阀座 Suction valve seat；6—液缸体 Liquid cylinder；7—柱塞 Plunger；
8—挡液环 Liquid retaining ring；9—蜗杆 Worm rod；10—蜗轮 Worm wheel

2.1.3.2 超高压引发剂双柱塞泵 High pressure initiator dual piston pump

1. 超高压引发剂双柱塞泵结构 High pressure initiator dual piston pump structure

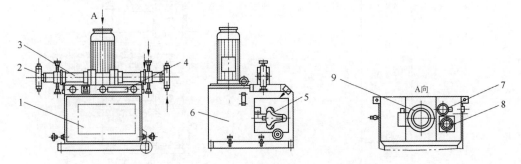

图 2.1.14 超高压引发剂双柱塞泵结构图
Structure drawing of high pressure initiator dual piston pump
1—液压驱动装置 Hydraulic drive；2—泵头 HP-pump head；
3—灯笼环 Lantern；4—铭牌 Nameplate；
5—气压驱动装置 Pneumatic drive；6—油箱 Oil tank；
7、9—电机 Motor；8—贮油器 Oil container

2 合成树脂 Synthetic resin

2. 超高压引发剂双柱塞泵主要零部件 Main parts of high pressure initiator dual piston pump

1) 换向器 Diverter

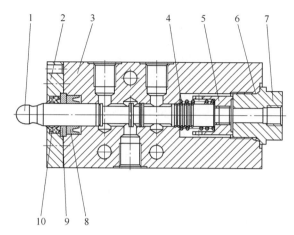

图 2.1.15 换向器结构图 Structure drawing of diverter

1—滑动杆 Slide stick；2—上阀盖 Upper cover；3—阀体 Valve body；4—压缩弹簧 Compression spring；
5—限位器 Stroke limit；6—O 形密封圈 O sealing ring；7—下阀盖 Cover；8—开槽环 Groove ring；
9—导向套 Guiding bush；10—刮油环 Oil scraper ring

2) 泵头 Pump head

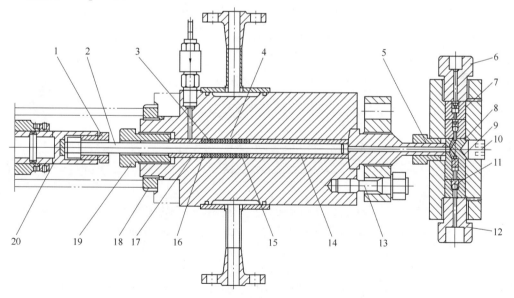

图 2.1.16 泵头结构图 Structure drawing of pump head

1、5、19—压紧螺母 Compression screw；2—柱塞 Plunger；3—填料环 Packing ring；4—支承环 Supporting ring；
6—出口接头 Discharge connection；7—阀座 Valve seat；8—Y 形件 Y-piece；9—球座 Ball cup；
10—螺纹销钉 Thread pin；11—球 Ball；12—入口接头 Suction connection；
13—双头螺栓 Stud bolt；14—定距套 Fixed pitch sleeve；15—压缩弹簧 Compression spring；16—压力环 Pressure ring；
17—导向套 Guiding bush；18—开槽圆螺母 Slotted round nut；20—联轴器 Coupling

2.1.3.3 高压润滑油柱塞泵 High pressure lubricant piston pump
1. C1 型高压润滑油柱塞泵 C1 – high pressure lubricant piston pump

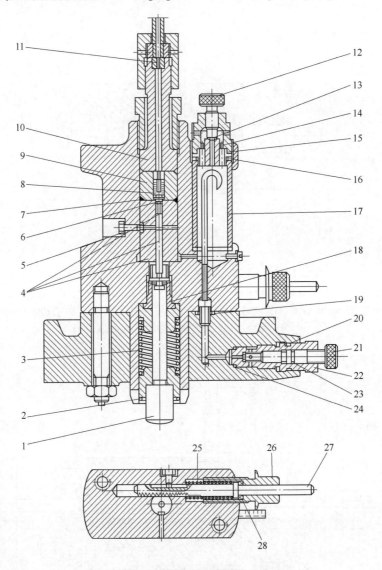

图 2.1.17　C1 型高压润滑油柱塞泵结构图
Structure drawing of C1-high pressure lubricant piston pump
1—推杆 Push rod；2—弹簧座 Spring seat；3、8、25—弹簧 Spring；
4—柱塞组件 Piston assembly；5、24—垫片 Gasket；6—阀头 Valve head；
7、14、15、16、19、20、23—O 形密封圈 O sealing ring；9—排出阀 Discharge valve；
10—连接件 Connector；11—透镜环 Lens ring；12、21—调节螺母 Adjusting nut；
13—压紧螺杆 Pressing screw；17—视镜 Sight glass；18—调节衬套 Adjusting bushing；
22—压紧盖 Tightening cap；26—调节螺杆 Adjusting screw；27—调节轴 Adjusting shaft；
28—卡环 Clamp ring

2. CH1 型高压润滑油泵 CH1 – high pressure oil pump

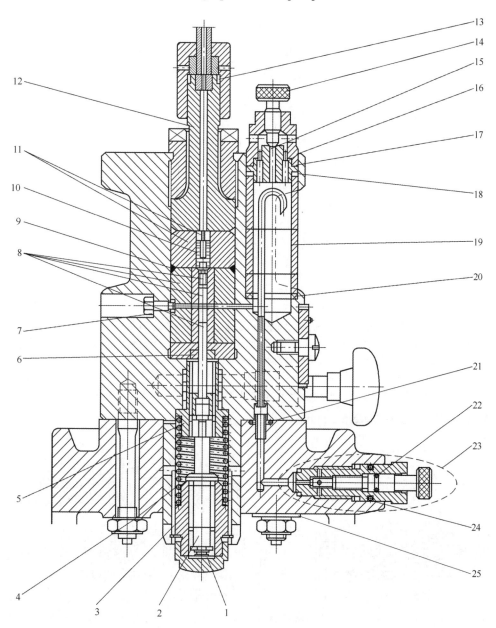

图 2.1.18 CH1 型高压润滑油泵结构图 Structure drawing of CH1 – high pressure oil pump
1—卡环 Clamp ring；2—推杆 Push rod；3—导向杆 Oriented rod；4、10—弹簧 Spring；
5—调节衬套 Adjusting bush；6—密封环 Seal ring；7、25—垫片 Gasket；8—柱塞组件 Piston parts；
9、15、16、17、18、20、21、22、24—O 形密封圈 O sealing ring；11—排出阀 Exhaust valve；
12—连接件 Connecting part；13—填料环 Packing ring；14—调节螺母 Adjusting nut；
19—视镜 Lens ring；23—切断阀 Shut-off valve

2.1.3.4 导热油屏蔽泵 Heat conduction oil shield pump

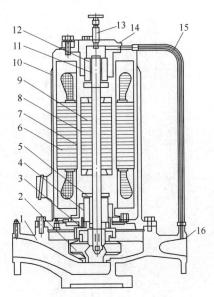

图 2.1.19 导热油屏蔽泵结构图
Structure drawing of heat conduction oil shield pump

1—泵体 Pump body；2—叶轮 Wheel；3—平衡端盖 Balance end cover；4—下轴承座 Bottom chuck；5—推力盘 Thrust disc；6—定子 Stator；7—定子屏蔽套 Stator shielding sleeve；8—转子屏蔽套 Rotor shielding sleeve；9—转子 Rotor；10—机座 Motor seat；11—轴套 Sleeve；12—石墨轴承 Graphite bearing；13—排水阀 Discharge valve；14—上轴承座 Upper bearing seat；15—循环管 Circulating pipe；16—过滤网 Filter

2.1.3.5 计量泵 Metering pump

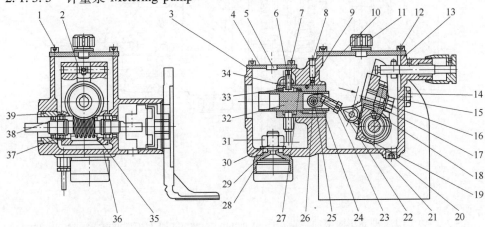

图 2.1.20 计量泵结构图 Structure drawing of metering pump

1、7、9、12、26、39—紧固螺钉 Tightening screw；2、27、29—O形密封圈 O sedling ring；3、13—密封垫 Seal pad；4、10—盖板 Cover；5、11、20—丝堵 Plug；6—排出口 Discharge ouflet；8—安全塞堵 Safety plug；14—油位指示计 Oil level indicator；15、30—垫圈 Washer；16、21、31—螺母 Nut；17—键 Key；18—隔板 Spacer；19—锥齿轮支架 Crank support；22—球形铰链接头 Ball-and-socket joint；23—十字头框架 Cross head support；24—连杆 Connecting rod；25—轴 Shaft；28—底阀 Foot valve；32—十字头 Cross head；33—紧固夹 Binding clip；34、38—唇形密封 Lip seal；35—蜗轮蜗杆 Worm gearing and worm shaft；36—箱体 Box；37—轴承 Bearing

2 合成树脂 Synthetic resin

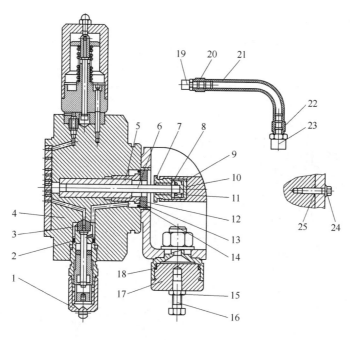

图 2.1.21 计量泵中体结构图 Structure drawing of metering pump midbody

1—进气阀 Admission valve；2、14、18、25—O 形密封圈 O sealing ring；3—止回球 Check ball component；
4—泵体 Pump body；5—柱塞衬里 Plunger liner；6—柱塞 Plunger；
7—柱塞连接螺母 Plunger hooking screw；8、9—垫圈 Casket；10—支架 Support；
11、16、24—螺栓 Screw；12—填料 Packing；13、15、17—螺母 Nut；19、23—衬套 Bush；
20、22—卡圈 Collar；21—管子 Tube

2.1.3.6 造粒水泵 Granulation pump

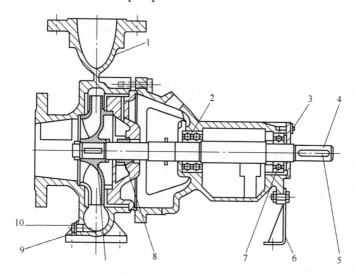

图 2.1.22 造粒水泵结构图 Structure drawing of granulation pump

1—泵体 Pump body；2、7—轴承 Bearing；3—轴承端盖 Bearing cover；4—轴 Shaft；5—键 Key；
6—支架 Support；8—密封函 Seal box；9—排液口 Drain hole；10—六角螺栓 Hexagon bolt

2.1.4 压缩机 Compressor

2.1.4.1 高密度聚乙烯循环气压缩机 High-density polyethylene recycle gas compressor

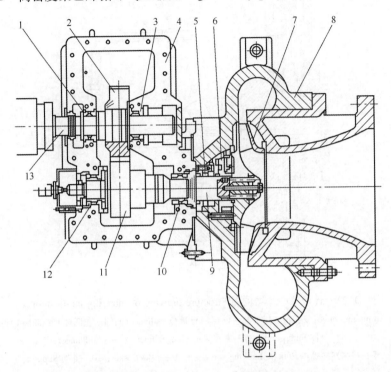

图 2.1.23 高密度聚乙烯循环气压缩机结构图
Structure drawing of high-density polyethylene recycle gas compressor

1、12—止推轴承 Thrust bearing；2、11—齿轮 Gear；3、10—径向轴承 Radial bearing；
4—箱体 Box body；5—浮环密封 Floating-ring seal；6—迷宫密封 Labyrinth seal；
7—叶轮 Wheel；8—机壳 Housing；9—主轴 Shaft；13—输入轴 Input shaft

2.1.4.2 线性低密度循环气压缩机 Linear low-density recycle gas compressor

1. 循环气压缩机外形 Recycle gas compressor outline

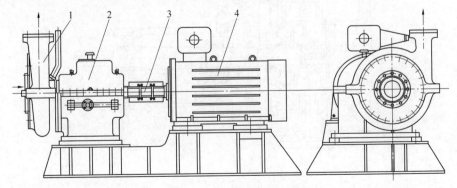

图 2.1.24 循环气压缩机外形图 Outline drawing of recycle gas compressor
1—机壳 Housing；2—齿轮箱 Gear case；3—联轴器 Coupling；4—电机 Motor

2. 循环气压缩机结构 Recycle gas compressor structure

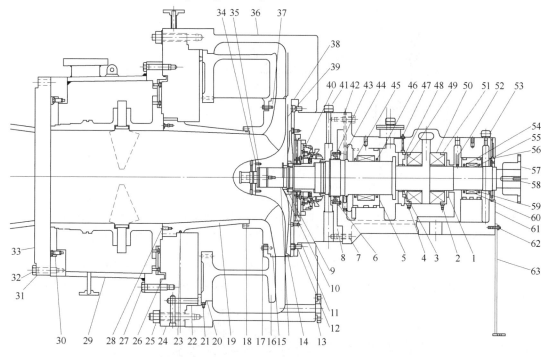

图 2.1.25　线性低密度循环气压缩机结构图
Structure drawing of linear low-density recycle gas compressor

1、3—油封 Oil seal；2、4—孔板 Orifice plate；5、59—轴承 Bearing；

6、12、17、27、32、47、62—六角头螺栓 Hex head screw；

7—锥销 Pin taper；8—密封箱 Seal housing；

9、10、11、15、18、20、22、23、26、28、30、31、

34、35—O 形密封圈 O sealing ring；13—盲法兰 Blind flange；

14、38、45、46、54、55—密封环 Seal ring；16—蜗壳 Scroll；

19—入口管 Inlet pipe；21—定位销 Dowd pin；24—端盖 End cover；

25—双头螺栓 Bolt stud；29—入口导叶管 Inlet guide pipe；33—进气管口 Inlet nozzle；

36—保护罩 Protective cover；37—叶轮护环 Impeller shroud ring；

39—旋转元件 Rotation element；40—密封组件 Seal group；41、53—排气罩 Vent cap；

42—管塞 Pipe plug；43、44—挡油板 Oil baffle；48—推力轴承支架 Thrust bearing retainer；

49、51—填充环 Filler ring；50—推力轴承 Thrust bearing；52—仪表 Instrument；

56—轴承箱 Bearing box；57—联轴器 Coupling；58—键 Key；

60—轴承垫 Bearing pad；61—轴承座 Bearing seat；

63—支座 Support

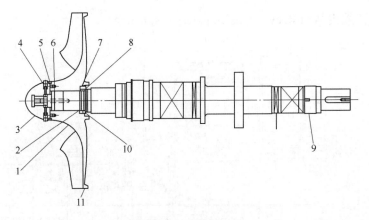

图 2.1.26 转子 Rotor

1—叶轮 Wheel；2、5、6、7、10—O 形密封圈 O sealing ring；3—锁紧螺母 Lock nut；4—紧固螺钉 Tightening screw；
8—液压释放环 Hydraulic release sealing ring；9—主轴 Shaft；11—J 型带状密封环 J-ribbon seal ring

2.1.4.3 高压聚乙烯一次压缩机 High pressure polyethylene first compressor

1. 高压聚乙烯一次压缩机外形 Outline of high pressure polyethylene first compressor

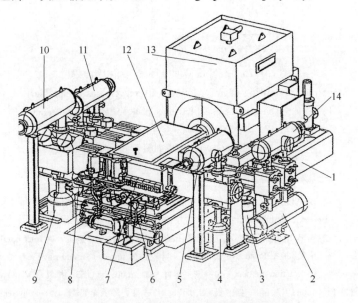

图 2.1.27 高压聚乙烯一次压缩机外形图
Outline drawing of high pressure polyethylene first compressor

1—五级吸入缓冲罐 Five-level suction buffer tank；2—五级排出缓冲罐 Five-level discharged buffer tank；
3—四级排出缓冲罐；4—二级排出缓冲罐 Two-level discharged buffer tank；
5—二级吸入缓冲罐 Two-level suction buffer tank；6—运动机构润滑油系统 Lube oil system of movement body；
7—气缸润滑油系统 Cylinder lubricating oil system；8—三级排出缓冲罐 Three-level discharged buffer tank；
9—一级排出缓冲罐 First level discharged buffer tank；10—一级吸入缓冲罐 First level suction buffer tank；
11—三级吸入缓冲罐 Three-level suction buffer tank；12—传动机构 Driving gear；
13—主电机 Main motor；14—四级吸入缓冲罐 Four-level suction buffer tank

2 合成树脂 Synthetic resin

2. 主要零部件 Main components

1) 曲轴箱 Crankcase

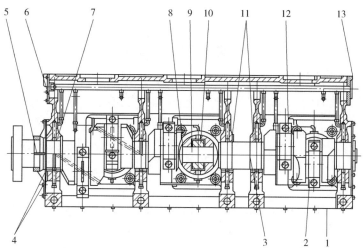

图 2.1.28 曲轴箱结构图 Structure drawing of crankcase

1—曲轴箱体 Crankshaft body；2、12—曲柄轴承 Crankshaft bearing；3—曲轴 Crankshaft；
4—轴导向环 Shaft guide ring；5—曲轴箱垫片 Crank case gaster；6、13—油槽垫片 Oil sump gasket；
7、11—曲轴轴承 Crankshaft bearing；8—十字头销簧环 Crosshead pin circlip；9—十字头销 Crosshead pin；
10—十字头销轴承 Crosshead pin bearing

2) 曲轴箱运动机构 Crankcase motion mechanism

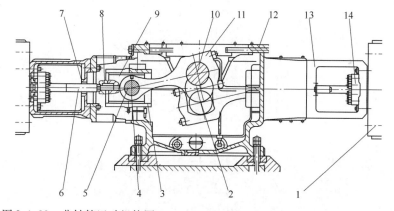

图 2.1.29 曲轴箱运动机构图 Structure drawing of crankcase motion mechanism

1—气缸 Cylinder；2—轴承 Bearing；3—十字头导轨 Crosshead guide rail；4—十字头销轴承 Crosshead pin bearing；
5—十字头销 Crosshead pin；6—活塞杆 Piston rod；7—刮油器 Oil wiper；8—联轴器 Coupling；
9—十字头 Crosshead；10—曲轴 Crankshaft；11—连杆 Connecting rod；12—曲轴箱 Crankcase；
13—定距块 Distance block；14—填料压盖 Stuffing gland

3) 刮油器 Oil wiper

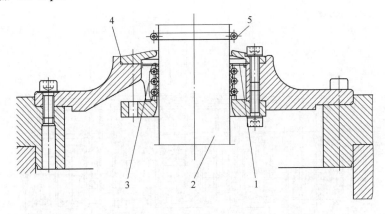

图 2.1.30　刮油器结构图 Structure drawing of oil wiper
1—弹簧板 Wiper spring board；2—活塞杆 Piston rod；3—刮油环 Oil scraper ring；
4—十字头端盖 Crosshead cover；5—甩油环 Oil slinger

4) 液压活塞杆联轴节 Hydraulic piston rod coupling

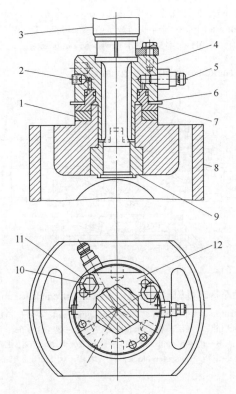

图 2.1.31　液压活塞杆联轴节结构图 Structure drawing of hydraulic piston rod coupling
1—密封环 Seal ring；2—泄放螺钉 Bleed screw；3—活塞杆 Piston rod；4—短接管 Hydraulic nut；
5—调节螺栓 Short Connecting pipe；6—定距环 Distance ring；7—调节垫 Adjusting pad；8—十字头 Crosshead；
9—十字头螺母 Crosshead nut；10—六角螺母 Hex nut；11—锁紧垫圈 Locking washer；12—锁定板 Locking plate

2 合成树脂 Synthetic resin

5) 增压机一级气缸 Structure of booster primory cylinder

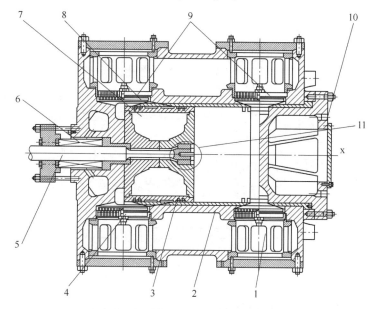

图 2.1.32　增压机一级气缸结构图 Structure drawing of booster primary cylinder

1—气阀固定罩 Fixed valve cover；2—气缸 Cylinder；3—活塞环 Piston ring；4—排出阀 Discharge valve；
5—活塞杆 Piston rod；6—填料 Packing；7—活塞 Piston；8—支承环 Support ring；9—吸入阀 Suction valve；
10—冷却水室 Cooling water chamber；11—活塞装配螺母 Piston assembly nut

6) 增压机二级气缸 Booster second cylinder

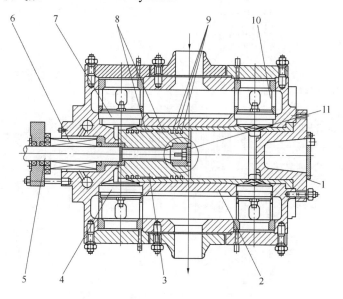

图 2.1.33　增压机二级气缸结构图 Structure drawing of booster seconda cylinder

1—冷却水室 Cooling water chamber；2—气缸 Cylinder；3—活塞 Piston；4—排出阀 Discharge valve；
5—活塞杆 Piston rod；6—填料 Packing；7—吸入阀 Suction valve；8—支承环 Support ring；
9—活塞环 Piston ring；10—气阀固定盖 Fixed valve cap；11—活塞装配螺母 Piston assembly nut

7) 一次机一级气缸 Primary structure of first compressor

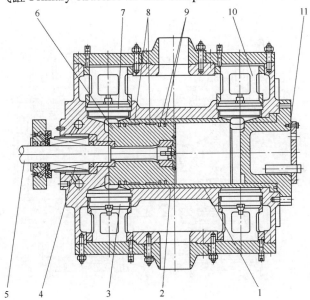

图 2.1.34　一次机一级气缸结构图 Structure drawing of first compressor primary structure
1—气缸衬套 Cylinder liner；2—O 型密封圈 O sealing ring；3—排出阀 Discharge valve；4—填料 Packing；
5—活塞杆 Piston rod；6—活塞 Piston；7—吸入阀 Suction valve；8—支承环 Support ring；
9—活塞环 Piston ring；10—气阀固定罩 Fixed valve cover；11—冷却水室 Cooling water chamber

8) 一次机二、三级气缸 Secondary and tertiary structure of first compressor

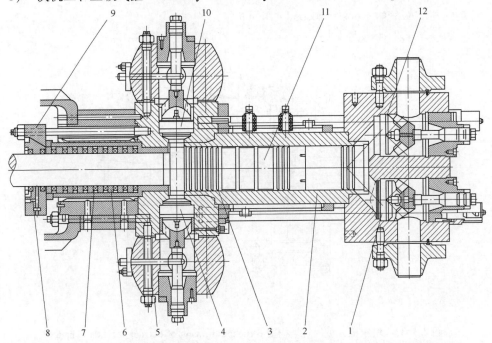

图 2.1.35　一次机二、三级气缸结构图 Structure drawing of first compressor secondary and tertiary structure
1、4—排出阀 Discharge valve；2—导向衬套 Guide bush；3—活塞环 Piston ring；5—螺栓 Bolt；
6—整体填料 Overall packing；7—填料环 Packing ring；8—压力缓冲环 Pressure buffer ring；
9—压紧法兰 Supported flange；10、12—吸入阀 Suction valve；11—活塞杆 Guide ring

9) 气阀 Gas valve

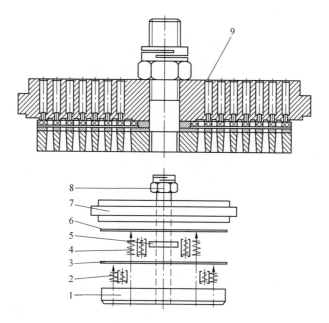

图 2.1.36 气阀结构图 Structure drawing of gas valve

1—升程限制器 Lift limiter；2—调节弹簧 Adjusting spring；3—调节板 Adjustment plate；
4—闭合弹簧 Closing spring；5—导向环 Guide ring；6—阀片 Valve plate；
7—阀座 Valve seat；8—螺母 Nut；9—气阀 Gas valve

10) 活塞杆连接螺母 Piston rod connecting nut

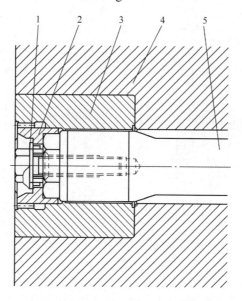

图 2.1.37 活塞杆连接螺母结构图 Structure drawing of piston rod connecting nut

1—活塞锁紧垫圈 Piston lock washer；2—活塞锁紧环 Piston locking ring；
3—活塞螺母 Piston nut；4—活塞 Piston；5—活塞杆 Piston rod

11）填料函 Packing box

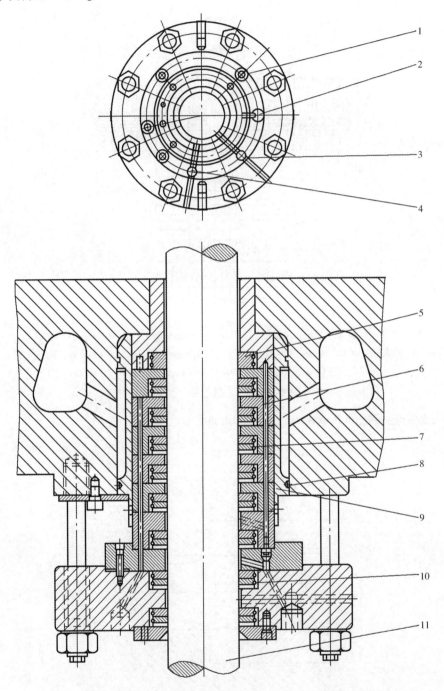

图 2.1.38 填料函结构图 Structure drawing of packing box
1—高压泄漏气孔 High-pressure leak hole；2—低压泄漏气孔 Low-pressure leak hole；
3—吹扫出口 Purge outlet；4—吹扫入口 Purge inlet；5—压力缓冲环 Pressure buffer ring；
6、7—填料杯 Packing cup；8—O 形密封圈 O sealing ring；9—冷却水夹套 Cooling water jacket；
10—侧装填料 Side Packing；11—活塞杆 Piston rod

2 合成树脂 Synthetic resin

12) 润滑油泵外形 Outline of lubrication pump

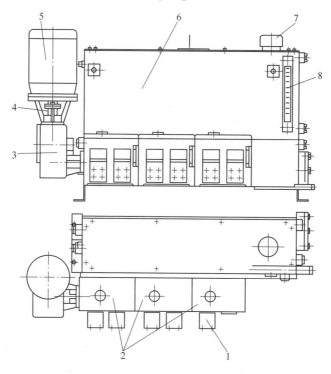

图 2.1.39 润滑油泵外形图 Outline drawing of cylinder lubricating oil pump
1—油泵 Oil pump；2—润滑油泵 Lubrication pump；3—齿轮箱 Gear；4—联轴器 Coupling；
5—电机 Motor；6—油箱 Oil tank；7—放空阀 Vent valve；8—液位计 Liquid level meter

13) 润滑油泵蜗轮蜗杆结构 Lubricating oil pump driving gear

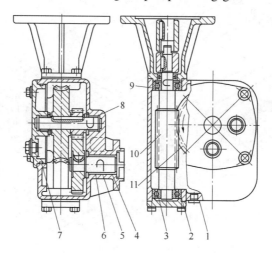

图 2.1.40 润滑油泵蜗轮蜗杆结构图 Structure drawing of lubricating oil pump driving gear
1—铜垫片 Copper gasket；2—环型垫片 Ring gasket；3、5、9—滚珠轴承 Ball bearing；
4、6、8—滚针轴承 Needle roller bearing；7—O 形密封圈 O sealing ring；
10—蜗杆 Worm rod；11—蜗轮 Worm wheel

· 125 ·

14）润滑油泵结构 Lubrication pump structure

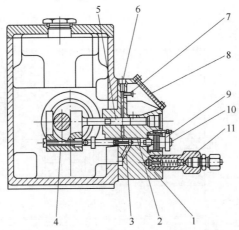

图 2.1.41　润滑油泵结构图 Structure drawing of lubrication pump
1—止回阀密封 Check valve seal；2—缸体 Cylinder liner；
3—冲程调节杆 Stroke adjustment rod；4—油泵 Oil pump；5—柱塞 Plunger；
6—放空螺栓 Vent screw；7—注油孔 Oil hole；8—视镜 Sight glass；
9—分度盘 Index plate；10—调节螺栓 Adjusting screw；11—止回阀 Check valve

2.1.4.4　高压聚乙烯二次压缩机 High pressure polyethylene second compressor

1. 高压聚乙烯二次压缩机外形 Outline of high pressure polyethylene second compressor

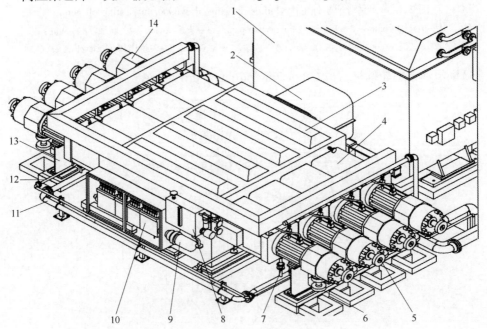

图 2.1.42　高压聚乙烯二次压缩机外形图 Outline drawing of high pressure polyethylene second compressor
1—主电机 Main motor；2—联轴器 Coupling；3—曲轴箱 Crankcase；4—中体 Midbody；
5—出口 Outlet；6—入口 Inlet；7—限流孔板 Restriction orifice；8—润滑油储罐 Lubricating oil tank；
9—润滑油电机 Lubricant motor；10—润滑油泵 Lubrication pump；11—泄漏气管线 Gas leak pipeline；
12—冷却油管线 Cooling oil pipeline；13—气缸支承板 Cylinder supporting plate；14—气缸 Cylinder

2 合成树脂 Synthetic resin

2. 二次压缩机曲轴箱 Crankcase of second compressor

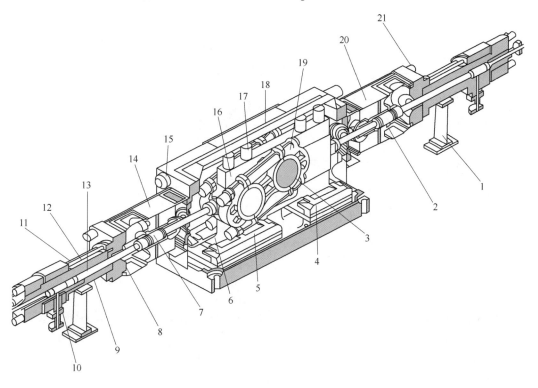

图 2.1.43 二次压缩机曲轴箱结构图 Structure drawing of second compressor crankcase

1—气缸支承板 Cylinder supporting plate；2—导向衬套 Guide bush；3—曲轴 Crankshaft；
4—副滑道 Secondary pipe slide；5—主滑道 Main pipe slide；6—十字头滑块 Crosshead slide；
7—弹性杆 Elastic rod；8—填料函 Stuffing box；9—金属填料 Metallic stuffing；
10—中心组合阀 Center combination valve；11—气缸紧固螺栓 Cylinder fastening bolt；12、14—气缸 Cylinder；
13—柱塞 Plunger；15、21—螺母 Nut；16—框架十字头 Frame cross；17—框架十字头螺栓 Frame cross bolt；
18—曲轴箱 Crankcase；19—连杆 Connecting rod；20—中体 Midbody

3. 主要零部件 Main components

1) 框架十字头 Frame cross

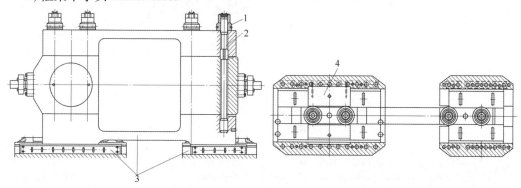

图 2.1.44 框架十字头结构图 Structure drawing of frame cross

1—螺母 Nut；2—螺栓 Bolt；3—十字头滑块 Crosshead slide；4—十字头 Crosshead body

2）连杆 Connecting rod

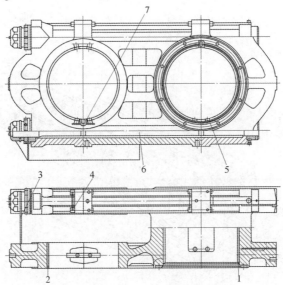

图 2.1.45　连杆结构图 Structure drawing of connecting rod

1—曲柄侧连杆轴承 Crank side rod bearings；2—十字头侧连杆轴承 Crosshead side connecting rod bearing；
3—连接板 Connecting plate；4—螺栓 Bolt；5、7—销 Pin；6—连杆螺栓 Connecting rod bolt

3）气缸 Cylinder

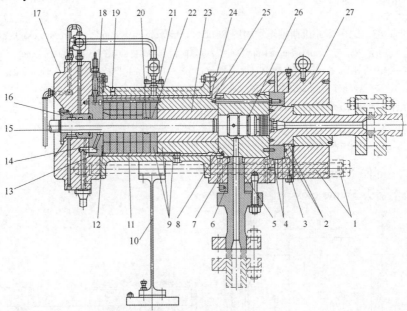

图 2.1.46　气缸结构图 Structure drawing of cylinder

1—螺栓 Bolt；2、4、5、12、14、24、25—O 形密封圈 O sealing ring；3—液压活塞 Hydraulic piston；6—透镜环 Lens ring；
7—弹簧 Spring；8—缸头内衬 Cylinder head lining；9—填料 Packing；10—支承架 Supporting plate；
11—对中衬套 Central alignment bushing；13—底部支承环 Bottom supporting ring；15—柱塞 Plunger；
16—整体填料函 Overall stuffing；17—缸体 Cylinder；18—注油单向阀 Oiling way valve；19—填料环 Packing ring；
20—导向衬套 Guide bushing；21—爆破膜 Rupture disk；22—压力缓冲环 Pressure buffer ring；
23—气缸衬套 Extrusion cylinder bushing；26—中心组合阀 Center combination valve；27—缸头盖 Clinder head cove

4）中心组合阀 Center combination valve

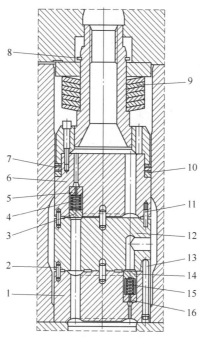

图 2.1.47 中心组合阀结构图 Structure drawing of center combination valve
1—吸气阀体 Suction valve；2、11—销 Pin；3、14—阀芯 Spool valve；4、8、15—弹簧 Spring；
5、16—定距盘 Spacer plate；6—排气阀体 Exhaust valve；7—密封环 Sealing ring；
9—盘状弹簧 Disc spring；10—支承环 Supporting ring；12—阀座 Valve seat；13—螺栓 Bolt

5）低压填料函 Low pressure packing box

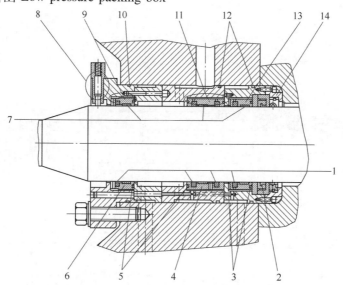

图 2.1.48 低压填料函结构图 Structure drawing of low pressure packing box
1—阶式密封 Step seal；2—填料环 Packing ring；3、5、10、12—O 形密封圈 O sealing ring；
4—滑动衬套 Sliding bushing；6、9、11、13—滑动环 Glide ring；7—导向环 Guide ring；
8—温度计 Temperature gauge；14—刮油器 Oil wiper

6) 高压喷射止回阀 High-pressure injection check valve

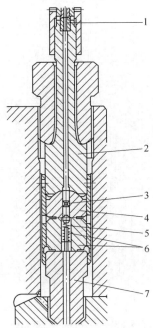

图 2.1.49　高压喷射止回阀结构图 Structure drawing of high-pressure injection check valve
1—透镜环 Lens ring；2、7—连接块 Connecting block；3—阀头 Valve head；
4—阀座 Valve seat；5—弹簧 Spring；6—排出阀外壳 Discharge valve housing

7) 中体 Midbody

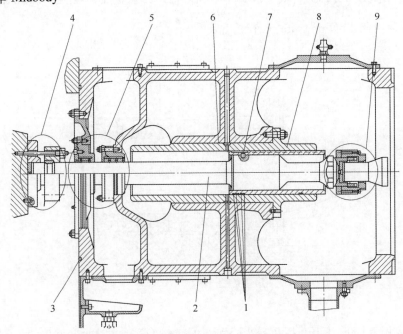

图 2.1.50　中体结构图 Structure drawing of midbody
1—导向环 Guide ring；2—弹性杆 Elastic rod；3、6—O 形密封圈 O sealing ring；4—联轴器 Coupling；
5—刮油器 Oil wiper；7—销 Pin；8—导向衬套 Guide bushing；9—柱塞联轴器 Plunger coupling

2 合成树脂 Synthetic resin

8) 柱塞联轴器 Plunger coupling

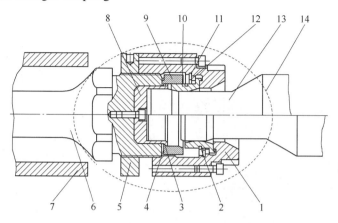

图 2.1.51 柱塞联轴器结构图 Structure drawing of plunger coupling
1—连接螺母 Coupling nut；2—弹簧 Spring；3、4—O 形密封圈 O sealing ring；
5—锁紧螺母 Lock nut；6—弹性杆 Elastic rod with brake blocks；7—导向衬套 Guide bushing；
8—中心环 Center ring；9—安全环 Safety ring；10—压紧块 Clamping block；11—弹簧垫 Spring pad；
12—定距环 Distance ring；13—柱塞 Plunger；14—柱塞联轴器 Plunger coupling

9) 刮油器 Oil wiper

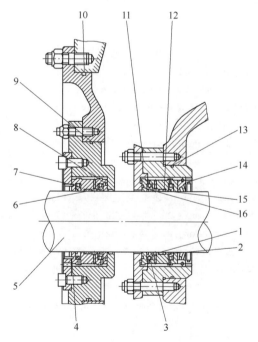

图 2.1.52 刮油器结构图 Structure drawing of oil wiper
1、6、12—导向环 Guide ring；2、7、15—带 O 形环的阶式密封 Step seal with O ring；
3—滑动衬套 Sliding bushing；4—滑动环 Glide ring；
5—弹性杆 Elastic rod；8、9、10、11、13、16—O 形密封圈 O sealing ring；
14—刮油环 Oil wiper ring

2.1.4.5 尾气回收压缩机 Exhaust gas recovery compressor

1. 迷宫式 Labyrinth type

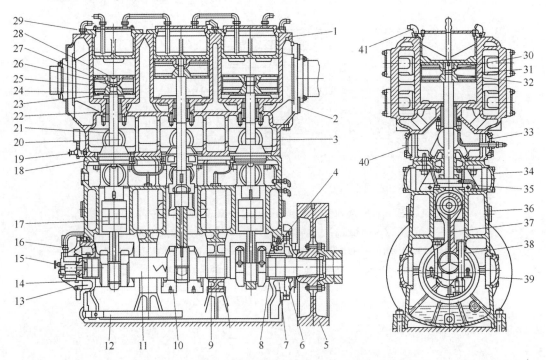

图 2.1.53 迷宫式尾气回收压缩机结构图
Structure drawing of gas recovery labyrinth compressor

1—气缸体 Cylinder block；2—侧盖 Side cover；3—中体 Midbody；
4—排放阀 Discharge valve；5—飞轮 Flywheel；6—轴封 Shaft seal；
7、13—轴承支架 Bearing bracket；8—主轴承 Main bearing；
9—轴承盖 Bering cover；10—平衡块 Counter balance；
11—曲轴 Crankshaft；12—过滤器 Filter；14—主油泵 Main pump；
15—油压力表 Oil pressure gauge；16—管接头 Pipe joints；
17—十字头 Crosshead；18—导向轴承及刮油环 Guide bearing and oil scraper ring；
19—压力表 Pressure gauge；20—挡油环 Oil slinger；21—活塞杆 Piston rod；
22—活塞杆填料 Piston rod packing；23—垫圈 Casket；24、27、30、32—活塞 Piston parts；
25、31—活塞裙 Piston skirt；26—销子 Pin；28—活塞螺母 Piston nut；
29—气缸盖 Cylinder cover；33、34、36、39、40—手孔盖 Hand hole cover；
35—过滤元件 Filter element；37—连杆 Connecting rod；
38—连杆螺栓及锁定销 Connecting rod bolt and its lock pin；
41—工艺管接口 Process pipe interface

2. 螺杆式 Screw type

1）螺杆式压缩机外形 Screw compressor outline

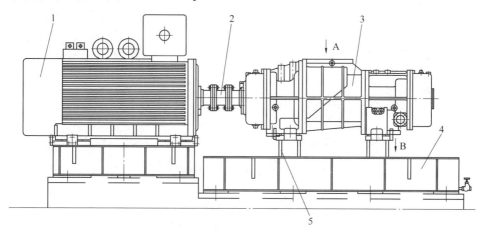

图 2.1.54　螺杆式压缩机外形图 Outline drawing of screw compressor
1—电机 Motor；2—联轴器 Coupling；3—螺杆压缩机 Screw compressor；
4—底座 Base；5—支架 Support；A—吸气 Suction；B—排气 Exhaust

2）螺杆式压缩机结构 Screw compressor structure

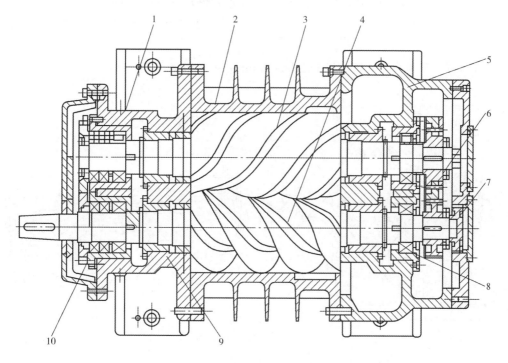

图 2.1.55　螺杆式压缩机结构图 Structure drawing of screw compressor
1—排气端座 Exhaust end seat；2—气缸体 Gas cylinder；3—阴转子 Female rotor；4—阳转子 Male rotor；
5—吸气端座 Suction end seat；6—同步齿轮 Synchronous gear；7—定位盘 Location disc；
8—径向轴承 Radial bearing；9—密封箱 Seal parts；10—止推轴承 Thrust bearing

2.1.4.6　排放气压缩机 Vent gas compressor

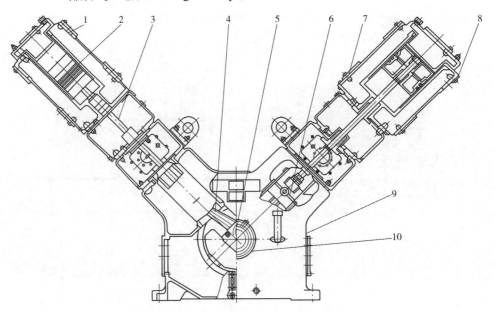

图 2.1.56　排放气压缩机结构图 Structure drawing of vent gas compressor

1—排气阀 Exhaust valve；2—活塞 Piston；3—活塞杆 Piston rod；4—连杆 Connecting rod；
5—曲轴 Crankshaft；6—十字头 Crosshead；7—吸气阀 Suction valve；8—排气阀 Exhaust valve
9—机壳 Housing；10—皮带轮 Belt pulley

2.1.4.7　隔膜式氮气压缩机 Diaphragm nitrogen compressor

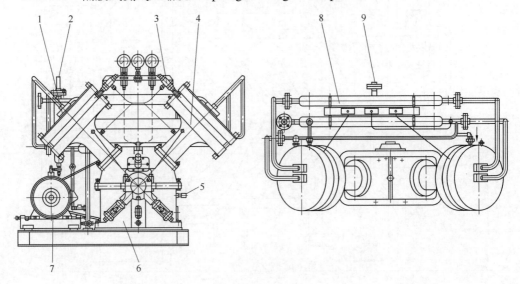

图 2.1.57　隔膜式氮气压缩机结构图 Structure drawing of diaphragm nitrogen compressor

1、4—气缸 Cylinder；2—排气口 Exhaust port；3—调压阀 Pressure regulating valve；
5—冷却水进口 Cooling water inlet；6—曲轴箱 Crankcase；7—电机 Motor；
8—冷却器 Cooler；9—进气口 Gas inlet

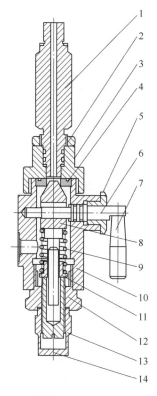

1—接头 Joint；2—螺母 Nut；3—阀盖 Valve cover；4—阀座 Valve seat；5—螺母衬套 Nut bushing；6—调节杆 Adjustment lever；7—手柄 Handle；8—阀芯 Spool valve；9—弹簧 Spring；10—阀体 Valve body；11—弹簧座 Spring seat；12—螺母 Nut；13—调节螺母 Adjusting screw；14—锁紧螺母 Lock nut

图 2.1.58 补偿油泵 Compensation pump

2.1.5 挤压造粒机 Extrusion granulator

2.1.5.1 Ⅰ型挤压造粒机 Ⅰ-extrusion granulator

1. Ⅰ型挤压造粒机构成 Configuration of Ⅰ-extrusion granulator

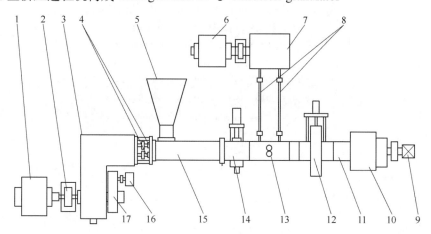

图 2.1.59 Ⅰ型挤压造粒机构成图 Configuration diagram of Ⅰ-extrusion granulator unit
1—主电机 Main motor；2—联轴器 Coupling；3—主减速箱 Main speed reduction box；4—螺杆 Screw；
5—进料斗 Feed hopper；6、9—电机 Motor；7—熔融泵减速箱 Melt pump speed reduction box；
8—同步联轴器 Synchronous coupling；10—切粒机 Pelletizer；11—模板 Die plate；12—换网器 Screen changer；
13—熔融泵 Melt pump；14—开车阀 Drive valve；15—筒体 Cylinder；16 开车电机 Drive motor；
17—辅助减速箱 Auxiliary speed reduction box

2. Ⅰ型挤压造粒机主要零部件 Main components of Ⅰ-extrusion granulator

1) 主减速箱 Main speed reduction box

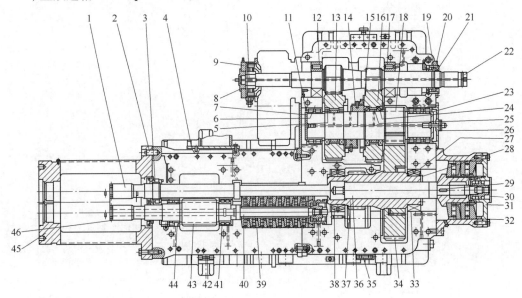

图 2.1.60 主减速箱结构图
Structure drawing of extruder main speed reduction box

1、36—主动轴 Main output shaft；2—销子 Pin；
3、5、6、12、17、19、23、24、33、38、41、44—轴承 Bearing；
4—观察孔 Observation hole；7、8、20、26、28、31—压盖 Gland；
9、30—端盖 End cover；10、29—键 Key；11、27—卡簧 Snap spring；
13—低速齿轮 Low speed gear；14—换挡器 Gear shifting；
15—高速齿轮 High speed gear；16、18、34—齿轮 Gear；
21—密封环 Seal ring；22—输入轴 Input shaft；
25—间隔环 Spacer ring；32—推力盘 Thrust disc；
35、42—找正平台 Alignment platform；37—轴套 Sleeve；
39—推力轴承 Thrust bearing；40—从动轴 Driven shaft；
43—从动齿轮；45—连接体 Connecting body；
46—专用测量件 Specific pieces for measurement

2）辅助减速箱 Auxiliary speed reduction box

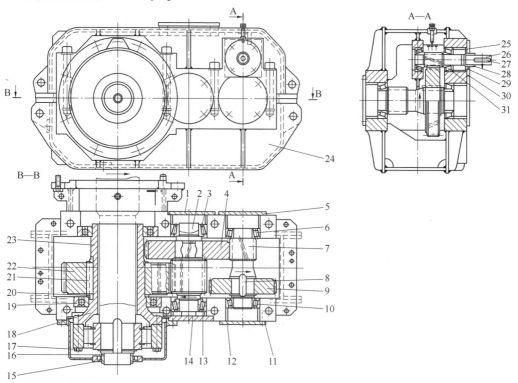

图 2.1.61　辅助减速箱结构图 Structure drawing of auxiliary speed reduction box

1、5、11、12、14、30—压盖 Gland；2、7、26—齿轮轴 Gear shaft；3、6、10、13、19、31—轴承 Bearing；4、9、22—齿轮 Gear；8、21、27—健 Key；15、29—轴封 Shaft seal；16—短轴 Minor axis；17—螺栓 Bolt；18、23、28—轴套 Sleeve；20—卡簧 Snap spring；24—壳体 Casing；25—端盖 End cover

3）联轴器 Coupling

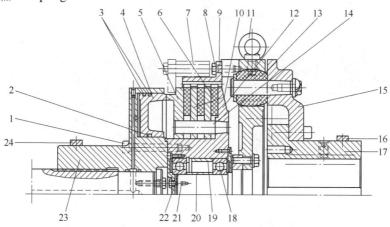

图 2.1.62　联轴器结构图 Structure drawing of extruder coupling

1、8、21—螺栓 Bolt；2、3—O 形密封圈 O sealing ring；4—活塞体 Piston body；5、10—摩擦片压板 Friction plate；6—摩擦片 Friction lining；7—压块 Briquetting；9—外齿套 External gear sets；11—吊耳 Lug；12—轮毂 Wheel hub；13—弹性柱销 Elastic pin；14—销子 Pin；15—传动盘 Driving disc；16、24—测速块 Velocimetry block；17—电机侧传动轴套 Motor side drive shaft sleeve；18、22—轴承 Bearing；19、20—间隔环 Spacer ring；23—减速箱侧传动轴套 Gearbox side drive shaft sleeve

4）筒体 Cylinder

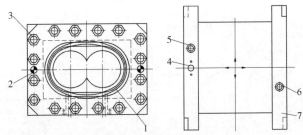

图2.1.63　筒体结构图 Structure drawing of cylinder

1—密封槽 Seal groove；2—销孔 Pin hole；3—螺栓孔 Bolt hole；4—预留孔 Reserve hole；
5—冷却水出口 Cooling water outlet；6—冷却水入口 Cooling water inlet；7—筒体法兰 Shell hange

5）螺杆 Screw

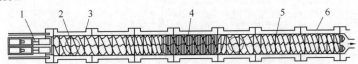

图2.1.64　螺杆结构图 Structure drawing of extruder screw

1—螺杆芯轴 Screw mandrel；2—输送啮合套 Conveying engagement block sets；
3、6—筒体 Cylinder；4—混合啮合套 Mix sliding sleeve；5—挤压啮合套 Squeeze sliding sleeve

6）熔融齿轮泵 Melt gear pump

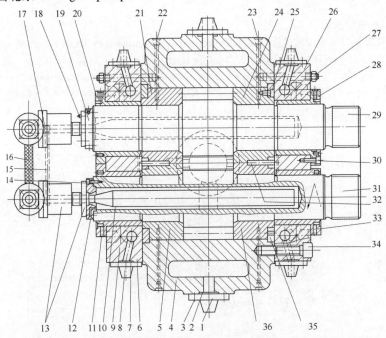

图2.1.65　熔融齿轮泵结构图 Structure drawing of melt gear pump

1、8、19—法兰 Flange；2、7—垫片 Gasket；3、6、18—螺栓 Bolt；4—泵体 Pump body；5、21、24、36—轴承 Bearing；
9、26—端盖 End cover；10、16—金属套管 Metal casing；11、20、28、33—黏性密封 Adhesive sealing；
12、14、25、30、34—内六角螺栓 Hex bolt；13—接头 Connector；15—支承螺栓 Stay bolt；17—螺母 Nut；
22、23—轴承温度探头 Bearing temperature probe；27—螺纹锥销 Thread taper pin；29、31—齿轮轴 Gear shaft；
32—花键 Spline；35—轴承止推环 Bearing thrust ring

2 合成树脂 Synthetic resin

7）开车阀 Driving valve

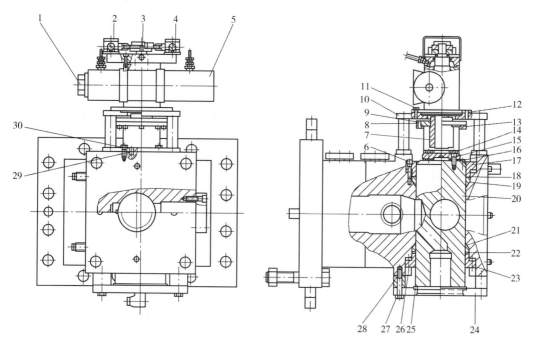

图 2.1.66 开车阀结构图 Structure drawing of driving valve

1—丝堵 Plug；2、4—限位开关 Limit switch；3—接触块 Contact block；5—液压油缸 Hydraulic cylinder；
6、11、13、14、27、30—螺栓 Bolt；7—传动轴套 Driving sleeve；8—传动环 Drive ring；9—销子 Pin；
10—支架 Support；12—油缸底座 Cylinder base；15—压板 Plate pinch；
16—调整垫片 Adjustable gasket；17—填料压盖 Stuffing gland；
18、23—压环 Compression ring；19、21—填料 Packing；
20、22—阀体 Valve body；24—支承板 Supporting plate；25—支承环 Supporting ring；
26、28—密封垫片 Sealing gasket；29—定位销 Location pin

8）换筛器 Screen changer

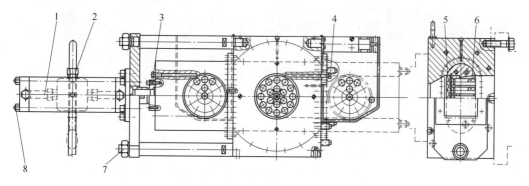

图 2.1.67 换筛器结构图 Structure drawing of screen changer

1—液压油缸 Hydraulic cylinder；2—液压油管 Hydraulic fluid pipe；3、4—顶丝 Jackscrew；
5、6—筛网 Screen cloth；7—支承架螺栓 Bracket bolt；8—螺栓 Bolt

· 139 ·

9) 模板 Die plate

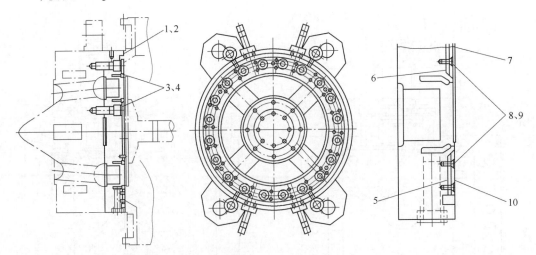

图 2.1.68 模板结构图 Structure drawing of die plate
1—模板 Die plate；2—模板连接体 Die plate drill；3—填料 Packing；
4—六角沉头螺栓 Hex socket bolt；5—隔热环 Insulation ring；
6—隔热盘 Insulation disk；7—卡盘 Retainer disk；
8—平头螺钉 Socket flat head cap screw；
9—带齿锁紧垫圈 Toothed lock washer；
10—卡环 Retainer ring

10) 水下切粒机结构 Underwater pelletizer structure

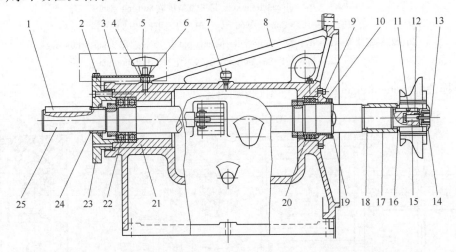

图 2.1.69 水下切粒机结构图 Structure drawing of underwater pelletizer
1、12—键 Key；2—紧固螺栓 Fastening bolt；3—支承板 Supporting plate；4—螺母 Nut；
5—紧固手柄 Tightening Handle；6—锁紧杆 Locking lever；7—排气孔 Gas vent；8—支架 Support；
9、11、15—螺栓 Bolt；10—迷宫密封 Labyrinth seal；13—紧固螺钉 Tightening screw；14、23—压盖 Gland；
16—间隔环 Spacer ring；17、24—轴套 Sleeve；18—O形密封圈 O sealing ring；19—轴封 Shaft seal；
20、22—轴承 Bearing；21—滑套 Sliding sleeve；25—轴 Shaft

11) 切刀及刀盘 Cutter and cutterhead

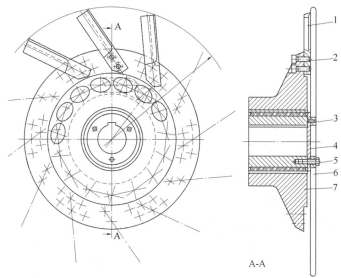

图 2.1.70　切刀及刀盘 Cutter and cutterhead

1—切刀 Cutter；2、5—螺栓 Bolt；3—紧固螺钉 Tightening screw；
4—间隔板 Spacer plate；6—防护板 Protection plate；7—刀盘 Cutterhead

2.1.5.2　L 形挤压造粒机 L-extrusion granulator

1. L 形挤压造粒机外形 L-extrusion granulator outline

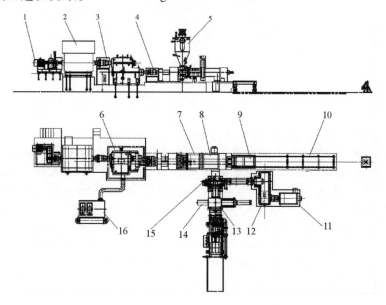

图 2.1.71　L 形挤压造粒机外形图 Outline drawing of L-extrusion granulator unit

1—启动电机 Start motor；2—主电机 Main motor；3、4—联轴器 Coupling；5—料斗 Hopper；
6—主减速箱 Main speed reduction box；7——段筒体 Cylinder of first section；8—二段筒体 Cylinder of seccond section；
9—托架 Bracket；10—导轨 Guide rail；11—熔融泵电机 Melt pump motor；
12—熔融泵减速箱 Melt pump speed reduction box；13—模板 Die plate；14—换筛器 Screen changer；
15—熔融泵 Melt pump；16—润滑油系统 Lube oil system

2. L形挤压造粒机主要零部件 Main components of L-extrusion granulator

1) 主减速箱 Main speed reduction box

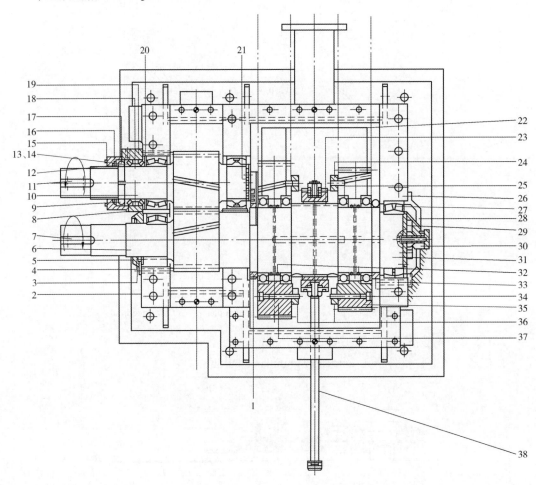

图 2.1.72 主减速箱结构图 Structure drawing of main speed reduction box

1、3、26、33、34—六角沉头螺栓 Hex socket bolt；2、16—油封盖 Oil seal cover；
4、9—油封 Oil seal；5、17、29—六角螺栓 Hexagon bolt；6、14、21—油封衬里 Oil seal Lining；
7、11—平键 Flat key；8、12、13、22、25、27—轴承 Bearing；10—输出轴 Output shaft；
15—挡块 Stopper；18—弹簧 Spring；19—轴承座 Bearing seat；20、32—隔板 Spacer；
23—移动花键 Shifting spline；24—平行销 Parallel pin；28—盖板 Cover；
30—油雾喷射器 Oil sprayer；31—输入轴 Input shaft；35、37—齿轮 Gear；
36—内花键 Internal spline；38—换挡器 Gear shifting

2 合成树脂 Synthetic resin

2）筒体 Cylinder

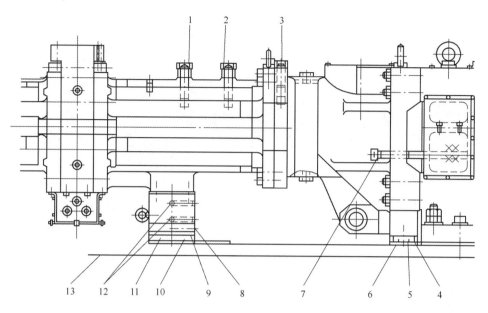

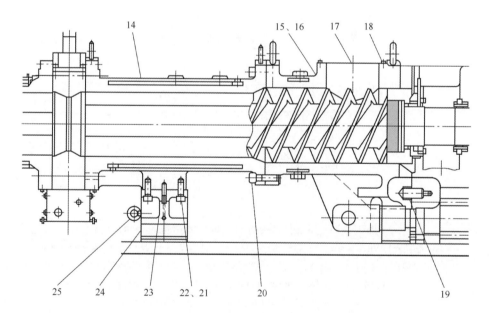

图 2.1.73 筒体结构图 Structure drawing of cylinder
1、2、3—销塞 Plug；4—带螺纹锥销 Pan screw-crossed；
5、10、21—弹簧销 Spring pin；6、11—滑板 Sliding plate；
7—定位螺栓 Jack bolt；8、22—加脂口 Grease nipple；
9—螺钉 Screw；12—插孔 Hub；13—底板 Base plate；
14—筒体 Cylinder；15、16—料斗部分 Hopper section；
17—盖板 Cover；18—六角螺栓 Hex bolt；19—套管 Bush；
20—六角沉头螺栓 Hex socket bolt；23—平行销 Parallel pin；
24—筒体支座 Cylinder support；25—吊环 Eye bolt

3）双螺杆 Twin screw

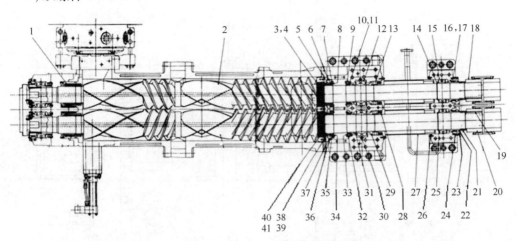

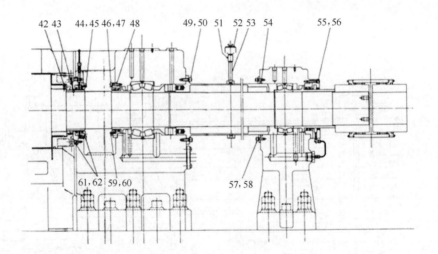

图 2.1.74 双螺杆结构图 Structure drawing of twin screw

1—粘性密封 Viscous seal；2—螺杆 Screw；3、39、41、42、43—定位螺丝 Set screw-socket；
4、40—螺杆密封 Screw seal；5、11、17—油封 Oil seal；6、26、36—油封护圈 Oil seal retainer；
7、9、24、32、35—油封箱 Oil seal housing；8、23、33、34—油封罩 Oil seal cover；
10、16—导向环 Guard ring；12、13、15—轴承 Bearing；
14—轴卡环 Ring-shaft；18、20—齿型联轴器 Gear type coupling；
19—联轴器键 Coupling key；21、25、29、31—轴承垫 Bearing pad；
22—毡圈 Felt；27—挡板 Retainer；28—弹簧 Spring；30—紧固销 Fastening pin；
37—油封隔板 Oil seal spacer；38—内环 Inner ring；
44、47、49、56、58、60、61—弹簧垫圈 Spring washer；
45、46、50、55、57、59、62—六角沉头螺栓 Hex socket bolt；
48—垫圈 Gasket；51—仪表 Gauge；52—呼吸阀 Breather valve；
53—接管 Connecting pipe；54—O 形密封圈 O sealing ring

4) 熔融泵 Melt pump

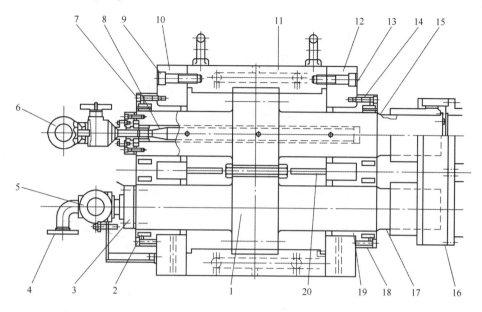

图 2.1.75　熔融泵结构图 Structure drawing of melt pump

1—齿轮 Gear；2、18—填料函密封 Packing box seal；3—配套接头 Coupling adapter；
4—管接口 Pipe interface；5、6—旋转接头 Rotary joint；7、9、13、14—紧固螺栓 Fastening bolt；
8—导流管 Diversion pipe；10、12—侧盖板 Side cover；11—泵体 Pump body；15—上转子 Top rotor；
16—联轴器 Coupling；17—下转子 Bottom rotor；19—轴承 Bearing；20—键 Key

5) 齿形联轴器 Gear type coupling

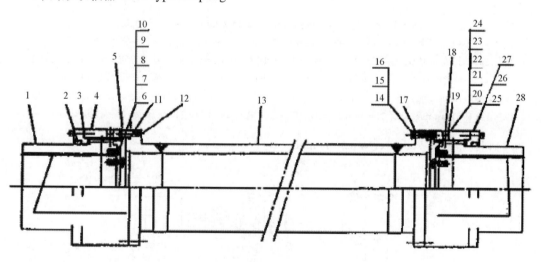

图 2.1.76　齿形联轴器结构图 Structure drawing of gear type coupling

1、28—轮毂 Hub；2、3、11、16、27—O 形密封圈 O sealing ring；4、18—齿套 Sleeve；5、17—注油口 Oil filler plug；
6、19—端板 End plate；7、20—垫圈 Gasket；8、14、21、25—六角螺栓 Hexagon bolt；
9、22—内六角螺栓 Socket cap screw；10、23—填料 Packing；12—平行销 Parallel pin；
13—隔板 Spacer；15、26—弹簧垫圈 Spring washer；24—盖板 Cover

5) 开车阀 Drive valve

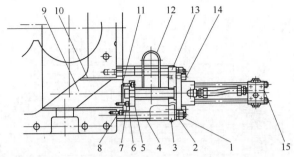

图 2.1.77　开车阀结构图 Structure drawing of drive valve

1、14—紧固螺栓 Fastening bolt；2—垫圈 Gasket；3—弹簧垫片 Spring shim；
4—导杆 Guide rod；5—夹板 Clamp plate；6—固定卡子 Guide bracket；7—六角沉头螺栓 Hex socket bolt；
8、11—调整垫片 Adjustable gasket；9—滑杆 Slide bar；10—衬套 Guide bush；12—止推块 Stopper；
13—导板 Guide plate；15—液压缸 Hydraulic cylinder

6) 换筛器 Screen changer

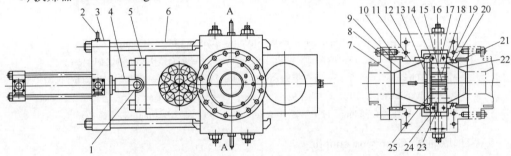

图 2.1.78　换筛器结构图 Structure drawing of screen changer

1—U 形夹销 Clevis pin；2—吊环 Eye bolt；3—液压缸 Hydraulic cylinder；4—叉杆 Clevis；
5—支架 Support；6—限制杆 Tie rod；7、8、22—O 形密封圈 O sealing ring；9、14—六角沉头螺栓 Hex socket bolt；
10—锁紧环 Locking ring；11—锁紧衬套 Locking bushing；12—换筛器本体 Screen changer body；
13—调整环 Adjust ring；15、18—筛板 Strainer plate；16—调整螺栓 Adjusting bolt；17—滑板支架 Plate holder；
19—止推环 Thrust ring；20、21—衬套 Bushing；23—销钉 Pin；24—滑板 Sliding plate；25—挡圈 Retainer ring

7) 切粒机水室 Pelletizer water chamber

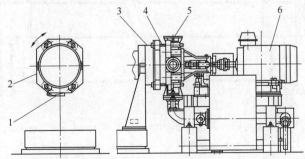

图 2.1.79　切粒机水室外形图 Outline drawing of pelletizer water chamber

1—气缸 Cylinder；2—夹紧环 Clamping ring；3—模头 Die-head；
4—液压夹紧气缸 Hydraulic clamping cylinder；5—水室 Water chamber；6—电机 Motor

8) 刀盘 Cutterhead

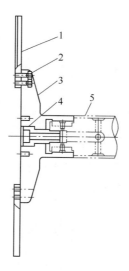

图 2.1.80 刀盘结构图 Structure drawing of cutterhead
1—切刀 Knife；2—六角沉头螺栓 Hex socket bolt；3—刀架 Knife holder；
4—专用螺栓 Special bolt；5—切刀轴 Knife shaft

2.1.5.3 单螺杆挤压造粒机 Single screw extrusion granulator

1. 单螺杆挤压造粒机外形 Outline of single screw extrusion granulator

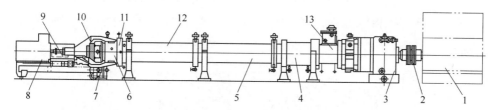

图 2.1.81 单螺杆挤压造粒机外形图 Outline drawing of single screw extrusion granulator
1—电机 Motor；2—联轴器 Coupling；3—齿轮箱 Gearbox；4、5、12、13—筒体 Cylinder；
6—分度板 Scale plate；7—模板 Die plate；8—切粒机电机 Pelletizer motor；
9—切粒机联轴器 Pelletizer coupling；10—切粒机 Pelletizer；11—模板法兰 Die plate flange

2. 单螺杆挤压造粒机主要零部件 Main components of single screw extrusion granulator

1) 单螺杆 Single screw

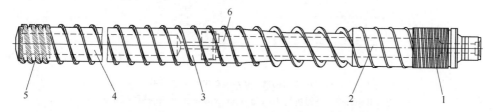

图 2.1.82 单螺杆结构图 Structure drawing of single screw
1—反向排气段 Reverse exhaust section；2—进料段 Feed section；3—压缩段 Compression segment；
4—计量段 Metering section；5—混合段 Mixing section；6—螺杆 Screw

2)减速箱 Speed reduction box

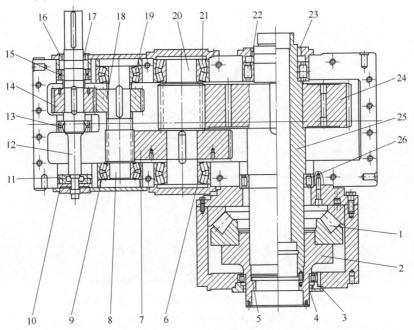

图 2.1.83 减速箱结构图 Structure drawing of speed reduction box

1、3、6、9、11、13、15、19、21、22、26—轴承 Bearing；
2—轴套 Sleeve；4、17、23—轴封 Shaft seal；5、10、16—O 形密封圈 O sealing ring；
7、14、18、24—齿轮 Gear；8、20—齿轮轴 Gear shaft；12—输入轴 Input shaft；25—输出轴 Output shaft

3)模板 Die plate

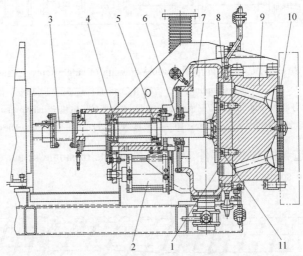

图 2.1.84 模板结构图 Structure drawing of die plate

1—切刀 Knife；2—进刀机构 Tool infeed mechanism；3—联轴器 Coupling；
4—圆柱滚子轴承 Cylindrical roller bearing；5—双列向心球轴承 Double column centripetal ball bearings；
6—填料 Packing；7—水室 Water chamber；8—模板 Die plate；9—分布板 Distribution plate；
10—筛网 Screen cloth；11—切刀盘 Cutter disc

4）切粒机 Pelletizer

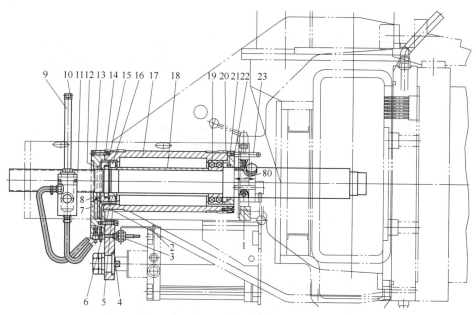

图 2.1.85 切粒机结构图 Structure drawing of pelletizer

1—旋转轴封 Rotary shaft sealing；2—支承板 Supporting plate；3—限位开关 Limit switch；4—垫板 Plate；
6、16—轴承 Bearing；5、17—轴承套 Bearing ring；7—防松垫 Locking pad；8—防松螺母 Lock nut；
9—螺纹管 Threaded pipe；10—螺母 Nut；11—刀柄 Shank；12、21—螺旋密封 Spiral seal；
13—内六角螺栓 Hex bolt；14、20—O形密封圈 O sealing ring；15—定位环 Locating ring；18—轴套 Sleeve；
19—向心止推轴承 Radial thrust bearings；22—轴承压盖；23—紧固螺栓 Fastening bolt

2.1.5.4 侧线挤压机 Lateral line extruder

1. 侧线挤压机外形 Lateral line extruder outline

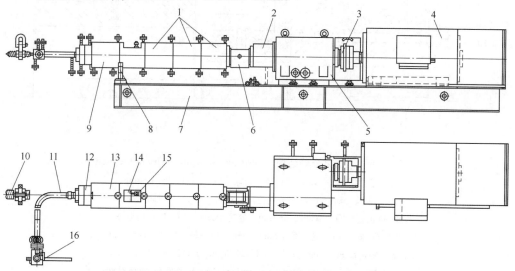

图 2.1.86 侧线挤压机外形图 Outline drawing of lateral line extruder

1、9—密闭筒体 Sealed cylinder；2—开放筒体 Open cylinder；3—摩擦联轴器 Friction coupling；4—电机 Motor；
5—齿轮箱 Gearbox；6—加热/冷却元件 Heating / cooling element；7—支架 Support；8—筒体支座 Cylinder support；
10—螺杆 Screw；11—加料管 Feeding tube；12—连接头 Connector；13—保护套 Protective sleeve；
14—注入阀盖 Injection valve cover；15—注入阀 Injection valve；16—三通阀 Three-way valve

2. 螺杆 Screw

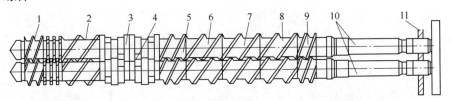

图 2.1.87　螺杆结构图 Structure drawing of screw

1、2、3、4、5、6、7、8、9—螺杆啮合元件 Screw engaging elements；10—螺杆 Screw；
11—调整板 Adjustment plate

3. 摩擦离合器 Friction clutch

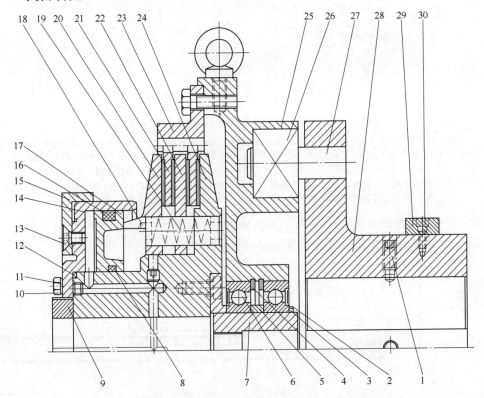

图 2.1.88　摩擦离合器结构图 Structure drawing of friction clutch

1—定位螺丝 Set screw；2、6—滚珠轴承 Needle roller bearing；
3、4—卡簧 Snap spring；5—定位环 Locating ring；7—轴套 Sleeve；
8—O 形密封圈 O sealing ring；9—套筒 Sleeve；10、12—调整垫；
11—垫片 Gasket；13—沉头螺钉 Sunk screw；14—轮毂气缸 Wheel cylinder；
15、29—固定块 Fixed block；16—密封环 Sealing ring；17—柱塞 Plunger；
18—销 Pin；19—压力弹簧 Pressure spring；20—压力盘 Pressure plate；
21—摩擦盘 Friction disc；22—内盘 Inner tray；23—齿形环 Gear ring；
24、28—轮毂 Hub；25—中间法兰 Spacer flange；
26—挠性块 Flexible block；27—插销 Plug pin；30—六角螺钉 Hex screw

2.1.6 加料器 Feeder

2.1.6.1 催化剂加料器 Catalyst feeder
2.1.6.2 添加剂加料器 Additive feeder

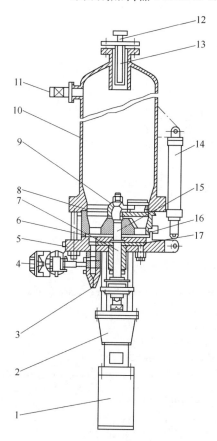

图 2.1.89 催化剂加料器结构图
Structure drawing of catalyst feeder

1—电机 Motor；2—减速器 Speed reducer；
3—催化剂出口 Catalyst outlet；
4—阀门驱动器 Valve actuator；
5—底部法兰 Bottom flange；
6—计量盘 Metering plate；
7—连接轴 Connecting sheet；
8—筛网 Screen cloth；
9—毛刷轮 Brush wheel；
10—储罐 Storage tank；
11—催化剂入口 Catalyst inlet；
12—排空口 Emptying port；
13—过滤器 Filter；
14—液压筒 Hydraulic cylinder；
15—搅拌器 Stirrer；
16—盖板 Cover plate；
17—耐磨板 Wear plate

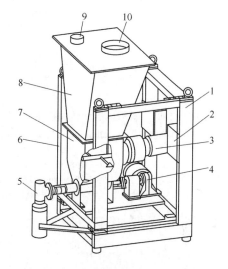

图 2.1.90 添加剂加料器结构图
Structure drawing of additive feeder

1—框架 Shell frame；2—接线盒 Junction box；
3—搅拌电机 Stirring motor；
4—加料电机 Feeding motor；
5—出料口 Discharge outlet；
6—称重支架 Weighing bracket；
7—加料漏斗 Feed hopper；8—连接槽 Link slot；
9—排气口 Exhaust port；10—进料口 Feed inlet

2.1.6.3 颗粒风送旋转加料器 Particles breeze rotary feeder

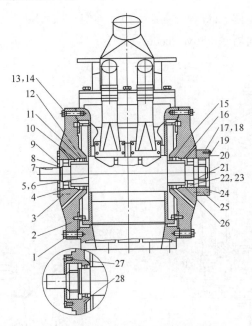

图2.1.91 颗粒风送旋转加料器结构图 Structure drawing of particles breeze rotary feeder

1—O形密封圈 O sealing ring；2、28—紧固板 Fastening plate；3、14、17、25、27—螺栓 Screw；
4、15、19、20—盖板 Cover plate；5—保护盘 Protection disc；6—转子 Closed rotor；
7—键 Key；8、11、12、23、26—阻隔环 Ring of arrest；
9、10—密封环 Seal ring；13、18、24—垫圈 Gasket；16—润滑器 Lubricator；
21—隔板 Partition；22—轴承 Bearing

2.1.6.4 旋转加料破块器 Rotary feeder block breaking device

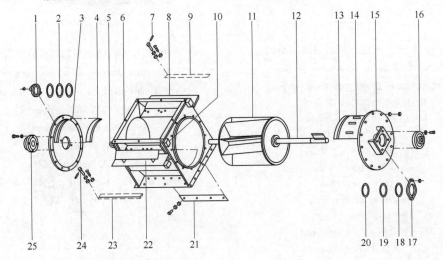

图2.1.92 旋转加料破块器结构图 Structure drawing of rotary feeder block breaking device

1、17—密封压盖 Shaft sealing gland；2、19—套环 Lantern ring；3、15—端盖 End cover；
4、13—密封环 Sealing ring；5—吊耳 Lifting lugs；6—箱体 Housing；7、24—螺杆组件 Screw post assemblies；
8，22—固定板 Fixed Plate；9、21、23—压板 Plate pinch；10—安全开关 Safety switch；11—转子 Rotor；
12—轴 Shaft；14—压缩弹簧 Compression springs；16、25—轴承 Bearings；18、20—填料环 Packing

2.1.7 振动筛 Vibrating screen

2.1.7.1 粉料振动筛 Powder vibrating screen

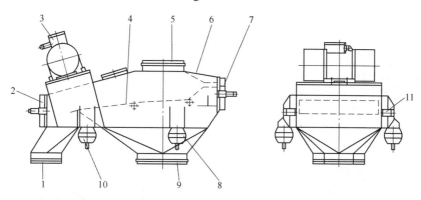

图 2.1.93 粉料振动筛外形图 Outline drawing of powder vibrating screen
1—排料口 Discharging port；2—前端盖 Front end cover；3—电机 Motor；
4—筛网 Screen cloth；5—进料口 Feed inlet；6—壳体 Casing；
7—后端盖 Rear end cover；8、10—橡胶弹簧 Rubber spring；
9—出料口 Discharge outlet；11—筛网压块 Screen press

2.1.7.2 颗粒振动筛 Particles vibrating screen

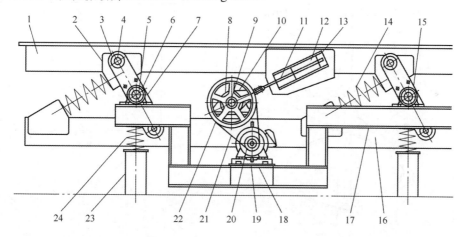

图 2.1.94 颗粒振动筛外形图 Outline drawing of particles vibrating screen
1—料槽 Trough；2—弹簧支架 Spring stent；3、5—橡胶套 Rubber sleeve；
4—夹持器 Clamp holder；6—销轴 Shaft；7—杠杆 Lever；
8—轴承座 Bearing seat；9—曲轴 Crankshaft；10—箱体 Cylinder；
11—弹性板 Elastic plate；12—橡胶弹簧 Rubber spring box；
13—橡胶弹簧箱 Rubber spring；14—弹簧 Spring；15—支座 Support；
16—传感器 Sensor；17—框架 Shell frame；18—电机座 Motor base；
19—电机 Motor；20、22—皮带轮 Belt pulley；
21—皮带 Belt；23—减振弹簧支座 Damping spring support；
24—减振弹簧 Damping spring

2.1.8 催化剂配制槽搅拌器 Catalyst preparation slot stirrer

2.1.9 料仓、干燥器及风机 Feed bin、dryer and blower

2.1.9.1 粉料料仓 Powder bin

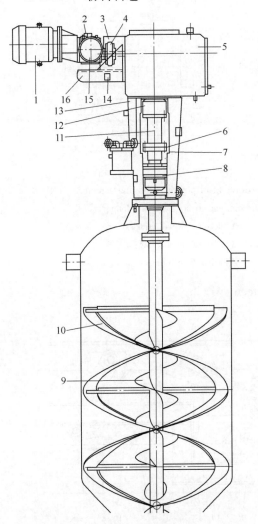

图 2.1.95 催化剂配制槽搅拌器结构图
Structure drawing of catalyst preparation slot stirrer

1—电机 Motor；2—齿轮箱 Gearbox；
3—联轴器防护罩 Coupling guard；4—联轴器 Coupling；
5—齿轮传动箱 Gear driving box；6—锥型联轴器 Conical coupling；
7—防护罩 Protective cover；8—机械密封 Mechanical seal；
9—螺旋器 Screw；10—螺带 Ribbon screw；
11—连接轴 Connecting sheet；12—齿形联轴器 Gear coupling；
13—支架 Support；14—接线盒 Junction box；
15—开关盒 Switch box；16—电机支座 Motor seat

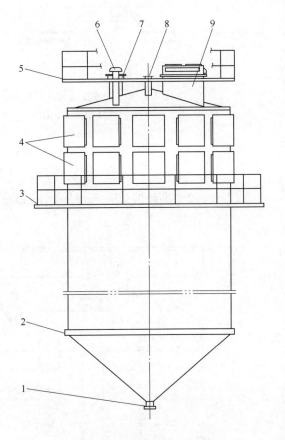

图 2.1.96 粉料料仓结构图
Structure drawing of powder bin

1—粉料出口 Powder outlet；2—支架 Support；
3、5—检修平台 Maintenance platform；
4—爆破片 Rupture disc；6—呼吸阀 Breathing valve；
7—人孔 Manhole；8—粉料入口 Powder inlet；
9—反吹过滤器 Reversed jet filter

2 合成树脂 Synthetic resin

2.1.9.2 粒料料仓 Pellet bin

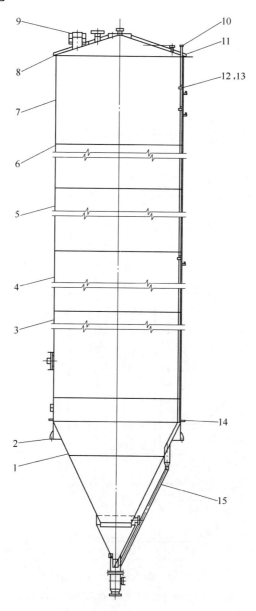

图 2.1.97 粒料料仓结构图 Structure drawing of pellet bin
1—料斗 Hopper；2—裙座 Skirt；3、4、5、6、7—筒体 Cylinder；
8—挡边 Rib；9—放空口 Vent；
10—水冲洗口 Water washing port；
11—锥顶 Cone top；12—挡板 Baffle；
13—加强筋 Reinforcing rib；
14—保温支承圈 Insulation supporting ring；
15—掺合件 Blending pieces

2.1.9.3 脱水颗粒干燥器 Dehydrated particle dryer

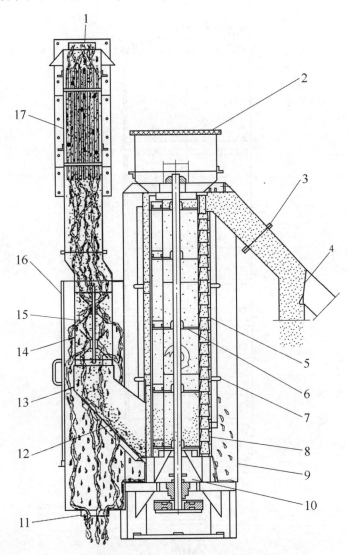

图 2.1.98 脱水颗粒干燥器结构图
Structure drawing of dehydrated particle dryer

1—料浆入口管 Slurry inlet pipe；2—空气入口管 Air inlet pipe；
3—产品出口管 Product outlet pipe；4—颗粒换向阀 Particle valve；
5—主滤网 Main filter；6—支架 Support；
7—中心支承环 Center support ring；8—提升叶片 Lifting blades；
9—干燥器箱体 Dryer casing；10—甩水环 Slinger ring；
11—水出口管 Water outlet pipe；12—进料网 Feed net；
13—进料斜槽 Feed chute；14—脱水网 Dehydration net；
15—水换向折板 Water reversing flap；
16—脱水箱体 Dehydration casing；
17—大块补集器 Chunk complement control

2 合成树脂 Synthetic resin

2.1.9.4 罗茨风机 Roots blower

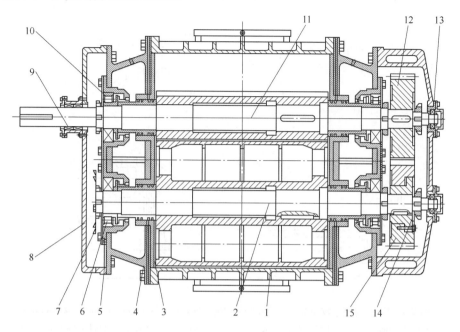

图 2.1.99 罗茨风机结构图 Structure drawing of Roots blower

1—壳体 Casing；2—从动转子 Driven rotor；3—侧板 Side plate；
4—盖板 Cover plate；5—轴承座 Bearing seat；6—圆柱滚子轴承 Cylindrical roller bearings；
7—甩油环 Oil slinger；8—油箱 Oil tank；9—机械密封 Mechanical seal；
10—轴承压盖 Bearing cover；11—主动转子 Main rotor；
12—主动齿轮 Driving gear；13—深沟球轴承 Deep groove ball bearing；
14—锁紧螺母 Locknut；15—从动齿轮 Driven gear

2.1.10 其它设备 Other equipments

2.1.10.1 电加热器 Electric heaters

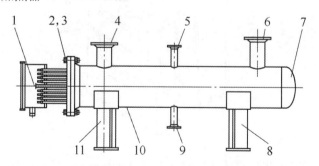

图 2.1.100 电加热器结构图 Structure drawing of electric heaters

1—热电偶 Thermocouple；2—螺栓 Bolt；3—垫片 Gasket；
4—介质进口 Media inlet；5—放空口 Vent；
6—介质出口 Media outlet；7—封头 Head cap；
8、11—支座 Support；9—排污口 Drain outlet；
10—筒体 Cylinder

2.1.10.2 造粒连续混合器 Continuous mixer granulator

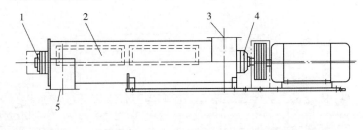
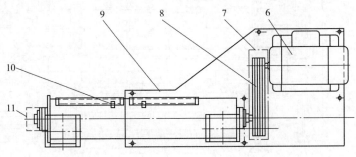

图 2.1.101　造粒连续混合器外形图 Outline drawing of continuous mixer granulator

1、4—轴承 Bearing；2—观察孔 Observing hole；3—进料口 Feed inlet；5—出料口 Discharge outlet；
6—电机 Motor；7—防护罩 Protective cover；8—皮带 Belt；9—基座 Base；
10—保护开关 Protective switch；11—轴承罩 Bearing cover

2.1.10.3 袋式过滤器 Bag filter

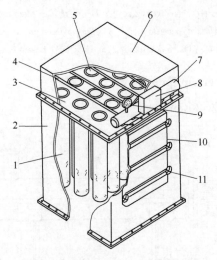

图 2.1.102　袋式过滤器结构图 Structure drawing of bag filter

1—过滤器 Filter；2—过滤器箱体 Filter box；3—管板 Tube sheet；
4—吹扫管 Scavenging conduit；5—文丘里管 Venturi tube；6—顶板 Roof；
7—顶部管 Top pipe；8—控制阀 Control valve；
9—压力表 Pressure gauge；10—检修口 Access hole；
11—手柄 Handle

2 合成树脂 Synthetic resin

2.1.10.4 反应器出料阀 Reactor outlet valve
2.1.10.5 反应器出料根部阀 Reactor outlet root valve

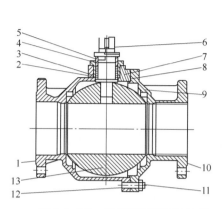

图 2.1.103 反应器出料阀结构图
Structure drawing of reactor outlet valve

1、10—阀体 Valve body；2—阀杆轴套 Stem sleeve；
3—辅助阀杆密封 Auxiliary stem seal；4—上部阀杆密封
Upper stem seal；5—压圈 Pressing ring；6—阀杆 Valve stem；
7—填料箱 Packing box；8—阀杆密封 Stem seal；9—密封座 Seal seat；
11—阀体螺母 Valve nut；12—阀体螺栓 Body bolt；
13—球体 Ball body

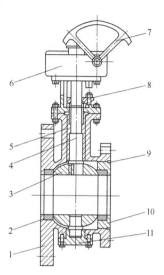

图 2.1.104 反应器出料根部阀结构图
Structure drawing of reactor outlet root valve

1—阀体 Valve body；2—阀座 Valve seat；
3—阀球 Valve ball；4—阀杆 Valve stem；
5—填料 Packing；6—蜗轮蜗杆减速箱
Worm reduction box；7—手轮 Hand wheel；
8、11—压盖 Gland；9—紧固螺钉 Tightening screw；
10—阀座支承环 Valve Seat supporting ring

2.1.10.6 风送三通阀 Wind sending three way valve

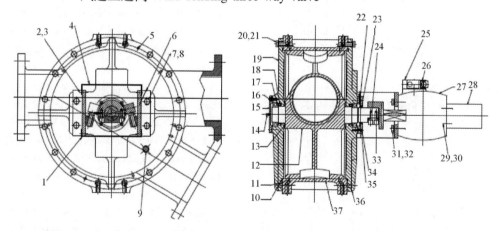

图 2.1.105 风送三通阀结构图 Structure drawing of wind sending three way valve

1、4—防护罩 Protective cover；2、7、20、29、31、34—六角螺栓 Hex head screw；3、8、21、30、32—弹簧垫圈
Spring washer；5—定位销 Centering pin；6—衬套 Bushing；9—销塞 Pin plug；10—硅橡胶绳 Silicone rope；11—硅胶密封
Silicone seal；12—转子 Rotor；13—定距块 Distance block；14、22—加脂口 Fat liquoring port；15、35—小盖板
Small cover plate；16、24、25—内沉六角螺栓 Socket cap screw；17—球轴承 Ball bearing；18—唇形密封环 Lip seal ring；
19—盖板 Cover plate；23—O 形密封圈 O sealing ring；26—电磁阀 Solenoid valve；
27—气动执行机构 Pneumatic actuator；28—限位开关 Limit switch junction box；33—防松板 Locking plate；
36—阀体 Valve body；37—阀盖 Valve cover

2.2 聚丙烯 Polypropylene

2.2.1 反应设备 Reaction equipment

2.2.1.1 环管聚合反应器 Loop polymerization reactor

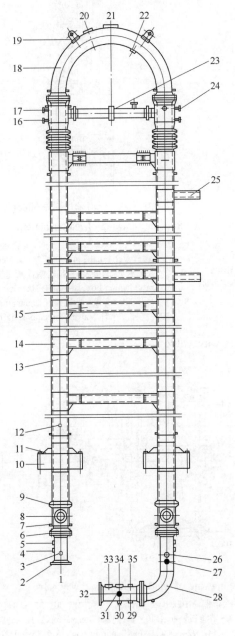

1—工艺排出管口 Process discharge nozzle；2—法兰 Flange；3—杀死剂注入口 Killer injection；4、5—温度计接口 Thermometer interface；6—垫片 Gasket；7—保温支承圈 Insulation supporting ring；8—夹套进/出水口 Jacket water inlet／outlet；9—轴向膨胀节 Axial expansion joint；10—支座 Support；11—锥形短节 Cone nipple；12—铭牌 Nameplate；13—夹套 Jacket；14—补强管 Reinforcing tube；15—连接梁 Connecting beam；16—水槽连接口 Water tank interface；17、24—夹套放空口 Jacket vent；18—180°弯管 180°bend tube；19—吊耳 Lug；20、30—放空口 Vent；21—安全阀接口 Safety valve interface；22—淤浆出口 Slurry outlet；23—横向膨胀节 Lateral expansion joint；25—管支承 Tube support；26、29—取样口 Sampling port；27—氮气入口 Nitrogen inlet；28—90°弯管 90°bend tube；31—排放口 Discharge outlet；32—淤浆入口 Slurry inlet；33—进料口 Feed inlet；34—压力计接口 Manometer interface；35—备用口 Alternate port

图 2.2.1 环管聚合反应器结构图
Structure drawing of loop polymerization reactor

2 合成树脂 Synthetic resin

2.2.1.2 立式搅拌流化床反应器 Vertical mixing fluidized bed reactor

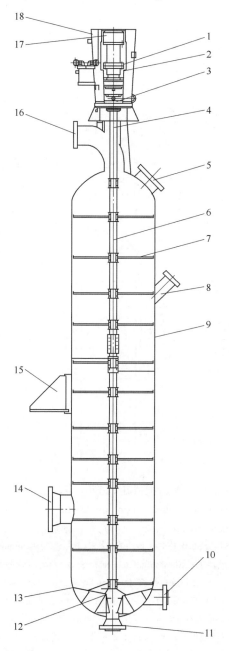

图 2.2.2 立式搅拌流化床反应器结构图 Structure drawing of vertical mixing fluidized bed reactor
1、17—联轴器 Coupling；2—防护罩 Protective cover；3—机械密封 Mechanical seal；
4—连接轴 Connecting sheet；5、14—人孔 Manhole；6—主轴 Shaft；7—刮板 Scraper；
8—物料入口 Material inlet；9—筒体 Cylinder；10—气体入口 Gas inlet；
11—物料出口 Material outlet；12—底部导料板 Bottom guide plate；
13—分布板 Distribution plate；15—支座 Support；
16—气体出口 Gas outlet；18—支架 Support

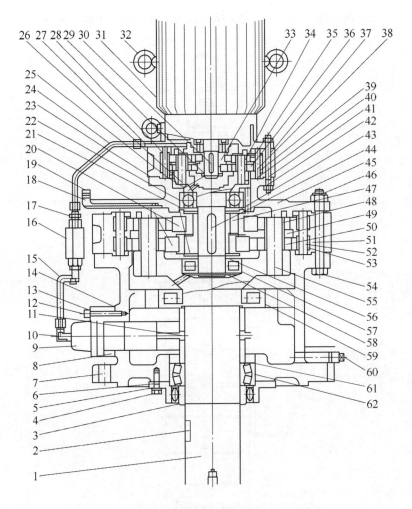

图 2.2.3 立式搅拌流化床反应器减速机结构图
Structure drawing of vertical mixing fluidized bed reactor speed reducer

1—低速轴 Slow speed shaft；2、11、30、44—键 Key；3—轴封 Shaft seal；
4、12、54—六角螺栓 Hex head bolt；5、13—弹簧垫 Spring washer；
6—低速端轴承座 Slow speed end seat；7—法兰套管 Flanged casing；
8—泵盖板 Pump plate；9—泵体 Pump body；10—弯头 Street elbow；
14、23、28、29—垫圈 Gasket；15、24—不锈钢管 Steel pipe；16—油视镜 Oil signal；
17—连接件 Connector；18—内外螺纹弯头 Street elbow；19—呼吸阀 Air vent；
20、57、59、62—滚柱轴承 Roller bearing；21—端板 End plate；22—球轴承 Ball bearing；
25、34、49、58—定位环 Locating ring；26、34、48、55—轴销 Shaft pin；
27—接头 Connector；31—电机 Motor；32—吊环 Suspension loop；
33—偏心轴承 Eccentric bearing；35、50—摆线啮合齿轮 Cycloid meshing gear；
36—齿轮 Gear；37、51—齿圈壳体 Ring gear housing；
38、52—齿环销 Ring gear pin；39、53—环形齿轮轴 Ring gear shaft；
40—传动轴 Mediate shaft；41—双头螺栓 Double end stud；42、56、61—垫片 Spacer；
43—弹簧挡圈 Retaining ring；45—中间盖 Inter mediate cover；
46—六角螺母 Hex nut；47—偏心盘 Eccentric disc；60—丝堵 Plug

· 162 ·

2.2.2 塔设备 Column equipment

2.2.2.1 洗涤塔 Scrubber column

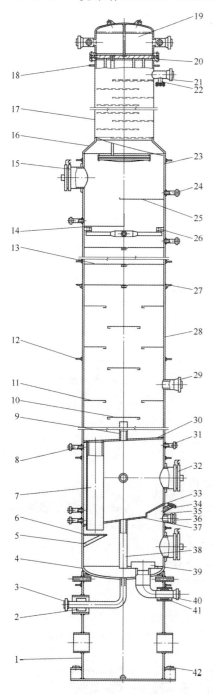

图 2.2.4 洗涤塔结构图
Structure drawing of scrubber column

1—裙座 Skirt；2—补强板 Support plate；3、8、24、31、34、36—接管 Connecting pipe；4—封头 Head；5—筋板 Reinforcing plate；6—防冲挡板 Impingement baffle；7—降液管 Downcomer；9、38—平衡管 Balance pipe；10、11、13—塔盘 Tray；12、18、27—保温支承圈 Insulating supporting ring；14、25—密封盘 Liquid receiving plate；15、32—人孔 Manhole；16—变径段 Reducers；17、23、28—筒体 Cylinder；19—换热器 Heat exchanger；20—筒体法兰 Shell hange；21、29、40—工艺管接口 Process pipe interface；22—旁路接管 Bypass pipe；26—紧固螺栓 Fastening bolt；30、33—弓形板 Arched plate；35—补强圈 Reinforcing ring；36—视镜 Sight glass；37—钢板支承 Steel plate support；39—防涡器 Vortex breaker；41—垫木 Skid；42—地脚螺栓 Anchor bolt

2.2.2.2 汽提塔 Stripping column

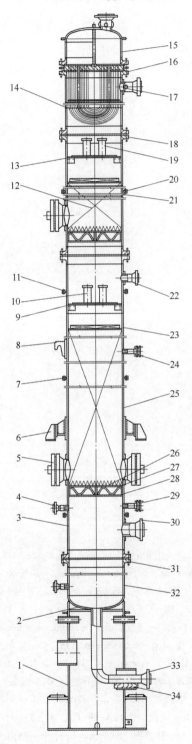

图 2.2.5 汽提塔结构图
Structure drawing of stripping column

1—裙座 Skirt；2—封头 Head；3—连接段筒体 Cylinder for connecting segment；4、17、22、24—接管 Connecting pipe；5—人孔 Manhole；6—支座 Support；7、11、20—保温支承圈 Insulation supporting ring；8—吊耳 Lug；9、13—支承件 Support；10、19—分配器 Distributor；12—填料 Packing；14、21、23、25、32—筒体 Cylinder；15—U形管换热器 U heat exchanger；16、18、31—筒体法兰 Shell hange；26—丝网除沫器 Screen mesh；27—支承盘 Supporting plate；28—支承圈 Supporting ring；29—工艺管接口 Process pipe interface；30—进料口 Feeding inlet；33—出料口 Discharge outlet；34—垫木 Skid

2.2.2.3 乙烯精制塔 Ethylene refining column

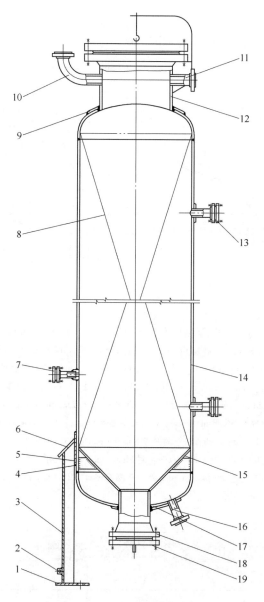

图 2.2.6 乙烯精制塔结构图
Structure drawing of ethylene refining column
1—底板 Foundation plate；2—接地板 Grounding plate；
3、6—支架 Support；4—下部筒体 Bottom cylinder；
5—补强板 Stiffening plate；7、10、11、13—工艺管接口 Process pipe interface；
8—催化剂 Catalyst；9、17—补强圈 Reinforcing ring；
12—上部筒体 Cylinder；14—筒体 Cylinder；
15—支承板 Support plate；16—接管 Connecting pipe；
18—泄料口 Discharge outlet；19—法兰盖 Flange cover

2.2.3 泵 Pump

2.2.3.1 循环轴流泵 Circulation axial pump

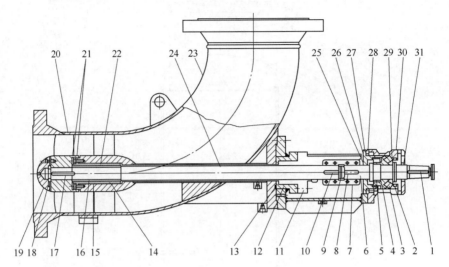

图 2.2.7 循环轴流泵结构图 Structure drawing of circulation axial pump

1、8、10、17—键 Key；2—弹簧 Spring；3—推力轴承 Thrust bearing；4—间隔环 Spacer ring；
5—径向轴承 Radial bearing；6—连接轴 Connecting shaft；7—联轴器 Coupling；
9、18—定位环 Positioning ring；11—双端面机械密封 12—密封腔 Sealed chamber；
Two-end balanced mechanical seal；13、28—O 形密封圈 O sealing ring；14—轴承 Bearing；
15—机械密封 Mechanical seal；16、22—轴套 Sleeve；19—固定帽 Fixed cap；
20—衬套壳体 Bushing shell；21—间距盘 Spacing plate；23—环管弯头 Ring tube elbow；
24—泵轴 Pump shaft；25、31—轴封 Shaft seal；26—轴承压盖 Bering cover；
27—调整垫 Adjusting gasket；29—垫圈 Gasket；30—轴承箱 Bearing box

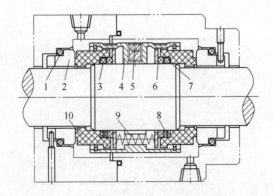

图 2.2.8 循环轴流泵双端面密封结构图
Structure drawing of two-end balanced mechanical seal
1—动环密封圈 Moving sealing ring；2—动环 Rotating ring；
3、7—静环密封圈 Static sealing ring；4—弹簧座 Spring seat；
5—锁紧螺钉 Set screw；6—推环 Push ring；8—卡环 Snap ring；
9—弹簧 Spring；10—静环 Static ring

2 合成树脂 Synthetic resin

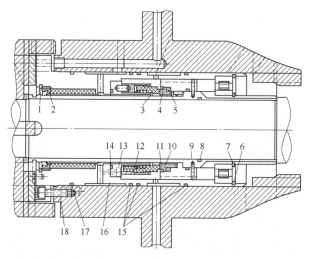

图 2.2.9 循环轴流泵单端面机械密封结构图 Structure drawing of typical mechanical seal
1—弹性挡圈 Snap ring；2—唇封 Lip seal；3—静环 Static ring；4—动环 Rotating ring；
5、8、14、15、17—O形密封圈 O sealing ring；6、7—挡圈 Retaining ring；
9—主轴套 Main shaft sleeve；10—垫片 Gasket；11—卡环 Snap ring；12—短轴套 Stub sleeve；
13、18—内沉六角螺栓 Socket cap screw；16—外轴套 Outer sleeve；

2.2.3.2 高速进料泵 High-speed feed pump

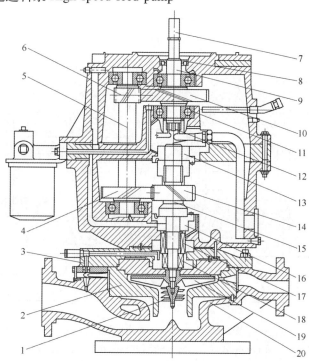

图 2.2.10 高速进料泵结构图 Structure drawing of high-speed feed pump
1—诱导轮 Inducer；2—扩压器 Diffuser；3—分离器 Extractor；4、10—驱动齿轮 Drive gear；
5—中速轴 Medium speed shaft；6、14—从动齿轮 Driven gear；7—驱动轴 Drive shaft；8—轴封 Shaft seal；
9—低速轴 Low speed shaft；11—滚动轴承 Rolling bearing；12—内部润滑泵 Internal lubrication pump；
13、16—滑动轴承 Sliding bearing；15—高速轴 High speed shaft；17、18—机械密封 Mechanical seal；
19—叶轮 Wheel；20—壳体 Pump casing

2.2.3.3 白油注入齿轮泵 White oil injection gear pump

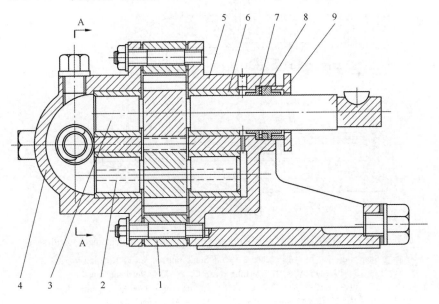

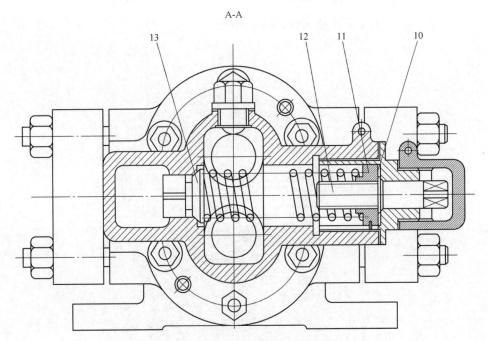

图 2.2.11 白油注入齿轮泵结构图
Structure drawing of white oil injection gear pump
1—泵体 Pump body；2—从动齿轮 Driven gear；3—主动齿轮 Driving gear；
4—阀体 Valve body；5—轴承座 Bearing seat；6—轴套 Sleeve；
7—密封圈 Sealing plug；8—挡圈 Retaining ring；9—压盖 Gland；
10—弹簧压盖 Spring gland；11—弹簧座 Spring seat；
12—调节螺栓 Adjusting bolt；13—阀芯 Valve core

2.2.3.4 助催化剂注入隔膜泵 Cocatalyst injection diaphragm pump

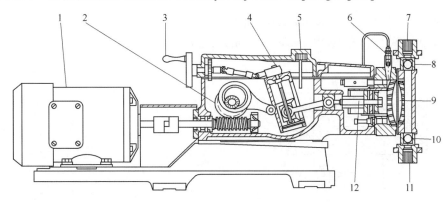

图 2.2.12 助催化剂注入隔膜泵结构图 Structure drawing of cocatalyst injection diaphragm pump

1—电机 Motor；2—曲轴连杆 Crankcase connecting rod；3—调节手柄 Adjusting handle；
4—调节机构 Adjusting mechanism；5—油视镜 Oil sight glass；6—排放阀 Drain valve；
7—出口接管 Outlet pipe；8—出口阀 Outlet valve；9—隔膜 Diaphragm；
10—入口阀 Inlet valve；11—入口接管 Inlet pipe；12—柱塞 Plunger

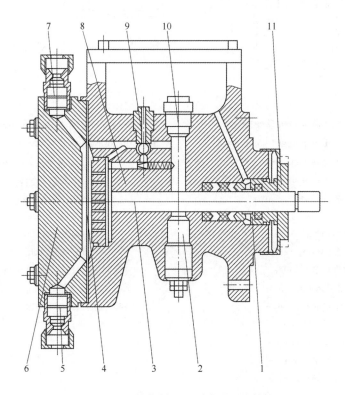

图 2.2.13 助催化剂注入隔膜泵泵头结构图
Structure drawing of pump head structure of cocatalyst injection diaphragm pump

1—填料密封 Packing seal；2—排放阀 Drain valve；3—柱塞 Plunger；
4—隔膜 Diaphragm；5—入口阀 Inlet valve；6—泵盖 Pump cover；
7—出口阀 Outlet valve；8—泵体 Pump body；9—充液阀 Prefill valve；
10—溢流阀 Relief valve；11—填料压盖 Stuffing gland

2.2.4 压缩机 Compressor

2.2.4.1 隔膜压缩机 Diaphragm compressor
2.2.4.2 循环气压缩机 Recycle gas compressor

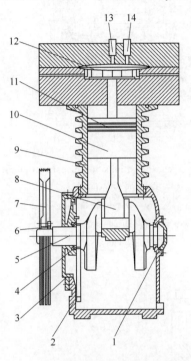

图 2.2.14 隔膜压缩机结构图
Structure drawing of diaphragm compressor

1—轴承压盖 Bearing cover；
2—底座 Base；
3—曲轴箱端盖 Crankcase cover；
4—轴承 Bearing；
5—曲轴 Crankshaft；
6—紧固套 Fixing sleeve；
7—皮带轮 Belt pulley；
8—连杆 Connecting rod；
9—缸体 Cylinder body；
10—活塞 Piston；
11—活塞环 Piston ring；
12—膜片 Membrane；
13—出口阀 Outlet valve；
14—入口阀 Suction valve

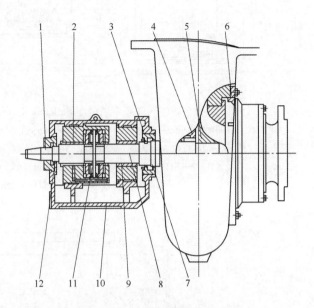

图 2.2.15 循环气压缩机结构图
Structure drawing of recycle gas compressor

1、3、7、12—轴封 Shaft seal；
2—轴套 Sleeve；
4—密封组件 Seal assembly；
5—叶轮 Wheel；
6—端盖 End cover；
8—主轴 Shaft；
9—径向轴承衬套 Radial bearing bush；
10—轴承箱 Bearing box；
11—止推轴承 Thrust bearing shell

2.2.5 挤压造粒机 Extrusion granulator

2.2.5.1 挤压造粒机外形 Extrusion granulator outline

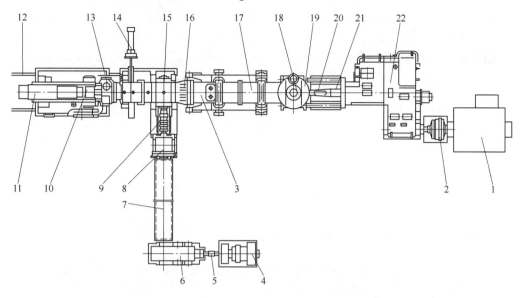

图 2.2.16 挤压造粒机外形图 Outline drawing of extrusion granulator

1—主电机 Main motor；2、5、9—联轴器 Coupling；3、17—筒体 Cylinder；4—电机 Motor；
6—减速箱 Reducer casing；7—万向联轴器 Universal Coupling；8—轴承箱 Bearing box；
10—切粒机 Pelletizer；11—切粒机电机 Pelletizer motor；12—导轨 Guide rail；
13—切粒机水室 Pelletizer water chamber；14—换网器 Screen changer；
15—熔融泵 Melt pump；16—开车阀 Drive valve；
18—下料管 Blanking pipe；19—下料斗 Lower hopper；
20—螺杆 Screw；21—主轴 Shaft；
22—主减速箱 Main reducer boxing

2.2.5.2 挤压造粒机结构 Extrusion granulator structure

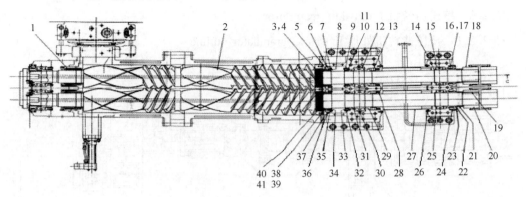

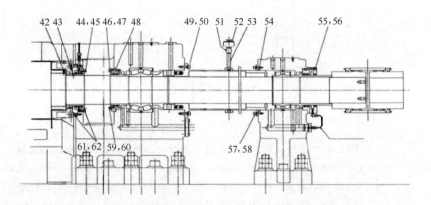

图 2.2.18 挤压造粒机结构图 Structure drawing of extrusion granulator

1—粘性密封 Visco seal；2—螺杆 Screw；3、39、41、42、43—定位螺丝 Set screw-socket；
4、40—螺杆密封 Screw seal；5、11、17—轴封 Shaft seal；6、26、36—轴封护圈 Shaft seal retainer；
7、9、24、32、35—油封箱 Oil sealing housing；8、23、33、34—油封罩 Oil sealing cover；
10、16—导向环 Guide ring；12、13、15—轴承 Bearing；14—轴卡环 Ring-shaft；
18、20—齿型联轴器 Gear coupling；19—键 Key；21、25、29、31—轴承垫 Bearing pad；
22—毡垫圈 Felt；27—挡板 Retainer；28—弹簧 Spring；30—紧固销 Fastening pin；
37—油封隔板 Oil seal spacer；38—内环 Inside ring；
44、47、49、56、58、60、61—弹簧垫圈 Spring washer；
45、46、50、55、57、59、62—六角沉头螺栓 Hex socket bolt；48—垫圈 Gasket；
51—仪表 Gauge；52—呼吸阀 Air breather；53—接管 Connecting pipe；
54—O 形密封圈 O sealing ring

2.2.5.3 减速箱 Reducer casing

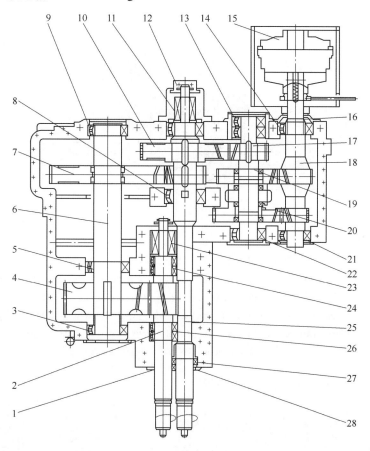

图 2.2.19 减速箱结构图 Structure drawing of reducer casing

1、28—轴封 Shaft seal；2、25—输出轴 Output shaft；

3、5、8、9、13、14、21、22—轴承 Bearing；

4—输出齿轮 Output gear；6—过渡轴 Transition shaft；

7、10—大齿轮 Large gear；11、23—止推轴承 Thrust bearing；

12—轴承挡圈 Bearing retainer；15—联轴器 Coupling；

16—轴封 Shaft seal；17、19—调速齿轮轴 Governor gear shaft；

18—输入轴 Input shaft；20—高速齿轮 High speed gear；

24、26、27—径向轴承 Radial bearing

2.2.5.4 离合器 Clutch

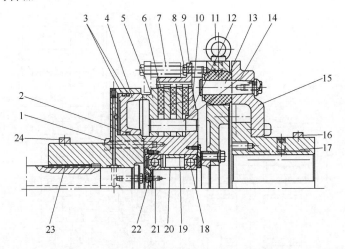

图 2.2.20 离合器结构图 Structure drawing of clutch

1、8、21—螺栓 Bolt；2、3—O 形密封圈 O sealing ring；4—活塞体 Piston body；5—摩擦压板 Friction plate；
6—摩擦片 Friction lining；7—压块 Briquetting；9—外齿套 External gear sets；10—摩擦板 Friction plate；
11—吊环 Lug；12—摩擦轮座 Friction wheel seat；13—弹性柱销 Elastic pin；14—销子 Pin；
15—传动盘 Driving disc；16、24—测速块 Velocimetry block；17、23—轴套 Sleeve；
18、22—轴承 Bearing；19、20—间隔环 Spacer ring

2.2.5.5 熔融齿轮泵 Melt gear pump

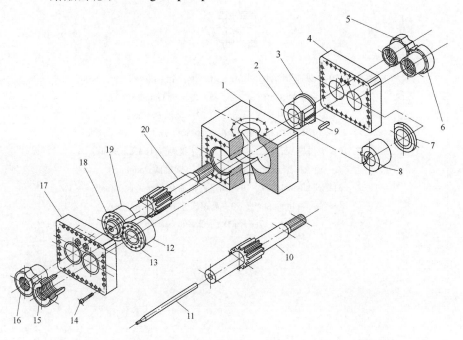

图 2.2.21 熔融齿轮泵结构图 Structure drawing of melt gear pump

1—泵体 Pump body；2、8、12、19—轴承 Bearing；3、7、13、18 轴承压盖 Bering cover；
4、17—侧盖 Pump side cover；5、6、15、16—螺旋密封 Spiral seal；9—键 key；
10、20—齿轮泵转子 Gear pump rotor；11—虹吸管 Siphon；14—紧固螺栓 Fastening bolt

2.2.5.6 换网器 Screen changer

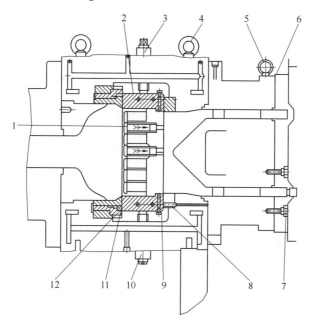

图 2.2.22 换网器结构图 Structure drawing of screen changer
1—过滤器 Filter；2—滑板 Sliding plate；3、7、9、10—紧固螺栓 Fastening bolt；
4、5—吊环 Lug；6—模板 Die plate；8—止回环 Check ring；
11—密封环 Seal ring；12—调节环 Adjusting ring

2.2.5.7 切粒机 Pelletizer

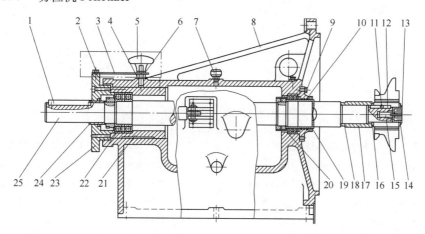

图 2.2.23 切粒机结构图 Structure drawing of pelletizer
1、12—键 Key；2、13—紧固螺栓 Fastening bolt；3—支承板 Supporting plate；4—螺母 Nut；
5—紧固手柄 Tightening Handle；6—锁紧杆 Locking lever；7—排气孔 Gas vent；
8—支架 Support；9、11、15—螺栓 Bolt；10—迷宫密封 Labyrinth seal；
14、23—压盖 Gland；16—间隔环 Spacer ring；17、24—轴套 Sleeve；
18—O 形密封圈 O sealing ring；19—轴封 Shaft seal；20、22—轴承 Bearing；
21—滑套 Sliding sleeve；25—轴 Shaft

2.2.6 搅拌混合器 Stirring mixer

2.2.6.1 磁力搅拌器 Magnetic stirrer
2.2.6.2 螺带混合器 Ribbon mixer

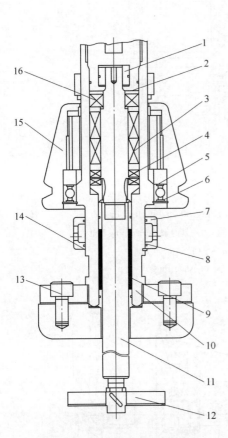

图 2.2.24 磁力搅拌器结构图
Structure drawing of magnetic stirrer

1—轴承压盖 Bering gland；2—止推弹簧 Thrust spring；
3—磁力系统 Magnetic system；4、5、16—轴承 Bearing；
6—定位环 Locating ring；7—夹套 Jacket；
8—O 形密封圈 O sealing ring；9—轴套 Sleeve；
10, 14—空心轴 Hollow shaft；11—搅拌轴 Stirring shaft；
12—叶片 Blade；13—紧固螺栓 Fastening bolt；
15—主动转动体 Active rotating body

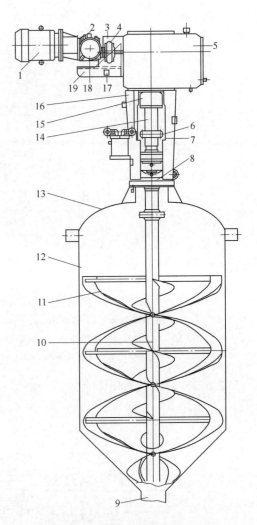

图 2.2.25 螺带混合器结构图
Structure drawing of helical ribbon mixer

1—电机 Motor；2—齿轮箱 Gearbox；
3—联轴节防护罩 Coupling guard；
4—联轴器 Coupling；5—齿轮变速箱 Gear box；
6—锥型联轴器 Conical coupling；7—防护罩 Protective cover；
8—机械密封 Mechanical seal；9—物料出口 Material outlet；
10—螺旋器 Screw；11—螺带 Helical ribbon；12—筒体 Cylinder；
13—封头 Head；14—连接轴 Connecting shaft；
15—齿型联轴节 Gear coupling；16—支架 Support；
17—接线盒 Junction box；18—开关盒 Switch box；
19—电机支座 Motor seat

2 合成树脂 Synthetic resin

2.2.7 其它设备 Other equipments

2.2.7.1 袋式过滤器 Bag filter

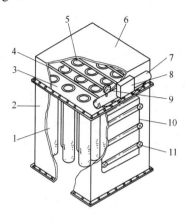

图 2.2.26 袋式过滤器结构图 Structure drawing of bag filter
1—过滤器 Filter；2—过滤器箱 Filter box；3—管板 Tube sheet；4—吹扫管 Scavenging conduit；
5—文丘里管 Venturi tube；6—顶板 Roof；7—顶部管 Top tube；8—控制阀 Control valve；
9—压力表 Pressure gauge；10—检修口 Access hole；11—手柄 Handle

2.2.7.2 烛式过滤器 Candle filter

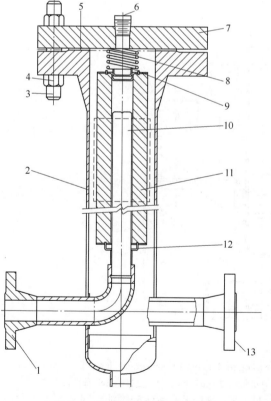

图 2.2.27 烛式过滤器结构图
Structure drawing of candle filter

1—出口法兰 Flange；2—筒体 Cylinder；3—螺栓 Studs；4—螺母 Hex nuts；5—垫片 Gasket；6—管塞 Pipe plug；7—压盖 Gland；8—弹簧 Spring；9—弹簧座 Spring seat；10—滤芯导向管 Tube guide；11—过滤芯 Filter tube；12—固定座 Set seat；13—入口法兰 Inlet flange

2.2.7.3 旋转加料器 Rotary feeder

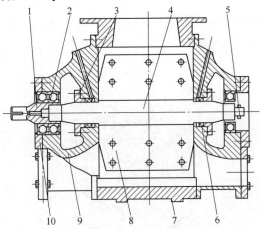

图 2.2.28 旋转加料器结构图 Structure drawing of rotary feeder
1—轴套 Sleeve；2—轴承 Bearing；3—填料压盖 Stuffing gland；4—轴 Shaft；5—锁紧套 Lock sleeve；
6—填料 Packing；7—壳体 Casing body；8—叶片 Vane；9—挡板 Baffle；10—轴承端盖 Bearing cover

2.2.7.4 离心干燥器 Centrifugal dryer

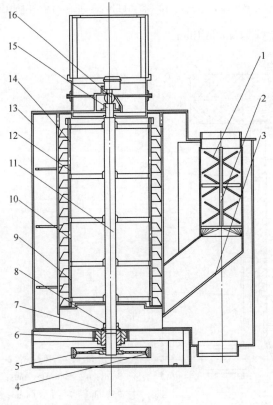

图 2.2.29 离心干燥器结构图 Structure drawing of centrifugal dryer
1—水翻转器 Water turner；2—轴 Shaft；3—进料斜槽 Feed chute；4—皮带轮 Belt pulley；
5—锁紧套 Lock sleeve；6、15—轴承 Bearing；7—锁紧垫圈 Lock washer；8—甩油环 Oil slinger；
9—底部过滤器 Bottom filter；10—中部过滤器 Middle filter；11—长轴 Long shaft；12—支架 Support；
13—干燥箱 Dryer box；14—顶部过滤器 Top filter；16—联轴节 Coupling

2.3 聚氯乙烯 Polyvinyl chloride

2.3.1 聚合釜 Polymerizers

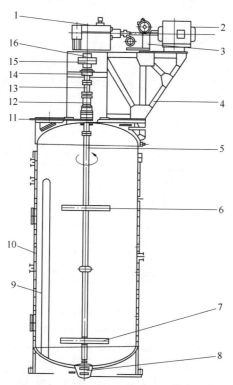

图 2.3.1 聚合釜结构图 Structure drawing of polymerizer
1—减速器 Speed reducer；2—电机 Motor；3—液力变矩器 Torque converter；
4—支架 Support；5—搅拌轴 Stirring shaft；6、7—搅拌桨叶 Stirring blade；
8—釜底轴承 Bottoms bearing；9—内冷挡板 Inner cold baffle；10—夹套 Jacket；
11—机械密封 Mechanical seal；12—输出轴 Input shaft；
13—联轴器短节 Coupling nipple；14—轴承箱 Bearing box；
15—联轴器 Coupling；16—输入轴 Output shaft

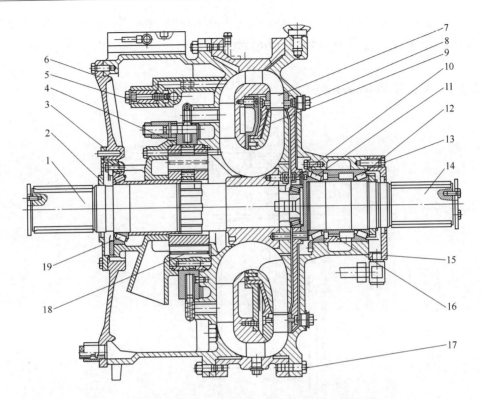

图 2.3.2 液力变矩器结构图 Structure drawing of torque converter

1—输入轴 Input shaft；2—轴封 Shaft seal；3、15—轴承压盖 Bearing cover；
4—齿轮泵 Gear pump；5—泄压阀 Pressure relief valve；6—大盖 Large cover；
7—涡轮 Turbine；8—丝堵 Plug；9—径向叶轮 Radial impeller；
10、13、17—紧固螺栓 Fastening bolt；11、12、19—轴承 Bearing；
14—输出轴 Output shaft；16—间隔环 Spacer ring；18—齿轮 Gear

2.3.2 流化床 Fluid bed

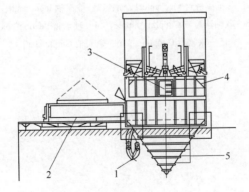

图 2.3.3 流化床结构图 Structure drawing of fluid bed

1—产品出口 Product outlet；2—加热板 Heating plate；3—清洗孔 Cleaning hole；
4—载气入口 Carrier gas inlet；5—气体入口 Gas inlet

2.3.3 螺旋板换热器 Spiral plate heat exchanger

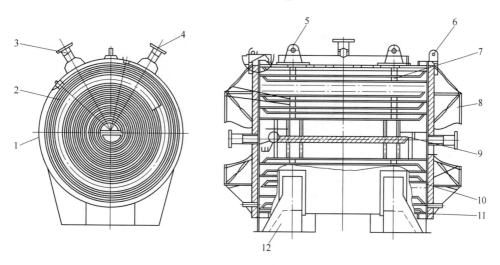

图 2.3.4 螺旋板换热器结构图 Structure drawing of spiral plate heat exchanger

1—压力套 Compression sleeve；2—螺旋板板片 Spiral plate sheet；3、4—接管 Connecting pipe；5、6—吊耳 Lug；7—定距柱 Spacer column；8—端盖 End cover；9—中间隔板 Intermediate partition；10—垫片 Gasket；11—卡箍 Hoop；12—支座 Support

2.3.4 分离器 Separator

2.3.4.1 旋风分离器 Cyclone separator

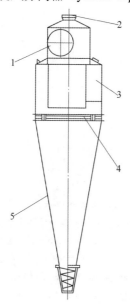

图 2.3.5 旋风分离器结构图
Structure drawing of cyclone separator

1—空气出口 Air outlet；2—清洗盖 Cleaning cover；3—空气入口 Air inlet；4—支架 Support；5—锥形筒体 Cone cylinder

2.3.4.2 磁性分离器 Magnetic separator

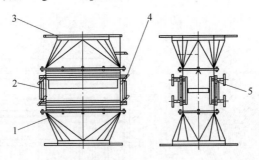

图 2.3.6 磁性分离器结构图 Structure drawing of magnetic separator
1—壳体 Casing；2、4—强磁材料 Magnetic material；3—接口法兰 Interface flange；
5—清洗孔 Cleaning hole

2.3.5 罐 Tank

2.3.5.1 汽提塔进料槽 Stripper feed channel

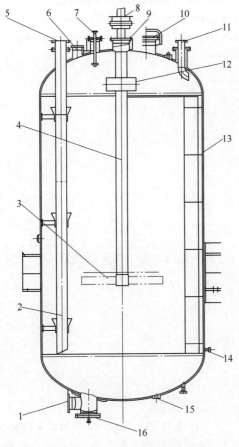

图 2.3.7 汽提塔进料槽结构图
Structure drawing of stripper feed channel

1—泵接口 Pump interface；2—浆料导流管 Slurry diversion pipe；3—搅拌浆叶 Mixing blades；4—搅拌轴 Mixing shaft；5—浆料进口 Slurry inlet；6—气体出口 Gas outlet；7—脱盐水喷淋口 Desalted water spray；8、12—联轴器 Coupling；9—连接法兰 Connecting flange；10—人孔 Manhole；11—浆料循环口 Slurry circulation port；13—筒体 Cylinder；14—温度计接口 Thermometer interface；15—接管 Connecting pipe；16—出料口 Feeding hole

2 合成树脂 Synthetic resin

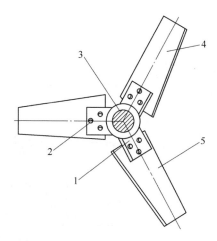

图 2.3.8　汽提塔进料槽搅拌桨叶 Stripper feed channel stirring impeller
1—轮毂 Hub；2—紧固螺栓 Fastening bolt；3—搅拌轴 Rotating shaft；
4、5—桨叶 Blade

2.3.5.2　汽提塔闪蒸槽 Stripper flash tank

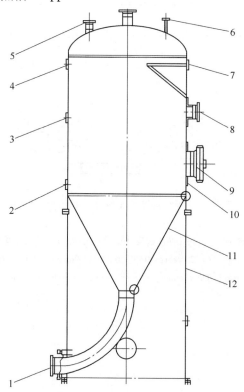

图 2.3.9　汽提塔闪蒸槽结构图
Structure drawing of stripper flash tank
1—浆料出口 Slurry outlet；2、3—液位计接口 Liquid level gauge interface；4—压力表接口 Pressure gauge interface；
5—手孔 Hand hole；6—安全阀接口 Safety valve interface；7—水喷淋口 Water spray opening；
8—浆料进口 Slurry inlet；9—人孔 Manhole；10—筒体 Cylinder；11—锥体 Cone；12—裙座 Skirt

2.3.5.3 气柜 Gas holder

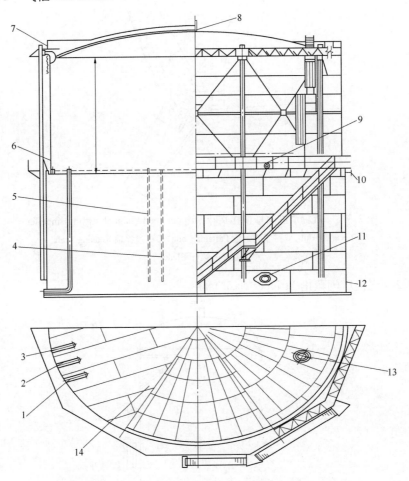

图 2.3.10 气柜结构图 Structure drawing of gas holder
1—气体出口 Gas outlet；2、3—气体入口 Gas inlet；4—供水管 Water supply pipe；
5—溢流管 Overflow pipe；6—防冻装置 Anti-freezing device；7—水喷嘴 Water spray；
8、9、13—放空口 Vent；10—压力表接口 Pressure gauge interface；
11—排放口 Drain outlet；12—筒体；14—气柜顶 Gas holder roof

2.3.6 离心机 Centrifuge

2.3.6.1 离心机外形 Centrifuge outline

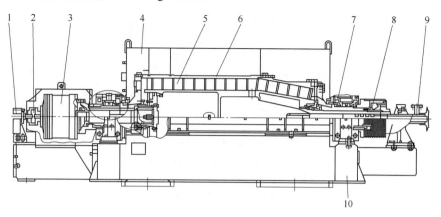

图 2.3.11 离心机外形图 Outline drawing of centrifuge

1—扭矩控制系统 Torque control system；2—联轴器 Coupling；3—差速器 Differentials；
4—机壳 Housing；5—螺旋输送器 Screw conveyor；6—转鼓 Rotor drum；
7—轴承箱 Bearing box；8—皮带轮 Belt pulley；9—进料口 Feed inlet；
10—机座 Machine base

2.3.6.2 离心机结构图 Centrifuge structure

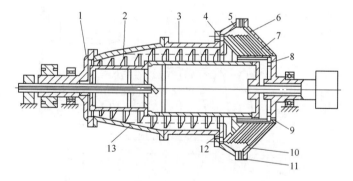

图 2.3.12 离心机结构图 Structure drawing of centrifuge

1—进料管 Feed pipe；2—螺旋 Spiral；3—转鼓 Rotor drum；
4—螺栓 Screw；5、6—倒空心锥 Inverted hollow cone；7—碟片 Disc；
8—空心套 Hollow sleeve；9—溢流口 Overflow port；10—手孔 Hand hole；
11—出渣口 Slag spraying hole；12—溢流堰 Weir；13—锥体 Cone

2.3.6.3 离心机差速器 Centrifuge differential

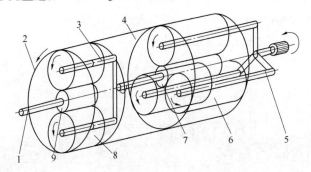

图 2.3.13 离心机差速器结构图 Structure drawing of centrifuge differential
1—扭矩限位开关 Torque limit switch;
2—一级内齿圈 First stage annulus gear(Input shaft);
3—一级行星架 First stage planet carrier;
4—二级内齿圈 Two-stage annulus gear(Input shaft);
5—二级行星架 Two-stage planet carrier(Output shaft);
6—二级行星轮 Two-stage planet gear; 7—二级太阳轮 Two-stage sun gear;
8—一级行星轮 First stage planet gear; 9—一级太阳轮 First stage sun gear

2.3.6.4 离心机转鼓 Centrifuge drum

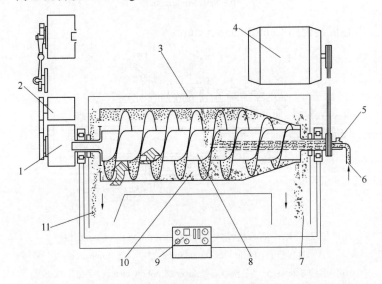

图 2.3.14 离心机转鼓结构图 Structure drawing of centrifuge drum
1—齿轮箱 Gearbox; 2—反向驱动电机 Reverse drive motor; 3—壳体 Casing;
4—主电机 Main motor; 5—冲洗液入口 Flushing fluid inlet; 6—进料口 Feed inlet;
7—固体出料口 Solid discharge outlet; 8—输送器 Conveyor;
9—润滑油系统 Lube oil system; 10—转筒 Revolving drum;
11—液体出料口 Liquid discharge outlet

2.3.7 振动筛 Vibrating screen

2.3.7.1 振动筛结构 Vibrating screen structure

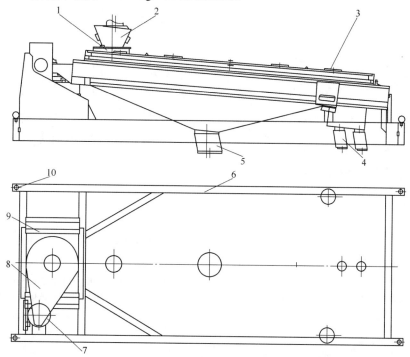

图 2.3.15 振动筛结构图 Structure drawing of vibrating screen

1—下垂入口连接器 Drooping inlet connector；2—滑动入口连接器 Sliding inlet connector；
3—观察口 Observing hole；4—大颗粒出口 Large particles outlet；5—产品出口 Product outlet；
6—底座 Base；7—电机 Motor；8—皮带防护罩 Belt guard；9—振动机座 Vibration stand；10—吊耳 Lug

2.3.7.2 振动筛主要部件 Main components of vibrating screen

1. 滑动入口连接器 Sliding inlet connector

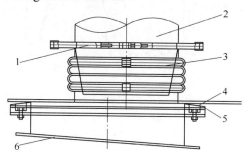

图 2.3.16 滑动入口连接器结构图 Structure drawing of sliding inlet connector

1—卡箍 Hoop；2—下料管 Filling tube；3—软联接 Soft connection；
4—耐磨盘 Wear-resistant plate；5—磨损指示器 Wear indicator；
6—振动筛入口 Vibrating screen inlet

2. 下垂入口连接器 Drooping inlet connector

2.3.8 振荡加料器 Oscillating feeder

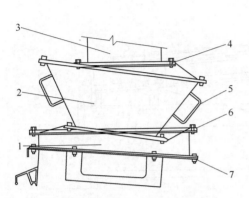

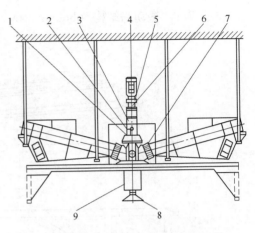

图2.3.17 下垂入口连接器结构图
Structure drawing of drooping inlet connector
1—连接器 Connector；2—料斗 Hopper；
3—下料口 Filling tube；
4、6、7—紧固螺栓 Fastening bolt；
5—手柄 Handle

图2.3.18 振荡加料器结构图
Structure drawing of oscillating feeder
1、3—轴承 Bearing；2—轴 Shaft；4—电机 Motor；
5—联轴器 Coupling；6—连接轴 Connecting shaft；
7—挠性套管 Flexible sleev；
8—分配器手轮 Distributor handwheel；
9—气缸 Cylinder

2.3.9 特种阀门 Special valves

2.3.9.1 换向阀 Reversing valve

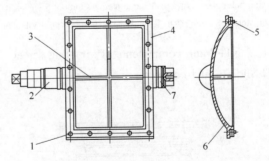

图2.3.19 换向阀结构图 Structure drawing of reversing valve
1—螺栓 Bolt；2、7—阀杆 Valve stem；3—阀挡板 Valve baffle；
4—阀体 Valve body；5—密封垫片 Sealing gasket；6—阀盖 Valve cover

2 合成树脂 Synthetic resin

2.3.9.2 旋转阀 Rotary valve

2.4 聚苯乙烯 Polystyrene

2.4.1 反应设备 Reaction equipment

2.4.1.1 预聚合反应釜 Prepolymerization reactor

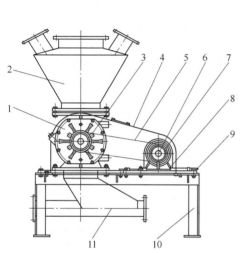

图 2.3.20 旋转阀结构图
Structure drawing of rotary valve

1—阀体 Valve body；
2—抽气室 Pumping chamber；
3、7—链轮 Sprocket；4—链轮罩 Chain cover；
5—链条 Chain；6—电机 Motor；
8—电机底座 Motor seat；
9—调整螺栓 Adjusting bolt；
10—阀座 Valve seat；
11—下料室 Filling chamber

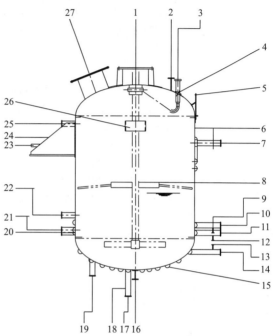

图 2.4.1 预聚合反应釜结构图
Structure drawing of prepolymerization reactor

1—搅拌器 Agitator；2—备用口 Spare opening；
3—进料口(来自预热器)Feed inlet(self preheater)；
4—封头 Head；5—吊耳 Lug；6—筒体 cylinder；
7—盘管出口 Coil outlet；
8—搅拌桨叶 Stirring blades；
9、12、13—排气口 Exhaust port；
10、11—盘管进口 Coil inlet；
14—夹套出口 Jacket outlet；
15—夹套盘管 Jacket coil；
16—排放口 Discharge outlet；
17—夹套进口 Jacket inlet；
18—夹套排放口 Jacket discharge outlet；
19—预聚物出口 Prepolymer outlet；
20—温度计接口 Thermometers interface；
21、25—液位计接口 Liquid level gauge interface；
22—进料口(来自进料罐)Feed inlet (Self feed tank)；
23—接地板 Grounding plate；24—支座 Support；
26—铭牌 Nameplate；27—人孔 Manhole

· 189 ·

2.4.1.2 聚合反应釜 Polymerization reactor

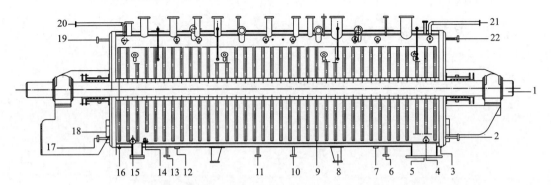

图 2.4.2 聚合反应釜结构图 Structure drawing of polymerization reactor

1—搅拌轴 Agitator；2、17—夹套热油入口 Jacketed hot oil inlet；
3、16—夹套排放口 Jacket discharge outlet；4、14、15—压力计接口 Manometer interface；
5—聚合物出口 Polymer outlet；6、10、13—夹套热油排放口 Jacketed hot oil discharge outlet；
7—取样口 Sampling port；8—支座 Support；9—桨叶 Blade；
11、12—聚合物取样口 Polymer sampling port；18—备用口 Alternate port；
19、22—热油出口 Hot oil outlet；20、21—热油入口 Hot oil supplement inlet

2.4.2 脱挥器 Devolatilizer

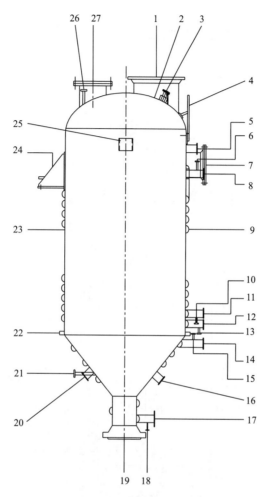

图 2.4.3 脱挥器结构图 Structure drawing of devolatilizer

1—气相出口 Gas phase outlet；2—封头 Head；
3—视镜 Sight glass；4—吊耳 Lug；5、16、20—液位计接口 Liquid level gauge interface；
6、15—夹套排气口 Jacket outlet；7—进料口 Feed inlet；8—盘管出口 Coil outlet；
9—夹套盘管 Tube coil；10、13、18—夹套排放口 Jacket drain port；
11、12、17—盘管入口 Coil inlet；14—底部盘管出口 Bottom coil outlet；
19—聚合物出口 Polymer outlet；21—温度计接口 Thermometers interface；
22—保温支承圈 Insulation supporting ring；
23—筒体 Cylinder；24—支座 Support；
25—铭牌 Nameplate；26—氮气入口 Nitrogen inlet；
27—人孔 Manhole

2.4.3 加热炉 Heating furnace

2.4.3.1 加热炉 Heating furnace

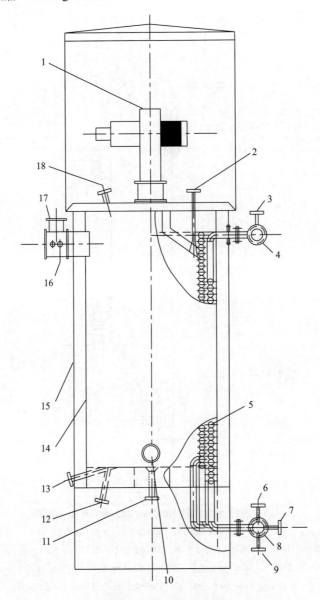

图 2.4.4 加热炉结构图
Structure drawing of heating furnace

1—燃烧器 Burner；2—烟气出口 Flue gas outlet；3—热油入口 Hot oil inlet；
4、17—出口集合管 Outlet manifold；5—加热盘管 Heating coil；
6—热油出口 Hot oil outlet；7、16—温度计接口 Thermometer interface；
8—入口集合管 Inlet manifold；9—排放口(集合管) Discharge outlet(Collection of tube)；
10—人孔 Manhole；11—排放口 Discharge outlet；12、13、18—观察口 Viewing port
14—隔热层 Heat insulation layer；15—炉体 Furnace body

2.4.3.2 加热炉燃烧器 Heating furnace burner

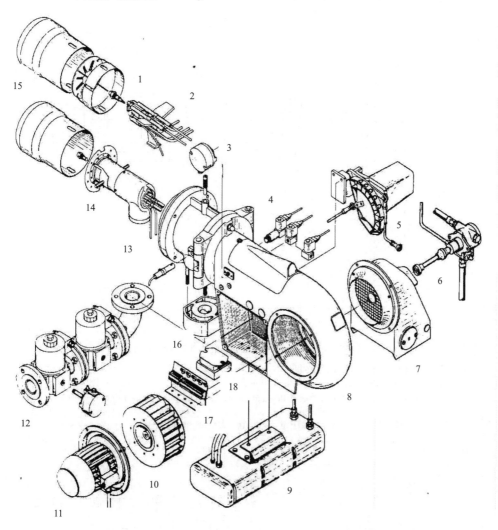

图 2.4.5 加热炉燃烧器结构图 Structure drawing of heating furnace burner
1—喷嘴 Nozzle；2—集装喷嘴 Manifold nozzle；3—燃气监控器 Gas monitor；
4—油电磁阀 Oil solenoid valve；5—伺服系统 Servo system；6—泵 Pump；
7—调节风门 Adjusting damper；8—燃烧器罩 Burner cover；9—油预热器 Oil preheater；
10—鼓风机 Blower；11—电机 Motor；12—气电磁阀 Gas solenoid valve；13—控制器 Controller；
14—烧嘴 Burner；15—主燃料气烧嘴 Main fuel gas burner；16—燃气风门 Gas damper；
17—终端箱 Terminal box；18—点火变压器 Ignition transformer

2.4.4 切胶机 Rubber cutting machine

2.4.4.1 切胶机 Rubber cutting machine

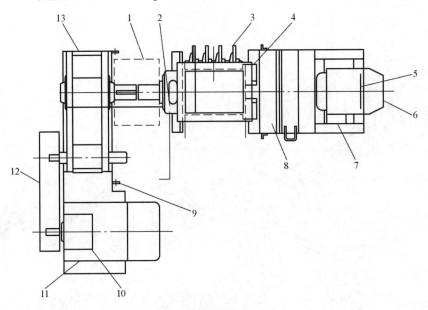

图 2.4.6 切胶机结构图 Structure drawing of rubber cutting machine

1—保护罩 Protective cover；2—联轴器 Coupling；3—可调节铁砧 Adjustable anvil；4—固定砧座 Fixed anvil；
5—电机转向标示 Motor turning mark；6、11—电机 Motor；7、10—接线盒 Wire box；8—铰接盖 Hinged cover；
9—吊耳 Lug；12—皮带轮防护罩 Pulley and protective cover case；13—变速箱 Gear box

2.4.4.2 切胶机刀垂 Rubber cutting machine knife hammer

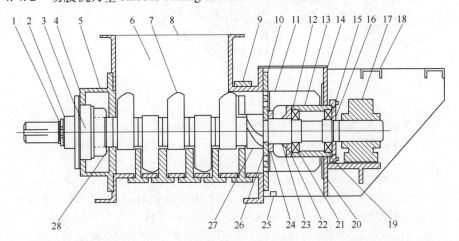

图 2.4.7 切胶机刀垂结构图 Structure drawing of rubber cutting machine knife hanging

1—锁紧螺母 Lock nut；2—锁紧垫圈 Lock washer；3、19、21—轴承 Bearing；4—垫片 Gasket；
5、13—轴承箱 Bearing box；6—下料口 Feed opening；7—单体刀锤 Monomer knife hammer；
8—箱体 Box；9—砧座 Anvil block；10—孔板 Orifice plate；11—切刀 Cutter；12—密封圈 Sealing ring；
14—端盖 Cover；15—止推环 Thrust ring；16、22—轴封 Shaft seal；17—皮带轮 Pulley；
18—电机座 Motor base；20—盖板 Cover plate；23—导向环 Guide ring；
24—加强环 Reinforcement ring；25—排放孔 Discharge hole；26—轴肩挡圈 Shoulder ring；
27—双体刀锤 Catamaran knife hammer；28—间隔套 Spacer sleeve

2.4.4.3 切胶机砧座 Rubber cutting machine anvil

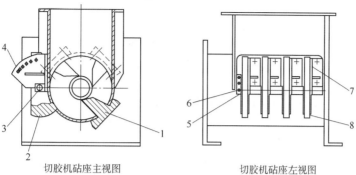

切胶机砧座主视图
Front view of rubber cutting machine anvil

切胶机砧座左视图
Left view of rubber cutting machine anvil

图 2.4.8 切胶机砧座结构图 Structure drawing of rubber cutting machine anvil
1—螺栓连接砧座 Bolted anvil；2、8—可调节砧座 Adjustable anvil；3—砧轴 Anvil shaft；
4、7—砧支架 Anvil stand；5—砧轴支架 Anvil shaft bracket；6—支架固定螺栓 Bracket screws

2.4.5 换热设备 Heat exchanger

2.4.5.1 反应器进料预热器 Reactor feed preheater

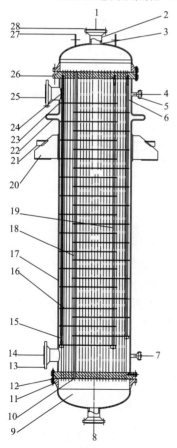

1—进料口 Feed inlet；2—接管 Connecting pipe；3—补强圈 Reinforcing ring；4—放空口 Vent；5、13、28—法兰 Flange；6、24—定距管 Spacer tube；7—排液口 Liquid outlet；8—出料口 Discharge outlet；9—管箱 Tube chamber；10、18—换热管 Heat exchange tube；11—螺栓 Bolt；12—下管板 Lower tube plate；14—热油出口 Hot oil outlet；15、19—拉杆 Tie rod；16—折流板 Baffle plate；17—筒体 Cylinder；20—支座 Support；21—接地板 Grounding plate；22—膨胀节 Expansion joint；23—防冲挡板 Impingement baffle；25—热油入口 Hot oil inlet；26—上管板 Upper tube plate；27—吊耳 Lug

图 2.4.9 反应器进料预热器结构图
Structure drawing of reactor feed preheater

2.4.5.2 预聚合真空冷凝器 Prepolymerization vacuum condenser

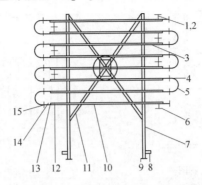

图 2.4.10 预聚合真空冷凝器结构图
Structure drawing of prepolymerization vacuum condenser
1—法兰 Flange；2—垫片 Gasket；3—U 形螺栓 U bolt；4、15—螺柱 Stud；
5—弯管 Bend tube；6—接管 Connecting pipe；7—立柱 Uprights；
8—接地板 Grounding plate；9—底板 Base plate；10、12—套管 Casing；
11—加强角钢 Stiffening angle steel；13—变径管 Adjustable tube；
14—内管 Inner tube

2.4.5.3 预聚合反应器冷凝器 Prepolymerization reactor condenser

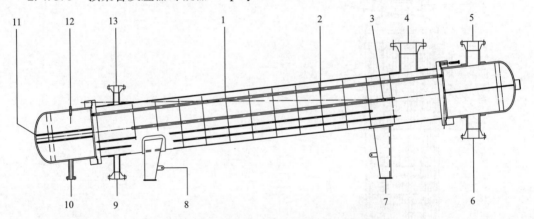

图 2.4.11 预聚合反应器冷凝器结构图
Structure drawing of prepolymerization reactor condenser
1—换热管 Heat exchange tube；2—筒体 Cylinder；3—拉杆 Tie rod；
4—蒸汽入口 Steam inlet；5—循环水出口 Circulating water outlet；
6—循环水入口 Circulating water inlet；7—支座 Support；
8—接地板 Grounding plate；9—液体出口 Liquid outlet；10—排放口 Drain port；
11—封头 Head；12—吊耳 Lug；13—蒸汽出口 Steam outlet

2.4.5.4 脱挥器气相冷凝器 Devolatilizer gas phase condenser

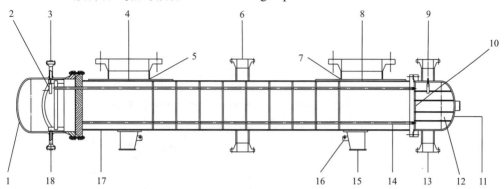

图 2.4.12 脱挥器气相冷凝器结构图 Structure drawing of devolatilizer gas phase condenser
1—外头盖 Outer cover；2—浮头 Floating head；3—排气口 Exhaust port；4—进气口 Gas inlet；
5、7—补强圈 Impingement baffle；6—接管 Connecting pipe；8—出气口 Outlet；
9—冷却水出口 Chilled water outlet；10—管板 Tube sheet；11—管箱封头 Tube box head；12—管箱 Channel box；
13—冷却水入口 Chilled water inlet；14—管束 Pipe bundle；15—支座 Support；
16—接地板 Grounding plate；17—筒体 Cylinder；18—排放口 Drain port

2.4.5.5 矿物油加热器 Mineral oil heater

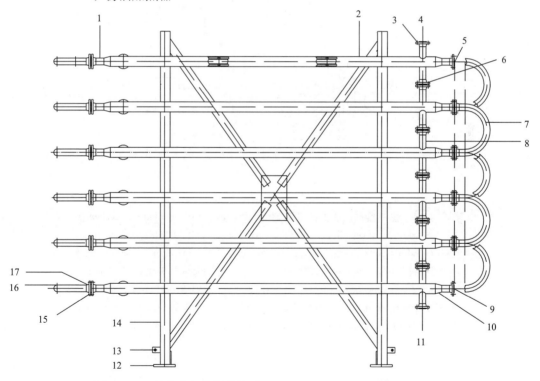

图 2.4.13 矿物油加热器结构图 Structure drawing of mineral oil heater
1—内管 Inner tube；2—套管 Casing；3、8、16—接管 Connecting pipe；4—热油入口 Hot oil inlet；
5—矿物油出口 Mineral oil outlet；6、15—螺栓 Bolt；7—弯管 Bend pipe；
9—矿物油入口 Mineral oil inlet；10—变径管 Reducing nipple；11—热油出口 Hot oil outlet；
12—固定底板 Fixed foundation plate；13—接地板 Grounding plate；14—支柱 Support；17—法兰 Flange

2.4.5.6 脱挥器预加热器 Devolatilizer preheater

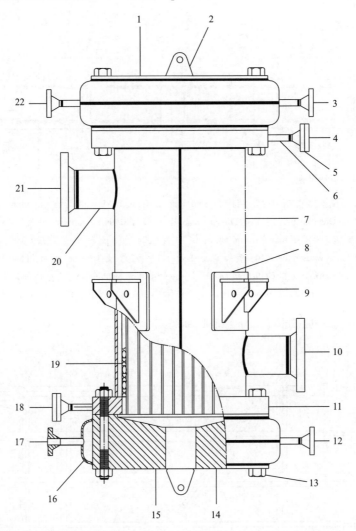

图 2.4.14 脱挥器预加热器结构图 Structure drawing of devolatilizer preheater

1—工艺介质入口 Process medium inlet；2—吊耳 Lug；3、12—夹套出口 Jacket outlet；
4—排气口 Exhaust port；5—法兰 Flange；6、20—接管 Connecting pipe；7—筒体 Cylinder；
8—补强板 Stiffening plate；9—支座 Support；10—壳程介质入口 Shell medium inlet；
11—管板 Tube sheet；13—螺栓 Bolt；14—封头 Head；15—工艺介质出口 Process medium outlet；
16—夹套封头 Jacket head；17、22—夹套入口 Jacket inlet；18—排液口 Liquid discharge outlet；
19—加热管 Heating pipe；21—壳程介质出口 Shell medium outlet

2.4.6 罐 Tank

2.4.6.1 苯乙烯储罐 Styrene storage tank

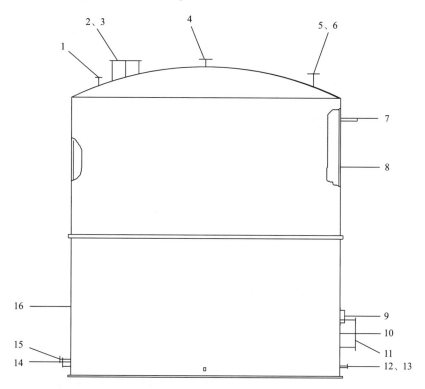

图 2.4.15 苯乙烯储罐结构图 Structure drawing of styrene storage tank

1—氮气接口 Nitrogen interface；2—法兰 Flange；3—放空口 Vent；4—呼吸阀 Breathing valve；
5—液位计接口 Liquid level gauge interface；6—压力计接口 Pressure gauge interface；
7—支承角钢 Angle steel support；8、16—筒体 Cylinder；9—支架 Support；
10—补强圈 Reinforcing ring；11—人孔 Manhole；12—苯乙烯入口 Styrene inlet；
13—温度计接口 Thermometer interface；14—苯乙烯出口 Styrene outlet；
15—苯乙烯循环口 Styrene circulation port

2.4.6.2 切胶溶解罐 Rubber cutting dissolving tank

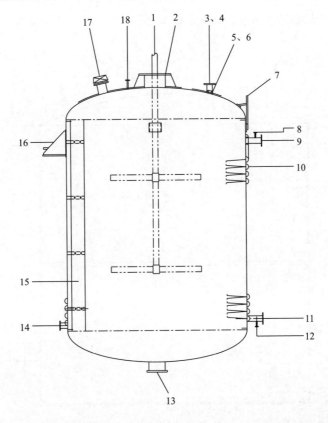

图 2.4.16 切胶溶解罐结构图 Structure drawing of rubber cutting dissolving tank
1—搅拌轴 Stirring shaft；2—固定支架 Fixed bracket；3—橡胶入口 Rubber inlet；
4—苯乙烯进料口 Styrene feed port；5—矿物油进料口 Mineral oil feed port；
6—抗氧化剂入口 Antioxidants inlet；7—吊耳 Lug；8、12—夹套排气口 Jacketed vent；
9—冷却水出口 Cooling water outlet；10—半管式夹套盘管 Semi-jacketed tube coil；
11—冷却水入口 Cooling water inlet；13—出料口 Discharge port；
14—液位计接口 Liquid level gauge interface；15—挡板 Baffle；16—支座 Support；
17—备用口 Alternate port；18—安全阀接口 Safety valve interface

2.4.6.3 预聚合进料罐 Prepolymerization feed tank

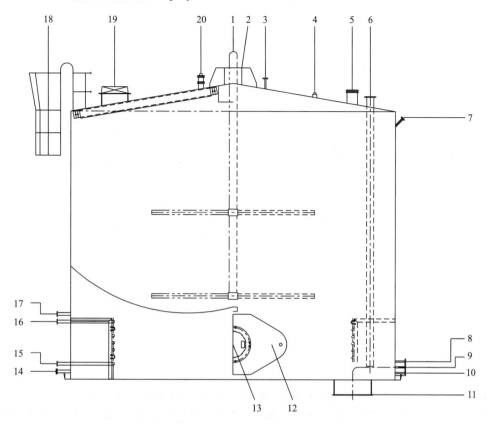

图 2.4.17 预聚合进料罐结构图 Structure drawing of prepolymerization feed tank

1—搅拌轴 Stirring shaft；2—固定支架 Fixed bracket；3—氮气入口 Nitrogen inlet；
4—压力计接口 Pressure gauge interface；5、13—人孔 Manhole；6—进料口 Feed inlet；
7—液位报警器接口 Level alarm interface；8—冷凝液出口 Condensate outlet；
9—排放口 Discharge outlet；10—液位计接口 Liquid level gauge interface；11—排污口 Sewage outlet；
12—补强板 Stiffening plate；14—备用口 Alternate opening；15—循环水入口 Circulating water inlet；
16—循环水出口 Circulating water outlet；17—温度计接口 Thermometers interface；
18—梯子 Ladder；19—紧急排放口 Emergency drain outlet；
20—安全阀接口 Safety valve interface

2.4.6.4 预聚合反应器凝液接收罐 Prepolymerization reactor condensate receiver tank

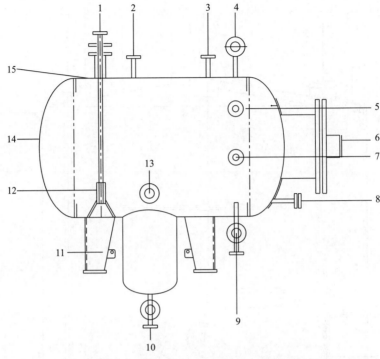

图 2.4.18 预聚合反应器凝液接收罐结构图
Structure drawing of prepolymerization reactor condensate receiver tank

1—插入管 Insertion tube；2—进料口 Feed inlet；3—蒸汽出口 Steam outlet；4、13—视镜 Sight glass；5、7—液位计接口 Liquid level gauge interface；6—人孔 Manhole；8—备用口 Alternate port；9—冷凝液出口 Condenser port；10—出料口 Exhaust port；11—支座 Support；12—支承管 Supporting tube；14—封头 Head；15—筒体 Shell

2.4.6.5 蓝料罐 Blue material tank

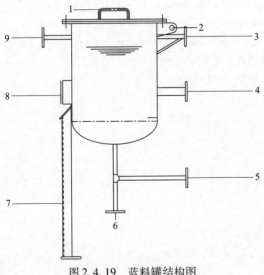

图 2.4.19 蓝料罐结构图
Structure drawing of blue material tank

1—把手 Handle；2—吊耳 Lug；3、5—视镜 Sight glass；4—低液位报警器接口 Low level alarm interface；6—出料口 Discharge port；7—支架 Support；8—铭牌 Nameplate；9—进料口 Feed inlet

2.4.6.6 引发剂罐 Initiator tank

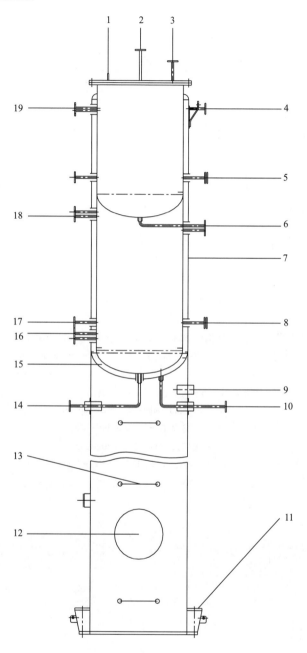

图 2.4.20 引发剂罐结构图 Structure drawing of initiator tank

1—盖板 Cover plate；2—排放口 Discharge outlet；3—矿物油入口 Mineral oil inlet；
4—冷却水出口 Cooling water outlet；5—温度报警器接口 Temperature alarm interface；
6—引发剂入口 Initiator inlet；7—筒体 Cylinder；8—夹套排放口 Jacket drain port；
9—裙座 Skirt；10—冷却水入口 Cooling water inlet；11—支座 Support；
12—人孔 Manhole；13—梯子 Ladder；14—引发剂出口 Initiator outlet；15—封头 Head；
16—低液位报警器接口 Low level alarm interface；17—手孔 Hand hole；18—视镜 Sight glass；
19—夹套排气口 Jacketed vent

2.4.6.7 脱挥器挥发物接收罐 Devolatilizer volatiles receiving tank

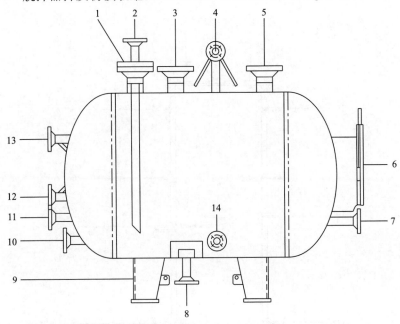

图 2.4.21 脱挥器挥发物接收罐结构图 Structure drawing of devolatilizer volatiles receiving tank

1—进料口 Feed inlet；2—气相物料入口 Gas phase material inlet；
3—液体入口 Liquid inlet；4、14—液位计接口 Liquid level gauge interface；
5—气体出口 Gas outlet；6—人孔 Manhole；7—备用口 Alternate opening；
8—液体出口 Liquid outlet；9—支座 Support；
10—温度计接口 Thermometer interface；
11—物料返回口 Material returning port；
12、13—液位控制口 Liquid level control port

2.4.6.8 仪表风罐 Instrument air tank

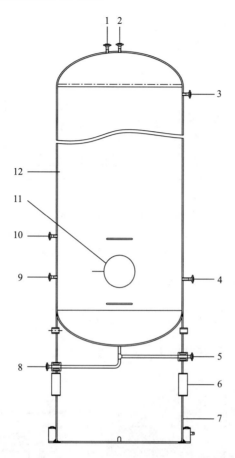

图 2.4.22 仪表风罐结构图 Structure drawing of instrument air tank
1—排放口 Vent nozzle；2—仪表风出口 Instrument air outlet；3—压力计接口 Pressure gauge interface；
4、5—液位计接口 Liquid level gauge interface；6、11—人孔 Manhole；7—裙座 Skirt；8—排放口 Discharge outlet；
9—备用口 Alternate opening；10—仪表风入口 Instrument air inlet；12—筒体 Cylinder

2.4.7 泵 Pump

2.4.7.1 预聚合进料泵 Prepolymerization feed pump

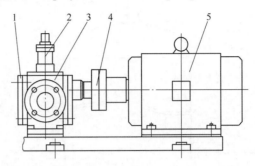

图 2.4.23 预聚合进料泵外形图 Outline drawing of prepolymerization feed pump

1—泵盖 Pump cover；2—安全阀 Safety valve；3—泵体 Pump body；
4—联轴器 Coupling；5—电机 Motor

2.4.7.2 硬脂酸锌泵 Zinc stearate pump

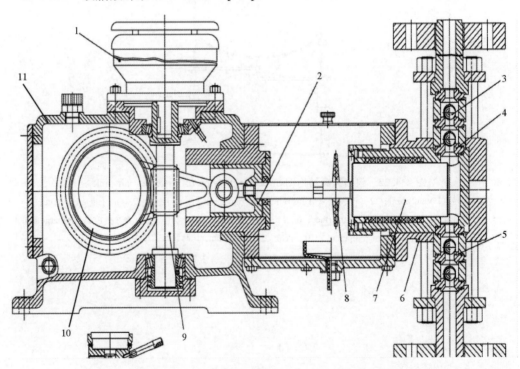

图 2.4.24 硬脂酸锌泵结构图 Structure drawing of zinc stearate pump

1—电机 Motor；2—十字头 Crosshead；3—阀球 Valve ball；4—排料阀 Outlet valve；
5—进料阀 Feed valve；6—液缸体 Liquid cylinder；7—柱塞 Plunger；
8—挡液圈 Liquid retaining ring；9—蜗杆 Worm rod；10—蜗轮 Worm wheel；11—传动箱 Transmission box

2.4.7.3 反应器屏蔽油泵 Jacketed reactor canned pump

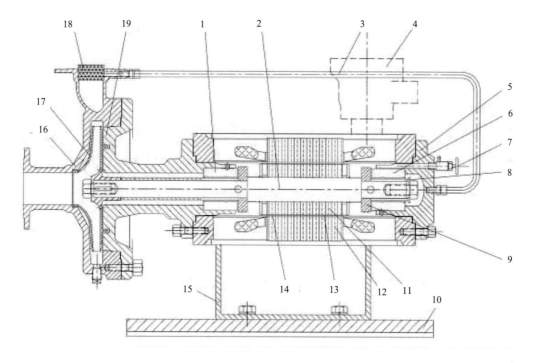

图 2.4.25 反应器屏蔽油泵结构图 Structure drawing of jacketed reactor canned pump
1、6—轴承 Bearing；2—泵轴 Pump shaft；3—循环管 Circulation pipe；
4—接线盒 Junction box；5—压盖 Gland；7—排气阀 Exhaust valve；8—轴套 Sleeve；
9—推力盘 Thrust disc；10—底座 Base；11—转子 Rotor；12—定子 Stator；
13—转子屏蔽套 Rotor shielding sleeve；14—定子屏蔽套 Stator shielding sleeve；
15—支架 Support；16—泵体 Pump body；17—叶轮 Wheel；
18—过滤器 Filter；19—端盖 End cover

2.4.7.4 苯乙烯进料泵 Styrene feed pump

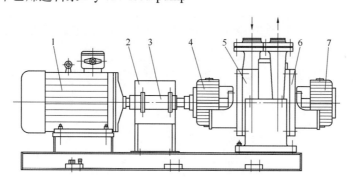

图 2.4.26 苯乙烯进料泵外形图 Outline drawing of styrene feed pump
1—电机 Motor；2—联轴器保护罩 Coupling guard；3—联轴器 Coupling；4—前轴承箱 Front bearing housing；
5—泵体 Pump body；6—泵盖 Pump cover；7—后轴承箱 Bearing housing

2.4.7.5 真空泵 Vacuum pump

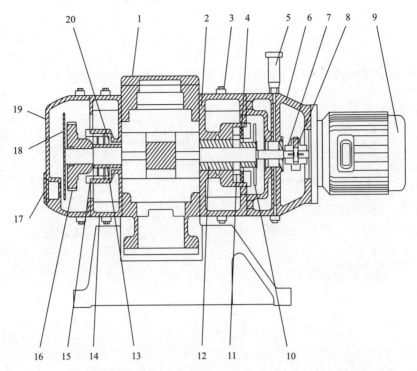

图 2.4.27 真空泵结构图 Structure drawing of vacuum pump

1—泵体 Pump body；2、14—侧盖 Side cover；3—丝堵 Screw plug；
4—波形弹簧 Wave spring；5—油杯 Oil cup；6、19—端盖 End cover；7、8—联轴器 Coupling；
9—电机 Motor；10、18—甩油环 Oil slinger；11—轴承 Bearing；12、13—轴套 Sleeve；
15—轴承压盖 Bearing cover；16—齿轮 Gear；17—视镜 Oil Lens；20—螺旋套 Spiral sleeve

2.4.8 造粒设备 Granulation equipment
2.4.8.1 切粒机 Pelletizer

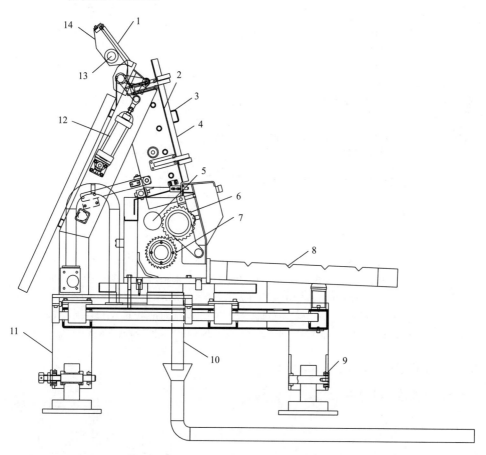

图 2.4.28 切粒机结构图 Structure drawing of pelletizer

1—溢流板 Overflow plate；2—导向槽 Guide groove；3—把手 Handle；4—挡板 Baffle；
5—压辊 Pressure roller；6—牵引辊 Traction rollers；7—切刀 Cutter；8—输送水槽 Transport sink；
9—螺栓 Bolt；10—排水管 Drain pipe；11—支座 Support；12—气缸 Cylinder；
13—溢流水入口 Overflow water inlet；14—翻板 Turning board

2.4.8.2 换网器 Screen changer

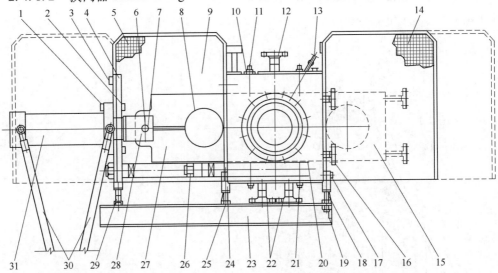

图 2.4.29 换网器结构图 Structure drawing of screen changer

1、11、17、18、21—螺栓 Bolt；2、3、10、19、25—螺母 Nut；4、15、27—挡板 Baffle；
5、9、14—防护网 Grid guard；6—销子 Pin；7—弹性环 Elastic ring；8—护栅 Grille；
12、16、22—法兰 Flange；13—转换器 Converter；20—机体 Compressor body；23—底座 Base；
24、29—支座 Support；26—排油口 Oil discharge outlet；28、31—活塞 Piston；30—管架 Pipe carrier

2.4.9 干燥设备 Drying equipment

2.4.9.1 干燥器 Dryer

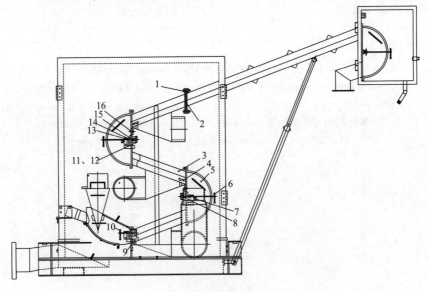

图 2.4.30 干燥器结构图 Structure drawing of

1、13—弹簧板 Spring plate；2—密封垫 Sealing gasket；3—防护罩 Protective cover；
4、9—角形支架 Angular supporting bracket；5—筛网 Screen cloth；6、10—操作杆 Operating rod；
7、8、11、15、16—六角螺栓 Hexagon bolt；12—卡环 Clamp ring；14—支架 Support

2.4.9.2 圆筒干燥器 Rotary drum dryer
1. 回转圆筒干燥器 Structgre of Rotary drum dryer

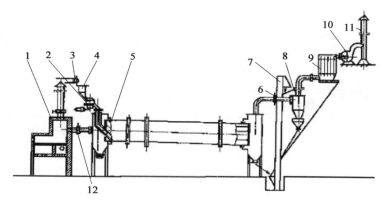

图 2.4.31　回转圆筒干燥器结构图

1—燃烧炉(或载热体加热器)Combustion tarnace(Heat carrier heater);
2—定量给料器 Metering feed; 3—湿料输送机 Wet material convegor; 4—料斗 Hopper; 5—回转圆筒干燥器 Rotary drum drger; 6、12—膨胀环 Expomsion joint; 7—斗式提升机 Bucket elevator; 8—旋风除尘器 Cyclone separator; 9—袋式除尘器 Bag dust collector; 10—引风机 Fan; 11—尾气排空烟囱 Chimneg

2. 直接传热转筒干燥器 Direct heat transter rotary drum dryer

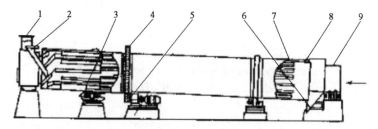

图 2.4.32　直接传热转筒干燥器结构图 Structure draming of direct heat tromsfer rotary drum dryer

1—空气出口 Vent; 2—加料口 Feed; 3—托轮与挡轮 Roller; 4—腰齿轮 Gear;
5—传动齿轮 Drive gear; 6—产品出口 Product; 7—抄板 Plate; 8—密封环 Sealing ring; 9—加热器 Heater

3. 外部加热转筒干燥器 Exterual heating rotary drum dryer

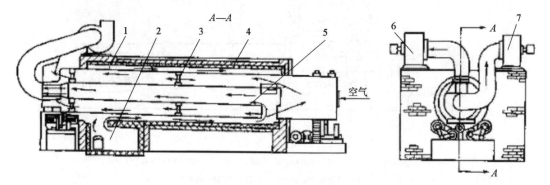

图 2.4.33　外部加热转筒干燥器结构图 Structure drawing of external henting rotary drum dryer

1—外转筒 Rotary drum; 2—炉膛 Furnace chamber; 3—内圆筒 Internal cylinder; 4—炉壁 Furnace wall;
5—连接管 Connecting pipe; 6, 7—风机 Fan

4. 内置加热转筒干燥器 Builtin heating rotary drum dryer

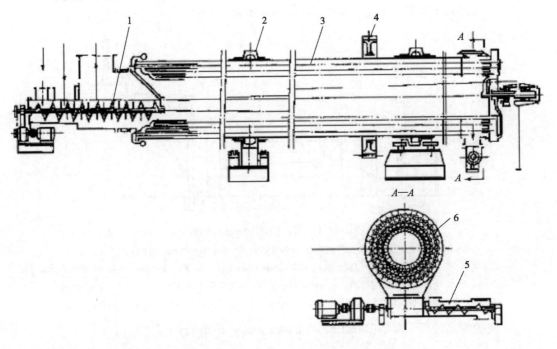

图 2.4.34　内置加热转筒干燥器结构图
Structure drawing of built in heating rotary drum dryer
1—进料螺旋输送器 Feed screw convegor；2—滚圈 Roll ring；3—回转圆筒 Rotary drum；
4—大齿轮 Large gear；5—卸料螺旋输送器 Unloading screw conveger；
6—加热管 Heating tube bumdle

5. 复式传热转筒干燥器

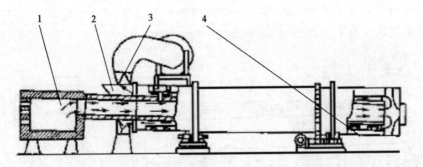

图 2.4.35　复式传热转筒干燥器结构图
Structure drawing of compound conduct heat rotary drum dryer
1—燃烧炉 Furnace；2—排风机 Blower；
3—外转筒 Rotarg drum；4—十字形管 Cross pipe

2.4.9.3 流化床干燥器 Fluidized bed dryer
1. 单层流化床干燥器 Single Layer fluidized bed dryer
2. 多层流化床干燥器

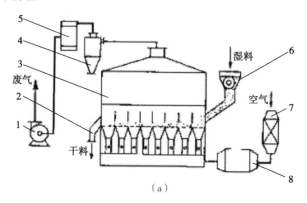

1—引风机 Blower；2—卸料管 Discharge tube；3—干燥器 Dryer；4—旋风分离器 Cyctoue；
5—袋式分离器 Bag dust collector；6—摇摆送料机 Feed；7—空气过滤器 Filter；8—加热器 Heater

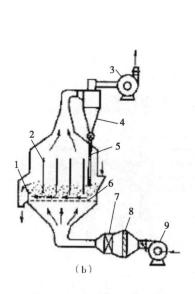

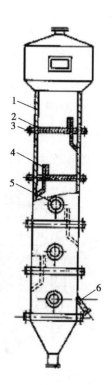

图 2.4.36 单层流化床干燥器结构图
Structure drowing of single layer fluidized bed dryer
1— 出口堰 Outlet weir；2—隔板 Baffle plate；
3—引风机 Blower；4—旋风分离器 Cyclone separator；
5—循环下料管 Discharge tube；6—流化床分布板 Oistributor；
7—空气加热器 Air heater；8—空气过滤器 Air filter；
9—鼓风机 Fan

图 2.4.37 多层流化床干燥器结构图
structure drawing of multilager fluidized bed dryer
1— 筒体 Cylinder；2—筛板 Sieve plate；
3—法兰 Flange；4—溢流装置 Overflow device；5—视镜 Sight glass；
6—卸料管 Discharge tube

2.4.10 输送设备 Conveying equipment

2.4.10.1 输送风机布置图 Conveying blower layout drawing

图 2.4.38 输送风机布置图 Layout drawing of conveying fan

1—传动底座 Conveying base；2—过滤式消音器 Filter-silencer；3—罗茨风机 Roots blower；4—皮带 Belt；5—皮带保护罩 Belt protective cover；6—消音罩 Silencer cover；7—电机 Motor；8—自启卸载阀 Self start unloading valve；9—减压阀 Pressure reducing valve；10—风扇 Fan；11—压力计 Pressure gauge；12—调节管箍 Adjustable hoop；13—连接罩 Connection cover；14—风机底座 Blower base

2.4.10.2 罗茨风机 Roots blower

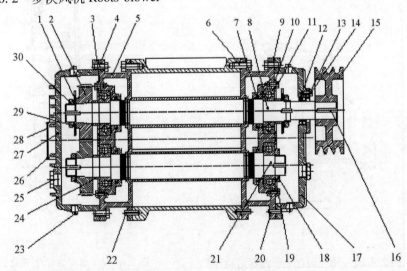

图 2.4.39 罗茨风机结构图 Structure drawing of Roots blower

1—甩油环 Oil slinger；2—齿轮 Gear；3—轴承 Bearing；4、6、11、14—螺栓 Bolt；5、17—轴承箱 Bearing box；7—O 形密封圈 O sealing ring；8—主动轴 Driving shaft；9—垫片 Gasket；10—轴承压盖 Bearing cover；12—轴封 Shaft seal；13—轴封压盖 Shaft sealing cover；15—皮带轮 Belt pulley；16—键 Key；18、30—圆螺母 Round nut；19、29—止退垫圈 Tab washer；20、22—销子 Pin；21—从动轴 Driven shaft；23—丝堵 Plug；24—轴套 Sleeve；25—油标 Grease pit；26—端盖 End cap 27—齿轮箱 Gear box；28—铭牌 Nameplate

2.4.10.3 罗茨风机自启卸载阀 Self start unloading valve of Roots blower

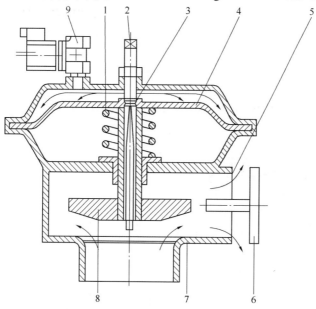

图 2.4.34 罗茨风机自启卸载阀结构图 Structure drawing of self start unloading valve of Roots blower
1—弹簧 Spring；2—阀杆 Valve stem；3—中空阀杆 Hollow stem；4—膜板 Diaphragm；5—排气口 Exhaust port；
6—挡板 Baffle；7—阀体 Casing；8—阀芯 Valve；9—电磁阀 Solenoid valve

2.4.10.4 罗茨风机减压阀 Pressure reducing valve of Roots blower

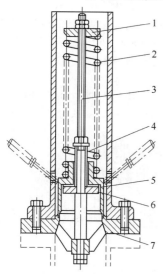

图 2.4.35 罗茨风机减压阀结构图
Structure drawing of pressure reducing valve of Roots blower
1—圆盘 Disk；2—弹簧 Spring；3—阀杆 Valve stem；4—导向衬套 Guide bushing；5—缠绕垫 Surrounding pad；6—活塞 Piston；7—阀座 Valve seat

2.5 丙烯腈－丁二烯－苯乙烯(ABS) Acrylonitrile-butadiene-styrene

2.5.1 聚合釜 Polymerizer

2.5.1.1 丁二烯反应聚合釜 Butadiene reaction polymerizer

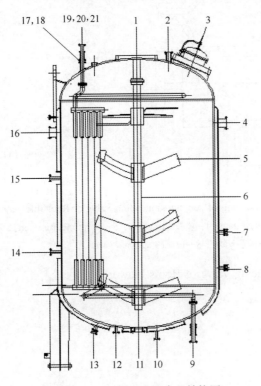

图 2.5.1 丁二烯反应聚合釜结构图
Structure drawing of butadiene reaction polymerizer

1—搅拌器 Agitator；2—蒸汽出口 Steam outlet；3—人孔 Manhole；4—夹套冷却水进口 Jacket cooling water inlet；5—搅拌桨叶 Mixing blades；6—搅拌轴 Mixing shaft；7、8—夹套液位计接口 Jacket level meter interface；9—冷却水进口 Cooling water inlet；10、13—夹套排放口 Jacket discharge outlet；11—物料出口 Material outlet；12—取样口 Sampling port；14、15—温度计接口 Thermometers interface；16—夹套冷却水出口 Jacket cooling water outlet；17—原料进口 Raw material inlet；18—添加剂进口 Additive inlet；19—催化剂进口 Catalyst inlet；20—安全阀接口 Safety valve interface；21—冷却水出口 Cooling water outlet

2.5.1.2 接枝反应聚合釜 Graft reaction polymerizer
2.5.2 凝聚罐 Cohesion tank

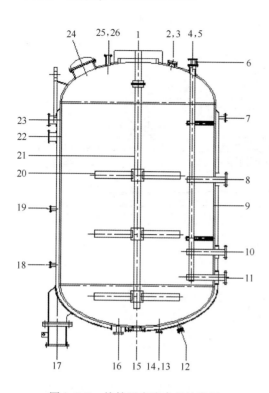

图 2.5.2 接枝反应聚合釜结构图
Structure drawing of graft reaction polymerizer

1—搅拌器 Agitator；2—透光口 Pervious to light mouth；
3—视镜 Sight glass；4—冷却水进口 Cooling water inlet；
5—支承板 Support plate；6、12、18、19、22—冷却水出口
Cooling water outlet；7—接管 Connecting pipe；8、10—热电偶接口
Thermocouple interface；9—夹套 Jacket；11—温度计接口
Thermometer interface；13—夹套排放口 Jacket drain port；
14—取样口 Sampling port；15—物料出口 Material outlet；
16—排放口 Drain port；17—支座 Support；
20—搅拌桨叶 Mixing blades；21—搅拌轴 Mixing shaft；
23—备用口 Alternate interface；24—人孔 Manhole；
25—胶乳进口 rubber latex inlet；
26—安全阀接口 Safety valve interface

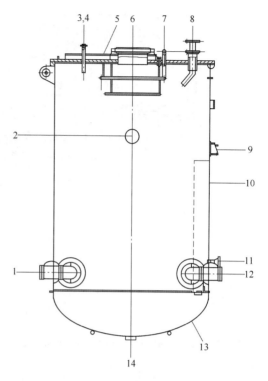

图 2.5.3 凝聚罐结构图
Structure drawing of cohesion tank

1、12—蒸汽入口 Steam inlet；
2—物料出口 Material outlet；
3—热水进口 Hot water inlet；
4—人孔 Manhole；
5—搅拌器底座 Agitator base；
6—凝结剂入口 Coagulant inlet；
7—蒸汽出口 Steam outlet；
8—物料入口 Material inlet；
9—支座 Support；
10—筒体 Cylinder；
11—热电偶接口 Thermocouple interface；
13—封头 Head；
14—排放口 Drain outlet

2.5.3 离心脱水机 Centrifugal hydroextractor

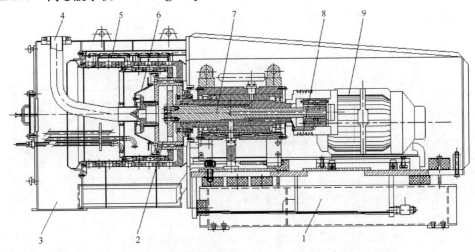

图 2.5.4 离心脱水机结构图 Structure drawing of centrifugal hydroextractor
1—油箱 Oil tank；2—推动盘 Push plate；3—湿粉料出口 Wet mash outlet；
4—胶乳入口 Rubber latex inlet；5—转毂 Revolving hub；
6—分散盘 Scattered disc；7—推动轴 Driving shaft；
8—推进器 Propeller；9—电机 Motor

2.5.4 流化床干燥器 Fluidized-bed dryer

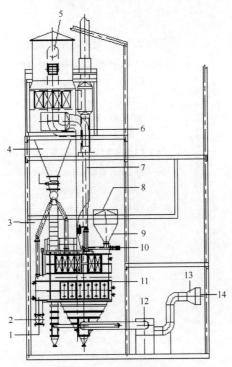

图 2.5.5 流化床干燥器结构图
Structure drawing of fluidized-bed dryer

1—干粉料出口 Dry mash outlet；2—旋转阀 Rotary valve；3—卸料旋转阀 Discharging rotary valve；4—袋式过滤器 Bag filter；5—氮气出口 Nitrogen outlet；6—增压风机 Booster fan；7—搅拌机 Mixer；8—湿粉料入口 Wet mash inlet；9—湿粉料斗 Wet powder hopper；10—螺旋给料器 Screw feeder；11—加热柜 Heating cabinet；12—预热器 Preheater；13—鼓风机 Blower；14—氮气入口 Nitrogen inlet

2.5.5 挤压造粒机 Extrusion granulator unit
2.5.5.1 TEX120α 型挤出机 TEX120αextruder

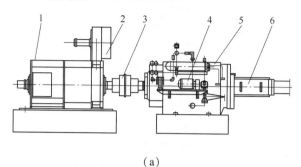

(a)

1—电机 Motor；2—冷却风机 Cooling fan；3—联轴器 Coupling；
4—润滑油泵 Lubrication pump；5—减速箱 Reduction gearbox；
6—螺杆连接器 Screw connector

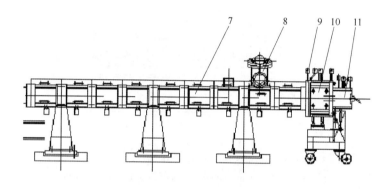

(b)

7—筒体 Cylinder；8—真空口 Vacuum port；
9—电加热器 Electric heater；10—换网器 Screen changer；11—模头 Die-head

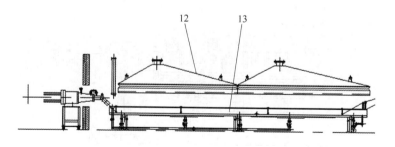

(c)

12—防护罩 Protective cover；13—拉条冷却器 Bracing wire cooler

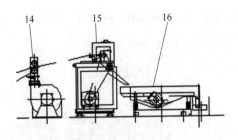

(d)

14—风机 Air wiper blower；15—切粒机 Pelletizer；
16—振动筛 Vibrating screen

图 2.5.6　TEX120α 型挤出机结构图 Structure of TEX120α extruder

2.5.5.2　ZSK133 型挤出机 ZSK133 extruder

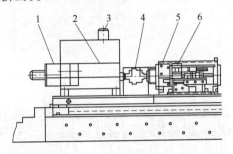

(a)

1、3—冷却风机 Cooling fan；2—电机 Motor；4—联轴器 Coupling；
5—减速箱 Reduction gearbox；6—润滑油泵 Lubrication pump

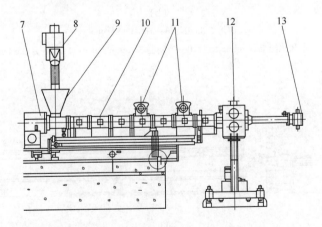

(b)

7—螺杆连接器 Screw connector；8—金属检测器 Metal detector；
9—下料料斗 Feed hopper；10—筒体 Cylinder；11—真空口 Vacuum port；
12—换网器 Screen changer；13—模头 Die-head

2 合成树脂 Synthetic resin

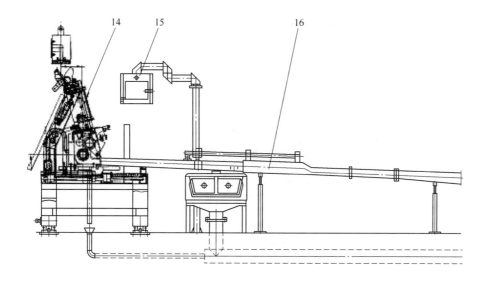

(c)

14—切粒机 Pelletizer；15—操作控制盘 Panel board；
16—输送水槽 Conveying water tank

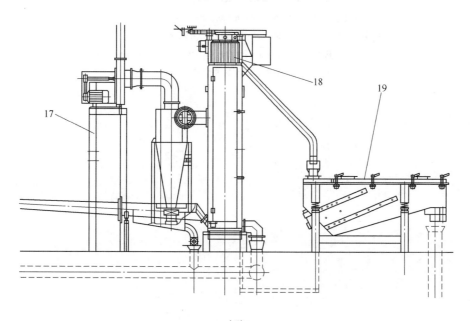

(d)

17—干燥风机 Drying blower；18—干燥机 Dryer；
19—振动筛 Vibrating screen

图 2.5.7　ZSK133 型挤出机结构图 Structure of ZSK133 extruder

2.5.6 计量秤 Measurement scale

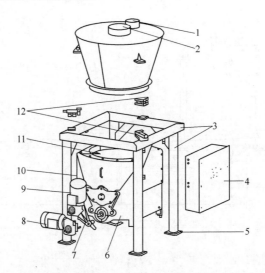

图 2.5.8 计量秤结构图 Structure drawing of measurement scale
1—放空口 Vent；2—进料口 Feed inlet；3—支架 Support；
4—电气接线箱 Electrical junction box；5—固定底座 Fixed base；
6—进料器支架 Feeder bracket；7—旋转板 Rotating plate；8—电机 Motor；
9—螺旋进料器电机 Screw feeder motor；10—后面板 Rear panel；
11—进料斗 Feed hopper；12—称重传感器 Weighing sensor

3 合成橡胶 Synthetic rubber

3.1 顺丁橡胶 Butadiene rubber

顺丁橡胶是顺式-1.4-聚丁二烯橡胶的简称,是由丁二烯聚合而成的结构规整的合成橡胶。顺丁橡胶生产的主要设备有聚合釜、凝聚釜、脱水干燥机组和包装设备等。

The major equipments to produce butadiene rubber includes the polymerization kettle, dehydration drying unit and the packaging equipment.

3.1.1 聚合釜 Polymerization kettle

3.1.1.1 釜体 Kettle

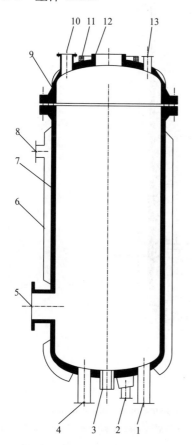

1—物料入口 Material inlet;2—壳层入口 Shell inlet;3—底座口 Bottom base hole;4—备用接口 Backup interface;5—人孔 Manhole;6—夹套 Jacket;7—筒体 Barrel;8—壳层出口 Shell outlet;9—封头 Head;10—物料出口 Material outlet;11—支架 Support;12—密封腔 Seal cavity;13—放空口 Vent

图 3.1.1 釜体结构图
Structure drawing of kettle

3.1.1.2 搅拌系统 Mixing system

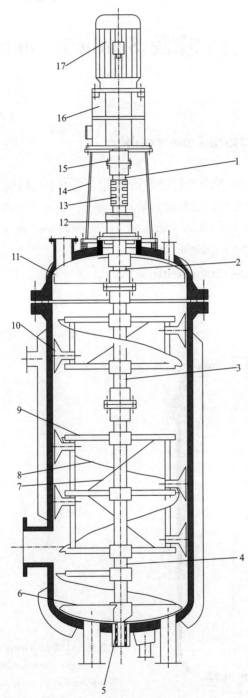

图 3.1.2 搅拌系统结构图 Structure drawing of mixing system

1、2、3、4—轴 Shaft；5—底部轴衬套 Bottom neck bush；6—下部刮板 Bottow scraper；7—小螺带 Small ribbon；
8—大螺带 Big ribbon；9—搅拌框 Stirring frame；10—刮刀 Scraper；11—上部刮板 Upper scraper；
12—机械密封 Mechanical seal；13—夹壳联轴器 Split coupling；14—支架 Support；
15—联轴器 Coupling；16—减速机 Reducer；17—电机 Motor

3 合成橡胶 Synthetic rubber

3.1.1.3 轴封 Shaft seal

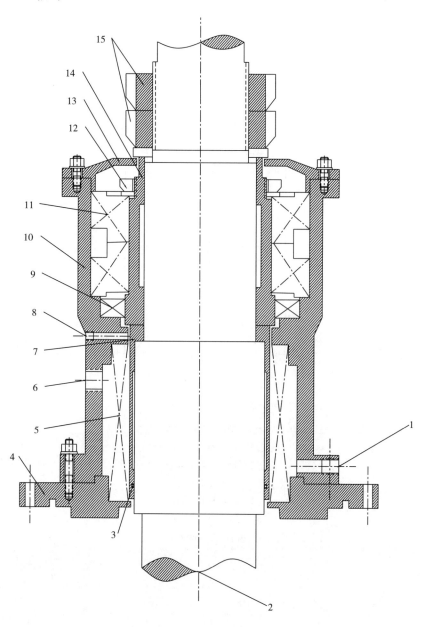

图 3.1.3　轴封结构图 Structure drawing of shaft seal

1—隔离液出口 Spacer fluid outlet；2—轴 Shaft；3—O 形密封圈 O sealing ring；4—下盖板 Lower cover；
5—双端面机械密封 Double mechanical seal；6—隔离液入口 Spacer fluid inlet；7—密封轴套 Seal cartridge；
8—泄漏检查口 Leakage inspection port；9—骨架油封 Framework oil seal；10—轴承箱 Bearing box；11—轴承 Bearing；
12—轴承螺母 Bearing nut；13—轴承压盖 Bearing cover；14—轴承套 Bearing sleeve；15—锁紧螺母 Gland nut

3.1.1.4 搅拌刮刀 Stirring scraper

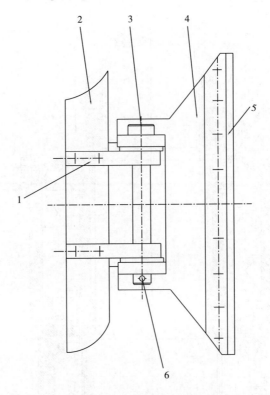

图 3.1.4 搅拌刮刀结构图
Structure drawing of stirring scraper
1—连接框架 Connect framework；
2—螺带框 Ribbon frame；
3—销轴 Pin；
4—刀架 Knife rest；
5—刀头 Cutting head；
6—销钉 Dowel

3.1.2 塔设备 Column equipment

3.1.2.1 丁二烯精馏塔 Butadiene distillation column

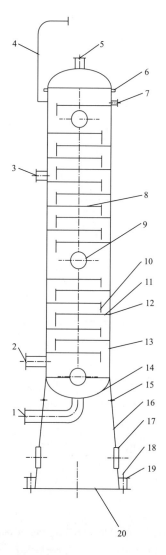

图 3.1.5 丁二烯精馏塔结构图 Structure drawing of butadiene distillation column

1—出料口 Liquid phase outlet；2—气相入口 Gas phase inlet；3—进料口 Feed inlet；
4—塔顶吊柱 Top davit；5—气相出口 Gas phase outlet；6—吊耳 Lifting lug；7—回流口 Reflux inlet；
8—塔盘(浮阀)Tray(float valve)；9—人孔 Manhole；10—降液板 Down-flow plate；
11—受液盘 Receiving tray；12—支承圈 Support ring；13—筒体 Barrel；14—封头 Head；
15—通气口 Vent hole；16—裙座 Skirt；17—检修口 Access hole；18—地脚螺栓 Anchor bolt；
19—地脚座 Foundation seat；20—基础环板 Base ring plate

3.1.2.2 丁二烯水洗塔 Butadiene washing column

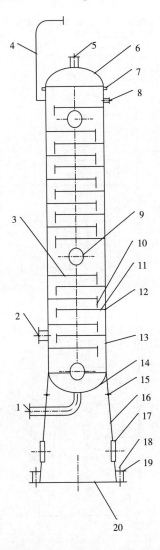

图 3.1.6 丁二烯水洗塔结构图 Structure drawing of butadiene washing column
1—洗涤水出口 Washing water outlet; 2—进料口 Feed inlet; 3—塔盘 Tray(sieve plate);
4—塔顶吊柱 Top davit; 5—出料口 Feed outlet; 6—上封头 Upper head; 7—吊耳 Lifting lug;
8—洗涤水入口 Washing water inlet; 9—人孔 Manhole; 10—降液板 Down-flow plqte;
11—受液盘 Receiuing tray; 12—支承圈 Support ring; 13—筒体 Barrel; 14—下封头 Lower head;
15—通气口 Vent hole; 16—裙座 Skirt; 17—检修口 Access hole; 18—地脚螺栓 Anchor bolt;
19—地脚座 Foundation seat; 20—基础环板 Base ring plate

3.1.3 塔顶冷凝器 Butadiene distillation overhead condenser

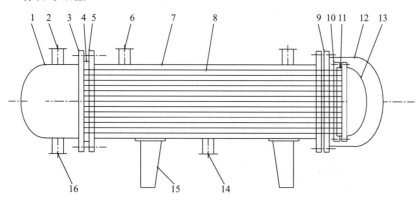

图 3.1.7 塔顶冷凝器结构图 Structure drawing of butadiene distillation overhead condenser
1—管箱 Channel；2—冷却液出口 Coolant outlet；3—管箱法兰 Channel flange；4—管箱侧管板 Channel side plate；
5—筒体法兰 Shell hange；6—气相入口 Gas phase inlet；7—筒体 Barrel；8—管束 Tube bundle；
9—大浮头法兰 Large floating head flange；10—小浮头钩圈 Small floating head backing device；
11—小浮头管板 Small floating head plate；12—大浮头 Large floating head；13—小浮头 Small floating head；
14—液相出口 Liquid phase outlet；15—支座 Support；16—冷却液入口 Coolant inlet

3.1.4 泵 Pump

3.1.4.1 丁二烯精馏塔回流泵 Butadiene distillation reflux pump

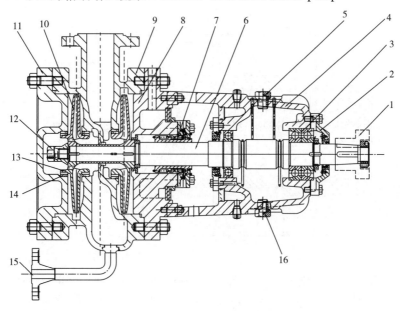

图 3.1.8 丁二烯精馏塔回流泵结构图 Structure drawing of butadiene distillation reflux pump
1—联轴器 Coupling；2—轴承压盖 Bearing cover；3—止推轴承 Thrust bearing；4—轴承箱 Bearing box；
5—加油口 Filler；6—轴 Shaft；7—机械密封 Mechanical seal；8—泵盖 Pump cover；9—泵体 Pump body；
10—叶轮 Wheel；11—端盖 End cover；12—叶轮锁紧螺母 Impeller lock nut；13—叶轮口环 Impeller ring；
14—泵体口环 Pump ring；15—泵体排液口 Pump condensate drain outlet；
16—轴承箱排液口 Bearing box drainage

3.1.4.2 胶液输送泵 Glue solution delivery pump

1. 胶液输送泵外形 Outline of glue solution delivery pump

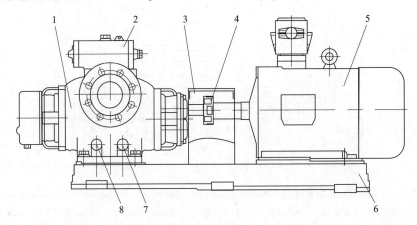

图 3.1.9　胶液输送泵外形图 Outside drawing of glue solution delivery pump

1—泵体 Pump body；2—出口安全阀 Outlet relief valve；3—联轴器防护罩 Coupling shield；
4—联轴器 Coupling；5—电机 Motor；6—底座 Base；7—排污口 Drain outlet；
8—泵体加热口 Pump heating gate

2. 胶液输送泵结构 Structure of glue solution delivery pump

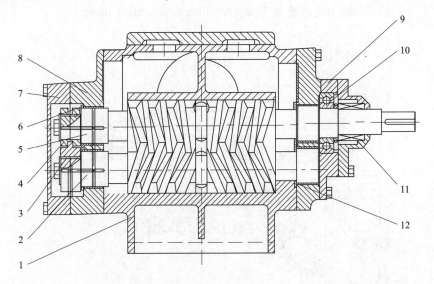

图 3.1.10　胶液输送泵结构图 Structure drawing of glue solution delivery pump

1—泵体 Pump housing；2—从动螺杆 Driven screw；3—从动齿轮 Driven gear；
4—主动齿轮 Driving gear；5—主动螺杆 Driving screw；6、12—螺杆轴衬套 Screw neck bush；
7—泵盖 Pump cover；8—泵体后端盖 Pump rear end cover；
9—轴承座 Bearing seat；10—轴承 Bearing；11—机械密封 Mechanical seal

3　合成橡胶 Synthetic rubber

3.1.4.3　催化剂计量泵 Catalyst metering pump

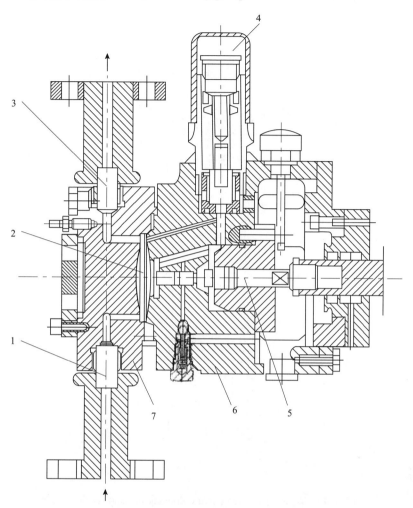

图 3.1.11　催化剂计量泵结构图
Structure drawing of catalyst metering pump
1—吸入口单向阀 Suction inlet valve；
2—隔膜 Membrane；
3—排出口单向阀 Discharge check valve；
4—隔膜腔补油器 Membrane filling-oil device；
5—柱塞 Plunger；
6—泵后腔体 Rear cavity of pump；
7—泵前腔体 Front cavity of pump

3.1.5 分离设备 Separation equipment

3.1.5.1 凝聚釜 Condensation kettle

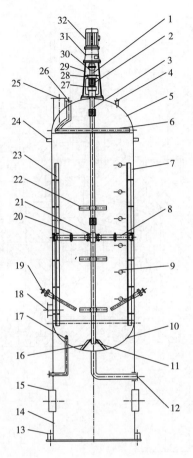

图 3.1.12 凝聚釜结构图 Structure drawing of condensation kettle

1、3—轴 Shaft；2—夹壳联轴器 Split coupling；4—接管 Connecting pipe；5—上封头 Upper head；
6—热水盘管 Hot water coil；7—筒体 Barrel；8—拉筋 Tiepiece；9—视镜 Sight glass；
10—下封头 Lower head；11—底部轴衬套 Bottom shaft sleeve；
12—排污口 Drain outlet；13—地脚螺栓 Anchor bolt；14—裙座 Skirt；15—裙座人孔 Skirt manhole；
16—底部支承架 Bottom supporting frame；17—蒸汽喷嘴 Steam nozzle；18—人孔 Manhole；
19—静态混合器 Static mixer；20—拉筋调节螺母 Tiepiece adjusting nut；
21—中间轴套 Intermediate shaft sleeve；22—搅拌桨叶 Stirring paddle；
23—折流板 Baffle plate；24—吊耳 Lifting lug；25—气相出口 Gas phase outlet；
26—热水入口 Hot water inlet；27—机械密封 Mechanical seal；
28—轴承座 Bearing seat；29—支架 Support；
30—柱销联轴器 Pin coupling；31—减速机 Reducer；
32—电机 Motor

3.1.5.2 洗胶釜 Washing gelatine kettle

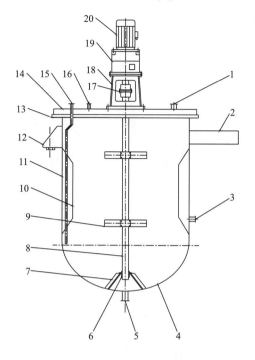

1—洗涤水入口 Washing water inlet；2—出料槽 Discharge chute；3—热电偶接口 Thermocouple interface；4—下封头 Lower head；5—排污口 Drain outlet；6—轴衬套 Neck bush；7、18—支架 Support；8—轴 Shaft；9—搅拌桨 Stirring paddle；10—折流板 Baffle plate；11—筒体 Barrel；12—支座 Support；13—盖板 Upper cover；14—搅拌支承 Soupport；15—蒸汽入口 Steam inlet；16—补水口 Moisturizing mouth；17—联轴器 Coupling；19—减速机 Reducer；20—电机 Motor

图 3.1.13 洗胶釜结构图
Structure drawing of washing gelatine kettle

3.1.5.3 振动脱水筛 Vibrating-dewatering screen

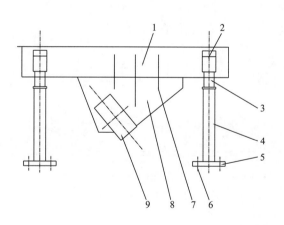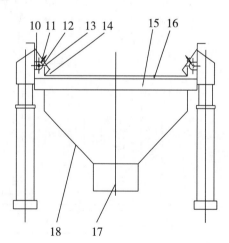

图 3.1.14 振动脱水筛结构图 Structure drawing of vibrating-dewatering screen

1—槽体 Cell body；2—支承座 Supporting seat；3—橡胶弹簧 Rubber spring；4—立柱 Upright；5—地脚座 Foundation seat；6—地脚螺栓 Anchor bolt；7—筋板 Rib plate；8—筛体 Screen box；9—激振电机 Shock excitation electric motor；10—铰接座 Hinged seat；11—柱销 Pin；12—压紧螺母 Gland nut；13—铰接拉杆 Hinged lever；14—筛面压板 Sieve plate；15—集水槽 Catch basin；16—筛面 Sieve；17—出水口 Water outlet；18—下水料斗 Water hopper

3.1.6 脱水干燥设备 Dehydration drying equipment

3.1.6.1 脱水挤压机 Dehydration extrusion machine

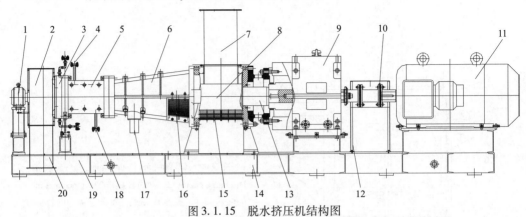

图 3.1.15 脱水挤压机结构图
Structure drawing of dehydration extrusion machine

1—机头轴承座 Handpiece bearing seat；2—出料口 Feed outlet；3—机头模板 Handpiece template；
4—筒体加热蒸汽入口 Cylinder heating steam inlet；5—直形脱水段 Straight dehydration section；
6—锥形脱水段 Conical dehydration section；7—进料料斗 Charging hopper；8—进料口 Feed inlet；
9—齿轮箱 Gear casing；10—膜片联轴器 Diaphragm coupling；11—电机 Motor；12—防护罩 Protective cover；
13—主轴 Shaft；14—地脚座 Foundation seat；15—进料口脱水板 Feed inlet dewatering plate；
16—锥段脱水板 Cone section dewatering plate；17—锥段脱水口 Dehydration mouth of small cone section；
18—筒体加热蒸汽出口 Cylinder heating steam outlet；19—联合底座 Joint base；20—出料料斗 discharging hopper

3.1.6.2 膨胀干燥机 Expansion dryer

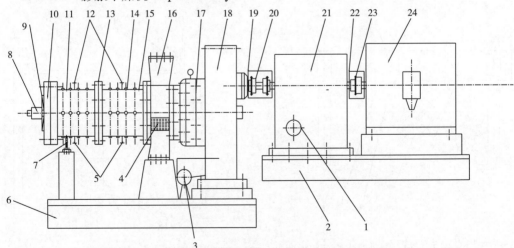

图 3.1.16 膨胀干燥机结构图 Structure drawing of expansion dryer

1、3—油泵 Mechanism；2、6—底座 Joint base；4—进料斗脱水板 Charging hopper dewatering plate；
5—夹套蒸汽出口 Jacket steam outlet；7—筒体支座 Cylinder bearing；8—造粒机轴承箱 Pelletizer bearing box；
9—造粒机切刀 Pelletizer parting tool；10—模头 Die head；11—二段筒体 2nd stage cylinder；
12—夹套蒸汽入口 Jacket steam inlet；13—筒体法兰 Shell hange；14—剪切螺栓 Shear screw；
15—一段筒体 1st stage cylinder；16—进料斗 Charging hopper；17—推力轴承座 Thrust bearing seat；
18—齿轮箱 Gear box；19、22—联轴器防护罩 Coupling shield；20、23—齿形联轴器 Gear-type coupling；
21—无级变速器 Continuously variable transmission；24—电机 Motor

· 234 ·

3 合成橡胶 Synthetic rubber

3.1.6.3 干燥箱 Drying oven

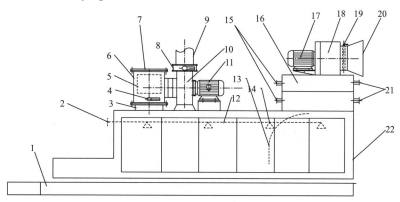

图 3.1.17 干燥箱结构图 Structure drawing of drying oven

1—水平输送筛 Horizontal conveying screen；2—消防蒸汽入口 Fire fighting steam inlet；3—引风筒 Suction cylinder；4—排风机入口风量调节阀 Exhaust fan inlet air flow adjusting valve；5—过滤器滤芯 Filter element；6—过滤器筒体 Filter cylinder；7—排风机过滤器盖 Exhaust fan inlet filter cover；8—排风机出口风量调节阀 Exhaust fan outlet volume regulating valve；9—引风筒 Uptake ventilator；10—排风机 Exhaust fan；11—排风机电机 Exhaust fan motor；12—消防蒸汽管 Fire fighting steam pipe；13—挡胶板 Stock guide；14—消防蒸汽喷嘴 Fire fighting steam nozzle；15—蒸汽出口 Steam outlet；16—空气加热器 Air heater；17—进风机电机 Intake air fan motor；18—进风机 Intake air fan；19—进风机入口风量调节阀 Intake air fan inlet volume damper；20—入口过滤网 Inlet filter；21—蒸汽入口 Steam inlet；22—干燥箱体 Drying cabinet

3.1.7 成形包装设备 Forming packaging equipment

3.1.7.1 电子定量秤 Electronic quantitative scale

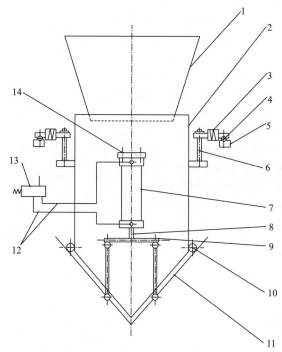

图 3.1.18 电子定量秤结构图
Structure drawing of electronic quantitative scale

1—料斗入口 Hopper inlet；2—料斗 Hopper；3—重量传感器 Weight sensor；4—铰接支承球 Bearing ball joint；5—支承座 Supporting Seat；6—秤体拉杆 Scale body pull rod；7—料门气缸 Feeding hopper cylinder；8—气缸活塞杆 Cylinder piston rod；9—料门执行机构 Feeding hopper performing machine；10—料门转轴 Feeding hopper rotor；11—下料门 Discharge outlet；12—空气风管 Air-duct；13—电磁换向阀 Magnetic exchange valve；14—料斗气缸座 Feeding hopper cylinder seat

3.1.7.2 压块机 Cuber

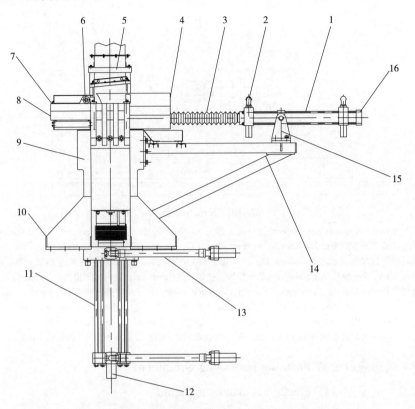

图 3.1.19 压块机结构图 Structure drawing of cuber

1—侧油缸 Side oil cylinder；2—侧油缸油管 Side oil cylinder pipe；
3—活塞杆护套 Piston rod guard；4—防护罩 Shield；
5—下料口 Discharge outlet；6—侧缸滚轮 Side cylinder wheel；
7—侧缸导轨 Side cylinder guide；8—主缸盖板 Main cylinder cover plate；
9—压块机腔体 Cuber cavity；10—底座 Base；
11—主油缸 Main oil cylinder；12—主油缸传感器 Main oil cylinder transducer；
13—主油缸油管 Main oil cylinder pipe；
14—侧油缸支架 Side oil cylinder stand；
15—侧油缸铰接架 Side oil cylinder articulated frame；
16—侧油缸传感器 Side oil cylinder transducer

3 合成橡胶 Synthetic rubber

3.1.7.3 薄膜包装机 Film packaging machines

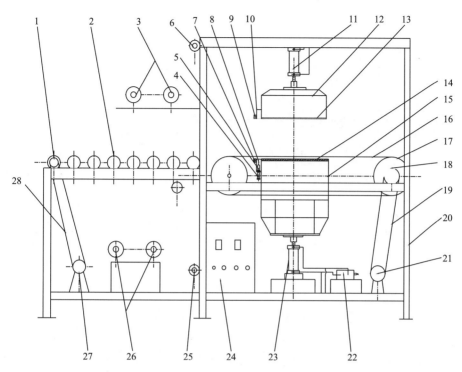

图 3.1.20 薄膜包装机结构图 Structure drawing of film packaging machines

1—驱动链轮 Drive sprocket；2—传送辊 Feed Roll；3—上部薄膜轴 Upper film shaft；
4—切刀气缸 Cutter cylinder；5—刀架 Knife rest；6—上部薄膜导向滚 Upper film guide roll；
7—直形段密封衬条 Straight sealing strip；8—切刀片 Cutting blade；9—直形段加热器 Straight heating body；
10—刀槽 Knife slot；11—上部气缸 Upper cylinder；12—上托架 Upper bracket；
13—U 形段加热器 U-type heating body；14—U 形段密封衬条 U-type sealing strip；
15—下托架 Bottom bracket；16—输送皮带 Conveyor belts；17—皮带轮 Belt wheel；
18—链轮 Chain wheel；19、28—链条 Chain；20—机体框架 Body frame；
21、27—电机 Motor；22—电磁控制阀 Solenoid electric valve；
23—下部气缸 Bottom cylinder；24—电气控制箱 Electric control box；
25—下部薄膜导向滚 Bottom film guide roll；26—下部薄膜轴 Bottom film shaft；

3.1.7.4　自动装袋机 Automatic baggers

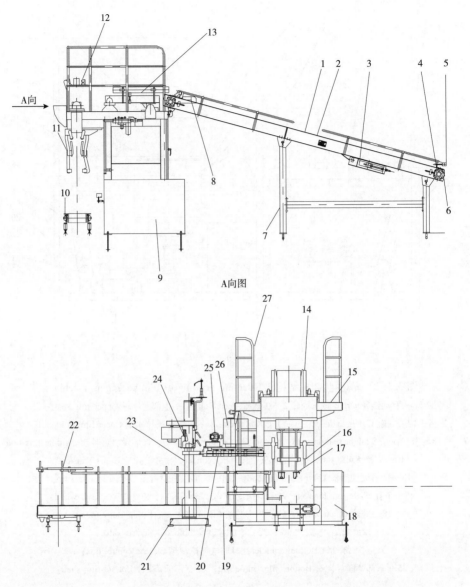

图 3.1.21　自动装袋机结构图 Structure drawing of automatic baggers

1—斜坡输送机 Slope conveyor；2—环形皮带 Endless-belt；3—皮带张紧装置 Belt tightening device；
4—轴承座 Bearing seat；5—光电开关 Optoelectronic switch；6—皮带滚子 Belt roller；7—机体支架 Body stents；
8—减速机及电机 Speed reducer and motor；9—真空泵 Vacuum pump；10、19—立袋输送机 Vertical bag conveyor；
11—夹袋装置 Bag clamping and taking device；12、14—垂直输送装置 Vertical conveyor device；
13—水平输送装置 Horizontal conveyor device；15—机顶平台 Roof platform；16—取袋装置 Bag taking device；
17—扶袋装置 Bag steadying device；18—电控箱 Electric control element box；
20—夹口整形皮带 Nip shaping belt；21—缝口机底座 Seam machine base；
22—倒袋调节架 Bag Pouring adjusting frame；23—缝口机立柱 Seam machine column；
24—缝口机 Seam machine；25—电机 Motor；
26—操作控制面板 Operation control panel；27—平台 Platform

· 238 ·

3 合成橡胶 Synthetic rubber

3.1.8 输送设备 Conveying equipment

3.1.8.1 螺旋提升机 Screw elevator

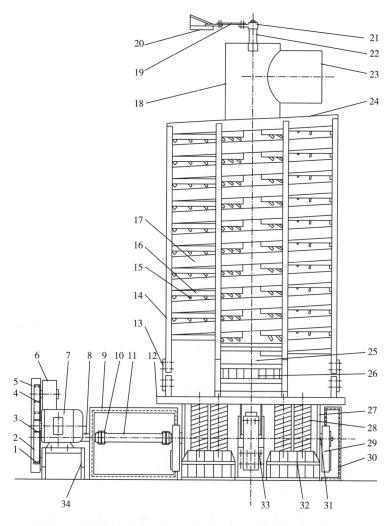

图 3.1.22 螺旋提升机结构图 Structure drawing of screw elevator

1—机体皮带轮 Body belt pulley；2—同步皮带 Timing belt；3—电机皮带轮 Motor belt pulley；
4—同步皮带轮 Timing belt pulley；5—皮带防护罩 Belt guard；6—同步齿轮箱 Timing gear box；
7—驱动电机 Drive motor；8—驱动轴 Drive shaft；9—联轴器防护罩 Coupling shield；
10—橡胶联轴器 Rubber coupling；11—联轴器长轴 Long axis of the coupling；12—机体底座 Body base；
13、19—连接板 Connecting plate；14—料槽立柱 Trough column；15—出风口 Air outlet；
16—通风横梁 Ventilation beam；17—料槽 Trough；18—中心柱(风筒) central post(air duct)；
20—顶部固定架 Top fixed frame；21—轴承座 Bearing block；
22—顶部固定柱 Top fixed post；23—风筒入口 Air duct inlet；24—出料口 Feed outlet；
25—入料口 Feed inlet；26—中心柱大法兰 Central post flange；27—底部轴承座 Base bearing block；
28—机体弹簧 Body spring；29—振动偏心块 Eccentric block vibration；30—偏心防护罩 Eccentric shield；
31—机体主轴 Body spindle；32—弹簧座 Spring seat；33—振动保护体 Vibration protection body；
34—电机底座 Motor base

· 239 ·

3.1.8.2 振动输送机 Vibrating conveyer

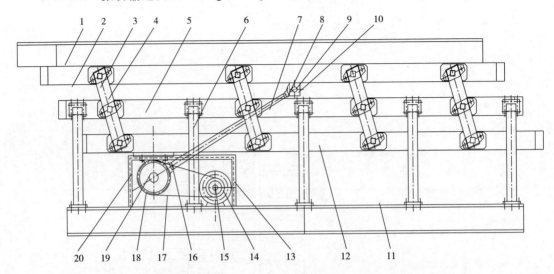

图 3.1.23 振动输送机结构图 Structure drawing of vibrating conveyer
1—槽体 Cell body；2—槽体座 Cell body base；3—扭簧轴 Torsional spring shaft；4—扭簧 Torsional spring；
5—支承框架 Supporting frame；6—支承柱 Bearing post；7—振动连杆 Connecting rod vibration；
8、16—连杆联接头 Connecting rod coupling head；9—销轴 Pin roll；10—销轴座 Pin roll base；
11—底座 Base；12—下质体 Lower plastid；13—皮带防护罩 Belt guard；14—电机 Motor；
15—电机皮带轮 Motor pulley；17—皮带 Belt；18—皮带轮 Belt pulley；19—曲轴 Crankshaft；
20—曲轴箱 Bent axle box

3.2 丁苯橡胶 Styrene butadiene rubber

丁苯橡胶是苯乙烯和丁二烯的共聚物,是最大的通用合成橡胶品种,也是最早实现工业化生产的橡胶品种之一。

Styrene buladiene rubber, also called a copolymer of styrene and butadiene, is the largest commonly used rubber variety, which is one of rubber varietiies produced earliest in industralization.

3.2.1 聚合反应器 Polymerization reactor

丁苯橡胶生产中使用的聚合反应器是釜式反应器,习惯上称为聚合釜。

The polymerization reactor which is used in the production of styrene butadiene rubber is the tank reactor, which is customarily called the polymerization kettle.

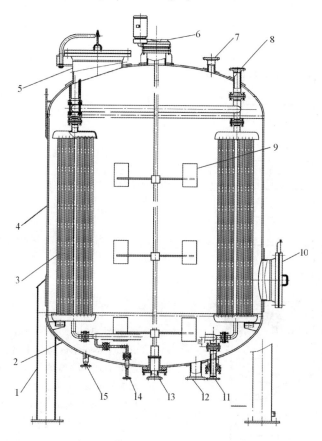

图 3.2.1 聚合反应器结构图 Structure drawing of polymerization reactor

1—支架 Support;2—封头 Head;3—氨列管 Ammonia column tube;4—筒体 Shell;5—上人孔 Upper manhole;
6—搅拌装置 Stirrer device;7—未脱气胶乳出口 Undegassed latex outlet;8—气氨出口 Ammonia gas outlet;
9—搅拌桨叶 Stirrer paddle;10—侧人孔 Side manhole;11—液氨回流 Liquid ammonia backflow mouth;
12—未脱气胶乳入口 Undegassed latex inlet;13—排液口 Liquid outlet;
14—液氨入口 Liquid ammonia inlet;15—氧化剂入口 Oxidizer mouth

3.2.2 塔设备 Column

3.2.2.1 苯乙烯脱气塔 Styrene degassing column

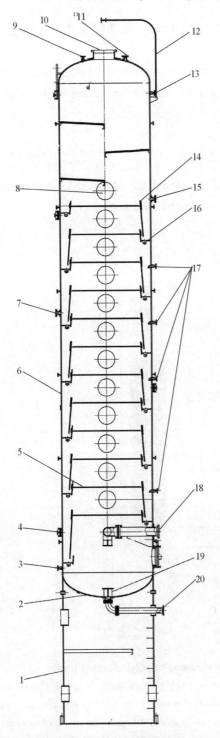

1—裙座 Skirt；2—封头 Head；3—测温口 Temperature measuring port；4—视镜 Sight glass；5—塔盘 Tray；6—筒体 Cylinder shell；7—回流口 Reflux inlet；8—人孔 Manhole；9、17—压力计接口 Pressure measurement mouth；10—气相苯乙烯出口 Gas phase styrene outlet；11—抽真空口 Vacuum orifice；12—塔顶吊柱 Top davit；13—泡沫检测口 Bubble detection mouth；14—溢流堰、降液板 Downflow weir、down-flow plate；15—胶乳入口 Latex inlet；16—受液盘 Tray；18—蒸汽入口 Steam inlet；19—挡胶板 Stock guide；20—胶乳出口 Latex outlet

图 3.2.2 苯乙烯脱气塔结构图
Structure drawing of styrene degassing column

3.2.2.2 煤油吸收塔 Kerosene absorption column

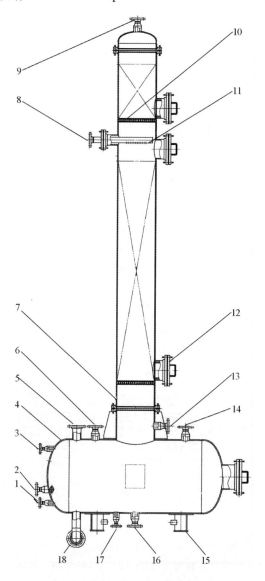

图 3.2.3 煤油吸收塔结构图
Structure drawing of kerosene absorption column

1、3、5、18—液位计接口 Liquid level gauge interface；2—温度计接口 Thermometer mouth；4—封头 Head；
6—压力计接口 Pressure gauge port；7—筒体 Barrel；8—煤油入口 Kerosene inlet；
9—放空口 Vent；10—栅板 Grid tray；11—喷淋管 Rainer；
12—手孔 Hand hole；13—丁二烯气体入口 Butadiene gas inlet；
14—安全阀接口 Safety valve interface；15—支座 Support；16—煤油出口 Kerosene outlet；
17—排放口 Discharge outlet

3.2.3 换热设备 Heat-exchange equipment
3.2.3.1 丁二烯进料换热器 Butadiene feed heat exchanger

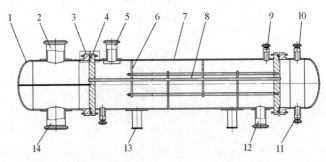

图 3.2.4 丁二烯进料换热器结构图 Structure drawing of butadiene feed heat exchanger
1—封头 Head；2—丁二烯出口 Butadiene outlet；3—法兰 Flange；4—管板 Tube sheet；
5—热水进口 Hot water inlet；6—折流板 Baffle plate；7—筒体 Cylinder；
8—换热管 Heat exchange tube；9—壳程放空口 Shell pass emptying outlet；
10—管程放空口 Tube pass emptying outlet；11—管程排放口 Tube pass discharge outlet；
12—热水出口 Hot water outlet；13—支座 Support；14—丁二烯入口 Butadiene inlet

3.2.3.2 排气冷凝器 Exhaust condenser

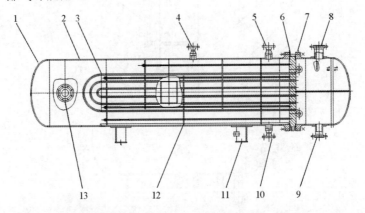

图 3.2.5 排气冷凝器结构图 Structure drawing of exhaust condenser
1—封头 Head；2—筒体 Cylinder；3—U 形管束 U tube bundle；
4—安全阀接口 Safety valve interface；5—丁二烯气相出口 Gas phase butadiene outlet；
6—管板 Tube sheet；7—法兰 Flange；8—气氨出口 Ammonia gas outlet；
9—液氨进口 Liquid ammonia inlet；10—丁二烯液相出口 Butadiene liquid outlet；
11—支座 Support；12—折流板 Baffle plate；13—丁二烯液相进口 Butadiene liquid inlet

3 合成橡胶 Synthetic rubber

3.2.3.3 冷胶进料冷却器 Cold glue feed cooler

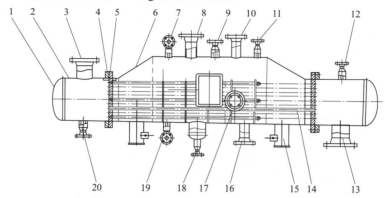

图 3.2.6　冷胶进料冷却器结构图 Structure drawing of cold glue feed cooler

1—封头 Head；2—管箱 Channel；3—管程进口 Gas phase inlet；4—法兰 Flange；5—管板 Tube sheet；
6—锥壳 Conical shell；7、10、16、19—液位计接口 Liquid level gauge interface；8—气氨出口 Ammonia gas outlet；
9—安全阀接口 Safety valve interface；11—压力计接口 Pressure gauge port；12—管程放空口 Tube pass vent outlet；
13—物料出口 Material outlet；14—换热管 Heat exchange tube；15—支座 Support；17—液氨进口 Liquid ammonia inlet；
18—壳程排液口 Shell side liquid discharge outlet；20—管程出口 Tube pass discharge outlet

3.2.4　泵 Pump

3.2.4.1　丁二烯输送屏蔽泵 Butadiene conveying canned motor pump

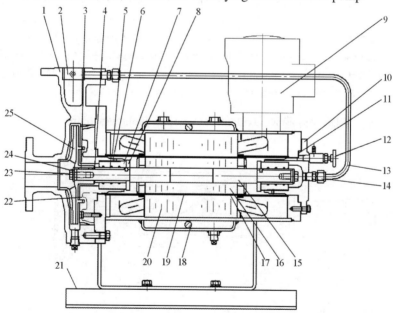

图 3.2.7　丁二烯输送屏蔽泵结构图 Structure drawing of butadiene conveying canned motor pump

1—泵体 Pump body；2—过滤器 Filter；3—键 Key；4—调整垫圈 Adjusting washer；5—轴套 Sleeve；
6—垫片 Gasket；7—轴承 Bearing；8—推力盘 Thrust collar；9—接线盒 Junction box；10、22—端盖 End cover；
11—密封垫圈 Seal washer；12—排气阀 Vent valve；13—循环管 Circulation tube；14—活结接头 Slipknot joint；
15—轴 Shaft；16—转子 Rotor；17—转子屏蔽套 Rotor can；18—水套 Water jacket；19—定子屏蔽套 Stator can；
20—定子 Stator；21—底座 Base；23—螺栓 Bolt；24—止动垫圈 Lock washer；25—叶轮 Wheel

3.2.4.2 活化剂输送旋涡泵 Activator conveying vortex pump

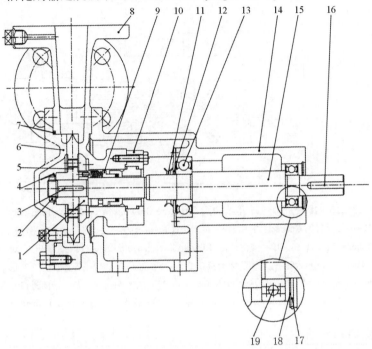

图 3.2.8　活化剂输送旋涡泵结构图 Structure drawing of activator conveying vortex pump

1—轴封锥衬 Push seal liner；2—叶轮螺母 Impeller nut；3—叶轮键 Impeller key；4—垫圈 Washer；
5—叶轮 Wheel；6—泵盖 Pump cover；7—O 形密封圈 O sealing ring；8—泵体 Pump body；9—机械密封 Mechanical seal；
10—密封压盖 Shaft sealing fastening ring；11—挡油环 Water retaining ring；12—轴承压盖 Bearing fastening ring；
13—后轴承 Rear bearing；14—轴承箱 Bearing box；15—泵轴 Pump spindle；16—键 Key；
17—甩油环 Oil flinger；18—轴承盖 Bearing cap；19—前轴承 Fore bearing

3.2.4.3 氧化剂输送隔膜泵 Oxidant conveying diaphragm pump

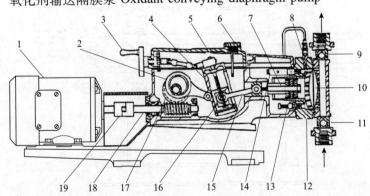

图 3.2.9　氧化剂输送隔膜泵结构图 Structure drawing of oxidant conveying diaphragm pump

1—驱动电机 Drive motor；2—曲柄 Crank；3—外部调节手轮 External regulation hand wheel；
4—振荡盒 Oscillation box；5—轴承箱 Bearing box；6—注油测试棒 Oil test bar；7—柱塞 Plunger piston；
8—自动排放阀 Automatic drain valve；9—排出阀 Discharge valve；10—隔膜 Membrane；11—吸入阀 Suction valve；
12—泵体 Pump body；13—液力补充阀 Hydraulic replenishing valve；14—液力旁通阀 Converter bypass valve；
15—连杆 Connecting rod；16—滑块 Slider；17—蜗轮 Worm wheel；18—蜗杆 Worm rod；19—联轴器 Cupling

3 合成橡胶 Synthetic rubber

3.2.4.4 胶乳输送螺杆泵 Latex conveying screw pump

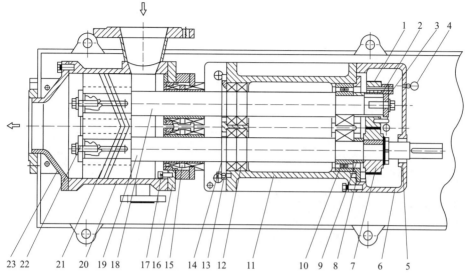

图 3.2.10 胶乳输送螺杆泵结构图 Structure drawing of latex conveying screw pump
1—从动齿轮 Driven gear；2—齿轮轴套 Gear sleeve；3、22—挡圈 Check ring；4—油标 Oil pointer；
5、14—油封 Oil seal；6—前盖 Front cover；7—主动齿轮 Driving gear；8、13—轴承压盖 Bearing cover；
9—挡环 Baffle ring；10、12—轴承 Bearing；11—轴承箱 Bearing box；15—机械密封 Mechanical seal；
16—连接板 Connecting plate；17—定位环 Locating ring；18—从动轴 Driven shaft；19—主动轴 Driving shaft；
20—泵体 Pump body；21—螺杆 Screw；23—后盖 Rear cover

3.2.4.5 丁二烯水环泵 Butadiene water ring pump

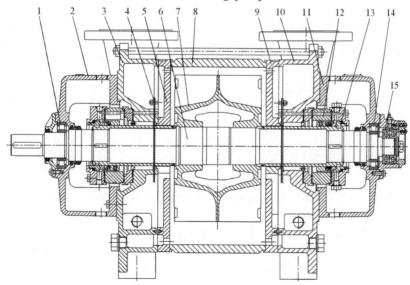

图 3.2.11 丁二烯水环泵结构图 Structure drawing of butadiene water ring pump
1、14—圆柱滚子轴承 Cylindrical rolling bearing；2—前轴承箱 Front bearing box；3—前侧盖 Front side cover；
4—柔性阀板 Flexible valve plate；5—前分配板 Front distribution plate；6—轴 Shaft；7—叶轮 Wheel；
8—泵体 Pump body；9—后分配板 Rear distribution plate；10—后侧盖 Rear side cover；
11—后轴承箱 Bearing housing；12—密封函 Mechanical seal body；
13—机械密封 Mechanical seal；15—推力球轴承 Thrust ball bearing

3.2.5 氨制冷压缩机 Ammonia refrigerating compressor

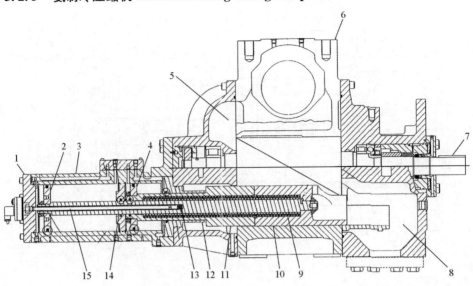

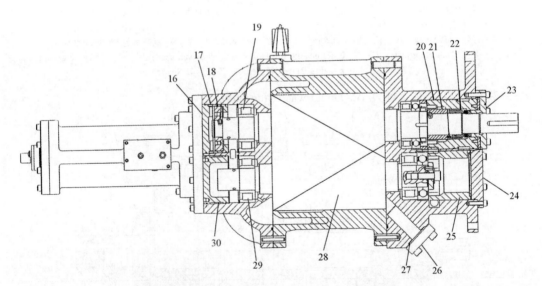

图 3.2.12 氨制冷压缩机结构图 Structure drawing of ammonia refrigerating compressor
1—油缸盖板 Oil cylinder cover；2—活塞 Piston；3—油缸 Oil cylinder；4—滑块活塞 Slide piston；
5—进气口 Gas inlet；6—转子外壳 rotor housing；7—键 Key；8—出气口 Gas outlet；
9—卸载弹簧 Unloading spring；10—滑阀 Slide valve；11—滑块 Sliding block；12—滑块外壳 Sliding block housing；
13—压簧 Compressed spring；14—内筒外壳 Internal cylinder cover；15—活塞杆 Piston rod；16—进气端盖 Inlet end cover；
17—平衡活塞套 Balance pistons sleeve；18—平衡活塞 Balance pistons；19、29—轴承 Bearing；
20—密封套筒 Seal sleeve；21—密封环 Seal ring；22—轴密封组件 Shaft seal assembly；
23—密封压盖 Sealing gland；24—压盖 Gland；25—隔离套筒 Female distance sleeve；
26—法兰 Flange；27—法兰垫片 Flange gasket；
28—转子 Rotor；30—轴承压盖 Bearing cover

3 合成橡胶 Synthetic rubber

3.2.6 其它设备 Other equipment

3.2.6.1 挤压脱水机 Extruding-desiccation machine

挤压脱水机用于将橡胶颗粒进行挤压脱水并切成薄片。

Extruding-desiccation machine is used to squeeze and dehydrate the rubber particles and cut them into thin slices.

1. 挤压脱水机结构 Structure of extruding-desiccation machine

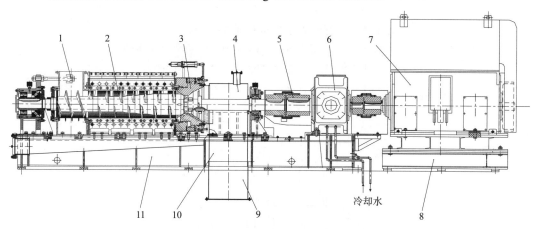

图 3.2.13 挤压脱水机结构图 Structure drawing of extruding-desiccation machine

1—进料口 Feed inlet；2—机体 Compressor body；3—调压锥体 Pressure regulating cone；4—出料口 Feed outlet；5—联轴器 Coupling；6—齿轮箱 Gear box；7—电机驱动装置 Motor drive device；8—电机底座 Motor base；9—活动接料筒 Mobile material barrel；10—过渡接料筒 Transition material barrel；11—机座 Machine base

2. 挤压脱水机主要零部件 Main components of extruding-desiccation machine

1) 机体 Machine body

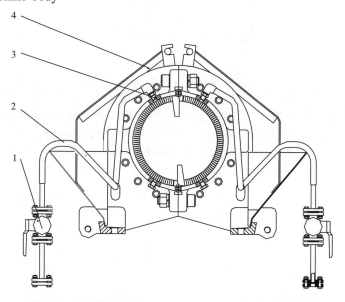

图 3.2.14 机体结构图 Structure drawing of machine body

1—阀门 Valve；2—蒸汽管路 Steam pipeline；3—接头 Joint；4—挡水帘组件 Retaining water component

2)螺旋 Spiral

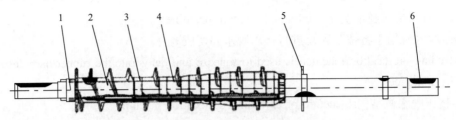

图 3.2.15　螺旋结构图 Structure drawing of spiral

1—主轴 Shaft；2—入料辊 Material input roller；3—隔套 Spacer bush；
4—螺旋 Spiral；5—破料器 Broken feeder；6—键 Key

3)调压锥体 Pressure regulating cone

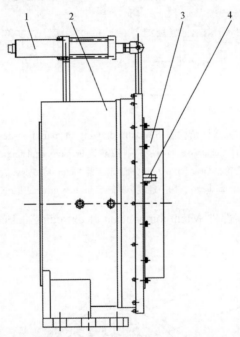

图 3.2.16　调压锥体结构图 Structure drawing of pressure regulating cone

1—传感器 Transducer；2—调压锥体 Pressure regulating cone；
3—导向环 Guide ring；4—防转螺杆 Non-rotating screw

3.2.6.2 压块机 Cuber

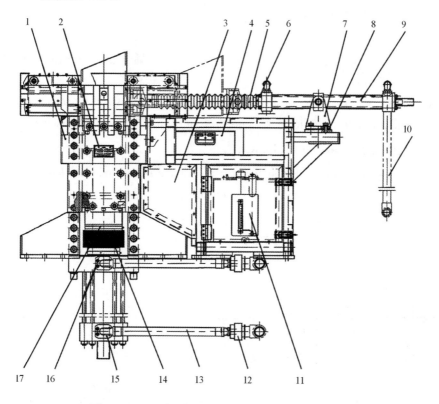

图 3.2.17 压块机结构图 Structure drawing of cuber
1—压块机机体 Cuber body；2—铭牌 Nameplate；3—大盖板 Big plate；
4—接料槽 Connecting material tank；5、14—防护套 Protecting cover；6、16、17—喉箍 Hose clamp；
7—铰轴支架 Hinged support；8—调整垫 Adjustable pad；9—液压缸 Hydraulic cylinder；
10、13—液压胶管 Hydraulic hose；11—硅油箱 Silicone oil tank；12—法兰 Flange plate；
15—对开法兰 Bisected flange

3.3 丁基橡胶 Butyl rubber

丁基橡胶是合成橡胶的一种，由异丁烯和少量异戊二烯合成。一般被应用在制作汽车轮胎以及汽车隔音用品。

Butyl rubber, one of synthesized rubber varieties, is synthesized by isobutene and a little isoprene, which is usually used to make auto tires and sound insulation products.

3.3.1 反应设备 Reaction equipment

3.3.1.1 丁基聚合反应釜 Butadiene styrene polymerization reaction kettle

1. 丁基聚合反应釜结构 Structure of butadiene styrene polymerization reaction kettle

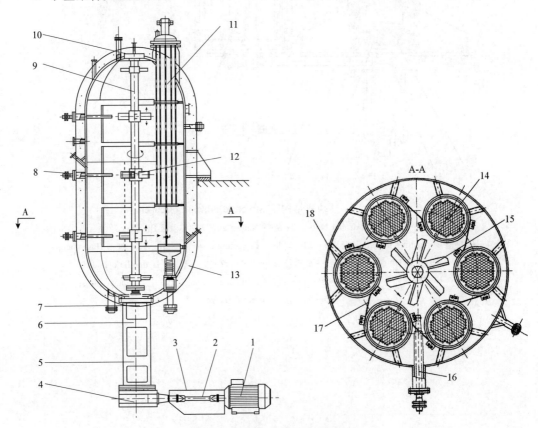

图 3.3.1 丁基聚合反应釜结构图

Structure drawing of butadiene styrene polymerization reaction kettle

1—电机 Motor；2—万向联轴器 Universal coupling；3—联轴器护罩 Coupling guard；
4—减速机 Reducer；5—推力轴承 Thrust bearing；6—径向轴承 Rodial bearing；
7—机械密封 Mechanical seal；8、16—喷嘴 Nozzle；9—搅拌轴 Mixer shaft；10—顶轴承 Top bearing；
11、14—管束 Tube bundle；12、15—搅拌桨叶 Propeller；13—保冷隔离层 Cold insulation；
17—折流板 Baffle plate；18—支承环 Supporting ring

3 合成橡胶 Synthetic rubber

2. 丁基聚合反应釜主要零部件 Main components of butadiene styrene polymerization reaction kettle

1）机械密封 Mechanical seal

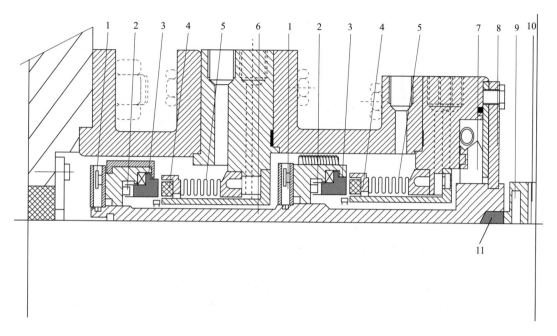

图 3.3.2 机械密封结构图 Structure drawing of mechanical seal

1、9—定位螺钉 Set screw；2—定位销 Locating pin；3—动环 Rotating ring；
4—静环 Stationary seal ring；5—波纹管 Bellows；6—轴套 Sleeve；
7—石墨密封环 Carbon gland ring；8—定压卡板 Constant pressure pallet；
10—轴套定位环 Shaft sleeve collar；
11—轴套密封圈 Shaft sleeve sealing ring

2)径向轴承 Radial bearing

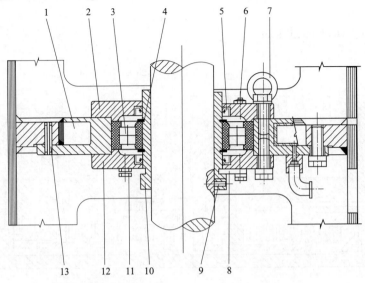

图 3.3.3　径向轴承结构图 Structure drawing of radial bearing

1—轴承支承 Bearing support；2—上盖 Upper cover；3—径向轴承 Radial bearing；4—弹簧挡圈 Circlip；
5—油封 Oil seal；6—排气阀 Vent valve；7—吊环螺栓 Eye bolt；8—定位卡板 Set card board；
9—定位顶丝 Set jackscrew；10—轴套 Sleeve；11—加脂口 Fatliquoring mouth；12—下盖 Lower cover；
13—弹性销 Elastic pin

3)推力轴承 Thrust bearing

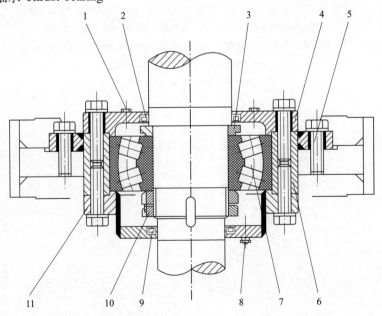

图 3.3.4　推力轴承结构图 Structure drawing of thrust bearing

1—排气阀 Vent valve；2、9—油封 Oil seal；3—拆卸环 Extractor；4—上盖 Upper cover；
5—弹性销 Elastic pin；6—轴承支承 Bearing support；7—轴承 Bearing；8—加脂口 Fatliquoring mouth；
10—圆螺母 Ring nut；11—下盖 Lower cover

3 合成橡胶 Synthetic rubber

4) 搅拌桨叶 Mixer turbine

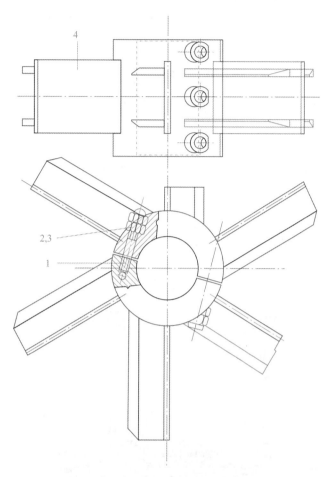

图 3.3.5 搅拌桨叶结构图 Structure drawing of mixer turbine
1—双头螺柱 Stud bolt；2—背螺母 Nut；3—盖形螺母 Cap nut；4—桨叶 Turbine

3.3.1.2 脱气釜/汽提釜 Degassing kettle/Stripping kettle

1. 脱气釜/汽提釜结构 Structure of degassing kettle/stripping kettle

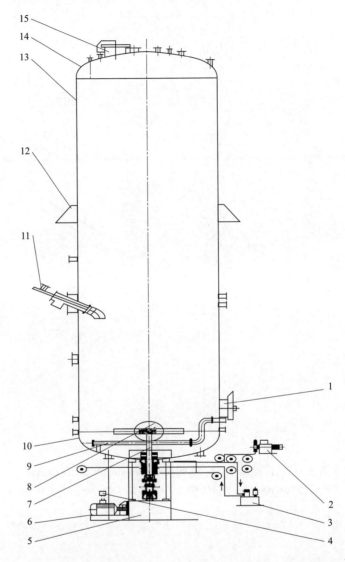

图 3.3.6 脱气釜/汽提釜结构图 Structure drawing of degassing kettle/stripping kettle

1—人孔 Man hole；2、3—注水泵 Water injection pump；4—变频调速器 Frequency modulating controller；5—减速机 Reducer；6—电机 Motor；7—搅拌轴 Agitator shaft；8—搅拌器 Agitator；9—蒸汽分布器 Steam distributor；10—下封头 Bottom head；11—三层套管 Three layer casing；12—支座 Support；13—筒体 Shell；14—上封头 Top head；15—上人孔 Top man hole

3 合成橡胶 Synthetic rubber

2. 脱气釜/汽提釜主要零部件 Main components of degassing kettle/stripping kettle

1) 三层套管 Three layer casing

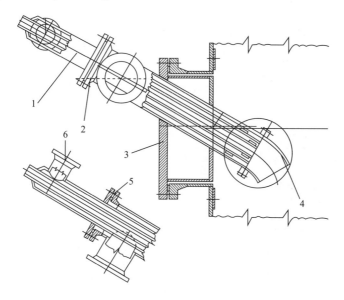

图 3.3.7 三层套管结构图 Structure drawing of three layer casing

1—三层套管 Three layer casing；2—套筒 Sleeve；3—与釜体连接法兰 Connecting flange with kettle body；
4—导流筒 Guide shell；5—螺栓 Bolt；6—法兰 Flange

2) 机械密封 Mechanical seal

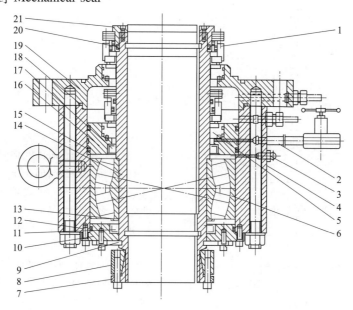

图 3.3.8 机械密封结构图 Structure drawing of mechanical seal

1—机械密封组件 Mechanical seal components；2、5—机封泄漏检测口 Machine sealing leak detection；
3—加脂口 Fat liquoring mouth；4—箱体 Box；6—轴承 Bearing；7、8—锁紧环 Locking ring；9—卡板 Nap-gauge；
10—压盖 Gland；11、17—油封 Oil seal；12、15、18、20—O 密封形圈 O ring；13—挡圈 Hubcap；14—轴套 Sleeve；
16—定位环 Set ring；19—底板 Base plate；21—滤筒 Filter sleeve

3.3.2 塔设备 Column equipment
3.3.2.1 氯甲烷精制塔 Methyl chloride refining column

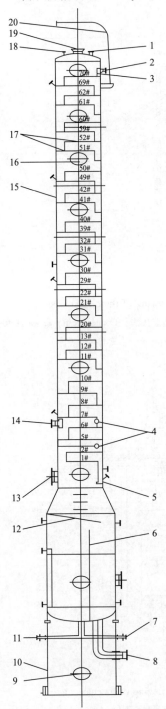

图 3.3.9 氯甲烷精制塔结构图
Structure drawing of methyl chloride refining column

1—真空口 Vacuum line；2—回流入口 Backflow inlet；3—防冲挡板 Impingement baffle；4—液相进料口 Liquid feed；5—受液盘 Fluid receive plate；6—隔板 Partition plate；7—排放口 Drain；8—再沸器进口 Reboiler feed inlet；9—检修口 Access hole；10—裙座 Skirt；11—塔底产品出口 Bottom column products outlet；12—斜隔板 Slant partition plate；13—再沸器返回口 Reboiler return；14—气相进料口 Gas phase feed；15—筒体 Shell；16—人孔 Manhole；17—塔盘 Tray；18—放空口 Vent；19—气体出口 Top column gas outlet；20—塔顶吊柱 Top davit

3.3.2.2 氯甲烷回收塔 Chloride methane recovery column

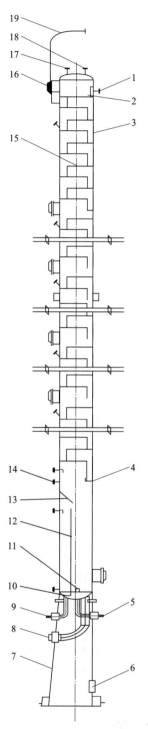

图 3.3.10 氯甲烷回收塔结构图
Structure drawing of chloride methane recovery column

1—进料口 Feed inlet；2—防冲挡板 Impingement baffle；3—筒体 Shell；4—受液盘 Liquid receive plate；5—排放口 Drain；6—检修口 Access hole；7—裙座 Skirt；8—再沸器进口 Reboiler feed inlet；9—塔底产品出口 Bottom column products outlet；10—防涡器 Vortex breaker；11—氮气吹扫口 Nitrogen purging mouth；12—塔釜隔板 Column kettle partition plate；13—斜膈板 Inclined diaphragm plate；14—再沸器返回口 Reboiler return；15—塔盘 Tray；16—人孔 Manhole；17—放空口 Vent；18—气体出口 Top column gas outlet；19—塔顶吊柱 Top davit

3.3.2.3 异丁烯回收塔 Isobutylene recovery column

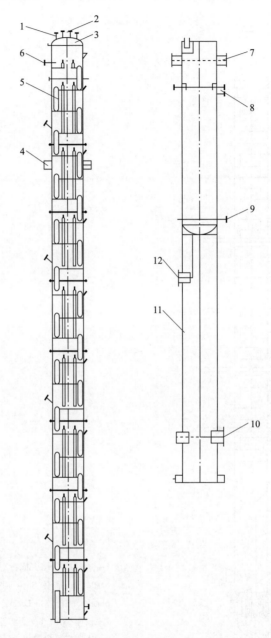

图 3.3.11 异丁烯回收塔结构图 Structure drawing of isobutylene recovery column
1—放空口 Vent；2—气体出口 Top column gas outlet；
3—顶封头 Top head；4—支座 Support；5—塔节 Column section；
6—回流入口 Backflow inlet；7—再沸器返回口 Reboiler return；
8—塔底泵返回口 Bottom column pump return；
9—蒸汽吹扫口 Steam purging mouth；
10—再沸器进口 Reboiler feed inlet；11—裙座 Skirt；
12—塔底产品出口 Bottom column products outlet

3 合成橡胶 Synthetic rubber

3.3.2.4 异丁烯精制塔 Isobutylene refining column

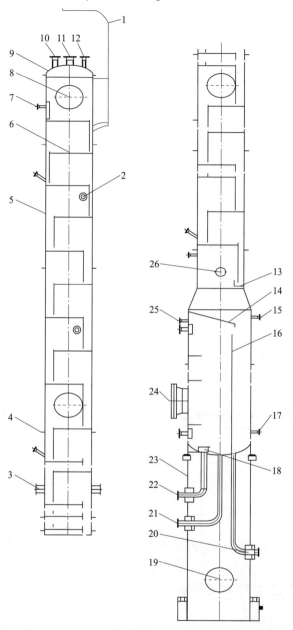

图 3.3.12 异丁烯精制塔结构图 Structure drawing of isobutylene refining column
1—塔顶吊柱 Top davit；2—进料口 Feed inlet；3—导向支架 Guide bracket；4—保温支承圈 Insulation supporting ring；
5—筒体 Shell；6—塔盘 Tray；7—回流入口 Backflow inlet；8、24—人孔 Manhole；9—封头 Head；
10—放空口 Vent；11—气体出口 Top column gas outlet；12—安全阀接口 Safety valve interface；
13—受液盘 Seal pan；14—斜隔板 Slant partition plate；15—塔底泵排气口 Bottom column pump gas outlet；
16—塔釜隔板 Column kettle partition plate；17—氮气吹扫口 Nitrogen purging mouth；
18—防涡器 Vortex breaker；19—检修口 Access hole；20—再沸器进口 Reboiler feed inlet；
21—排放口 Reboiler feed inletUnloading mouth；22—塔底产品出口 Unloading mouth；23—裙座 Skirt；
25—塔底泵回流入口 Bottom column pump backflow inlet；
26—再沸器返回口 Reboiler return

· 261 ·

3.3.2.5 氯甲烷干燥塔 Chloromethane drying column

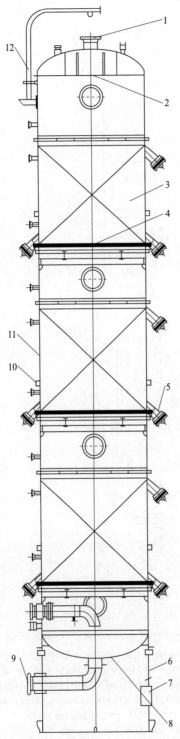

图 3.3.13 氯甲烷干燥塔结构图
Structure drawing of chloromethane drying column

1—气体进口 Gas inlet；2—防冲挡板 Anti-impact damper；3—干燥剂 Desiccant；4—填料 Packing；5—出料口 Discharge outlet；6—裙座 Skirt；7—检修口 Access hole；8—封头 Head；9—气体出口 Gas outlet；10—保温支承圈 Insulation supporting ring；11—筒体 Shell；12—塔顶吊柱 Top davit

3.3.3 换热设备 Heat-exchange equipment

3.3.3.1 乙二醇低温水冷却器 Ethylene glycol low temperature water cooler

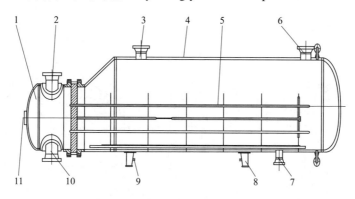

图 3.3.14 乙二醇低温水冷却器结构图 Structure drawing of ethylene glycol low temperature water cooler

1—管箱 Channel；2—低温水出口 Low temperature water outlet；3、6—壳程出口 Shell outlet；
4—筒体 Shell；5—管束 Tube bundle；7—壳程入口 Shell inlet；8—支座 Support；
9—接地板 Ground plate；10—低温水入口 Low temperature water inlet；11—铭牌 Nameplate

3.3.3.2 再生氮气加热器 Regeneration nitrogen gas heater

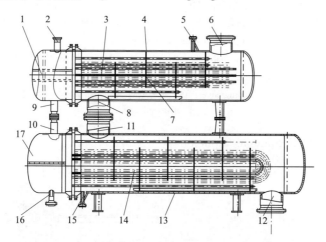

图 3.3.15 再生氮气加热器结构图 Structure drawing of regeneration nitrogen gas heater

1—上管箱 Upper channel；2—蒸汽入口 Steam inlet；3—上管束 Upper tube bundle；4—上筒体 Upper shell；
5—放空口 Vent；6、11—氮气出口 Nitrogen outlet；7—折流板 Baffle plate；8、12—氮气入口 Nitrogen inlet；
9—蒸汽出口 Steam outlet；10—蒸汽入口 Steam inlet；13—下筒体 Lower shell；14—下管束 Lower tube bundle；
15—排放口 Unloading mouth；16—凝液出口 Condensate outlet；17—下管箱 Lower channel

3.3.4 泵 Pump

3.3.4.1 胶粒水输送泵 Colloidal particles water delivery pump

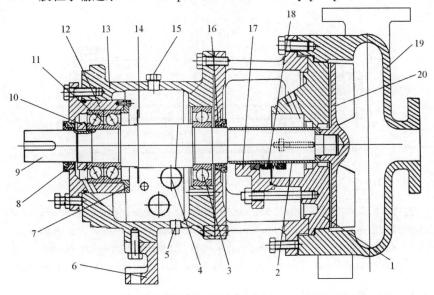

图 3.3.16 胶粒水输送泵结构图 Structure drawing of colloidal particles water pump

1—泵盖 Pump cover；2—轴套 Sleeve；3—径向轴承 Radial bearing；4—视镜 Sight glass；
5—放油口 Oil drain；6—轴承箱支架 Bearing housing support；7—轴承压盖 Bearing gland；
8、16—迷宫密封 Labyrinth seal；9—泵轴 Shaft；10—轴承锁紧螺母 Bearing locknut；11—止推轴承 Thrust bearing；
12—轴承箱压盖 Bearing housing gland；13—轴承箱 Bearing box；14—甩油环 Oil flinger；15—加油口 Oil fill；
17—机封压盖 Mechanical clamping cover；18—机械密封 Mechanical seal；19—泵体 Pump body；
20—叶轮 Wheel

3.3.4.2 配置泵 Configuration pump

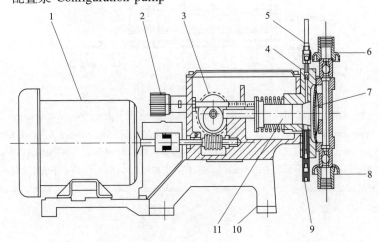

图 3.3.17 配置泵结构图 Structure drawing of configuration pump

1—电机 Motor；2—冲程调节柄 Stroke control handle；3—比例齿轮 Gear ratio assembly；
4—泵头 Pump head；5—自动排放阀 Automatic bleed valve；6—出口阀 Discharge valve；
7—隔膜 Diaphragm；8—入口阀 suction valve；9—可调液压阀 Adjustable hydraulic make-up valve；
10—支座 Support；11—活塞 Piston

3 合成橡胶 Synthetic rubber

3.3.4.3 添加剂输送泵 Additive delivery pump

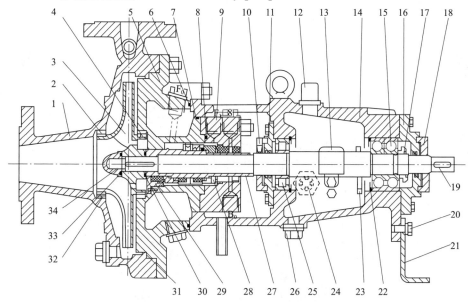

图 3.3.18 添加剂输送泵结构图 Structure drawing of additive delivery pump

1—泵体 Pump body；2—泵体密封环 Pump body sealing ring；3—叶轮 Wheel；
4—叶轮密封环 Impeller labyrinth seal ring；5—泵盖 Pump cover；6、7、8、30、33—O 形密封圈 O sealing ring；
9—密封压盖 Sealing cover；10—防尘盖 Dust shield；11—轴承盖 Bearing cap；12—加油口 Filler；
13—恒位油杯 Constant level oil cup；14—轴承架 Bearing bracket；15、25—轴承 Bearing；
16—轴承盖 Bearing cover；17—轴承螺母 Bearing nut；18—端盖 Bearing cap；
19—轴 Shaft；20、32—螺栓 Screw；21—轴承箱支架 Bearing housing support；22—挡圈 Ring；
23—甩油环 Oil flinger；24—油窗 Oil window；26—丝堵 Plug；
27—轴套 Sleeve；28—机械密封 Mechanical seal；29—静环座 Stationary seal ring seat；
31—平垫片 Flat gasket；34—叶轮螺母 Impeller nut

3.3.5 压缩机 Compressor
3.3.5.1 氯甲烷压缩机 Chloride methane compressor

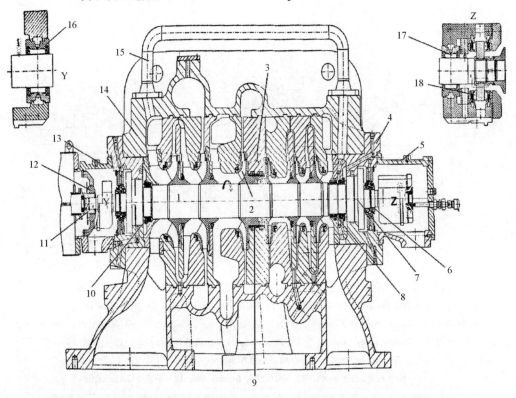

图 3.3.19 氯甲烷压缩机结构图 Structure drawing of chloride methane compressor
1—轴 Shaft；2—叶轮 Wheel；3—隔板 Diaphragm；4—密封箱 Seal box；5、13—轴承箱 Bearing box；
6、12—油封 Oil seal；7—干气密封 Gas seal；8、10—迷宫密封 Labyrinth seal；9—级间密封 Interstage seal；
11—联轴器 Coupling；14—机体 Compressor body；15—平衡管 Balance pipe；
16、17—可倾瓦径向轴承 Tilting pad journal bearing；18—推力轴承 Thrust bearing

3 合成橡胶 Synthetic rubber

3.3.5.2 乙烯压缩机 Ethylene compressor

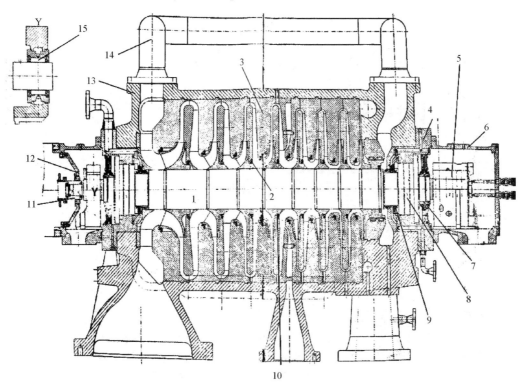

图 3.3.20 乙烯压缩机结构图 Structure drawing of ethylene compressor
1—轴 Shaft；2—叶轮 Wheel；3—隔板 Diaphragm；4—密封箱 Seal box；
5—可倾瓦径向轴承及推力轴承 Tilting pad journal bearing；6—轴承箱 Bearing box；
7、12—油封 Oil seal；8—干气密封 Gas seal；9—迷宫密封 Labyrinth seal；10—级间密封 Interstage seal；
11—联轴器 Coupling；13—机体 Compressor body；14—平衡管 Balance pipe；
15—可倾瓦径向轴承 Tilting pad journal bearing

3.3.5.3 再生氮气压缩机 Regeneration nitrogen compressor

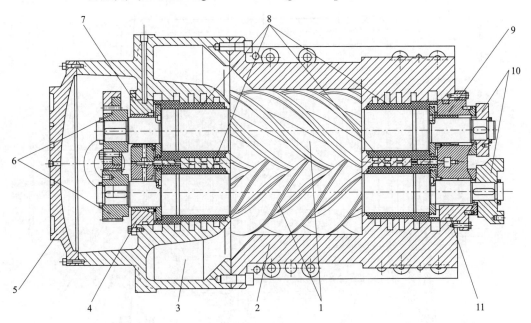

图 3.3.21 再生氮气压缩机结构图 Structure drawing of regeneration nitrogen compressor
1—螺杆 Screw；2—机体 Compressor body；3—机体入口 Inlet casing；4、11—主动螺杆轴承 Drive shaft bearing；
5—端盖 End cover；6—同步齿轮 Synchromesh gear；7、9—从动螺杆轴承 Driven shaft bearing；
8—机械密封 Mechanical seal；10—推力盘 Thrust disc

3 合成橡胶 Synthetic rubber

3.3.6 干燥设备 Drying equipment

3.3.6.1 脱水挤压机 Dehydration extrusion machine

1. 脱水挤压机结构 Structure of dehydration extrusion machine

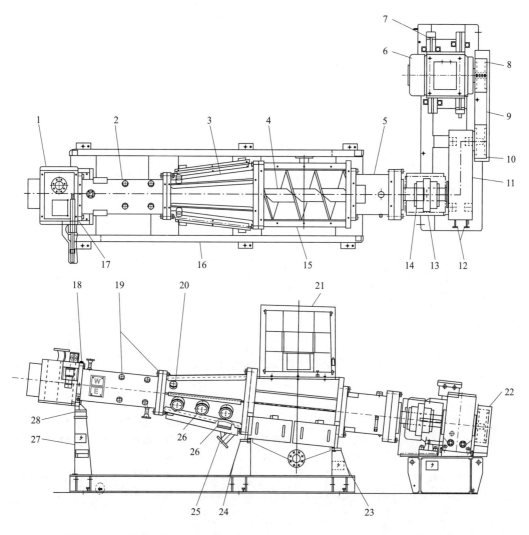

图 3.3.22 脱水挤压机结构图 Structure drawing of dehydration extrusion machine

1—切刀箱 Cutting box；2—圆柱形筒体 Cylindrical barrel；3—过渡筒体 Transition barrel；
4—螺杆 Screw；5—推力轴承箱 Thrust block；6—电机 Motor；7—滑轨 Slide rail；
8、10—皮带轮 Pulley；9—皮带 Belt；11—减速机 Reducer；12—冷却水接管 Cooling water conneting pipe；
13—联轴器防护罩 Coupling guard；14—联轴器 Coupling；15—进料箱 Feed barrel；16—底座 Base；
17—温度计 Temperature gauge；18—压力计 Pressure gauge；19、20—剪切螺栓 Shear screw；
21—进料斗 Feed hopper；22—皮带防护罩 Belt guard；23—进料箱后支座 Feed barrel rear support；
24—进料箱前支座 Feed barrel front support；25—集水管 Water collector；26—筛滤管 Screen pipe；
27—前端筒体支座 Front cylinder support；28—滑动支座 Sliding support

2. 脱水挤压机主要零部件 Main components of dehydration extrusion machine

1) 切刀箱 Cutting box

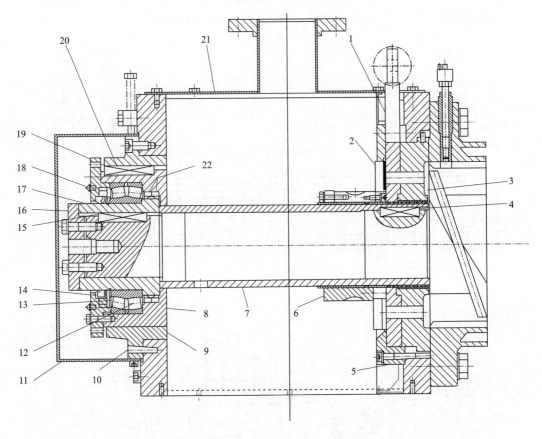

图 3.3.23　切刀箱结构图 Structure drawing of cutting box

1—可调模板 Adjustable die；2—切刀 Cutter；3—固定模板 Fixed die；4、15、20—键 Key；5—压紧圈 Retainer ring；
6—切刀夹具 Knife holder；7—模板定距套 Die spacer sleeve；8—轴承箱 Bearing box；9—外环 Outer ring；
10—锥形销 Taper pin；11—防护罩 Guard；12—滚柱轴承 Roller bearing；13—弹性挡圈 Circlip；14、22—油封 Oil seal；
16—挡板 Retainer；17—内环 Inner ring；18—加油口 Grease nipple；19—压盖 Cover；21—顶盖 Top cover

3 合成橡胶 Synthetic rubber

2）推力轴承 Thrust bearing

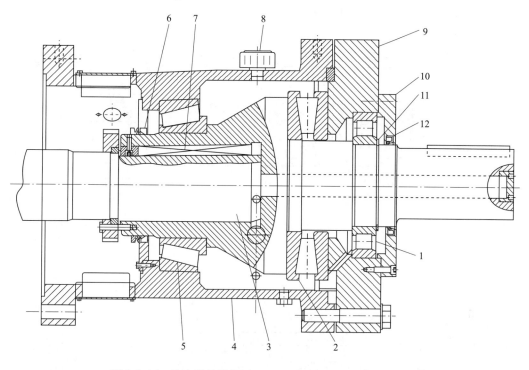

图 3.3.24 推力轴承结构图 Structure drawing of thrust bearing
1—滚柱轴承 Roller bearing；2—推力轴承 Thrust bearing；3—轴 Shaft；
4—推力轴承箱 Thrust block housing；5—圆锥轴承 Conical bearing；
6、12—油封 Oil seal；7—键 Key；8—加油口 Oil filling port；9—轴承箱压盖 Bearing housing gland；
10—轴承压盖 Bearing gland；11—弹性挡圈 Circlip

3.3.6.2 流化床 Fluidized bed
1. 流化床结构 Structure of fluidized bed

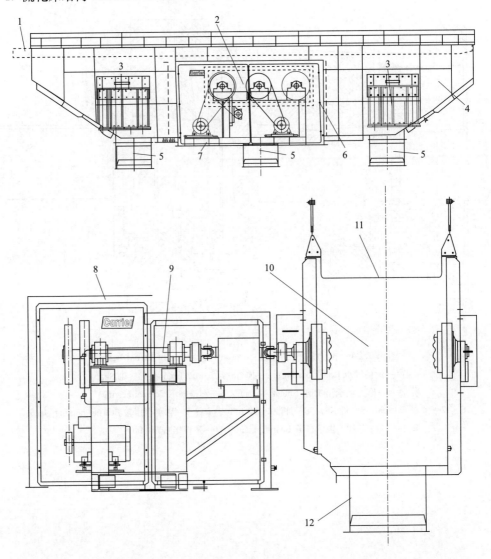

图 3.3.25 流化床结构图 Structure drawing of fluidized bed

1、11—带孔床层 Perforated deck；2、9—传动装置 Drive unit；3—隔振弹簧 Isolation spring；
4、10—气室 Plenum；5、12—连接风筒 Connecting sleeve；6、8—传动装置防护罩 Drive guard；
7—支座 Support

3 合成橡胶 Synthetic rubber

2. 流化床传动装置 Driving medium of fluidized bed

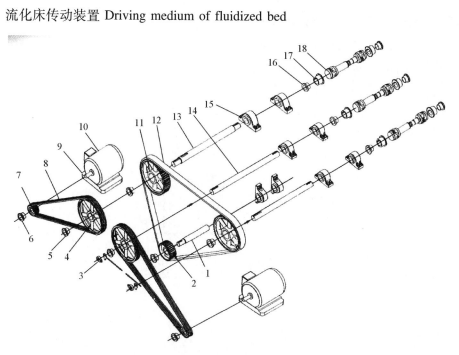

图 3.3.26 流化床传动装置组成图 Structure drawing of driving medium of fluidized bed

1—从动轴 Driven shaft；2、11—齿形皮带轮 Gear belt pulley；3—编码器 Encoder；4—V 形皮带轮 V-belt sheave；5、6、16—锥形轴套 Tapered bushing；7—电机皮带轮 Motor sheave；8—V 形皮带 V-belt；9—电机滑动底座 Motor slide base；10—电机 Motor；12—齿形皮带 Gear belt；13、14—轴 AShaft；15—轴承座 Support bearing；17—对轮 Hub；18—U 形关节驱动轴 U-joint drive shaft

3.3.7 包装设备 Packaging equipment
3.3.7.1 压块机 Briquetting machine
1. 压块机结构 Structure of briquetting machine

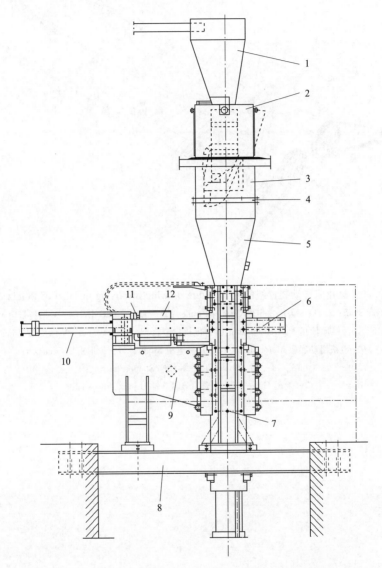

图 3.3.37 压块机结构图 Structure drawing of briquetting machine
1—进料斗 Hopper；2—秤 Weight scale；3、5—进料槽 Discharge chute；4—软连接 Flexible connection；
6—滑轨 Guide rail；7—主缸 Master cylinder unit；8—地下支承梁 Support beams in pit；
9—侧缸支承 Side cylinder support；10—侧缸 Side cylinder unit；11—硅油喷雾管 Spray pipes；
12—侧缸料槽 Side cylinder chute

2. 压块机主要零部件 Main components of briquetting machine

1）侧缸 Side cylinder

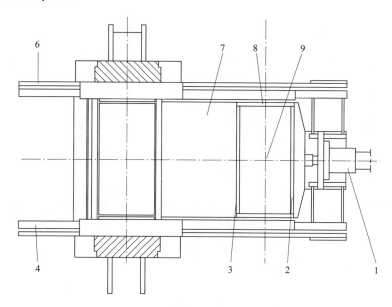

图 3.3.28　侧缸结构图 Structure drawing of side cylinder

1—水平油缸 Leveling cylinder；2、3、8—侧缸料槽密封条 Side cylinder seal strip before and after the chute；
4—左滑轨 Guide rail L. H；5—压块箱顶防磨条 Briquetting box top wear strip；6—右滑轨 Guide rail R. H；
7—侧缸滑块 Side cylinder sliding block；9—侧缸料槽 Side cylinder chute

2）水平油缸 Leveling cylinder

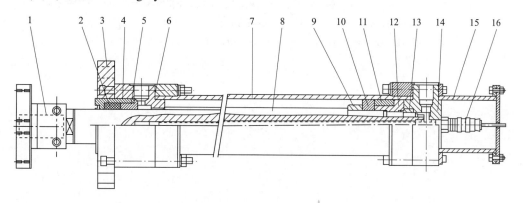

图 3.3.29　水平油缸结构图 Structure drawing of leveling cylinder

1—侧缸头 Side cylinder head；2—活塞杆密封 Rod seal；3—法兰 Flange；4—填料压盖 Stuffing gland；
5—杆轴承 Rod bearing；6—前压盖 Front cover；7—筒体 Barrel；8—活塞杆 Rod；
9、13—缓冲套 Cushion bush；10—前活塞 Front piston；11—活塞 Piston；
12—后活塞 Rear piston；14—后压盖 Rear cover；
15—传感器护罩 Transducer cover；16—传感器 Transducer

3）主缸 Master cylinder

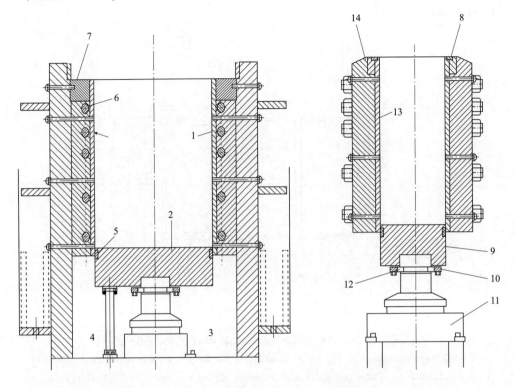

图 3.3.30　主缸结构图 Structure drawing of master cylinder
1—腔体侧衬板 Cavity side liner；2、9—支承台 Platen；3、11—主缸 Master cylinder；4—导向杆 Guide rod；
5—箱顶防磨条 Top box wear strip；6—腔体紧固螺栓 Cavity tie bolt；7—侧衬板 Side liner block；
8—前衬板 Front liner block；10—外环 Outer ring；12—内开口环 Inner split ring；
13—主缸前衬板 Master cylinder front and rear liner；14—主缸后衬板 Rear liner block

3.3.7.2 薄膜包装机 Film packaging machine

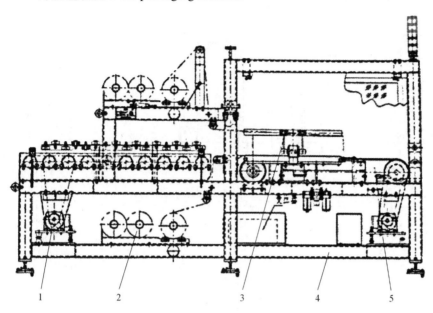

图 3.3.31 薄膜包装机结构图 Structure drawing of film packaging machine
1—胶块输入装置 Gommures input device；2—薄膜供应装置 Film supply device；
3—薄膜热装装置 Film hot charging device；4—机座 Rack；
5—胶块输出装置 Gommures output device

3.3.7.3 金属检测器 Metal detector

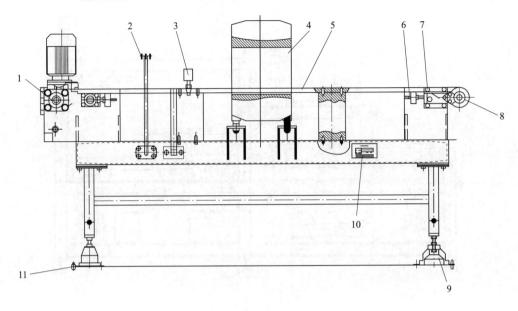

图 3.3.32 金属检测器结构图 Structure drawing of metal detector
1—主动滚筒 Driving roller；2—报警灯 Alarm light；3—反射板 Baffle-board；
4—金属检测器 Metal detector；5—环形皮带 Ring belt；6—张紧杆 Tension rod；
7—张紧滑板 Tension sliding plate；8—从动滚筒 Driven roller；
9—橡胶隔振器 Rubber vibration isolators；10—铭牌 Nameplate；
11—膨胀螺栓 Expansion bolt

3.4 丁腈橡胶 Nitrile rubber

丁腈橡胶是由丁二烯和丙烯腈经乳液聚合法制得的，主要用于制造耐油橡胶制品。

Nitrile rubber obtained by the emulsion polymerization polymerization of butadiene and acrylonitrile is mainly used to make oil resistant rubber products.

生产丁腈橡胶设备主要有聚合釜、丙烯腈汽提塔、凝聚槽、高压鼓风机、脱水挤压机和自动包装设备等。

The Main equipments for producing the nitrile butadiene rubber contain polymeric kettle, acrylonitrile stripper, coagulation, high pressure blower, dehydration extrusion machine and automatic packaging equipment.

3.4.1 聚合釜 Polymeric kettle

3.4.1.1 釜体 Kettle body

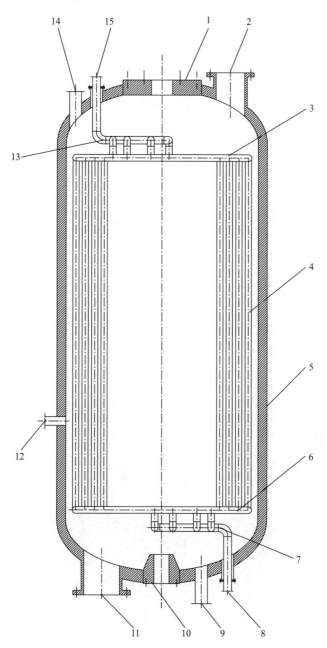

图 3.4.1　聚合釜釜体结构图 Structure drawing of polymeric kettle body

1—凸缘 Flange；2、11—人孔 Manhole；3—上环管 Upper ring pipe；4—冷却管束 Cooling tube bundle；
5—筒体 Shell；6—下环管 Lower ring pipe；7—盐水入口分布管 Salt water inlet distribution pipe；
8—盐水入口 Salt water inlet；9—下部物料口 Bottom feed outlet；10—底部轴承座 Bottom bearing seat；
12—温度计接口 Thermometer mouth；13—盐水出口分布管 Salt water outlet distribution pipe；
14—上部物料口 Upper feed mouth；15—盐水出口 Salt water outlet

3.4.1.2 搅拌系统 Mixing system

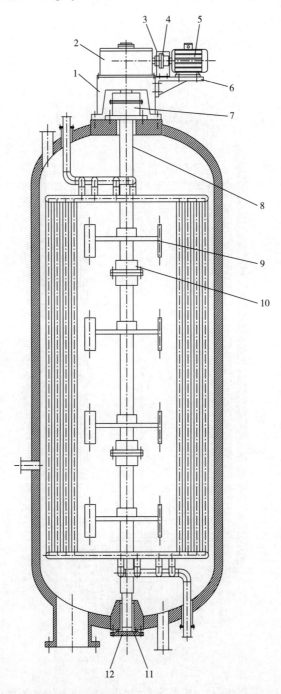

图 3.4.2 聚合釜搅拌系统结构图 Structure drawing of polymeric kettle mixing system
1—减速机支架 Reducer supporting bracket；2—减速机 Reducer；3—联轴器防护罩 Coupling shield；
4、10—联轴器 Coupling；5—电机 Drive motor；6—电机支架 Motor supporting bracket；
7—轴封 Shaft seal assembly；8—搅拌轴 Mixing shaft；9—搅拌桨 Agitating paddle；
11—底部轴套 Bottom shaft sleeve；12—轴套压盖 Shaft sleeve gland

3.4.1.3 轴封 Shaft seal

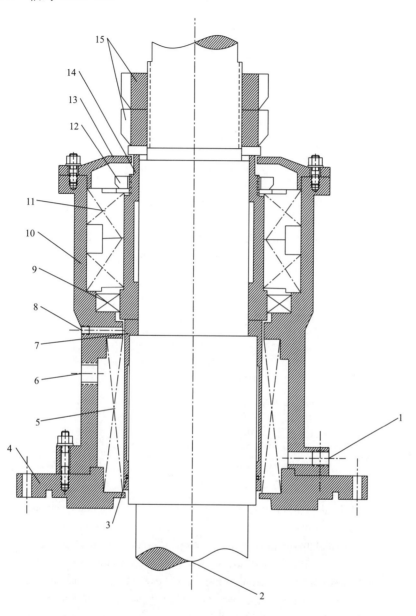

图 3.4.3 轴封结构图 Structure drawing of shaft seal
1—隔离液出口 Spacer fluid outlet；2—轴 Shaft；3—O 形密封圈 O sealing ring；4—下盖板 Bottom cover plate；
5—双端面机械密封 Dual-face mechanical seal；6—隔离液入口 Spacer fluid inlet；
7—密封轴套 Seal sleeve；8—泄漏检查口 Leakage mouth；9—骨架油封 Framework oil seal；
10—轴承箱 Bearing box；11—轴承 Bearing；12—轴承螺母 Bearing nut；13—轴承压盖 Bearing gland；
14—轴承套 Bearing sleeve；15—锁紧螺母 Locking nut

3.4.2 丙烯腈汽提塔 Acrylonitrile stripper

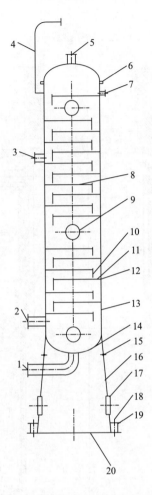

图 3.4.4　丙烯腈汽提塔结构图 Structure drawing of acrylonitrile stripper
1—液相出口 Liquid phase outlet；2—气相入口 Gas phase inlet；3—进料口 Feed inlet；
4—塔顶吊柱 Top davit；5—气相出口 Gas phase outlet；6—吊耳 Lifting lug；
7—回流口 Reflux inlet；8—塔盘 Tray；9—人孔 Manhole；
10—降液板 Down-flow plate；11—受液盘 Liquid tray；12—支承圈 Support ring；
13—筒体 Shell；14—封头 Head；15—通气孔 Venthole；16—裙座 Skirt；
17—检修口 Access hole；18—地脚螺栓 Anchor bolt；19—地脚座 Foundation seat；
20—基础环板 Foundation ring plate

3 合成橡胶 Synthetic rubber

3.4.3 汽提塔顶冷凝器 Stripping top column condenser

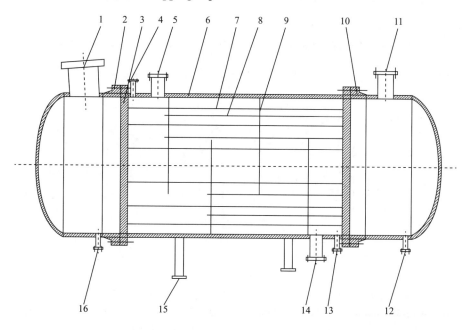

图 3.4.5 汽提塔顶冷凝器结构图 Structure drawing of stripping top column condenser
1—物料入口 Material inlet；2—左浮头 Left floating head；3—管板法兰 Tube plate flange；
4—排气口 Exhaust port；5—循环水出口 Circulating water outlet；6—筒体 Shell；
7—管束 Tube bundles；8—定距管 Spacer tube；9—折流板 Baffle plate；
10—右浮头 Right floating head；11—物料出口 Feed outlet；12—乙腈水出口 Acetonitrile-water outlet；
13、16—排放口 Evacuation mouth；14—循环水入口 Circulating water inlet；15—支座 Support

3.4.4 机泵 Blower and pump

3.4.4.1 丁二烯水环真空泵 Butadiene water ring vacuum pump

1. 丁二烯水环真空泵外形 Shape of the butadiene water ring vacuum pump

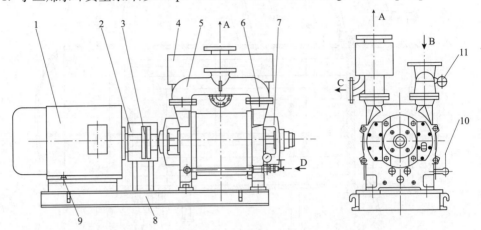

图 3.4.6 丁二烯水环真空泵外形图 Structure drawing of butadiene water ring vacuum pump

1—电机 Motor；2—联轴器防护罩 Coupling shield；3—联轴器 Coupling；4—气水分离器 Gas-water separator；
5—进气管 Intake-tube；6—密封垫片 Seal gasket；7—水环真空泵 Water ring vacuum pump；
8—底座 Base；9—电机地脚螺栓 Motor anchor bolt；10、11—真空表 Vacuum meter
A—排气口 Exhaust port；B—进气口 Air inlet；C—排水口 Outfall；D—进水口 Working fluid mouth

2. 丁二烯水环真空泵结构 Structure of the butadiene water ring vacuum pump

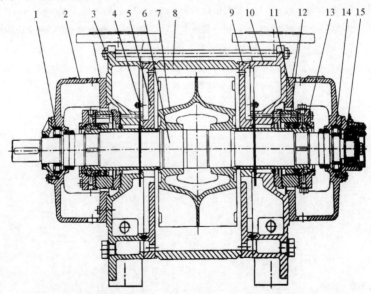

图 3.4.7 丁二烯水环真空泵结构图 Structure drawing of butadiene water ring vacuum pump

1—前轴承 Fore bearing；2—前部轴承箱 Fore bearing box；3—前侧盖 Fore side cover；
4—柔性阀板 Flexible valve plate；5—前分配板 Fore distribution board；6—泵轴 Shaft；7—叶轮 Wheel；
8—泵体 Pump body；9—后分配板 Rear distribution board；10—后侧盖 Rear side cover；
11—后部轴承箱 Rear bearing box；12—机械密封函体 Mechanical seal body；
13—机械密封组件 Mechanical seal component；14—后轴承 Rear bearing；15—止推轴承 Thrust bearing

3 合成橡胶 Synthetic rubber

3.4.4.2 高压鼓风机 Pressure blower

1. 高压鼓风机外形 Shape of the pressure blower

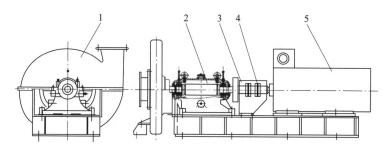

图 3.4.8 高压鼓风机外形图 Structure drawing of pressure blower
1—壳体 Shell；2—轴承箱 Bearing box；3—联轴器防护罩 Coupling shield；
4—联轴器 Coupling；5—电机 Motor

2. 高压鼓风机结构 Structure drawing of pressure blower

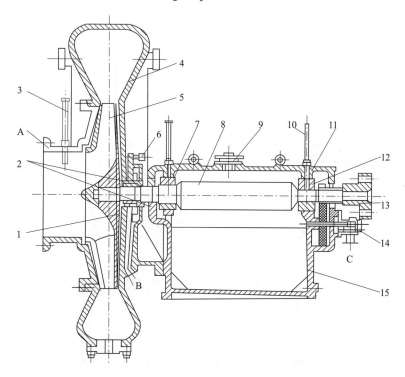

图 3.4.9 高压鼓风机结构图 Structure drawing of pressure blower
1—排气口 Exhaust port；2—轴封 Shaft seal；3、10—温度计 Thermometer；4—壳体 Shell；
5—叶轮 Wheel；6—油杯 Oil ring；7—止推轴承 Thrust bearing；8—轴 Shaft；
9—呼吸阀 Breather valve；11—轴承箱 Bearing box；12—径向轴承 Rodial bearing；
13—联轴器 Coupling；14—主油泵 Main oil pump；15—底座 Base
A—进气口 Air inlet；B—平衡口 Balance hole；
C—润滑油泵出口 Lubrication pump outlet

3.4.5 分离设备 Parting device
3.4.5.1 凝聚槽 Coagulating bath

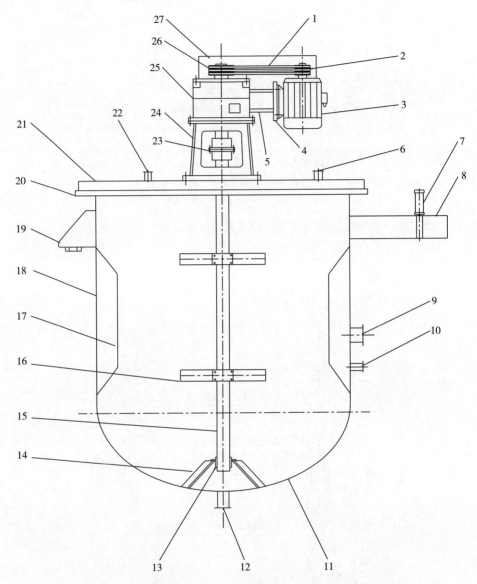

图 3.4.10 凝聚槽结构图 Structure drawing of coagulating bath

1—皮带 Belt；2—电机皮带轮 Motor belt pulley；3—电机 Motor；4—电机底座板 Motor base plate；
5—电机支架 Motor bearing support；6—胶乳入口 Latex inlet；7—出料控制口 Discharge control gate；
8—出料槽 Discharge chute；9—蒸汽入口 Steam inlet；10—热电偶接口 Thermocouple jack；
11—封头 Head；12—排污口 Drain outlet；13—轴套 Bush；14—底部轴套支架 Bottom shaft support bracket；
15—搅拌轴 Shaft；16—搅拌桨 Agitating paddle；17—折流板 Baffle plate；18—筒体 Shell；
19—支座 Support；20—上盖板 Upper cover plate；21—搅拌支承型钢 Agitating bearing steel；
22—补水口 Replenishment mouth；23—联轴器 Coupling；24—支架 Support；
25—减速机 Reduction drive；26—减速机皮带轮 Reduction drive belt pulley；27—皮带防护罩 Belt guard

3.4.5.2 洗胶釜 Rubber washing kettle

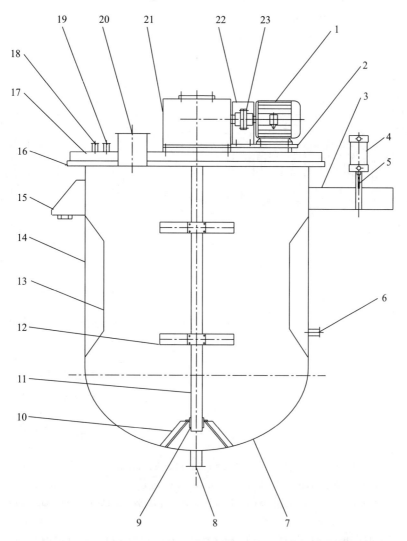

图 3.4.11 洗胶釜结构图 Structure drawing of rubber washing kettle

1—电机 Motor；2—电机支座 Motor bearing support；3—出料槽 Discharge chute；
4—闸板气缸 Wedge disc air cylinder；5—出料槽闸板 Discharge chute wedge disc；
6—热电偶接口 Thermocouple jack；7—封头 Head；8—排污口 Drain outlet；
9—底部轴套 Bottom shaft sleeve；10—底部轴套支架 Bottom shaft support bracket；
11—搅拌轴 Shaft；12—搅拌桨 Agitating paddle；13—折流板 Baffle plate；14—筒体 Shell；
15—支座 Support；16—上盖板 Upper cover plate；
17—搅拌支承型钢 Agitating bearing steel；18—补水口 Replenishment mouth；
19—洗涤水入口 Washings inlet；20—物料入口 Material inlet；21—减速机 Reduction drive；
22—联轴器防护罩 Coupling shield；23—联轴器 Coupling

3.4.6 脱水挤压机 Dehydration extrusion machine

3.4.6.1 脱水挤压机外形 Shape of the dehydration extrusion machine

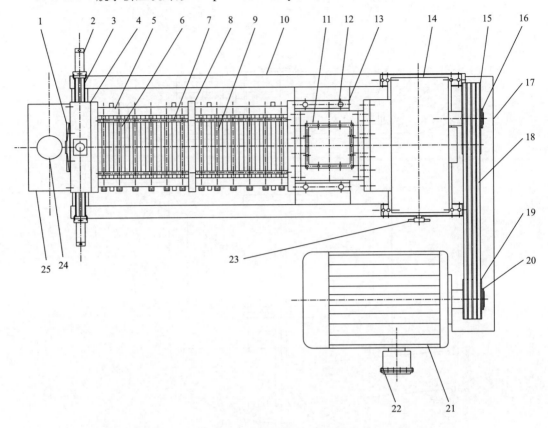

图 3.4.12 脱水挤压机外形图 Shape of the dehydration extrusion machine

1—机头切刀 Handpiece cutter；2—模头气缸 Die head air cylinder；3—气缸支架 Air cylinder stand；
4—模头 Die head；5—横梁 Beam；6—横向夹紧压板 Horizontal clamping pressure plate；
7—轴向拉杆 Axial pull rod；8—机架 Rack；9—横向夹紧螺栓 Horizontal clamping pressure bolt；
10—底座 Base；11—进料斗 Feed hopper flange；12—螺栓 Feed hopper foundation bolt；
13—进料斗法兰 Feed hopper flange；14—齿轮箱 Gear box；15、19—皮带轮 Input shaft belt pulley；
16、20—锁紧螺母 belt pulley cap；17—皮带防护罩 Belt guard；18—皮带 Belt；
21—电机 Motor；22—电机接线盒 Motor terminal box；
23—润滑油控制板 Lubricating oil control panel；
24—顶部视窗 Top window；25—出料斗 Outlet hopper；

3　合成橡胶 Synthetic rubber

3.4.6.2　脱水挤压机结构 Structure drawing of dehydration extrusion machine

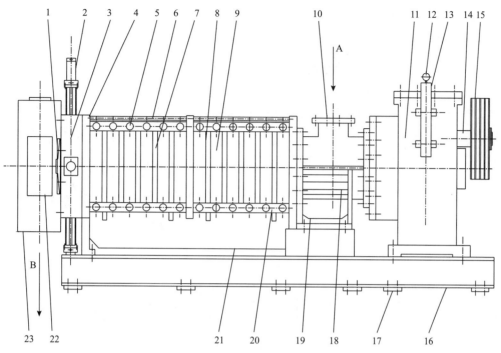

图 3.4.13　脱水挤压机结构图 Structure drawing of dehydration extrusion machine

1—机头切刀 Handpiece cutter；2—模头气缸 Die head air cylinder；3—模头 Die head；
4—机架 Rack；5—横向夹紧螺栓 Horizontal clamping pressure bolt；6—轴向拉杆 Axial pull rod；
7—横向夹紧压板 Horizontal clamping pressure plate；8—笼条固定板 Cage fixed plate；
9—笼条 Cage；10—进料斗 Feed hopper；11—齿轮箱 Gear box；12—油压表 Oil pressure gauge；
13—润滑油控制板 Lubricating oil control panel；14—齿轮箱输入轴 Gear box input shaft；
15—皮带轮 Belt pulley；16—底座 Base；17—地脚螺栓 Anchor bolt；
18—进料斗脱水笼条 Feed hopper dehydration cage；19—进料斗接水槽 Feed hopper water catcher；
20—横梁 Beam；21—机身接水槽 Frame water catcher；
22—出料视窗 Discharge Window；23—出料斗 Outlet hopper；
A—进料口 Inlet hopper；B—出料口 Discharge outlet

3.4.7 包装设备 Forming packaging equipment
3.4.7.1 压块机 Cuber

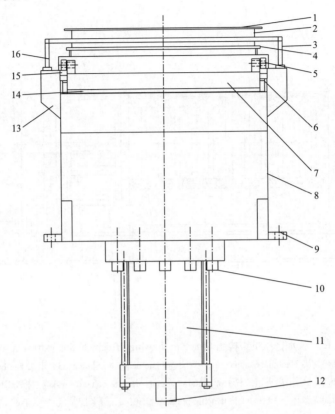

图 3.4.14 压块机结构图 Structure drawing of cuber

1—下料口 Discharge outlet flange；2—下料斗 Hopper；
3—下料斗支架 Hopper support；4—入料口 Inlet hopper；
5—上盖轮滚 Upper cover wheel rolling；6—上盖滑块 Upper cover sliding block；
7—上盖主体 Upper cover main body；8—压块机壳体 Cuber shell；
9—底座 Base；10—主油缸联接螺栓 Main oil cylinder connecting bolt；
11—主油缸 Main oil cylinder；12—主油缸传感器 Main oil cylinder transducer；
13—轮滚导轨支架 Wheel rolling slideway support；
14—上盖主体滑板 Upper cover main body slide plate；
15—轮滚导轨 Wheel rolling slideway；16—防护罩 Shield

3.4.7.2 薄膜包装机 Film packing machine

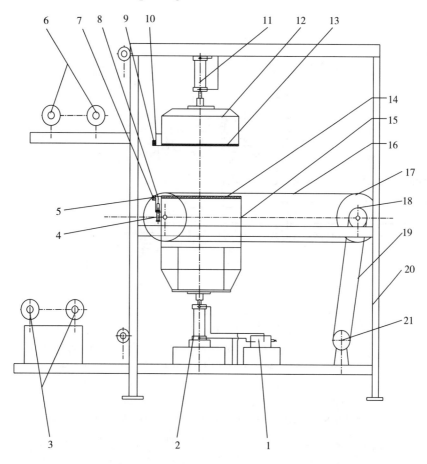

图 3.4.15 薄膜包装机结构图 Structure drawing of film packing machine

1—电磁控制阀 Solenoid electric valve；2—下部气缸 Bottom air cylinder；
3—下部薄膜轴 Lower film shaft；4—切刀气缸 Cutter air cylinder；5—刀架 Knife rest；
6—上部薄膜轴 Upper film shaft；7—直形段密封衬条 Straight sealing strip；
8—切刀片 Cutting blade；9—直形段加热器 Straight heating element；
10—刀槽 Knife slot；11—上部气缸 Upper air cylinder；12—上托架 Upper bracket；
13—U形段加热器 U heating element；14—U形段密封衬条 U sealing strip；
15—下托架 Bottom bracket；16—输送皮带 Conveying belt；17—皮带轮 Belt pulley；
18—链轮 Chain wheel；19—链条 Chain；
20—机体框架 Body frame；21—电机 Motor

3.4.7.3 自动装袋机 Automatic bagging machine

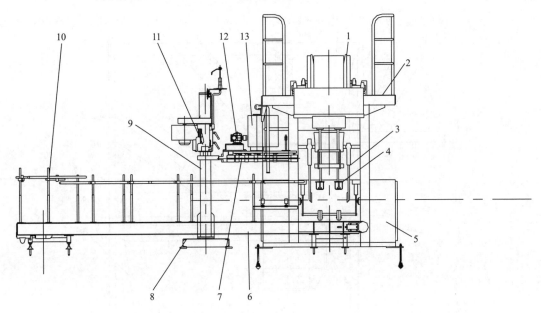

图 3.4.16 自动装袋机结构图 Structure drawing of automatic bagging machine
1—垂直输送装置 Vertical conveying equipment；2—机顶平台 Roof platform；
3—取袋装置 Fetching bag equipment；4—扶袋装置 Taking bag equipment；
5—电控箱 Electric control element box；6—立袋输送机 Vertical bag conveyor；
7—夹口整形皮带 Clip reshaping belt；8—缝口机底座 Seam machine base；
9—缝口机立柱 Seam machine column；10—倒袋调节架 Pouring bag adjusting frame；
11—缝口机 Seam machine；
12—电机 Motor；
13—操作控制面板 Operation control panel

3.5 SBS 橡胶 SBS(Styrene-butadiene-styrene) rubber

SBS 橡胶是以苯乙烯、丁二烯为单体，通过阴离子聚合而成的嵌段共聚物，主要生产设备主要有聚合釜、凝聚釜、后处理挤压脱水机和膨胀干燥机等。

SBS is the block copolymer which is formed by the anionic polymerization based on styrene and butadiene monomer, whose main production equipment contain contains polymerizing pot, condensation kettle, post-processing squeeze dewatering machine, expansion dryer and so on.

3 合成橡胶 Synthetic rubber

3.5.1 聚合釜 Polymerizing pot

3.5.2 闪蒸罐 Flash drum

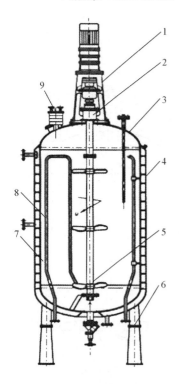

图 3.5.1 聚合釜结构图
Structure drawing of polymerizing pot
1—传动装置 Transmission device；
2—机械密封 Mechanical seal；
3—封头 Head；
4—冷却夹套 Cooling jacket；
5—搅拌轴 Stirring paddle；
6—支座 Support；
7—釜体 Kettle；
8—内冷管 Inner cooling tube；
9—隔离罐 Isolation tank

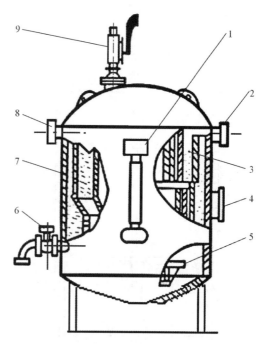

图 3.5.2 闪蒸罐结构图
Structure drawing of flash drum
1—液位传感器 Liquid level sensor；
2—进料口 Feed nozzle；
3—调压装置 Regulator device；
4—人孔 Manhole；
5—冷却盘管 Cooling coil；
6—排出口 Outlet；
7—冷却管束 Cooling tube bundle；
8—氮气接口 Nitrogen gas nozzle；
9—安全阀 Safety valve

3.5.3 凝聚釜 Condensation kettle

3.5.4 洗涤水罐/洗胶罐 Washing tank/ Washing glue tank

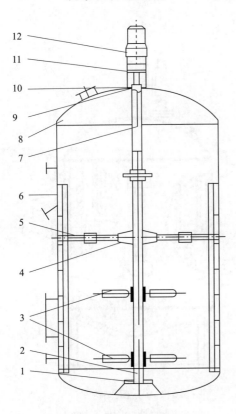

图 3.5.3 凝聚釜结构图
Structure drawing of condensation kettle

1—底轴承座 Bottom bearing seat；
2—下轴 Lower shaft；
3—搅拌桨 Stirring paddle；
4—中间轴承座 Middle bearing seat；
5—拉杆 Tie rod；6—釜体 Kettle；
7—上轴 Upper shaft；8—封头 Head；
9—凸缘 Flange；
10—机械密封 Mechanical seal；
11—轴承座 Bearing seat；
12—传动机构 Transmission mechanism

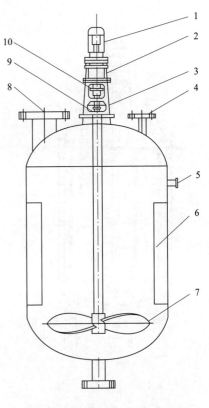

图 3.5.4 洗涤水罐/洗胶罐结构图
Structure drawing of washing tank/
washing glue tank

1—电机 Motor；2—减速机 Reducer；
3—支架 Support；
4—加入口 Feed mouth；
5—出料口 Discharge port；
6—折流板 Baffle plate；
7—搅拌桨 Stirring paddle；
8—进料口 Feed inlet；
9—轴承座 Bearing seat；
10—联轴器 Coupling

3.5.5 泵 Pump

3.5.5.1 胶液泵/出料泵 Glue pump / Discharge pump

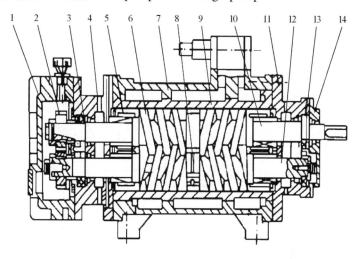

图 3.5.5 胶液泵/出料泵结构图 Structure drawing of glue pump and discharge pump
1—齿轮箱 Gear box；2—齿轮 Gear；3、13—滚动轴承 Rolling bearing；4—后支架 Back bracket；
5—密封 Seal；6—螺杆 Screw；7—泵体 Pump body；8—调节螺栓 Adjusting bolt；
9—衬套 Sleeve；10—主动轴 Driving shaft；11—前支架 Front support；12—从动轴 Driven shaft；
14—压盖 Gland

3.5.5.2 丁二烯泵/苯乙烯泵/热水泵 Butadiene pump/ Styrene pump/ Hot water pump

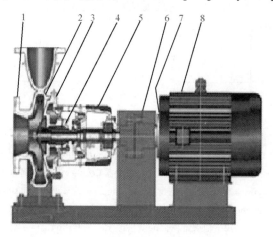

图 3.5.6 丁二烯泵/苯乙烯泵/热水泵结构图
Structure drawing of butadiene pump/ styrene pump/ hot water pump
1—泵体 Pump body；2—叶轮 Wheel；3—冷却腔 Cooling chamber；
4—机械密封 Mechanical seal；5—轴承箱 Bearing box；
6—变速箱 Reducer；7—联轴器 Coupling；8—电机 Motor

3.5.5.3 引发剂/防老剂/终止剂加料泵 Initiator/ Antiager / Terminator charge pump

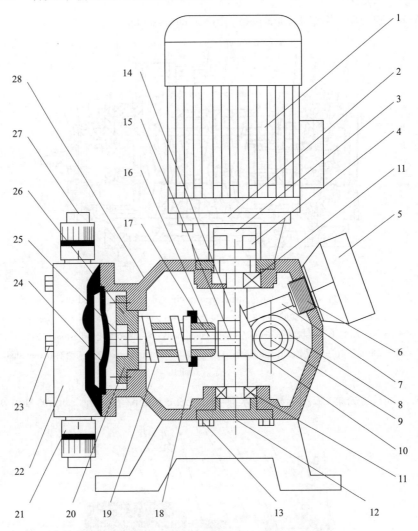

图 3.5.7 引发剂/防老剂/终止剂加料泵结构图
Structure drawing of initiator/ antiager / terminator charge pump

1—电机 Motor；2—电机座 Motor seat；3—联轴器 Coupling；4—缓冲块 Damper block；
5—调节手轮 Adjusting handle；6—密封圈 Seal ring；7—调节顶杆 Adjusting knockout pin；
8—偏心轮 Eccentric shaft bearing；9—蜗轮轴 Main shaft；10—蜗轮 Worm wheel；
11—轴承 Bearing；12—轴承盖 Bearing cover；13—螺栓 Bolt；14—蜗杆 Worm rod；
15—顶杆 Knockout pin；16—导向轴套 Guide sleeve；17—销钉 Pin；
18—弹簧座 Spring seat；19—弹簧 Spring；20—密封圈 Seal ring；
21—进水阀 Inlet valve assembly；22—泵头 Pump head；
23—端盖螺栓 End cover bolt；24—膜片 Membrane；25—油封 Oil seal；
26—油封座 Oil seal seat；27—出水阀 Outlet valve assembly；
28—泵体 Pump body

3 合成橡胶 Synthetic rubber

3.5.5.4 加料泵/环烷油泵 Antiadherent/ Charge pump/ Naphthenic oil pump

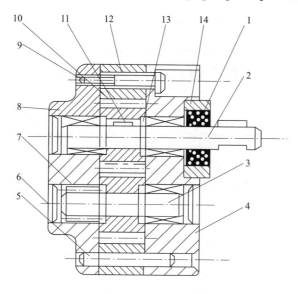

图 3.5.8 加料泵/环烷油泵结构图

Structure drawing of antiadherent/ charge pump/ naphthenic oil pump

1—油封 Oil seal；2—传动轴 Transmission shaft；3—短轴 Minor axis；4—前端盖 Front end cover；
5—圆柱销 Cylindrical pin；6—压盖 Gland；7—轴承 Bearing；8—后端盖 Rear end cover；
9—端盖螺栓 End cover screw；10—齿轮 Gear；11—平键 Flat lsey；12—泵体 Pump body；
13—弹性挡圈 Flexible retaining rings for shaft；14—油封座 Oil seal seat

3.5.5.5 真空泵 Vacuum pump

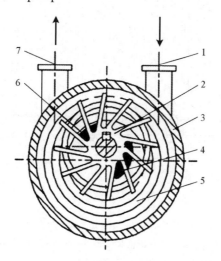

图 3.5.9 真空泵结构图 Structure drawing of vacuum pump

1—吸气口 Air entry；2—叶轮 Wheel；3—泵体 Pump body；4—吸气孔 Air intake；
5—液环 Liquid ring；6—排气孔 Exhaust vent；7—排气口 Air outlet

3.5.6 换热设备 Heat-exchange equipment

3.5.6.1 热风加热器 Hot blast heater

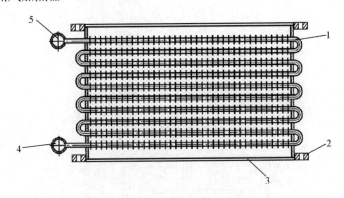

图 3.5.10 热风加热器结构图 Structure drawing of hot blast heater
1—翅片管 Finned tube；2—固定基座 Permanent seat；
3—加热器壳体 Heat exchanger shell；
4—出风口 Outlet；5—进风口 Inlet

3.5.6.2 润滑油冷却器 Lubricating oil cooler

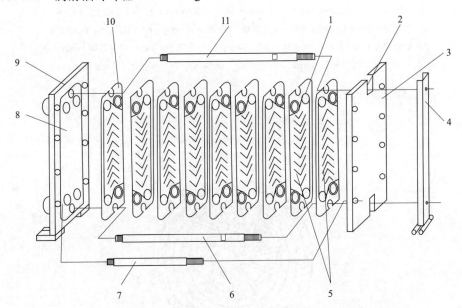

图 3.5.11 润滑油冷却器结构图 Structure drawing of lubricating oil cooler
1—密封垫 Sealing gasket；2—滚轮 Roller；3—活动夹紧板 Mobile clamp plate；
4—支柱 Column；5、10—换热板片 Sheet bar；6—下导杆 Bottom guide bar；
7—夹紧螺杆 Clamping bolt；8—橡胶板 Rubber plate；
9—固定夹紧板 Fixing clamp plate；
11—上导杆 Upper guide bar

3 合成橡胶 Synthetic rubber

3.5.7 振动输送设备 Vibrating conveyer equipment

3.5.7.1 振动脱水筛 Vibrating-dewatering screen

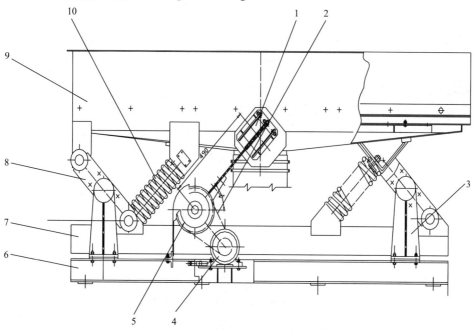

图 3.5.12 振动脱水筛结构图 Structure drawing of vibrating-dewatering screen
1—橡胶弹簧 Rubber spring；2—三角带 Triangular belt；3—支座 Support；4—电机 Motor；
5—皮带轮 Rotor；6—机座 Engine base；7—平衡架 Balancing stand；
8—导向杆 Guide rod；9—槽体 Cell body；10—螺旋弹簧 Spiral spring

3.5.7.2 水平输送机 Horizontal screw conveyer

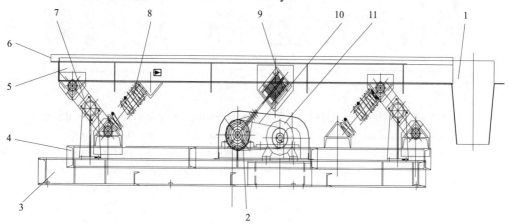

图 3.5.13 水平输送机结构图 Structure drawing of horizontal screw conveyer
1—出料口 Discharge port；2—传动装置 Rotor assembly；3—机座 Engine base；
4—平衡架 Balancing stand；5—输送槽 Conveyor chute；6—加强板 Stiffening plate；
7—导向杆 Guide rod；8—螺旋弹簧 Spiral spring；9—橡胶弹簧 Rubber spring；
10—三角带 Triangular belt；11—电机 Motor

3.5.8 干燥设备 Dehydration drying equipment
3.5.8.1 挤压脱水/膨胀干燥机 Extruding-desiccation / Expansion dryer

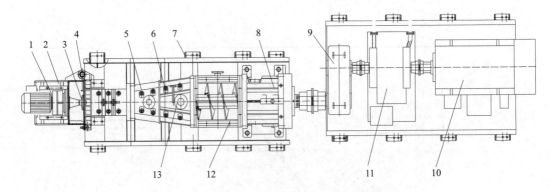

图 3.5.14 挤压脱水/膨胀干燥机结构图
Structure drawing of extruding-desiccation / expansion dryer
1、9—减速机 Reducer；2、7—机座 Engine base；3—切刀 Cutter；
4—机头 Handpiece hinge；5—筒体 Shell；6—剪切螺栓 Shearing screw；
8—轴承座 Bearing seat；10—电机 Motor；11—液力耦合器 Hydraulic couplers；
12—底筛 Bottom screen；13—管筛 Tube screen

3.5.8.2 双螺旋搅拌干燥机 Double helix mixing dryer

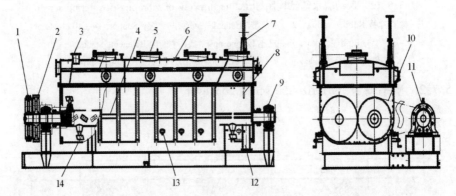

图 3.5.15 双螺旋搅拌干燥机结构图 Structure drawing of double helix mixing dryer
1—链轮 Chain wheel；2—齿轮 Gear；3—进料口 Feed inlet；4—搅拌轴 Stirring shaft；
5—人孔 Manhole；6—上盖 Upper cover；7—调节杆 Regulating stem；8—挡板 Baffle；
9—轴承座 Bearing seat；10—链条 Chain；11—电机 Motor reducer；
12—出料口 Feed outlet；13—辅助风口 Auxiliary tuyere；14—搅拌桨叶 Paddle

4 合成纤维 Synthetic fiber

4.1 聚酯 Polyester

4.1.1 反应设备 Reactor equipment

4.1.1.1 酯化反应器 Esterification stage

第一、第二酯化反应器都是全夹套带搅拌的立式反应器，反应器内有液相热媒加热，夹套内是气相热媒保温。搅拌器是由上下两层多个叶片组成的推进式搅拌器，物料通过搅拌器混合搅拌，进行酯化反应。

The first and second esterification reactors are all jacketed and stirred vertical reactor, which contain heated liquid phase heat medium and gas phase heat medium preservation in the jacket. The agitator is consist of the push-blender which is made of the multi-blades by upper and lower layers. The materials are mixed and stirred by an agitator for esterification.

1. 酯化反应器结构 Structure esterification stage

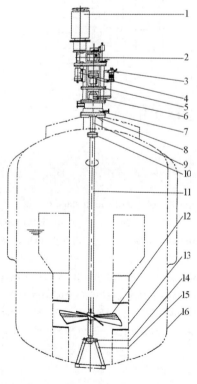

图 4.1.1 酯化反应器结构图
Structure drawing of esterification stage

1—电机 Motor；2—减速机 Speed reducer；3—密封液罐 Storage vessel；4—提升旋转装置 Lift and swiveling device；5—搅拌架 Agitator lantern；6—机械密封 Mechanical seal；7—搅拌基座 Agitator base；8—连接盘 Joint tray；9—连接轴 Mian shaft；10—联轴器 Flange coupling；11—搅拌轴 Agitator shaft；12—搅拌桨叶 Impeller；13—挡板 Plate；14—轴承 Bearing；15—轴承支架 Bearing support；16—筒体 Shell

2. 酯化反应器筒体 Esterification stage casing

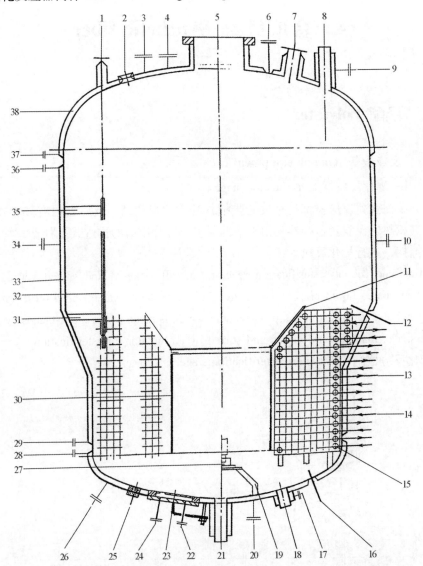

图 4.1.2　酯化反应器筒体结构图 Structure drawing of esterification stage casing

1—液位计接口 Liquid level gauge interface；2—视镜 Sight glass；3、10、24、26—气相热媒入口 HTM vapour inlet；
4、28、34、36—气相热媒出口 HTM vent；5—搅拌器接口 Agitator interface；6—乙二醇回流口 Glycol reflux；
7—物料入口 Material inlet；8—气相出口 Vapour outlet；9、17、20、22、25、29、37—接管 Connecting pipe；
11、30—挡板 30—Plate；12、16—温度计接口 Thermometer；13—液相热媒出口 HTM liquid outlet；
14—液相热媒入口 HTM liquid inlet；15—热媒盘管 HTM pipe；18—物料回流口 Product reflux；
19—轴承支架 Steady bearing support；21—物料出口 Product outlet；23—人孔 Manhole；27—轴承 Bearing；
31、35—液位计支架 Level transmitter support；32—液位计 Level transmitter；33—筒体 Shell；38—封头 Head

4.1.1.2　预缩聚反应器 Precondensation stage

1. 第一预缩聚反应器 Precondensation stage I

第一预缩聚反应器是一个全夹套的立式反应器，反应器内有液相热媒加热盘管加热，夹套内是气相热媒保温，反应器不需搅拌，而是靠物料自身的沸腾进行混合，反应器内酯化反应和缩聚反应是同时进行的。

The precondensation stage Ⅰ is a full-jacketed vertical reactor, which contains the liquid phrace heat medium heating coil and heating tube and the vapor heat medium heat preservation in the jacket. The reactor is not need being stired, but being mixed by the material itself boiling. The lactonization and polycondensation of the stage are carried out simultaneously.

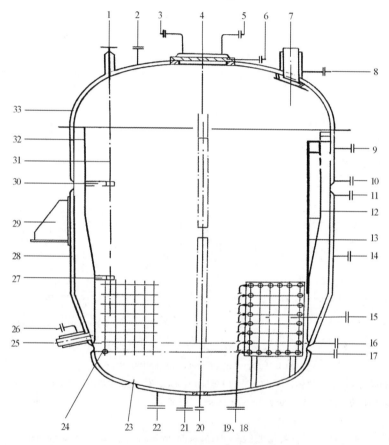

图 4.1.3 第一预缩聚反应器结构图

Structure drawing of precondensation stage Ⅰ

1—液位计接口 Liquid level gauge interface；2、5、11、17—气相热媒出口 HTM vapour vent；
3、9、14、22—气相热媒入口 HTM vapour inlet；4—人孔 Manhole；
6、8、10、16、21—冷凝气相热媒出口 HTM condensate outlet；7—气相出口 Vapour outlet；
12、13、32—挡板 Plate；15—物料入口 Material inlet；18—液相热媒入口 HTM liquid inlet；
19、26—液相热媒出口 HTM liquid outlet；20—排放口 Product drain；23—温度计接口 Thermowell interface；
24—热媒盘管 HTM pipe；25—物料出口 Product outlet；27、30—液位计支架 Level transmitter support；
28—筒体 Shell；29—支座 Support；31—液位计 Level transmitter；33—封头 Head

2. 第二预缩聚反应器 Precondensation stage Ⅱ

第二预缩聚反应器是一个全夹套卧式单轴环盘反应器，反应器内分设5个室，每个室内有一组环盘，环盘固定在搅拌轴上，环盘上分布有圆孔。

Precondensation stage Ⅱ is full jacketed horizontal reactor with the single shaft ring plate, which contains five chamber, one of them has a group of ring plates. The ring plate which has circle holes is fixed on the stirring shaft.

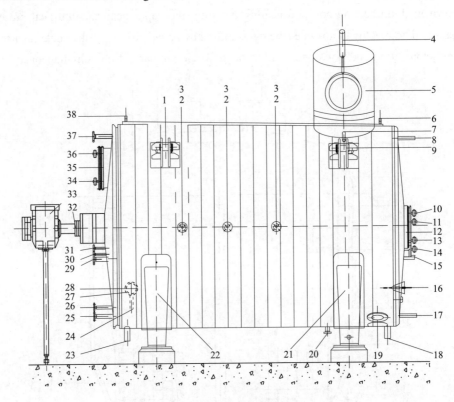

图 4.1.4　第二预缩聚反应器外形图
Outline drawing of precondensation stage Ⅱ

1、9—吊耳 Lug；2—挡板用热媒入口 HTM inlet for baffle plate；3—挡板用热媒出口 HTM outlet for baffle plate；
4—蒸汽圆顶室用热媒出口 HTM outlet for vapour dome；5—蒸汽圆顶室 Vapour dome；6、38—出气口 HTM vent；
7—蒸汽圆顶室用热媒入口 HTM inlet for vapour dome；8—封头Ⅱ用热媒出口 HTM outlet for cover Ⅱ；
10、11、36—人孔用热媒出口 HTM outlet for manhole；12—人孔 Manhole；
13、14、34—人孔用热媒入口 HTM inlet for manhole；15、17—封头Ⅱ用热媒入口 HTM inlet for cover Ⅱ；
16—温度计接口 Thermowell interface；18、23—热媒入口 HTM inlet；19—物料出口 Product outlet；
20—液位计接口 Liquid level gauge interface；21、22—支座 Support；
24—物料入口用热媒出口 HTM outlet for product inlet；25—加热盘管用热媒入口 HTM inlet for heating pipe；
26—加热盘管用热媒出口 HTM outlet for heating pipe；
27—物料入口用热媒加热夹套 HTM heating jacket for product inlet；28—物料入口 Material inlet；
29、31—封头Ⅰ用热媒入口 HTM inlet for cover Ⅰ；30、37—封头Ⅰ用热媒出口 HTM outlet for cover Ⅰ；
32—轴封 Shaft sealing；33—减速机 Speed reducer；35—氦检口 Helium-test nozzle

4.1.1.3　终缩聚反应器 Ringscheiben reactor

终缩聚反应器是一个全夹套卧式单轴环盘反应器，反应器内分设 8 个室，32 块挡板，55 块环盘，环盘固定在搅拌轴上，前四个室的环盘是组合型环盘，后四个室的环盘是单盘，环盘均无开孔。

4 合成纤维 Synthetic fiber

Ringscheiben reactor is a full-jacketed horizontal uniaxial ring disk reactor, which contain 8 chambers, 32 baffles, 55 ring disks, which fixed on the agitator shaft. The ring disks of the front four chambers are the combination of the ring disks, the ring disks of the rear four chambers are the single disks, the ring disks are all without the holes.

1. 终缩聚反应器外形 Ringscheiben reactor

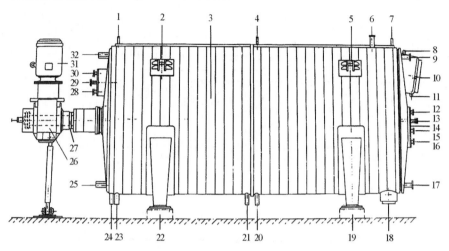

图 4.1.5 终缩聚反应器外形图
Outline drawing of outside of ringscheiben reactor

1、7—热媒出气口 HTM vent；2、5—吊耳 Lug；3—反应釜 Reactor vessel；4、6—热媒出口 HTM outlet；8—蒸汽用热媒出口 HTM outlet for vapour；9—封头Ⅱ用热媒出口 HTM outlet for cover Ⅱ；10—蒸汽出口 Vapour outlet；11—蒸汽用热媒入口 HTM inlet for vapour；12、13、15、16、28、30—人孔用热媒出口 HTM outlet for manhole；14—人孔 Manhole；17—封头Ⅱ用热媒入口 HTM inlet for cover Ⅱ；18—物料出口 Product outlet；19、22—支座 Support；20、21、24—热媒入口 HTM inlet；23—物料入口 Material inlet；25—封头Ⅰ用热媒入口 HTM inlet for cover Ⅰ；26—减速机 Speed reducer；27—轴封 Shaft sealing；29—氦检口 Helium-test nozzle；31—电机 Motor；32—封头Ⅰ用热媒出口 HTM outlet for cover Ⅰ

2. 反应釜 Reactor vessel

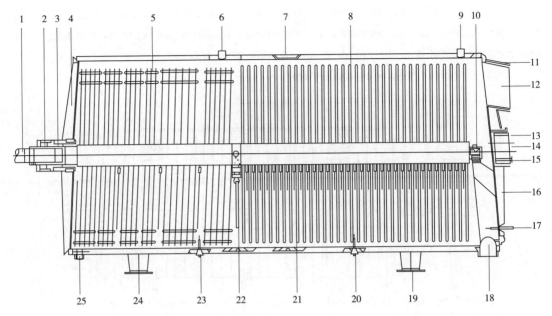

图 4.1.6　反应釜结构图 Structure drawing of reactor vessel

1—轴 Shaft；2—滚动轴承 Rolling bearing；3—轴封 Shaft sealing；4、16—封头 Head；5、8、21—环盘 Annular plate；
6、9—热媒出口 HTM outlet；7、11—加热夹套 Heating jacket；10—滑动轴承 Sliding bearing；
12—蒸汽出口 Vapour outlet；13—人孔用热媒出口 HTM outlet for manhole；14—人孔 Manhole；
15—人孔用热媒入口 HTM inlet for manhole；17—温度计接口 Thermowell interface；
18—物料出口 Product outlet；19、24—支座 Support；
20、23—液位计接口 Liquid level gauge interface；
22—挡板 Baffle plate；25—物料入口 Material inlet

3. 加热夹套 Heating jacket

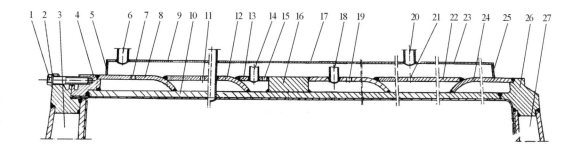

图 4.1.7 加热夹套结构图 Structure drawing of heating jacket structural drawing

1—六角螺钉 Hex screw；2—垫片 Gasket；3、27—封头 Head；4、26—法兰环 Flange ring；5、25—挡板 Plate；
6、9、14、17、18、20、23—热媒管 HTM pipe；7、11、15、19、21—热媒孔 HTM hole；
8、12、13、22、24—加热夹套 Heating jacket；10—釜内壳 Inside shell；16—加强圈 Reinforcement ring

4.1.2 塔设备 Column equipment

4.1.2.1 乙二醇精馏塔 Ethylene glycol distillation column

精馏是将第一、二酯化反应器中的乙二醇和生成的水共同蒸发形成的混合蒸汽在塔内进行精馏分离。

The distillation is separation about the mixed steam from the ethylene glycol in the first and second esterification reactor and the generated water.

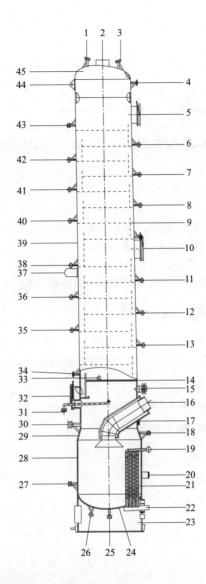

图4.1.8 乙二醇精馏塔结构图 Structure drawing of ethylene glycol distillation column

1、4、34—压力计接口 Pressure gauge interface；2—蒸汽出口 Steam outlet；3、6、7、8、11、12、13、26、35、36、38、40、41、42—温度计接口 Thermometer interface；5、10、32—人孔 Manhole；
9、14—塔盘 Tray；15—安全阀接口 Safety valve interface；16—蒸汽入口 Steam inlet；
17—视镜 Sight glass；18—乙二醇入口 Ethylene glycol inlet；19—热媒出口 Heat meadium outlet；
20—铭牌 Nameplace；21—热媒盘管 Heat meadium coiler；22—热媒入口 Heat meadium inlet；
23—裙座 Skirt support；24、45—封头 Head；25—排放口 Discharge outlet；
27、30—液位计接口 Liquid level gauge interface；28—下筒体 Lower shell；29—筒体过渡段 Cylinder transition section；
31—喷淋乙二醇入口 Spray glycol inlet；33—泡罩 Bubble cap；37—支座 Support；
39—上筒体 Upper shell；43—乙二醇回流口 Ethylene glycol backflow opening；44—吊耳 Lifting lug

4.1.2.2 尾气洗涤塔 Exhaust gas scrubber column

尾气洗涤塔是将酯化、缩聚反应中产生的不凝气体用工业水洗涤后放空。

Exhaust gas scrubber column is the equipment of using the industrial water to wash the non-condensable gas generated from the esterification and the polycondensation reaction and then vent.

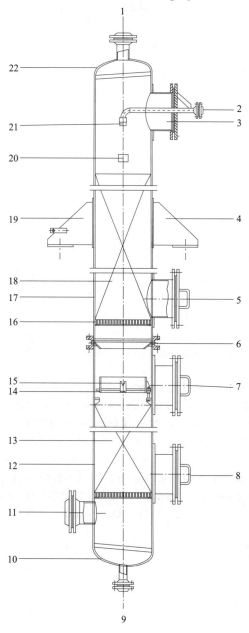

图 4.1.9 尾气洗涤塔结构图 Structure drawing of exhaust gas scrubber column

1—排气口 Exhaust port；2—水入口 Water inlet；3、5、7、8—人孔 Manhole；
4、19—支座 Support；6—集液器 Liquid trap；9—水出口 Water outlet；10、22—封头 Head；
11—尾气入口 Exhaust gas inlet；12、17—筒体 Shell；13、18—填料 Packing；14—堰板 Weir plate；
15—堰升器 Weir liters machine；16—栅板 Grid tray；20—铭牌 Nameplate；21—喷头 Sprinkler

4.1.3 料仓 Hopper

4.1.3.1 精对苯二甲酸贮存料仓 Purified terephthalic acid storage hopper

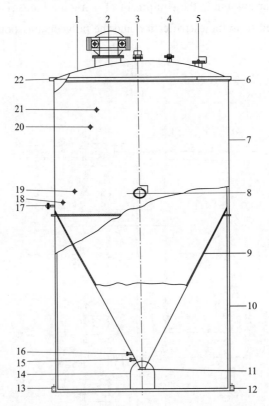

图 4.1.10 精对苯二甲酸贮存料仓结构图 Structure drawing of purified terephthalic acid storage hopper
1—拱顶 Arch crown；2—过滤器 Filter；3—安全阀 Safety valve；4—物料进口 Material inlet；5、8—人孔 Manhole；
6—拱顶加强圈 Arch crown reinforcing ring；7—筒体 Shell；9—锥形筒体 Conical barrel；10—裙座 Skirt；
11—物料出口 Material outlet；12、13—地脚螺栓 Anchor bolt；14—检修口 Access hole；
15、16—锥体反吹管 Cone blowback pipe；17—筒体反吹管 Shell blowback pipe；
18—超低料位报警器接口 Super low level alarm interface；19—低料位报警器接口 Low level alarm interface；
20—高料位报警器接口 High level alarm interface；21—超高料位报警器接口 Super high level alarm interface；
22—护栏底座 Guardrail base

4.1.3.2 切片分析料仓 Slice analysis hooper

生产出来的成品切片送至切片分析料仓取样并分析,划分等级,再送至切片储存料仓。

Put the finished slice to the slice analysis hooper, sample and analylize, divide the level and put it to the slice analysis hooper.

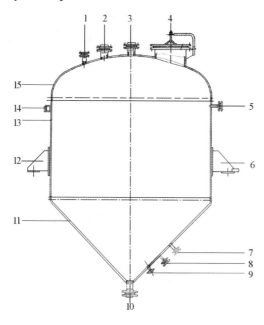

图 4.1.11 切片分析料仓结构图 Structure drawing of slice analysis hooper

1—呼吸阀接口 Respiration valve interface;2—排气口 Exhaust port;
3—切片入口 Slice inlet;4—人孔 Manhole;
5—高料位报警器接口 High level alarm interface;
6、12—支座 Support;7—低料位报警器接口 Low level alarm interface;
8—超低料位报警器接口 Super low level alarm interface;
9—取样口 Sample connection;10—切片出口 Slice outlet;11—锥体 Cone;
13—筒体 Shell;14—铭牌 Nameplace;15—封头 Head

4.1.3.3 切片贮存料仓 Slice storage hooper

切片贮存料仓用于贮存切片，而后进行包装。

Slice storage hooper is for the storage of the slice and for packaging.

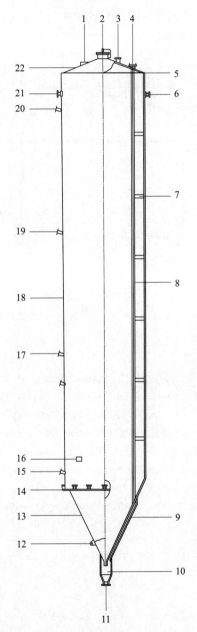

图 4.1.12 切片贮存料仓结构图 Structure drawing of slice storage hooper

1—呼吸阀接口 Respiration valve interface；2—人孔 Manhole；3—切片入口 Slice inlet；
4—配料管接口 Ingredient pipe interface；5—锥顶加强圈 Cone stiffening ring；6、12、21—吊耳 Lifting lug；
7—配料管支承 Ingredient pipe bearing；8—配料管 Ingredient pipe；9—混合板 Mixing plate；
10—混合室 Mixing chamber；11—切片出口 Slice outlet；13—锥体 Cone；14—反吹管 Blowback pipe；
15、17、19、20—料位计接口 level gauge interface；16—铭牌 Nameplace；18—筒体 Shell；22—锥顶 Cone top

4.1.4 换热设备 Heat exchange equipment

4.1.4.1 乙二醇管壳式冷却器
Ethylene glycol tube shell cooler

乙二醇管壳式冷却器的作用是冷却用于喷淋的乙二醇,管程的介质是乙二醇,壳程的介质是冷却水。

The function of the ethylene glycol tube shell type cooler is to cool the ethylene glycol using for spraying, the medium in the tube pass is the ethylene glycol, the medium in the shell pass is the cooling water.

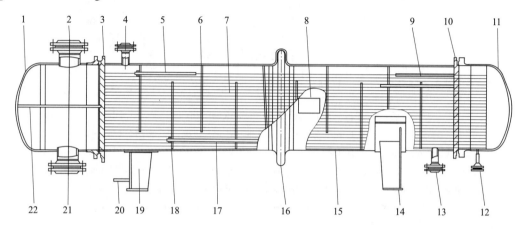

图 4.1.13 乙二醇管壳式冷却器结构图 Structure drawing of ethylene glycol tube shell cooler
1—前管箱 Front cover pipe box;2—乙二醇出口 Ethylene glycol outlet;3、10—管板 Tube sheet;
4—冷却水出口 Cooling water outlet;5、9、17—拉杆 Tie rod;6、18—折流板 Baffle plate;
7—换热管 Heat exchange tube;8—铭牌 Nameplace;11—后管箱 Rear cover pipe box;
12—排放口 Discharge outlet;13—冷却水进口 Cooling water inlet;14、19—支座 Support;
15—筒体 Shell;16—膨胀节 Expansion joint;20—接地板 Earth plate;
21—乙二醇进口 Ethylene glycol inlet;22—分程隔板 Pass partition plate

4.1.4.2 乙二醇板式冷却器 Ethylene glycol plate cooler

乙二醇板式冷却器用于冷却第一、二预缩聚反应器以及蒸汽喷射泵喷淋系统中喷淋用的乙二醇。

Ethylene glycolplate cooler is for cooling the ethylene glycol for sprying in the first and second pre-polycondensation reactor and steam injection pump spray sprinkler system.

1. 乙二醇板式冷却器外形 Outline of ethylene glycol plate heat exchanger

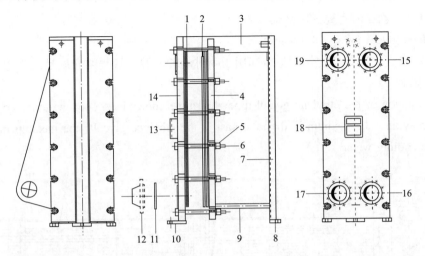

图 4.1.14　乙二醇板式冷却器外形图 Structure drawing of ethylene glycol plate heat exchanger

1—板片 Plate；2—4 角垫片 Corners gasket；3—顶部导架 Top guide；4—移动架 Moveable frame；
5—锁紧螺母 Lock nut；6—拉杠 Tie-rod；7—固定支承 Fixed support；8、10—地脚螺栓 Anchor bolt；
9—底部导架 Bottom guide；11—垫片 Gasket；12—法兰 Flange；13、18—铭牌 Nameplate；
14—固定架 Fixed frame；15—冷却水入口 Cooling water inlet；16—乙二醇入口 Ethylene glycol inlet；
17—冷却水出口 Cooling water outlet；19—乙二醇出口 Ethylene glycol outlet

2. 乙二醇板式冷却器板片 Plate of ethylene glycol plate heat exchanger

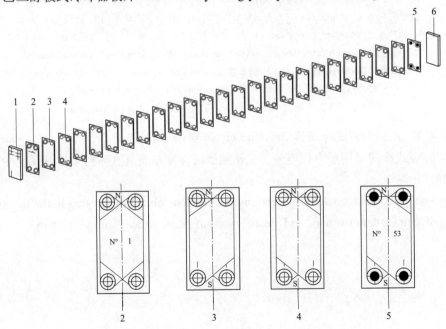

图 4.1.15　乙二醇板式冷却器板片图 Structure drawing of plate of ethylene glycol plate heat exchanger drawing

1—固定架 Fixed frame；2—第一块板片 1st plate；3—偶数板片 Even plate；
4—奇数板片 Odd plate；5—最后一块板片 Special plate；6—移动架 Moveable frame

4.1.5 切粒设备 Grain-sized dicing equipment

切粒设备的作用是将来自终缩聚反应器聚合物熔体加工成最终产品——聚酯切片,而后送至切片储存料仓。

The function of the grain-sized dicing equipment is making the polymer melt come from the final polycondensation reactor into the final product that is the polyester chips and sending it to the slice storage hooper.

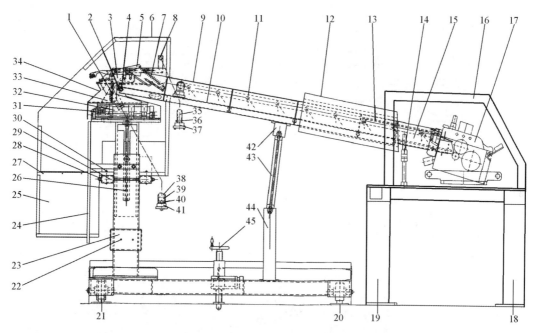

图 4.1.16 切粒设备外形图 Structure drawing of pelletizing machinery

1、22—开槽沉头螺钉 Slotted countersunk head screw;2、29—螺杆 Threaded rod;3、23—夹板 Holder;4—调节环 Adjusting ring;5—括刮器 Scraper;6、24、25—挡板 Plate;7—弹簧 Spring;8—喷水管 Spray tube;9—钢槽 Steel trough;10—导流板 Strand guide section;11—喷嘴 Spray nozzles;12—隔音设备 Sound insulation;13—盖板 Cover plate;14、26—调节丝杆 Adjusting screw;15—水下切粒机进口 Inlet part of underwater pelletizing machinery;16—防护罩 Protection cover;17—水下切粒机 Underwater pelletizing machinery;18、19—底座 Machine base;20、21—导轨 Guide;27—盘头自攻螺钉 Pan head tapping screw;28—六角螺母 Hexagon nut;30—内六角圆柱头螺钉 Hexagon socket head cap screw;31—保护格栅 Protection grid with sliding door;32—气缸 Pneumatic cylinder;33—水箱 Water box;34—起动设备 Starting device;35—蒸汽软管 Rubber-steam-hose;36、40—软管夹子 Hose clamp;37—软管接头 Hose coupling with viton-sealing;38—软管 Hose;39—软管套 Hose liner;41—连接管 Coupling with sealing ring;42、44—支架 Support;43—螺旋千斤顶 Screw jack;45—调节手轮 Adjusting device

4.1.5.1 铸带头 Die head

铸带头的作用是将聚合物熔体进行铸带。The function of the die head is casting the polymer melt.

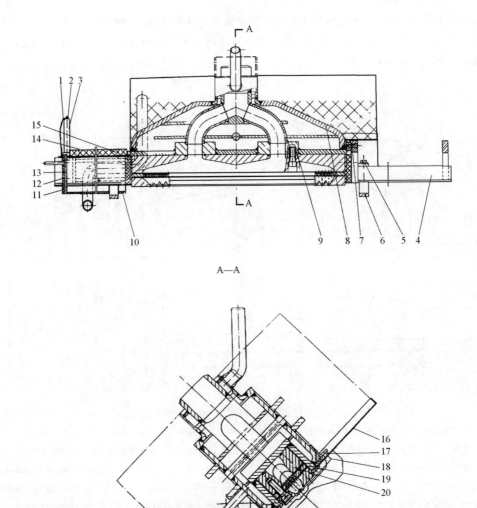

图 4.1.17 铸带头结构图 Strueture drawing of die head

1—链 Chain；2、5—六角螺栓 Hexagon cap screw；3—螺栓 Hexagon bolt；4—卡具 Holder；
6、13—挡板 Plate；7—内六角圆柱头螺钉 Hexagon socket head cap screw；
8—加热箱 Heating box；9—插销螺套 Keensert；10、11、15—保温板 Insulating plate；
12、14、19、21—沉头螺钉 Slotted countersunk head screw；16—底板 Base plate；
17—定距环 Distance ring；18、22—肋形导架 Ribbed guide；20—铸带机构 Die pack；
23—隔热垫 Insulating mat

4 合成纤维 Synthetic fiber

4.1.5.2 水下切粒机 Underwater pelletizing machinery

水下切粒机的作用是将带状熔体切断成粒，它由螺旋切割转子(动刀部分)、可调节的机架刀(定刀部分)和前、后进料辊组成。

The function of the underwater pelletizing machinery is to cut the strip of melt granulation, which consists of helical cutting rotor (ro tating blades portion), adjustable rack knives (fixed blade portion) and the front and backward feed rollers.

1. 水下切粒机 Underwater pelletizing machinery

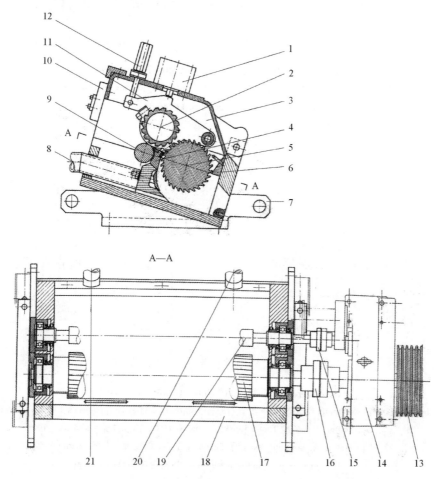

图 4.1.18 水下切粒机结构图 Strueture drawing of underwater pelletizing machinery

1—气缸 Pneumatic-cylinder；2—前进料辊 Ahead feed roller；3—翻板 Flap；4—机架刀 Bed knife；
5、17—旋转切粒刀 Cutting rotor；6—挡水板 Flashing；7—底座 Base；8、20、21—进水管 Water inlet tube；
9、19—后进料辊 Latter feed roller；10—盖板 Cover plate；11—摆轴臂 Rocker arm with bearing cover；
12—调节丝杆 Adjusting screw；13—皮带轮 Pulley；14—齿轮箱 Gear box；
15、16—联轴器 Coupling；18—机壳 Housing

· 317 ·

2. 旋转切粒刀 Cutting rotor

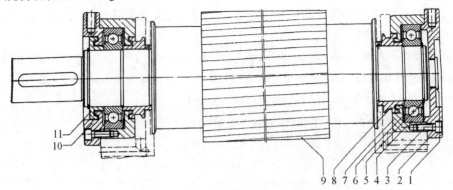

图 4.1.19　旋转切粒刀结构图 Structure drawing of cutting rotor

1—盲密封盖 Dummy gland；2、10—轴承压盖 Bearing cover；
3—内六角圆柱头螺钉 Hexagon socket head cap screw；
4—球轴承 Grooved ball bearing；5—轴挡圈 Retaining ring for shafts；
6—星型弹簧 Star spring；7—轴承座 Bearing housing；
8、11—迷宫密封 Labyrinth ring；9—旋转切粒刀 Cutting rotor

3. 前进料辊 Ahead feed roller

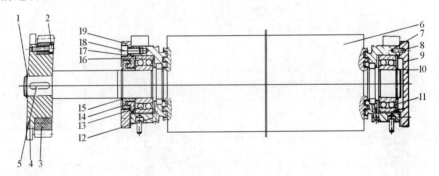

图 4.1.20　前进料辊结构图 Structure drawing of ahead feed roller

1、10、15—轴挡圈 Retaining ring for shafts；2、18—内六角圆柱头螺钉 Hexagon socket head cap screw；
3—齿轮 Gear wheel；4—齿盘 Toothed disk；5—键 Key；6—前进料辊 Ahead feed roller；
7、12—轴承座 Bearing housing；8—销子 Parallel pin；9、19—轴承压盖 Bearing cover；
11、13—轴承 Self-aligning bearing；14—迷宫密封 Labyrinth ring；
16—垫片 Gasket；17—O 形密封圈 O sealing ring

4. 后进料辊 Latter feed roller

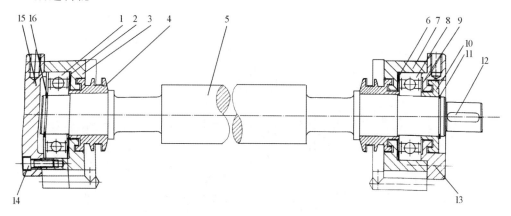

图 4.1.21　后进料辊结构图 Structure drawing of latter feed roller

1、8—球轴承 Ball bearing；2、6—星型弹簧 Star spring；3、7—轴承座 Bearing housing；
4、10—迷宫密封 Labyrinth ring；5—后进料辊 Latter feed roller；9、15—轴承压盖 Bearing cover；
11、16—轴挡圈 Retaining ring for shafts；12—键 Key；
13、14—内六角圆柱头螺钉 Hexagon socket head cap screw

4.1.5.3　切片干燥器 Pellet drier

切片干燥器的作用是去除来自水下切粒机含水切片中的水分，使切片水含量达到允许值。

The function of the pellet drier is to remove moisture from the underwater pelletizer aqueous slice in order to make the water content of the slices to the permissible value.

1. 切片干燥器外形 Outline of pellet drier

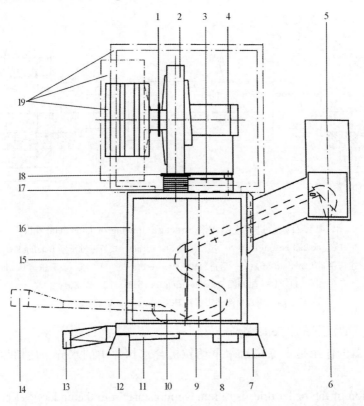

图 4.1.22　切片干燥器外形图 Outline drawing of pellet drier

1、18—法兰 Flange；2—鼓风机 Blower；3—鼓风机底座 Blower base；4—电机 Motor；5—风出口 Wind outlet；6—切粒出口 Pellet outlet；7、12—支座 Support；8、15—弧形筛 Curved sieve；9—切片干燥器底座 Pellet drier base；10—水分离器 Water pre-separator；11—水收集槽 Water collecting tray；13—水出口 Water outlet；14—物料入口 Material inlet；16、19—吸声罩 Sound absorbing cover；17—软连接 Soft connection

2. 切片干燥器结构 Drier housing structure

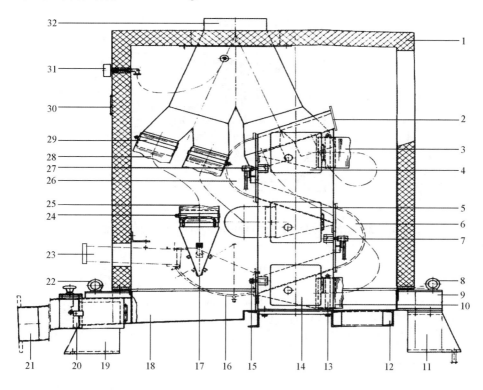

图 4.1.23　切片干燥器结构图 Structure drawing of drier housing

1—吸声罩 Sound absorbing cover；2—物料出口 Product outlet；3、10、25、28—软管 Hose；
4、5、14—手孔 Hand hole；6、26—弧形筛 Curved sieve；7、15、27—自锁卡 Self-locking holder；
8、22—吊环螺栓 Eye bolt；9、12—底座 Base frame；11、19—支座 Support；13、24、29—软管夹子 Hose clamp；
16—水分离器 Water pre-separator；17—喷嘴 Air-flow nozzle；18—水收集槽 Water collecting tray；
20—滤网 Filter；21—水出口 Water outlet；23—物料入口 Material inlet；30—铭牌 Nameplate；
31—压力表 Pressure meter；32—分配管 Distributor tube

4.1.6　泵 Pump

4.1.6.1　熔体出料泵 Product feed pump

熔体出料泵采用的是齿轮泵，作用是将聚合物熔体由第二预缩聚反应器输送到终缩聚反应器，再由终缩聚反应器输送到切粒系统。

The product feed pump is using the gear pump in order to send the polymer melt from the second pre-polymer reactor to the final polycondensation reactor, then from the the final polycondensation reactor to the pelletizing system.

1. 熔体出料泵外形 Outline of product feed pump

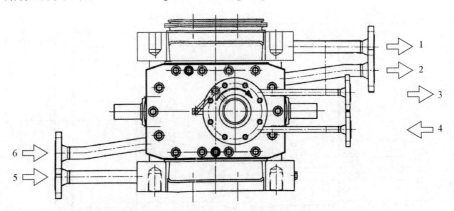

图 4.1.24　熔体出料泵外形图 Outline drawing of product feed pump
1—泵壳热媒出口 HTM outlet for pump housing；
2—驱动侧热媒出口 HTM outlet for cover drive-side；
3—密封液出口 Sealing liquid outlet；4—密封液入口 Sealing liquid inlet；
5—泵壳热媒入口 HTM outlet for pump housing；
6—驱动侧热媒入口 HTM inlet for cover drive-side

4 合成纤维 Synthetic fiber

2. 熔体出料泵结构 Product feed pump structural

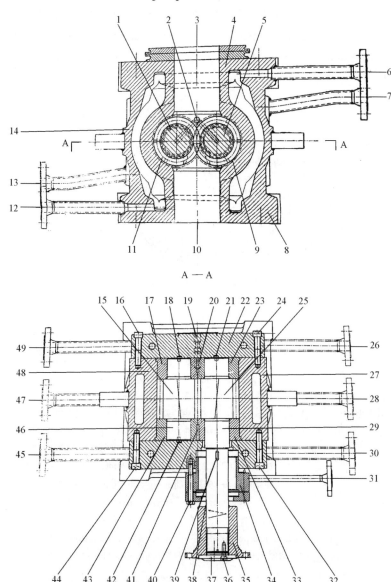

图 4.1.25 熔体出料泵结构图 Structure drawing of product feed pump

1、46—从动轴 Driven shaft；2、18、21、42—V形槽 V-trough；3—物料入口 Material inlet；4—入口端盖 Cover inlet；
5、29—驱动轴 Drive shaft；6、28—泵壳热媒出口 HTM outlet for pump housing；7、26、30—端盖热媒出口 HTM outlet for cover；
8—出口端盖 Outlet cover；9、25—驱动齿轮 Drive gear；10—物料出口 Product outlet；11、15—从动齿轮 Driven shoft；
12、47—泵壳热媒入口 HTM inlet for pump housing；13、45、49—端盖热媒入口 HTM inlet for cover；
14、27、48—泵体 Pump body；16、24、32、40、44—六角螺栓 Bolt；
17、23、33、43—滑动轴承 Sliding bearing；19—对中销 Pin；20、39、41—键 Key；22—从动侧端盖 Cover non-drive-side；
31—密封液入口 Sealing liquid inlet；34—密封 Sealing；35—轮毂 Hub；36—内六角圆柱头螺栓 Allen cap screw；
37—固定板 Retaining plate；38—O形密封圈 O sealing ring

4.1.6.2 黏度计泵 Viscosimeter pump

黏度计泵设在熔体出料泵和切粒系统之间,实现聚合物熔体黏度的在线控制。

The viscosimeter pump exists between the melt discharging pump and the pelletizing system, which realize the on-line control to the viscosity of the polymer melt.

1. 黏度计泵外形 Outside of viscosimeter pump

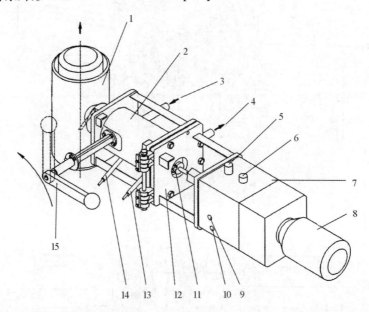

图 4.1.26 黏度计泵外形图 Structure drawing of outside of viscosimeter pump

1—热物料管道 Heated product pipeline;
2—加热夹套 Heating jacket; 3—热媒入口 HTM inlet;
4—热媒出口 HTM outlet; 5—排气帽 Vent;
6—注油口 Oil inlet; 7—齿轮箱 Gear box; 8—电机 Motor;
9—油位计 Oil level gauge; 10—排油口 Oil drain plug;
11—压力计或温度计接口 TI, PI interface;
12—连接盘 Connecting plate; 13—压力传感器 Pressure sensor;
14—温度传感器 Temperature sensor; 15—切断装置 Shutoff device

2. 黏度计泵结构 Viscosimeter pump structural

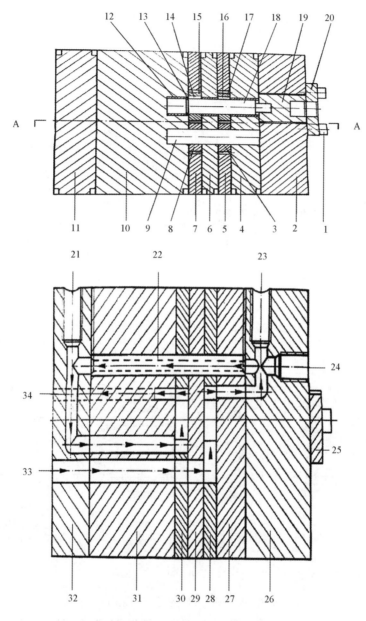

图 4.1.27　黏度计泵结构图
Structure drawing of viscosimeter pump

1—内六角螺栓 Cylindrical screw；2、26—顶板 Top plate；3、8—从动齿轮 Driven metering gear；
4、27—轮毂板 Hub plate；5、7、28、30—中心板 Center plate；6、10、29、31—中间板 Intermediate plate；
9—从动轴 Driven shaft；11、32—背板 Back plate；12—轴套 Sleeve；13—开口环 Snap ring；
14、16—圆柱销 Cylindrical pin；15、17—驱动齿轮 Driving metering gear；18—驱动轴 Drive shaft；
19—滑动轴承 Sliding bearing；20、25—轮毂头 Hub top；21、23—温度传感器接口 Temperature sensor interface；
22—毛细管 Capillary tube；24—压力传感器接口 Pressure sensor interface；33—物料入口 Material inlet；
34—物料出口 Product outlet

4.1.6.3 液环式真空泵 Liquid ring vacuum pump

1. 真空泵系统 Vacuum pump system

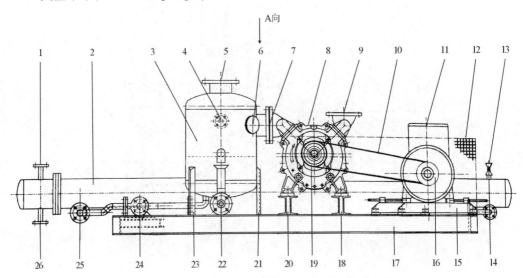

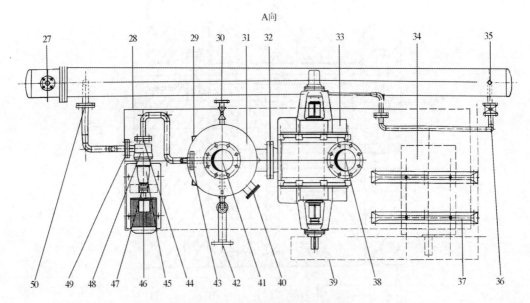

图 4.1.28 真空泵系统布置图
Structure drawing of outside of vacuum pump system

1、27—冷却水出口 Cooling water；2—换热器 Heat exchanger；3、31—分离器 Separator；
4、30—乙二醇进口 Fresch glycol inlet；5、41—废气出口 Off gas outlet；6、40—手孔 Hand hole；
7、32—分离器入口 Vapour and Glycol separator inlet；8、33—液环泵 Liquid ring pump；9、38—蒸汽入口 Vapour inlet；
10—V 形皮带 V belt；11、34、46—电机 Motor；12、39—防护罩 shield；13—排气阀 Air vent valve；
14、36—乙二醇出口 Glycol outlet；15、37—电机底座 Motor base；16—驱动皮带轮 Driving pulley；
17、28—底座 Common structural steel base；18、20—液环泵底座 Liquid ring pump base；19—从动皮带轮 Driven pulley；
21、23、29—分离器支座 Separator support；22、42—分离器溢流口 Glycol separator overflow；
24、48—循环泵 Electro-recirculating pump；25、50—换热器进口 Heat exchanger inlet；
26—冷却水进口 Cooling water inlet；35—排气口 Exhaust port；43—分离器进口 Glycol separator inlet；
44—循环泵入口 Glycol electro-recirculating pump inlet；45—轴承箱 Bearing box；47—联轴器 Coupling；
49—循环泵出口 Glycol electro-recirculating pump outlet

2. 液环式真空泵结构 Structure of liquid ring vacuum pump

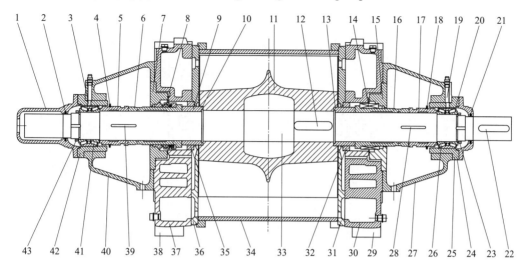

图 4.1.29 液环式真空泵结构图 Structure drawing of liquid ring vacuum pump

1—非驱动侧轴承盖 Non-drive-side bearing cover；2、21—毛毡密封 Felt seal；3、19—销 Pin；
4—V形密封圈 V-ring seal；5、17—定位圈 Piston ring holding ring；6、16—轴套 Sleeve；
7、15—压盖 Mechanical seal cover；8、14—机械密封 Mechanical seal；9、13—耐磨套 Wear sleeve；
10、18—O形密封圈 O sealing ring；11—叶轮 Wheel；12、22、28、39—键 Key；
20—驱动侧轴承盖 Drive-side bearing cover；23、43—锁紧螺母 Lock nut；24—驱动侧轴承箱 Drive-side bearing housing；
25—垫片 Spacer petting lateral clearance；26—驱动侧轴承 Drive-side bearing；27—驱动侧轴承箱支架 Bearing box support；
29、38—底座 Liquid ring pump base；30—驱动侧侧盖 Drive-side cover；31—驱动侧分配板 Drive-side distribution plate；
32、35—密封环 Sealing ring；33—轴 Shaft；34—泵体 Pump body；36—非驱动侧分配板 Non-drive-side distribution plate；
37—非驱动侧侧盖 Non-drive-side cover；40—非驱动侧轴承箱支架 Non-drive-side bearing box support；
41—非驱动侧轴承 Non-drive-side bearing；42—非驱动侧轴承箱 Non-drive-side bearing housing

4.1.7 其它设备 Other equipments

4.1.7.1 热媒加热炉 HTM furnace

热媒加热炉通过燃烧燃料油，对热媒循环系统中的热媒进行加热。

The HTM furnace is used to heat the heating medium in the HTM circulation system by burning the fuel oil.

1. 热媒加热炉 HTM heaters

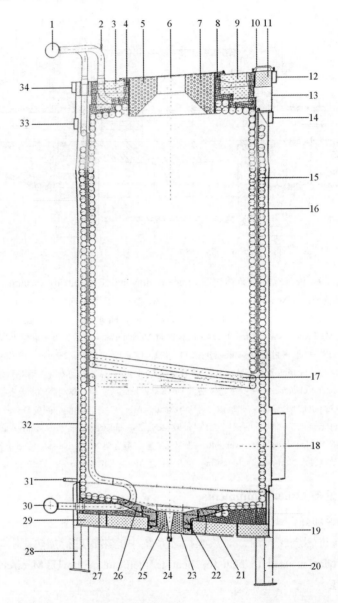

图 4.1.30 热媒加热炉结构图 Structure drawing of HTM heaters

1—热媒出口 HTM outlet；2—温度计接口 Thermowell interface；3、9、11、19、29—陶瓷纤维毡 Ceramic fiber blanket；4、8—密封层 Sealing layer；5、7、23、25、26、27—浇注料 Castable；6—燃烧器插入口 Burner insert；10—清洗口 Cleaning opening；12、14、33、34—加强圈 Reinforcement ring；13—炉顶 Heaters arch；15、16、17—加热炉管 Heater coils；18—烟气出口 Flue gas outlet；20、28—支腿 Support；21—炉底 Heaters floor；22—人孔 Manhole；24—视镜 Sight glass；30—热媒入口 HTM inlet；31—清洗排水管 Drain nozzle cleaning；32—炉壳体 Heaters casing

4 合成纤维 Synthetic fiber

2. 燃烧器 Burner

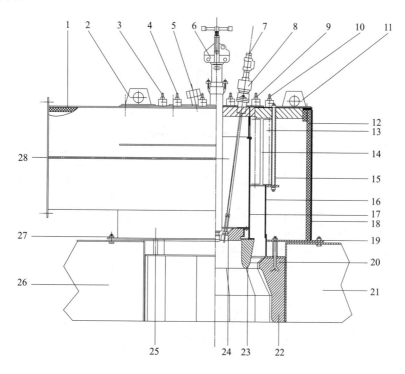

图 4.1.31 燃烧器结构图 Structure drawing of burner

1、12—衬里 Lining；2、11—吊耳 Lug；3、4、9、10—调风机构 Air adjusting setup；5—观火孔 Sight hole；6—油枪 Oil gun；7—燃气入口 Fuel gas inlet；8—气枪 Gas gun；13—调风挡板 Air adjusting baffle；14—调风转轴 Air adjusting axes；15—拉杆 Tie rod；16—稳焰罩外圈 Flame stabilized cover outwrd circle；17—稳焰罩内圈 Flame stabilized cover inner circle；18—壳体 Shell；19、27—螺栓 Bolt；20、25—Y形锚钉 Y Clip anchor；21、26—炉顶 Heaters Arch；22—大火盆砖 Big firebrick；23、24—小火盆砖 Small firebrick；28—风道入口 Air inlet

4.1.7.2 二氧化钛离心机 TiO_2 centrifuge

二氧化钛离心机的作用是将乙二醇、二氧化钛悬浮液进行离心分离，除掉未分散开的二氧化钛大颗粒。

The function of the TiO_2 centrifuge is making the centrifugal separation to the suspension liquid of the ethylene glycol and the titanium dioxide and to remove large particles of the disperse titanium dioxide.

1. 二氧化钛离心机 TiO₂ centrifuge

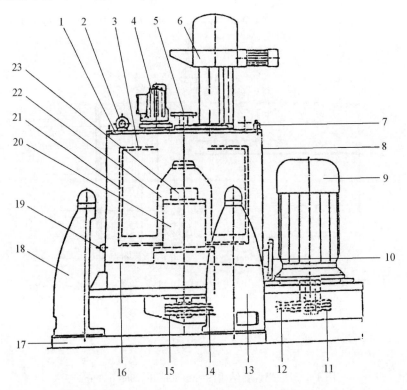

图 4.1.32 二氧化钛离心机外形图 Structure drawing of outside of TiO₂ centrifuge

1—机盖 Cover；2—吊环 Eye bolt；3—挡液板 Lip ring；4—检查灯 Inspection light；5—物料入口 Material inlet；6—旋转刮刀卸料装置 Whirling peeler device；7—六角螺栓 Hexagon bolt；8—机壳 Housing；9—电机 Motor；10—物料出口 Product outlet；11—驱动皮带轮 Driving pulley；12、14—V形皮带 V-belt；13、18—摇摆式支座 Pendulum column；15—从动皮带轮 Driven pulley；16—底盘 Chassis；17—机座 Base；19—螺旋塞 Screw plug；20—轴承箱 Bearing box；21—转鼓体 Durm；22—转鼓底 Durm base；23—主轴 Shaft

2. 旋转刮刀卸料装置 Whirling peeler device

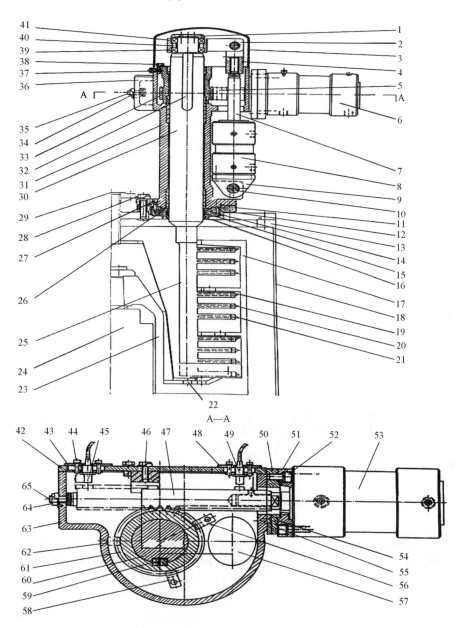

图 4.1.33 旋转刮刀卸料装置结构图 Structure drawing of whirling peeler device structural drawing

1—轴承压盖 Bearing cover；2—连杆 Articulated lever；3—机罩 Cap；4—U形夹 U Clevis；5、47—齿条 Rack；
6、8、53—液压缸 Hydraulic cylinder；7、57—推杆 Push rod；9—带孔底座 Eye mount；10、35、65—销 Pin；
11—法兰环 Flange ring；12—密封垫 Teflon seal；13—机盖 Cover；14—耐磨片导向环 Wear-resisting guide ring；
15、16—O形密封圈 O sealing ring；17—转鼓体 Durm；18—离心机壳体 Centrifuge shell；19、20、21—刮刀刀片 Peeler blade；
22—物料出口 Product outlet；23—转鼓底 Durm base；24—轴承箱 Bearing box；25—转鼓主轴 Durm main shaft；
26—压盖 Cover；27、44、52、54、55—内沉六角螺栓 Cylindrical screw；28、37、46—六角螺栓 Hexagon bolt；
29—物料入口 Material inlet；30、60—卸料刀轴 Peeler shaft；31、36、63—壳体 Shell；32、59—键 Key；
33、61—齿轮 Gear；34、64—六角螺母 Hex nut；38、56、58、62—压紧凸耳 Hold-down lug；
39、41—球轴承 Grooved ball bearing；40—定位环 Positioning ring；42—端板 End plate；43、48—盖板 Plate；
45、49—开关 Approximation switch；50—中间法兰 Intermediate flange；51—液压缸法兰 Cylinder flange

4.1.7.3 二氧化钛研磨机 TiO₂ pearl mill

二氧化钛研磨机的作用是将消光剂二氧化钛进行研细，使其在乙二醇中分散均匀。二氧化钛研磨机主要包括研磨桶、搅拌器、机械密封、研磨介质、驱动装置、控制装置、轴承、底座和螺杆泵等。

The function of the TiO₂ pearl mill is porphyrizing the titanium dioxide and making it disperse uniform.

The TiO₂ pearl mill contains the grinding barrel, the stirrier, the mechanical seal, the grinding medium, the driving device, the control device, the bearing assembly, the base supporting and the subsidiary screw pump.

1. 二氧化钛研磨机 TiO₂ pearl mill

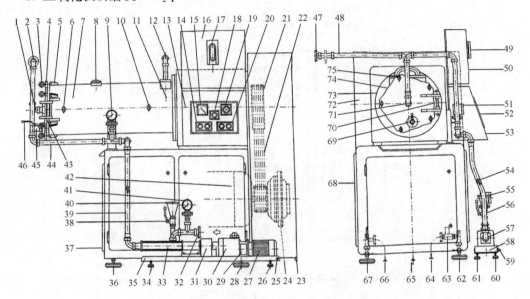

图 4.1.34 二氧化钛研磨机外形图 Structure drawing of outside of TiO₂ pearl mill

1、71—物料入口 Material inlet；2、44、48—接管 Connecting pipe；3—吊环 Eye bolt；4、73—研磨筒底部 Grinding vessel bottom；
5—加强筋 Stiffeners；6、10、75—吊耳 Lug；7—研磨筒 Grinding vessel；8—注入塞 Beads filling plug；
9、41、51—压力表 Pressure transmitter with pressure measuring device；11、74—筛筒 Screen cartridges；
12、47—物料出口 Medium-outlet；13—机械密封 Mechanical seal；14—控制面板 Push-button unit；
15—电流表 Amperemeter；16、37、50、53、68、72—防护罩 Protection cover；
17、49—热虹吸罐 Thermo-syphon-vessel；18—紧急停机按钮 Emergency-stop；19—控制灯 Control lamp；
20—搅拌器按钮 Push-button agitator；21—驱动皮带轮 Pulley；22—V 形皮带 V-belt；
23、25、59—接地板 Earthing connection；24—起动离合器 Starting clutch；26、28、34、36、60、61、62、67—地脚 Foot；
27—电机 Motor；29—联轴器 Coupling；30—减速机 Speed reducer；31—轴承箱 Bearing box；
32、40、55—泵物料入口 Pump product inlet；33、57—螺杆泵 Screw pump；35、58—底座 Base；
38、56—球阀 Ball cock；39、54—软管 Hose；42—搅拌器电机 Agitator motor；43、70—旋转支承 Rotational support；
45、69—排放球阀 Ball cock including discharge；46—阀柄 Valve handle；52—操作台 Operating desk；
63—流量控制器 Discharge controler；64—冷却水出口 Cooling water-outlet；65—电缆接口 Cable-inlet；
66—冷却水入口 Cooling water-entry

2. 搅拌盘 Agitator disk

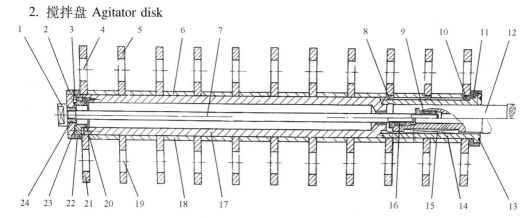

图 4.1.35　搅拌盘结构图 Structure drawing of agitator disk

1—螺帽 Nut；2、23—压盖 Cover disk；3—内沉六角螺栓 Hexagon socket cylindrical screw；4、19—孔 Hole；5、21—圆盘 Disk；6、18—定距套筒 Distance sleeve；7—螺杆 Threaded rod；8、10、15—键 Key；9、16—填料函 Stuffing box；11—定距环 Distance ring；12—主轴 Shaft；13、20、22、24—密封环 Sealing ring；14、17—轴套 Sleeve

4.2　涤纶 Polyester

4.2.1　搅拌釜式反应器 Stirred tank reactor

搅拌釜式反应器由釜体（包括封头、法兰、支座、接管等）、搅拌器、密封装置和传动装置等组成。

The basic structure of strirred tank reactor is made up of kettle body (including head, flanges, supports, adapter, etc.), agitator, sealing devices, transmission devices and other components.

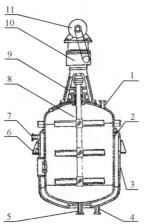

图 4.2.1　搅拌釜式反应器结构图 Structure drawing of horizontal polycondensation kettle

1—出料口 Discharge pipe；2—筒体 Barrel；3—夹套 Jacket；4—热水进口 Hot water inlet；5—进料口 Feed pipe；6—支座 Support；7—热水出口 Hot water outlet；8—搅拌器 Agitator；9—轴封 Shaft seal；10—传动装置 Transmission device；11—电机 Motor

4.2.2 卧式预缩聚釜 Horizontal pre-polycondensation kettle

卧式反应器分为卧式预缩聚釜和卧式后缩聚釜，主要由筒体、搅拌装置、密封装置、传热装置和传动装置等组成。

In polyester production, horizontal reactor is divided into horizontalpre-polycondensation kettle and the horizontal post-polycondensation kettle, mainly composed of shell, mixing device, sealing device, heat transfer device and transmission device, etc.

4.2.2.1 卧式预缩聚釜结构 Structure horizontal pre-polycondensation kettle

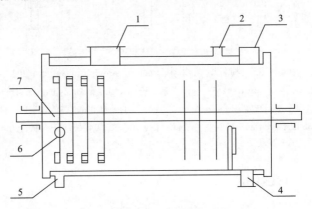

图 4.2.2　卧式预缩聚釜结构图

Structure drawing of the horizontal pre-polycondensation kettle

1、3—抽气口 Bleeding point；2—热载体进口 Heat carrier inlet；4—出料口 Discharge port；
5—热载体出口 Heat carrier outlet；6—进料口 Feed port；7—搅拌轴 Stirring shaft

4.2.2.2 预缩聚釜传动装置 Pre-polycondensation kettle gearing

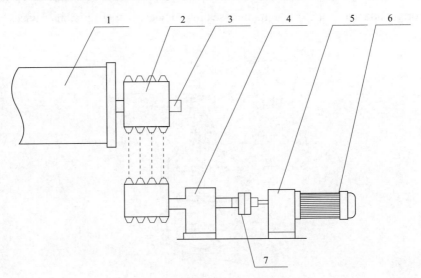

图 4.2.3　预缩聚釜传动装置结构图

Structure drawing of pre-polycondensation kettle gearing

1—预缩聚釜 Pre-polycondensation kettle；2—四排链轮 Four row chain wheel；3—轴 Shaft；
4—行星摆线减速器 Cycloid planetary gear speed reducer；5—变速器 Transmission；6—电机 Motor；7—联轴器 Coupling

4.2.2.3 预缩聚釜轴封 Pre-polycondensation kettle shaft seal

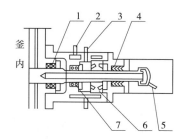

图 4.2.4 预缩聚釜轴封结构图 Structure drawing of pre-polycondensation kettle

1—填料 Packing；2—冷却水出口 Cold water outlet；3—抽真空管 Vacuum tube；4—V 形密封环 V-ring；
5—冷却水进口 Cold water inlet；6—推力轴承 Bearing；7—径向轴承 Radial bearing

4.2.3 卧式后缩聚釜 Horizontal post-polycondensation kettle

4.2.3.1 卧式后缩聚釜结构
Structure horizontal post-polycondensation kettle

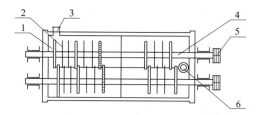

图 4.2.5 卧式后缩聚釜结构图 Structure drawing of horizontal post-polycondensation kettle

1—光板 Light board；2—孔板 Orifice；3—进料口 Feed inlet；4—螺杆 Screw；
5—链轮 Chain wheel；6—出料口 Discharge outlet

4.2.3.2 卧式后缩聚釜传动装置 Horizontal post-polycondensation kettle gearing

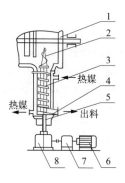

图 4.2.6 卧式后缩聚釜传动装置结构图
Structure drawing of the horizontal post-polycondensation kettle gearing

1—后缩聚釜 Post-polycondensation kettle；2—引料螺旋带 Cited material spiral band；
3—出料螺杆 Outlet screw；4—机械密封 Machanical seal；5—巴氏合金轴承 Babbitt bearing；
6—电机 Motor；7—齿轮变速箱 Gear box；8—蜗轮蜗杆减速器 Worm gear reducer

4.2.4 纺前设备 Pre-spinning equipment

4.2.4.1 纺前干燥设备 Pre-spinning drying equipment

1. 真空转鼓干燥机 Vacuum drum drying machine

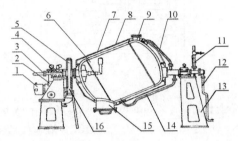

图 4.2.7 真空转鼓干燥机结构图 Structure drawing of vacuum drum drying machine

1—电机 Electric motor；2—减速器 Reducer；3—抽真空管 Vacuum tube；4—滑动轴承 Sliding bearing；
5—大齿轮 Big gear wheel；6—转鼓 Barrate；7—加热夹套 Heating jacket；8—保温层 Insulating layer；
9—进、出料口 Feed inlet&outlet；10—蒸汽管 Steam pipe；11—热载体进入管 Heat carrier inlet pipe；
12—热载体回流管 Heat carrier return pipe；13—支座 Support；14—冷凝水管 Condensate water pipe；
15—人孔 Manhole；16—小齿轮 Pinion

2. 真空转鼓干燥机轴头 Vacuum drum dryer shaft head

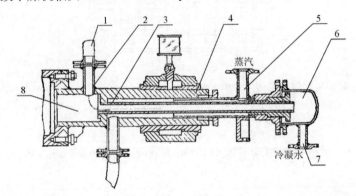

图 4.2.8 真空转鼓干燥机轴头结构图 Structure drawing of vacuum drum dryer shaft head

1—接管 Connecting pipe；2—回流管 Return pipe；3—轴 Spindle head；4—轴套 Sleeve；
5—进汽管 Steam inlet pipe；6—收集室 Collection chamber；7—排水管 Drain pipe；
8—回流室 Return chamber

3. 回转圆筒干燥机 Rotary drum dryer

回转圆筒干燥机由筒体、纱板、旋风分离器和传动装置组成。

Rotary drum dryer consists of barrel, shovelling plate, cyclone separator group and transmission parts.

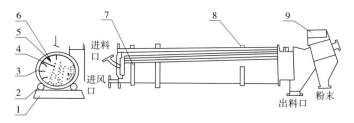

图 4.2.9　回转圆筒干燥机结构图 Structure drawing of Rotary drum dryer

1—底座 Pedestal；2—传动齿轮 Transmission gear；3—大齿轮 Big gear wheel；4—切片 Cutting disk；5—纱板 Plate；6—回转筒体 Rotary barrel；7—止推轮 Thrust wheel；8—托轮 Riding wheel；9—回风口 Return air inlet

4. 料量控制器 Feed quantity controller
5. 填充式干燥机 Pacded dryer

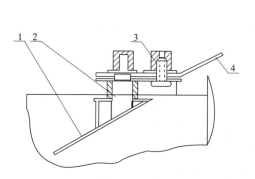

图 4.2.10　料量控制器结构图
Structure drawing of feed quantity controller

1—控制板 Control board；2—斜面螺栓 Cant bolt；
3—固定螺栓 Fastening screw；4—手柄 Hand shank

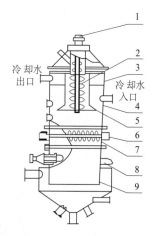

图 4.2.11　填充式干燥机结构图
Structure drawing of pacded dryer

1—传动装置 Transmission equipment；
2—立式搅拌器 Vertical agitator；
3—上筒体 Upper shell；
4—中间筒体 Middle shell；
5—套筒 Sleeve；
6—炉栅搅拌器 Grid melter agitator；
7—炉栅搅拌筒体 Grate stirring barrel；
8—支座 Support；9—下筒体 Lower shell

6. 预结晶器 Pre-crystallizer

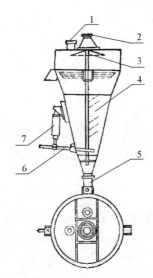

图4.2.12 预结晶器结构图 Structure drawing of pre-crystallizer

1—进料口 Feed port；2—排气口 Exhaust port；3—锥形罩 Conical cap；
4—撞击棒 Impact bar；5—出料口 Feed port；6—杠杆 Lever；7—气缸 Air cylinder

4.2.4.2 熔体过滤器 Melt filter

熔体过滤器是由许多整齐排列的烛芯元件装在圆筒中构成的。

Melt filter is composed of many candlewick components those lined up in the cylinder.

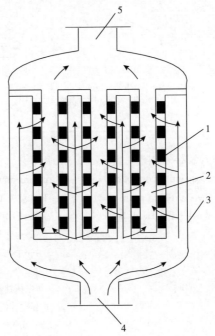

图4.2.13 熔体过滤器结构图 Structure drawing of melt filter

1—过滤介质 Filtering medium；2—烛芯 Candlewick；
3—筒体 Barrel；4—熔体进口 Melt inlet；5—熔体出口 Melt outlet

4.2.5 纺丝设备 Spinning equipment

4.2.5.1 螺杆挤出机 Screw extruder

螺杆挤出机有卧式和立式两种结构。螺杆在空间呈水平安装的为卧式,呈垂直安装的为立式。

Screw extruder is divided to two structures, horizontal type and vertical type. For horizontal type, the screw is installed horizontally in space, while for the other type, the screw is installed vertically.

1. 卧式螺杆挤出机 Horizontal screw extruder

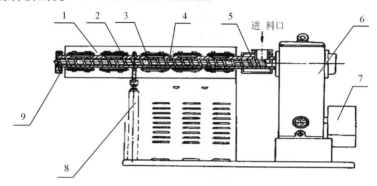

图 4.2.14　卧式螺杆挤出机结构图 Structure drawing of horizontal screw extruder
1—保温套 Cosy;2—加热器 Heater;3—套筒 Sleeve;4—螺杆 Screw;
5—冷却夹套 Cooling jacket;6—变速箱 Transmission case;
7—电机 Electric motor;8—支柱 pillar;9—出料滤板 Discharge filter plate

2. 立式螺杆挤出机 Vertical screw extruder

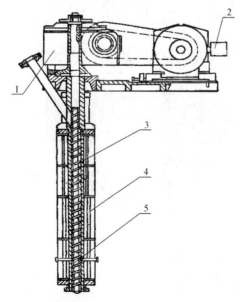

图 4.2.15　立式螺杆挤出机结构图 Structure drawing of vertical screw extruder
1—传动箱 Transmission case;2—电机 Electric motor;3—套筒 Sleeve;4—加热器 Heater;5—螺杆 Screw

3. 压力传感器 Pressure sensor

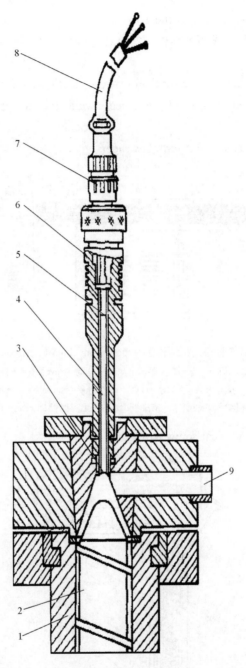

图 4.2.16　压力传感器结构图 Structure drawing of pressure sensor

1—套筒 Sleeve；2—螺杆 Screw；3—薄膜片 Film sheet；4—应变杆 Strain pole；5—应变片 Strain sheet；6—外壳 Outermost shell；7—压紧螺母 Compact sheet；8—连接导线 Connecting lead；9—熔体出口 Melt outlet

4 合成纤维 Synthetic fiber

4.2.5.2 齿轮增压泵 Gear booster pump

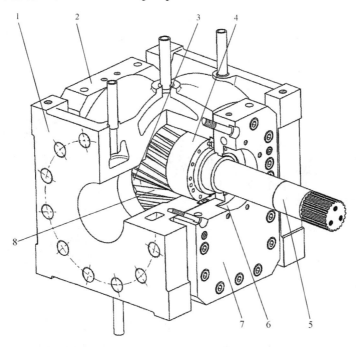

图 4.2.17 齿轮增压泵结构图 Structure drawing of gear booster pump

1—泵体 Pump body；2—非传动侧泵盖 Non-drive side-pump cover；3—非传动侧滑动轴承 Non-drive side-sliding bearing；4—传动侧滑动轴承 Drive side-sliding bearing；5—主动轴 Transmission shaft；6—密封盘 Sealing plate；7—传动侧泵盖 Drive side-pump cover；8—从动轴 Driven shaft

4.2.5.3 纺丝箱体 Spinning box

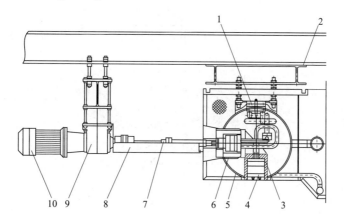

图 4.2.18 纺丝箱体结构图 Structure drawing of spinning box

1—测试法兰 Testing flange；2—传送架 Transmission frame；3—喷丝保护板 Spinning fender；4—纺丝组件 Spinning subassembly；5—纺织束管壳 Spinning beam tube shell；6—纺丝泵 Spinning pump；7—驱动轴 Drive shaft；8—保护架 Guard；9—变速器 Transmission；10—电机 Motor

4.2.5.4 计量泵 Metering pump

计量泵又称纺丝泵，一般采用结构简单的齿轮泵。

Metering pumps, also known as spinning pump, generally a simple gear pump structure.

1. 纺丝计量泵 Spinning metering pump

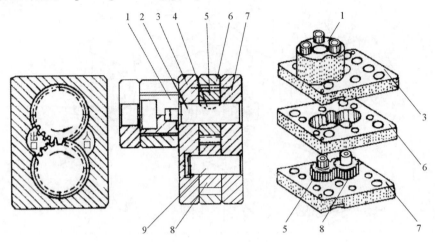

图 4.2.19 纺丝计量泵/熔纺泵结构图
Structure drawing of spinning metering pump/melt spinning pump
1—联轴器 Coupling；2—主动轴 Driving shaft；3—上盖板 Upper cover plate；4—键 Key；
5—主动齿轮 Driving gear；6—中间板 Middle plate；7—下盖板 Lower cover plate；
8—从动齿轮 Driven gear；9—从动轴 Driven shaft

2. 三齿轮计量泵 Three-gear metering pump

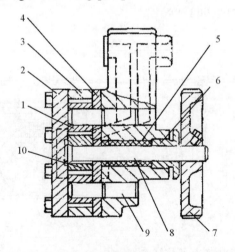

图 4.2.20 三齿轮计量泵结构图 Structure drawing of three-gear metering pump
1—从动齿轮 Driven gear；2—外盖板 Outer cover plate；3—中间板 Middle plate；
4—内盖板 Inner cover plate；5—填料 Packing；6—填料压盖 Stuffing gland；
7—传动齿轮 Driven gear；8—主动轴 Driving shaft；9—从动轴 Driven shaft；
10—主动齿轮 Driving gear

4 合成纤维 Synthetic fiber

3. 计量泵传动轴 Transmission shaft of metering pump

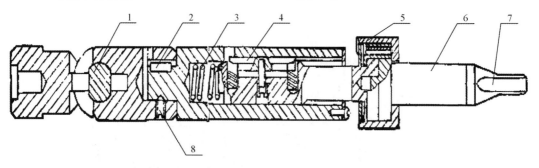

图 4.2.21　计量泵传动轴结构图 Structure drawing of transmission shaft of metering pump
1—万向联轴节 Universal coupling；2—键 Key；3—弹簧 Spring；4—滑键 Sliding key；
5—安全销 Safety pin；6—传动轴 Transmission shaft；7—键槽 Keyway；8—顶丝 jackscrew

4.2.5.5　纺丝组件 Spinning subassembly

1. 短纤维高压纺丝组件 Short fiber high-pressure spinning subassembly

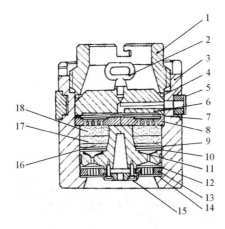

图 4.2.22　短纤维高压纺丝组件结构图 Structure drawing of short fiber high-pressure spinning subassembly
1—压紧螺母 Gland nut；2—吊环 Ring；3—喷丝板座 Spinneret holder；4、5—O 形密封圈 O sealing ring；
6—扩散板 Diffuser plate；7、12、14—密封垫片 Sealing gasket；8—分配板 Distributing plate；
9、10—过滤网 Filter screen；11—耐压板 Pressure resistant plate；13—喷丝板 Spinneret plate；
15—压板 Press plate；16、17、18—滤砂 Filter sand

2. 长丝高压纺丝组件
Filament high-pressure spinning subassembly

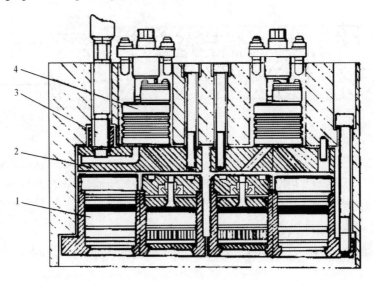

图 4.2.23　长丝高压纺丝组件结构图 Structure drawing of filament high-pressure spinning subassembly
1—纺丝头 Spinning head；2—泵座 Pump base；3—针形阀 Needle valve；4—计量泵 Metering pump

3. 复合纺丝组件（下装式）Composite spinning subassembly（Bottom installation type）

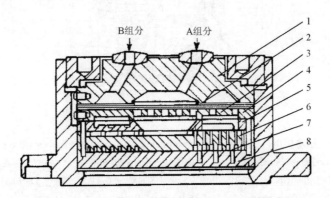

图 4.2.24　复合纺丝组件（下装式）结构图
Structure drawing of composite spinning subassembly（Bottom installation type）
1—扩散板 Diffuser plate；2—密封垫片 Sealing gasket；3—滤网 Filter screen；
4—耐压板 Pressure resistant plate；5—上分配板 Upper distributing plate；
6—下分配板 Lower distributing plate；7—喷丝孔 Spinneret hole；8—喷丝板 Spinneret plate

4 合成纤维 Synthetic fiber

4. 复合纺丝组件(上装式) Composite spinning subassembly (Top installation type)

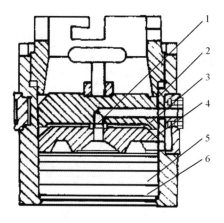

图 4.2.25　复合纺丝组件(上装式)结构图
Structure drawing of composite spinning subassembly

1—密封垫片 Sealing gasket；2—O 形密封圈 O seal ring；3—上扩散板 Upper diffuser plate；
4—下扩散板 Lower diffuser plate；5—分配板 Distributing plate；6—喷丝板 Spinneret plate

5. 皮芯型复合纺丝组件 Core-sheath composite spinning subassembly

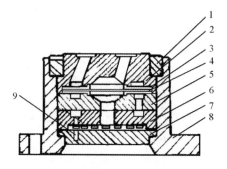

图 4.2.26　皮芯型复合纺丝组件结构图
Structure drawing of core-sheath composite spinning subassembly

1—防松螺钉 Backing-up screw；2—扩散板 Diffuser plate；3—垫圈 Gasket；4—滤网 Filter screen；
5—上分配板 Upper distributing plate；6—下分配板 Lower distributing plate；7—喷丝板 Spinneret plate；
8—喷丝板座 Spinneret plate base；9—喷丝孔 Spinneret orifice

6. 普通短纤维纺丝组件 Ordinary staple spinning subassembly

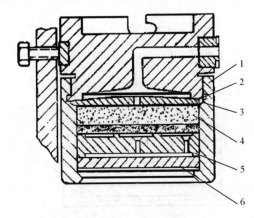

图 4.2.27　普通短纤维纺丝组件结构图 Structure drawing of ordinary staple spinning subassembly
1—喷丝板座 Spinneret plate base；2—分配板 Distributing plate；3—压紧螺母 Gland nut；
4—过滤层 Filter layer；5—耐压板 Pressure resistant plate；6—喷丝板 Spinneret plate

7. 湿法纺丝组件 Wet spinning subassembly

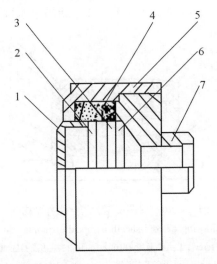

图 4.2.28　湿法纺丝组件结构图 Structure drawing of wet spinning subassembly
1—喷丝帽 Spinneret cap；2、3、6—垫片 Gasket；4—滤布 Filter cloth；
5—上螺母 Upper nut；7—下螺母 Lower nut

8. 板式纺丝组件 Plate type spinning subassembly

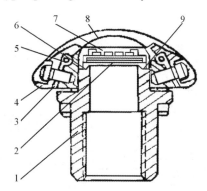

图 4.2.29 板式纺丝组件结构图 Structure drawing of plate type spinning subassembly

1—分配板座 Distributing plate support；2—滤网 Filter screen；3—扇形板 Sector plate；
4—压板 Press plate；5—压棒 Press bar；6、9—密封圈 Seal ring；7—分配板 Distributing plate；
8—喷丝帽 Spinneret cap；

4.2.5.6 卷绕机构 Winding mechanism

卷绕机构可按运动规律分成两部分，完成往复运动的导丝机构和完成回转运动的卷取机构。

According to the requirement of the motion way, winding mechanism could be divided into two part: one is thread guide mechanism, which complete reciprocating motion, other one is winding down frame, which complete rotational motion.

1. 导丝机构 Thread guide mechanism

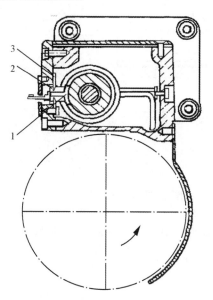

图 4.2.30 导丝机构结构图 Structure drawing of thread guide mechanism

1—滑梭 Shuttle；2—导丝器 Thread guide；3—导丝压板 Thread guide press plate

2. 筒管夹头 Bobbin chuck

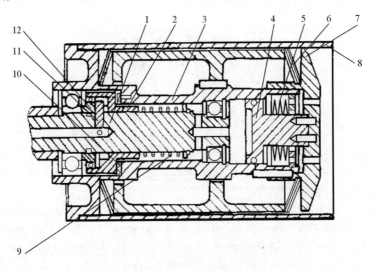

图 4.2.31 筒管夹头结构图 Structure drawing of bobbin chuck
1—芯轴 Pivot；2—滑套 Sliding sleeve；3—筒管座 Bobbin seat；4—活塞 Piston；5—锥形弹簧 Conical spring；
6—碟形弹簧 Disk spring；7—弹簧座 Spring seat；8—筒管 Bobbin；9—压缩弹簧 Compression spring；
10—制动片 Brake pad；11—定位销 Locating pin；12—套环 Lanter ring

3. 直径式控制卷取机构 Diameter control winding mechanism

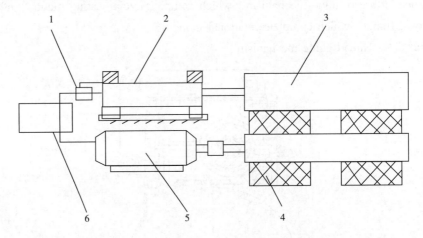

图 4.2.32 直径式控制卷取机构结构图 Structure drawing of diameter control winding mechanism
1—电位器 Potential device；2—同步电机 Synchronous motor；3—控制辊 Control roller；4—卷辊 Winding up roller；
5—变速电机 Variable speed motor；6—调速器 Control governor

4 合成纤维 Synthetic fiber

4. 张力式控制卷取机构 Coiling tension control mechanism

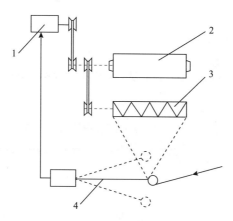

图 4.2.33　张力式控制卷取机构结构图
Structure drawing of coiling tension control mechanism
1—电机 Electric moter；2—卷辊 Winding roller；
3—导丝凸轮 Thread guide cam；4—张力检测杆 Tension detection lever

5. 热拉伸盘 Hot stretch disk

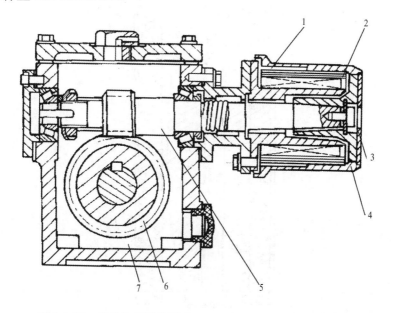

图 4.2.34　热拉伸盘结构图 Structure drawing of hot stretch disc
1—测温元件 Temperature measurer；2—拉伸盘 Stretch disc；3—端盖 End cap；
4—加热线圈 Load coil；5—蜗杆 Worm rod；6—蜗轮 Worm wheel；7—蜗轮箱 Turbine case

6. 小转子 Small rotor

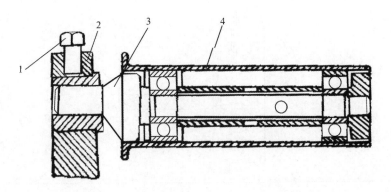

图 4.2.35 小转子结构图 Structure drawing of small rotor

1—调节螺栓 Adjusting screw；2—机架 Rack；3—小转子 Small rotor；4—套筒 Small sleeve

7. 加热板 Heating plate

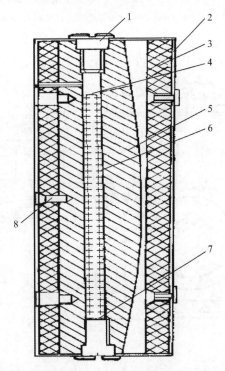

图 4.2.36 加热板结构图 Structure drawing of heating plate

1—螺塞 Plug；2—外壳 Shell；3—保温层 Heat preservation cover；
4—电热丝 Heating wire；5—热板 Hot plate；6—绝缘瓷环 Insulating ceramic ring；
7—云母垫片 Mica gasket；8—测温孔 Thermometer hole

8. 槽式加热箱 Slot type heating box

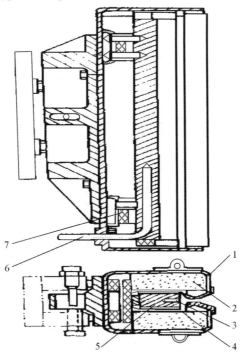

图 4.2.37 槽式加热箱结构图 Structure drawing of slot type heating box
1—壳体 Shell；2—保温层 Insulating layer；3—电阻丝通道 Resistance wire channel；
4—铝板 Aluminum plate；5—电阻丝加热器 Resistance wire heating；
6—测温孔 Thermometer hole；7—电源接点 Power connection point

9. 锭子变速机构 Spindle speed change mechanism

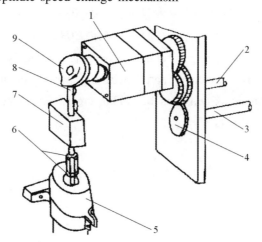

图 4.2.38 锭子变速机构结构图 Structure drawing of spindle speed change mechanism
1—减速器 Reducer；2—变速传动轴 Variable speed transmission shaft；3—复位传动轴 Restoration transmission shaft；
4—复位传动齿轮 Restoration transmission gear；5—差动变速器 Differential transmission；
6—动程调整机构 Stroke adjustment mechanism；7—支架 Support；8—顶杆 Ejector rob；
9—变速凸轮 Variable speed cam

10. 平移式摩擦辊支承 Translational friction roller bearing

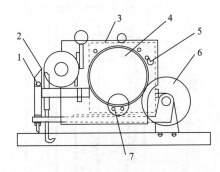

图 4.2.39 平移式摩擦辊支承结构图
Structure drawing of translational friction roller bearing

1—斜轴 Tilting axis；2—定位凹槽 Positioning groove；3—箱体 Box；4—摩擦辊 Friction roller；
5—导丝器 Thread guide；6—卷辊 Winding roller；7—滚轮 Idler wheel

4.2.6 纺后设备 Post-spinning equipment

由常规纺(或高速纺)纺出的成形丝(初生纤维)的特点是强力低、伸度高，不具备纺织加工的性能。因此，这种初生纤维必须经过一系列后道工序的处理，如拉伸加捻、假捻变形和热定形等。

The formed wire(nascent fiber) which is span out by conventional spinning (or high-speed spinning) is characterized by low strength, stretch high. But it does not have the properties of textile processing. Thus, this nascent fibers must be treated after a series of processes, such as tensile twisting, false-twist texturing, heat setting, etc.

4.2.6.1 牵伸加捻机 Draft twisting machine

牵伸加捻机的主要作用是拉伸，其次是加捻。牵伸加捻机分为喂入、拉伸、加捻、卷绕、成形和传动等部分，它包括筒子架、导丝杆和横动导丝机构等。

The main role of draft twisting machine isstretching, secondly, twisting. Draft twisting machine includes feed part, stretching part, twist ing part, winding part, forming part, transmission part, etc. And it is consisted of bobbin creel, thread guide bar, traverse thread guide mechanism and other parts.

1. 筒子架 Bobbin creel

1)直放固定式筒子架 Vertically fixed bobbin creel

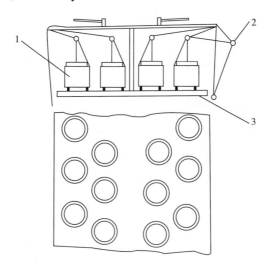

图 4.2.40 直放固定式筒子架结构图 Structure drawing of vertically fixed bobbin creel
1—筒子丝 Bobbin wire；2—导丝钩 Thread guide hook；3—托板 Pallet

2)斜放固定式筒子架 Diagonal fixed bobbin creel

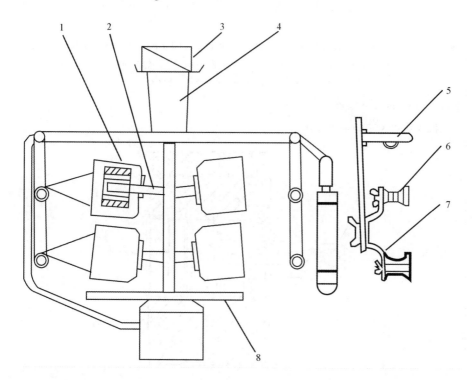

图 4.2.41 斜放固定式筒子架结构图 Structure drawing of diagonal fixed bobbin creel
1—筒子丝 Bobbin wire；2—木板座 Plank support；3—风管 Air hose；4—排风管 Exhaust duct；
5—导丝钩 Thread guide hook；6—导丝棒 Thread guide bar；7—转盘 Turntable；8—隔热板 Thermal baffle

2. 横动导丝机构 Traverse thread guide mechanism

1）偏心式横动装置 Eccentric type traverse device

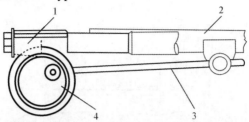

图 4.2.42　偏心式横动装置结构图 Structure drawing of eccentric type traverse device

1—蜗杆 Worm rod；2—横动导杆 Traverse guide bar；3—连杆 Connencting rod；4—蜗轮 Worm wheel

2）凸轮式横动装置 Cam type traverse device

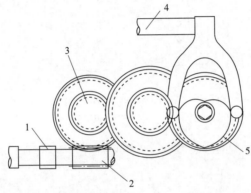

图 4.2.43　凸轮式横动装置结构图 Structure drawing of cam type traverse device outletre

1—喂丝辊 Wire-feeding roller；2—蜗杆 Worm rod；3—蜗轮 Worm wheel；
4—横动导杆 Traverse guide rod；5—凸轮 Cam

3. 压辊 Compression roller

1）上压辊 Upper compression roller

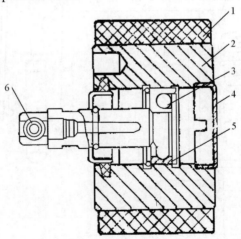

图 4.2.44　上压辊结构图 Structure drawing of upper compression roller

1—丁腈橡胶套 Nitrile rubber sleeve；2—铁芯 Core；3—滚动轴承 Antifriction bearing；
4—端盖 Seal cover；5—弹簧挡圈 Spring collar；6—固定轴 Fixed shaft

4 合成纤维 Synthetic fiber

2）扭簧加压压辊 Torsion spring inflating compression roller

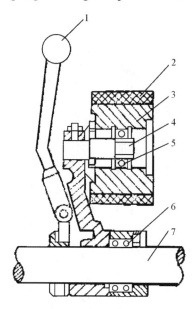

图 4.2.45 扭簧加压压辊结构图 Structure drawing of torsion spring inflating compression roller

1—手柄 Handle；2—丁腈橡胶套 Nitrile rubber sleeve；3—铁芯 Core；4—固定轴 Fixed shaft；

5、6—滚动轴承 Antifriction bearing；7—支承轴 Supporting shaft

4. 牵伸盘 Drafting plate

1）牵伸盘结构 Drafting plate

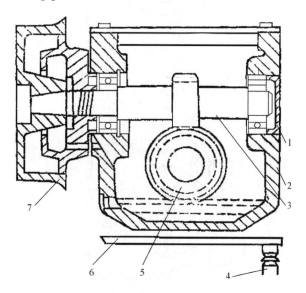

图 4.2.46 牵伸盘结构图 Structure drawing of drafting plate

1—端盖 End cover；2—壳体 Shell；3—蜗杆 Worm rod；4—排油管 Oil discharge pipe；

5—蜗轮 Worm wheel；6—接油盘 Oil reciver；7—牵伸盘 Drafting disc

2) 热箱 Heating case

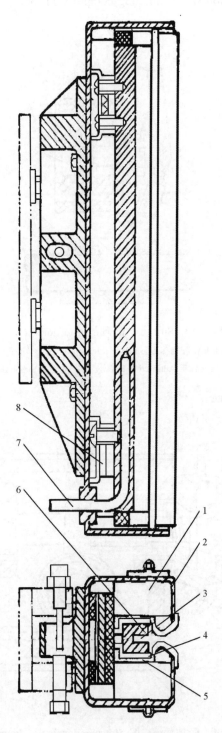

图 4.2.47 热箱结构图 Structure drawing of heating case
1—保温层 Insulating layer；2—壳体 Shell；3—电阻丝通道 Resistance wire channel；
4—电阻丝 Resistance wire；5—绝缘层 Insulating layer；6—铝板 Aluminum plate；
7—测温孔 Temperature measurement hole；8—电源接点 Power connection point

4.2.6.2 加捻变形机 Twist texturing machine

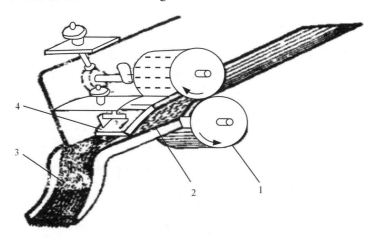

图 4.2.48 加捻变形机结构图 Structure drawing of twist texturing machine
1—卷曲辊 Crimp roller；2—填塞箱 Stuffing box；3—平面卷曲变形丝 Plane crimped wire；
4—弹性活门 Elasticity valve

1. 喂给装置 Feeding device

1）喂丝装置 Feeding device

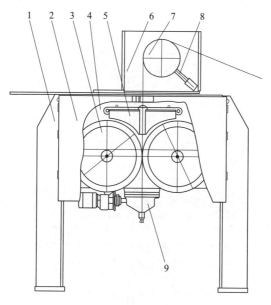

图 4.2.49 喂丝装置结构图 Structure drawing of feeding device
1—机架 Rack；2—转轮保护罩 Wheel protect cover；3—喂入轮 Down beater；4—齿轮箱 Gear box；
5—进口保护器 Entrance protector；6—导盘 Guide plate；7—导向辊 Guide roller；
8—绕辊检测器 Around roll detector；9—转动轮调整器 Rotating wheel adjuster

2)筒子架 Bobbin creel

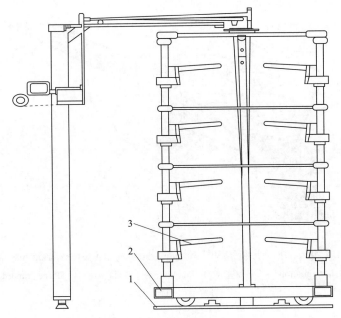

图 4.2.50　筒子架结构图 Structure drawing of bobbin cree
1—地面 Ground；2—旋转轮 Swiveling wheel；3—插座 Socket

2. 握持装置 Holding device

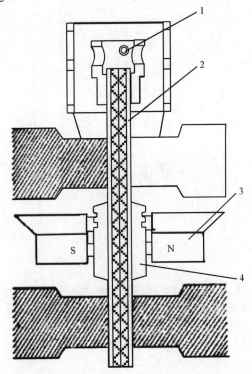

图 4.2.51　握持装置结构图 Structure drawing of holding device
1—横销 Lateral pin；2—套筒 Sleeve；3—磁铁 Magnet；4—矽钢片 Silicon steel sheet

3. 叠盘型加捻器 Folded plate twisting

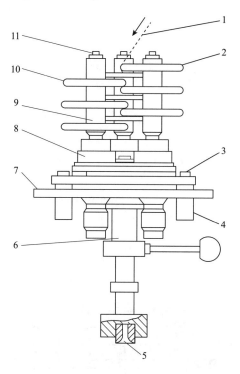

图 4.2.52 叠盘型加捻器结构图 Structure drawing of folded plate twisting

1—丝条 Thread line；2、10—摩擦盘 Friction plate；3—螺母 Nut；4—销轴 Pin roll；5—丝条出口 Sliver outlet；6—导丝管 Thread guide pipe；7—底板 Baseplate；8—轴承 Bearing；9—轴套 Sleeve；11—螺栓 Bolt

4. 球盘型加捻器 Ball-disc twisting

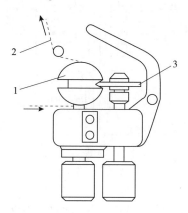

图 4.2.53 球盘型加捻器结构图 Structure drawing of ball-disc twisting

1—摩擦球体 Friction sphere；2—丝条 Thread line；3—摩擦圆盘 Friction disc

5. 双套筒型加捻器 Double-sleeve twisting

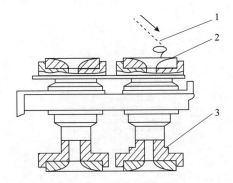

图 4.2.54 双套筒型加捻器结构图 Structure drawing of double-sleeve twistin
1—丝条 Thead line；2—环形导丝管 Annular thread guide pipe；3—传动管 Drive pipe

6. 可调式摩擦加捻器 Variable friction twisting

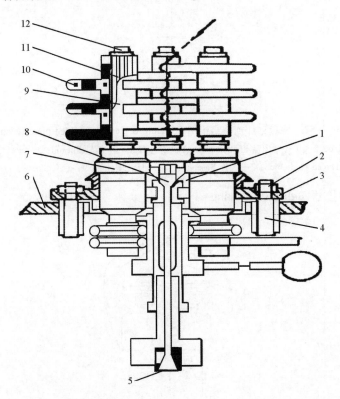

图 4.2.55 可调式摩擦加捻器结构图 Structure drawing of variable friction twisting
1—圆锥轮 Conical wheel；2—螺母 Nut；3—转臂 Tumbler；4—销轴 Pin roll；5—丝条出口 Thread line outlet；
6—底板 Baseplate；7—轴承 Bearing；8—圆锥体导丝管 Cone thread guide pipe；9—轴套 Sleeve；
10—摩擦盘 Friction disc；11—定距套 Fixed pitch sleeve；12—螺栓 Screw

7. 双增速轮式加捻器 Double-multiplying wheel twisting

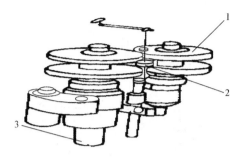

图 4.2.56　双增速轮式加捻器结构图 Structure drawing of double-multiplying wheel twisting
1—增速轮 Multiplying wheel；2—小转子 Small rotor；3—管形皮带轮 Tubular belt pulley

8. 小转子 Small rotor
9. 摩擦式加捻器 Friction twisting

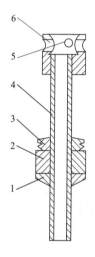 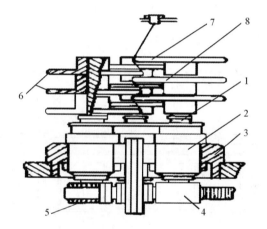

图 4.2.57　小转子结构图
Structure drawing of small rotor
1—下定位圈 Lower locating ring；
2—硅钢片 Silicon steel sheet；
3—上定位圈 Upper locating ring；
4—传动管 Drive pipe；
5—转子头 Rotor head；
6—横销 Lateral pin

图 4.2.58　摩擦式加捻器结构图
Structure drawing of friction twisting
1—轴 Axle；2—轴承 Bearing；
3—轴承座 Bearing pedestal；
4—齿形皮带轮 Tooth profile belt pulley；
5—齿形皮带 Toothed belt；
6、8—摩擦盘 Friction disc；
7—导向盘 Positioning disk；

10. 电器式防凸装置 Electric-type anti-convex device

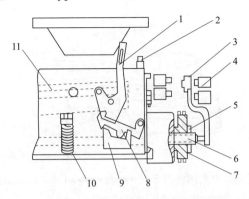

图 4.2.59 电器式防凸装置结构图
Structure drawing of electric-type anti-convex device

1—导丝器 Thread guide；2—成形杆 Forming bar；3、4—行程限位开关 Lead limit switch；
5—螺母 Nut；6—往复螺杆 Reciprocate screw；7—链轮 Chain wheel；8—L形弯杆 L-curved rod；
9—横板 Tabula；10—压簧 Compressed spring；11—变幅导板 Luffing guide

4.2.6.3 集束架 Creel

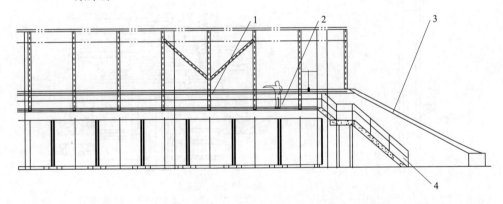

图 4.2.60 集束架结构图 Structure drawing of creel

1—集束架 Creel；2—操作台 Operating floor；3—丝束 Tow；4—楼梯 Stairs

4.2.6.4 整经上浆机 Warping sizing machine

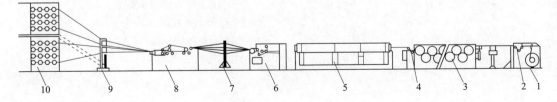

图 4.2.61 整经上浆机结构图 Structure drawing of warping sizing machine

1—轴架 Shaft bracket；2—上油辊 Oiling roller；3—干燥定形辊 Drying stereotypes roller；
4—张力调节辊 Dance roller；5—干燥器 Dryer；6—上浆槽 Sizing groove；7—网络板 Network board；
8—拉伸机构 Stretching mechanism；9—孔板架 Pore plate frame；10—筒子架 Bobbin creel

4.2.6.5 导丝机 Thread guide

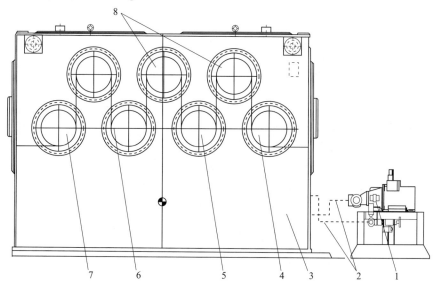

图 4.2.62 导丝机结构图 Structure drawing of thread guide

1—润滑油站 Lubricating oil station; 2—润滑油线 Lubricating oil line;
3—机壳 Housing; 4、5—轴承辊 Bearing roller; 6、7、8—导丝辊 Thread guide roller

4.2.6.6 卷曲装置 Crimping device

1. 牵引机 Tractor

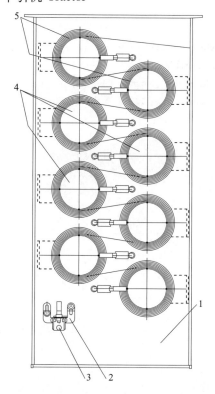

1—机架 Rack; 2—牵引导向装置 Traction guide;
3—连接装置 Connecting device; 4、5—导丝盘 Thread guide plate

图 4.2.63 牵引机结构图 Structure drawing of tractor

2. 上油辊 Oiling roller

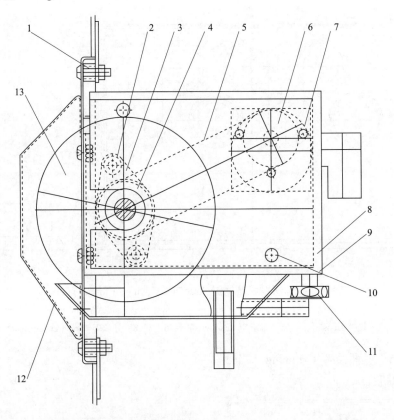

图 4.2.64　上油辊结构图 Structure drawing of oiling roller

1—导管 Conduit；2、7—六角头螺栓、螺母 Hexagonal head screw-nut；3—法兰 Flange；
4、6—皮带轮 Belt pulley；5—齿形皮带 Toothed belt；8—支架 Support；9—角型板 Angle plate；
10—保护器 Protector；11—星型手柄螺栓 Star handle screw；12—引擎罩 Engine cover；
13—润湿盘 Moistening tray

3. 浸油槽 Leaching tank

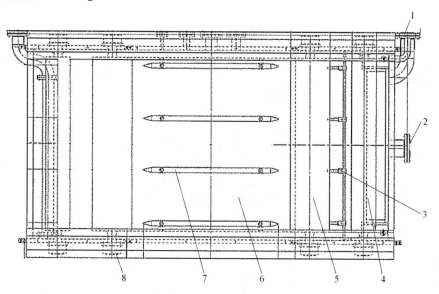

图 4.2.65 浸油槽结构图 Structure drawing of leaching tank

1—油剂入口 Oil inlet；2—油剂出口 Oil outlet；3—分丝棒 Devillicate rod；4—轴密封 Shaft seal；
5—导辊 Guide roller；6—槽头 Trough；7—导流板 Guide plate；8—轴承 Bearing

4. 牵伸槽 Drawing groove

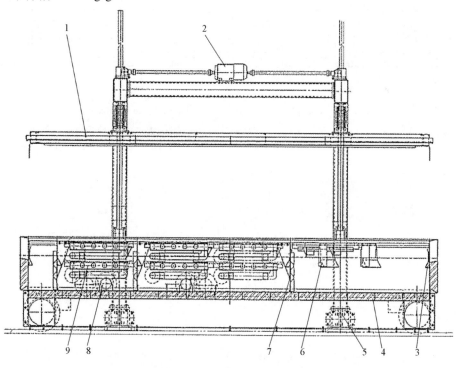

图 4.2.66 牵伸槽结构图 Structure drawing of drawing groove

1—上盖板 Upper cover plate；2—电机 Motor；3—导丝器 Thread guide；4—槽头 Trough；
5—机架 Rack；6—刮削器 Scrapper；7—隔板 Clapboard；8—导管 Conduit；9—喷涂架 Spray frame

5. 牵伸机 Drawing machine

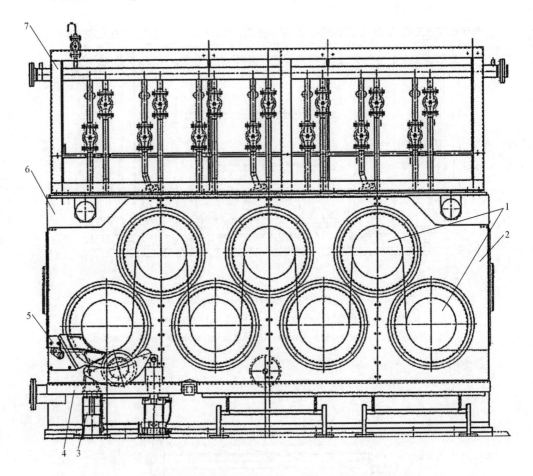

图 4.2.67 牵伸机结构图 Structure drawing of drawing machine
1—导盘 Guide disc；2—机罩 Hood；3—夹丝辊 Clip wire rolls；4—收集缸 Collecting cylinder；
5—进口保护器 Import protection；6—中心润滑器 Center lubricator；7—热水管线 Hot water pipeline

4.2.6.7 纺丝定形设备 Spinning sizing device

1. 蒸汽箱 Steam box

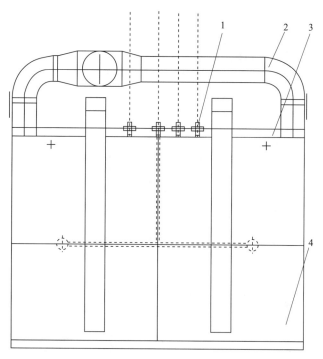

图 4.2.68 蒸汽箱结构图 Structure drawing of steam box

1—蒸汽管线 Steam pipeline；2—排汽管 Steam discharge pipe；3—上盖 Upper cover；4—箱体 Tank

2. 热定形机 Heat-setting machine

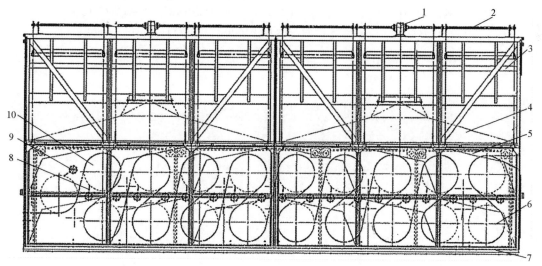

图 4.2.69 热定形机结构图 Structure drawing of heat-setting machine

1—电机 Motor；2—轴 Shaft；3—机外罩 Outer cover；4—排气罩 Exaust stream cover；
5—机前罩 Machine front cover；6—导盘 Guide plate；7—冷凝回流器 Condense return channel；
8、9—绕丝探测器 Tie wire detector；10—轴承辊 Bearing roller

3. 冷却机 Cooler

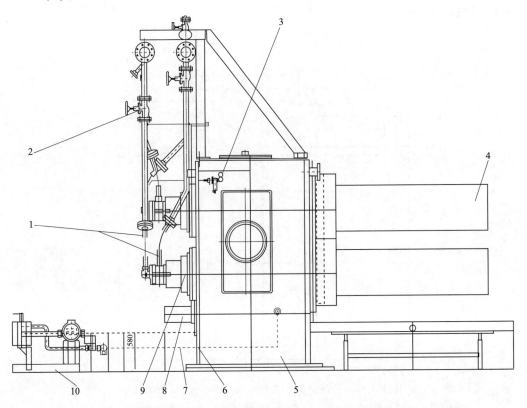

图 4.2.70 冷却机结构图 Structure drawing of cooler

1—软管 Hose；2—阀门 Valve；3—压缩空气供给口 Compressed air supply port；
4—轴辊 Bearing roller；5—传动箱 Transmission case；6—机壳 Housing；7—油线 Oil line；
8—驱动轴 Driving shaft；9—轴承座 Bearing set；10—润滑油站 Lubricant oil station

4 合成纤维 Synthetic fiber

4. 立式上油机 Vertical oiling machine

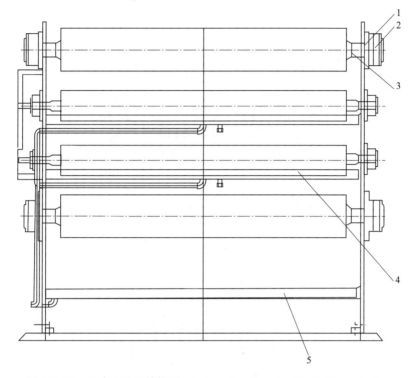

图 4.2.71 立式上油机结构图 Structure drawing of vertical oiling machine

1—O 形箍 O-hoop；2—轴承座 Bearing seat；3—自调节轴辊 Self-adjusting roller；
4—导辊 Guide roller；5—油剂槽 Oil groove

5. 叠丝机 Folding machine

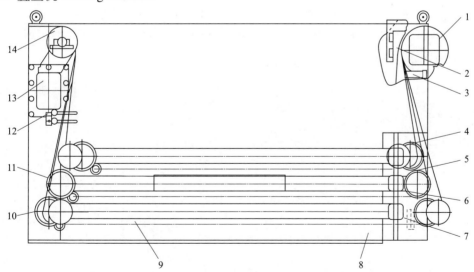

图 4.2.72 叠丝机结构图 Structure drawing of folding machine

1—导辊 Guide roller；2—夹辊 Pinch roll；3—气动系统 Air-operated system；4、6、7—旋转件 Rotating part；
5—调节装置 Regulating device；8—机架 Rack；9—履带 Crawler；10—旋转轴 Rotation shaft；
11—导辊 Guide roller；12—导丝器 Thread guide；13—托架 Bracket；14—进口辊 Inlet roller

6. 蒸汽预热箱 Steam preheat box

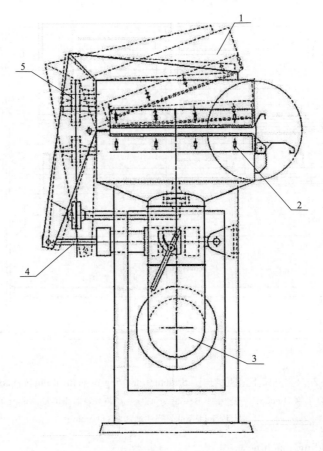

图 4.2.73 蒸汽预热箱结构图 Structure drawing of steam preheat box
1—上盖 Cover；2—喷嘴 Nozzle；3—排汽管 Exhaust pipe；4—汽缸 Cylinder；5—蒸汽管 Steam pipe

7. 松弛定形机 Relaxation setting machine

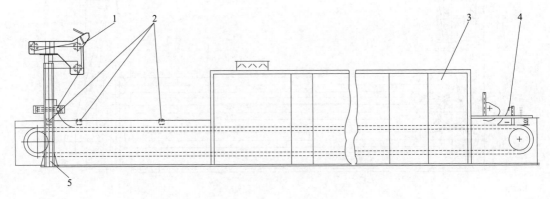

图 4.2.74 松弛定形机结构图 Structure drawing of relaxation setting machine
1—牵引导向装置 Traction guiding device；2—灯光栅 Lamp grating；3—干燥箱 Dryer box；
4—填充盒 Packing box；5—机架 Rack

4 合成纤维 Synthetic fiber

8. 铺丝机 Thread paving machine

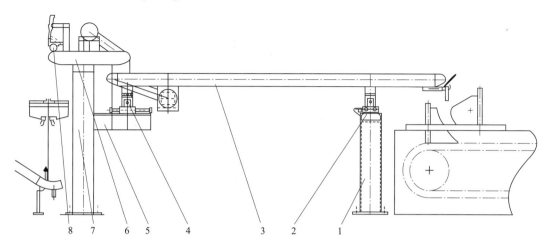

图 4.2.75 铺丝机结构图 Structure drawing of thread paving machine

1—机架 Rack；2—横动装置 Traverse equipment；3、6—输送带 Conveyor；4—旋转轴 Rotating shaft；
5—支架 Support；7—压缩空气通道 Compressed air channel；8—输送辊 Conveyor roller

9. 电阻丝加热器 Resistance wire heater

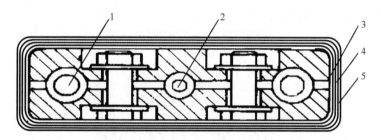

图 4.2.76 电阻丝加热器结构图
Structure drawing of resistance wire heater

1—加热管 Heating pipe；2—测温、控温管 Thermometry&temperature control pipe；3—绝缘漆 Insulating paint；
4—电阻丝缠绕层 Resistance wire winding layer；5—玻璃丝带包覆层 Glass bobbin coating layer

· 371 ·

10. 热媒加热器 Heating medium heater

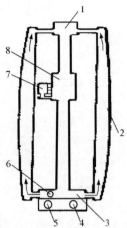

图 4.2.77 热媒加热器结构图 Structure drawing of heating medium heater
1—集气管 Gas collecting pipe；2—热板 Hot plate；3—热媒 Heating medium；
4—调节加热电热棒 Adjusted electric heating rod；5—基本加热电热棒 Fundamental electric heating rod；
6—测温口 Thermometry hole；7—压力控制器 Pressure controller；8—冷凝器 Condenser

4.2.6.8 切断设备 Cutting device

1. 跳辊张力机 Jump roller tension machine

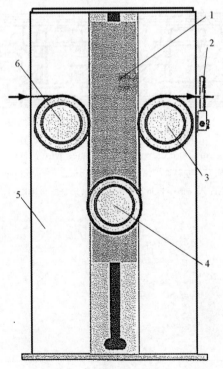

图 4.2.78 跳辊张力机结构图 Structure drawing of jump roller tension machine
1—传送器 Transporter；2—导丝器 Thread guide；3—出口辊 Outlet roller；
4—跳动辊 Jump roller；5—机壳 Housing；6—进口辊 Inlet roller

4 合成纤维 Synthetic fiber

2. 切断张力机 Cutting tension machine

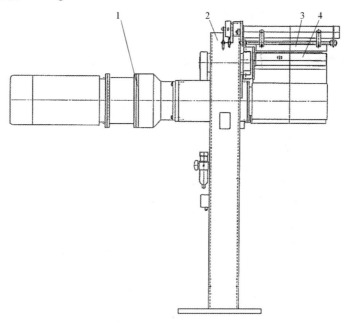

图 4.2.79 切断张力机结构图 Structure drawing of cutting tension machine
1—驱动装置 Driving device；2—支架 Support；3—顶部辊 Top roller；4—底部辊 Bottom roller

3. 切断机 Cutting machine

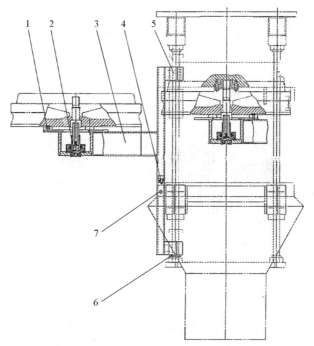

图 4.2.80 切断机结构图 Structure drawing of cutting machine
1—螺栓 Bolt；2—耦合器 Coupler；3—支承架 Support frame；4—夹头 Chuck；
5—轴承 Bearing；6—垫圈 Washer；7—六角头螺栓 Hexagonal head screw

4.3 腈纶 Acrylic fibers

4.3.1 纺丝原液制备设备 Spinning solution preparation equipment

4.3.1.1 聚合反应釜 Polymerization reactor

1. 湿法聚合反应釜 Wet process polymerization reactor

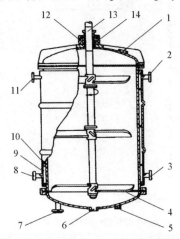

图 4.3.1 湿法聚合反应釜结构图
Structure drawing of wet process polymerization reactor

1—出料口 Discharge port；2—夹套水溢流口 Jacket water overflow port；3—夹套排水口 Jacket drain；4—搅拌桨叶 Stirring paddle；5—排放口 Discharge outlet；6—底阀；7—进料口 Feed port；8—夹套水进口 Jacket water inlet；9—夹套 Jacket；10—釜体 Autoclave body；11—夹套水出口 Jacket water outlet；12—填料密封 Packing seal；13—搅拌轴 Mixing shaft；14—轴套 SSleeve

2. 干法聚合反应釜 Dry process polymerization reactor

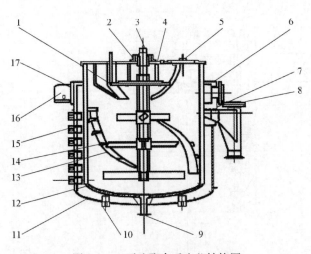

图 4.3.2 干法聚合反应釜结构图
Structure drawing of dry process polymerization reactor

1—环管喷淋 Loop spray；2—水封套 Water jacket；3—搅拌轴 Mixing shaft；4—上端盖 Upper cover；5—人孔 Manhole；6—冷却水出口 Cooling water outlet；7—方法兰 Square flange；8—溢流管 Overflow pipe；9—底阀 Bottom valve；10—夹套水进口 Jacket water nozzle；11—夹套 Jacket；12—釜体 Autoclave body；13—折流板 Baffles；14—搅拌桨叶 Stirring blades；15—侧向夹套水进口 Lateral jacket water nozzle；16—支座 Support；17—补强板 Stiffener

· 374 ·

3. 聚合反应釜主要零部件 Polymerization reactor parts

1）搅拌轴 Mixing shaft

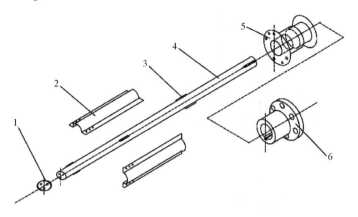

图 4.3.3　搅拌轴结构图 Structure drawing of mixing shaft

1—搅拌轴底部端盖 Mixing shaft bottom cover；2—搅拌轴保护套 Mixing shaft protective sleeve；
3—桨叶连接键 Blade linkage；4—搅拌轴 Mixing shaft；5—水封套 Water jacket assembly；6—联轴器 Coupling

2）夹套水喷嘴 Jacket water nozzle

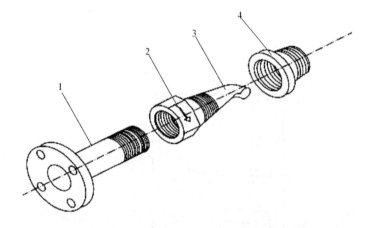

图 4.3.4　夹套水喷嘴结构图
Structure drawing of jacket water nozzle

1—连接短管 Junction pipe；2—喷嘴安装指向 Nozzle mounting direction；
3—喷嘴 Nozzle；4—丝套 Wire set

3）放空喷淋器 Unloading spray thrower

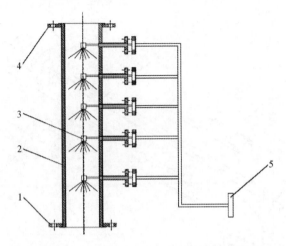

图 4.3.5　放空喷淋器结构图 Structure drawing of unloading spray thrower
1、4、5—法兰 flange；2—筒体 Cylinder；3—喷嘴 Nozzle

4.3.1.2　试剂混合槽 Reagent mixing tank

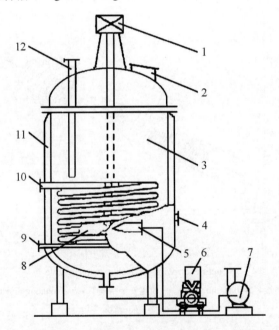

图 4.3.6　试剂混合槽结构图 Structure drawing of reagent mixing tank
1—电机 Motor；2—人孔 Manhole；3—筒体 Barrel；4—主料入口 Main ingredient inlet；
5—试剂出口 Reagent outlet；6—比例泵 Proportional pump；7—循环泵 Circulating pump；
8—搅拌器 Agitator；9—冷却水入口 Cooling water inlet；10—冷却水出口 Cooling water outlet；
11—夹套 Jacket；12—配料入口 Ingredients entrance

4.3.1.3 脱单体塔 Demonomerization tower
4.3.1.4 脱泡塔 Deaeration tower

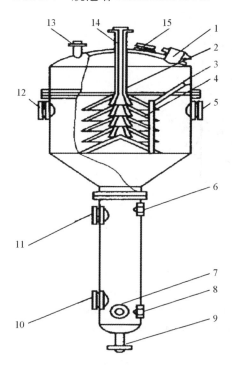

图 4.3.7 脱单体塔结构图
Structure drawing of demonomerization tower

1—抽真空口 Evacuation port；
2—锥形套管 Tapered sleeve；
3—固定架 Supporting structure；
4—伞面 Umbrella plate；
5、10、11、12—视镜 Sight glass；
6、8—液位计接口 Liquid level gauge interface；
7—阀接口 Safety valve interface；
9—出料口 Discharge opening；
13—真空表 Vacuum table；
14—进料口 Feed opening；
15—视窗 Sight window

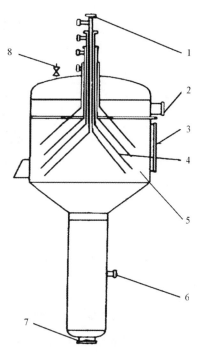

图 4.3.8 脱泡塔结构图
Structure drawing of deaeration tower

1—进料管 Feed pipe；
2—视镜 Sight glass；
3—视窗 Sight window；
4—伞面 Umbrella plare；
5—筒体 Barrel；
6—液位视镜 Level sight glass；
7—出料口 Discharge port；
8—抽真空口 Vacuum orifice

4.3.1.5　多级混合器 Multistage mixer

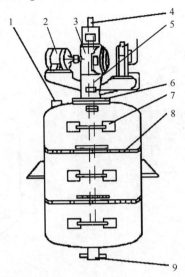

图4.3.9　多级混合器结构图 Structure drawing of multistage mixer

1—进料口 Feed inlet；2—电机 Motor；3—减速箱 Reducer box；
4—搅拌轴 Mixing shaft；5—机械密封 Mechanical seal；6—补强板 Sealing device；
7—搅拌桨叶 Stirring paddle；8—挡板 Baffle；9—出料口 Discharge port

4.3.1.6　板框压滤机 Plate-and-frame filter press

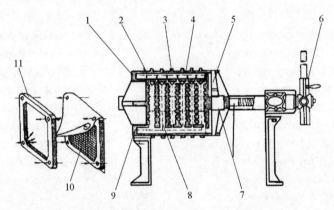

图4.3.10　板框压滤机结构图 Structure drawing of plate-and-frame filter press

1—滤液出口 Filtrate outlet；2—固定端板 Fixed end plate；3、11—滤框 Filter box；
4、10—滤板 Filter plate；5—压紧板 Pressing plate；6—压紧手轮 Compress hand wheel；
7—滑轨 Sliding rail；8—滤布 Filter cloth；9—滤液进口 Filtrate inlet

4.3.1.7 勺管调速型液力耦合器 Spoon speed regulation hydraulic coupler

1. 勺管调速型液力耦合器结构 Scoop speed regulation hydraulic coupler

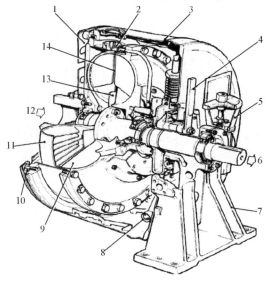

图 4.3.11　勺管调速型液力耦合器结构图 Structure drawing of scoop control fluid coupler

1—驱动板 Resilient driving plate；2—喷嘴 Leak off nozzle；3—储油凹管 Reservoir casing；
4—勺管操纵杆 Scoop housing bracket；5—注油漏斗 Filling tundish；6—输出轴 Output shoft；
7—支架 Support；8—勺管 Scoop；9—蜗轮 Worm wheel；10—壳体 Back casing assembly；
11—泵轮 Runner；12—输入轴 Input shaft；13—隔板 Baffle plate；14—工作油回路 Working oil circuit

2. 勺管室 housing

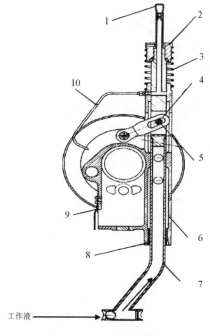

图 4.3.12　勺管室结构图
Structure drawing of housing

1—油吸管 Oil dipper；2—止退环 Collar stop；3—弹簧 Spring；4—控制杠杆 Control lever；5—勺管销 Scoop pin；6—勺管室 Scoop housing；7—勺管 Scoop tube assembly；8—底部衬套 Bottom bush；9—节流挡板 Deflector plate；10—喂入管 Feed tube

4.3.1.8 真空转鼓过滤机 Vacuum drum filter

1. 真空转鼓过滤机外形 Vacuum drum filter

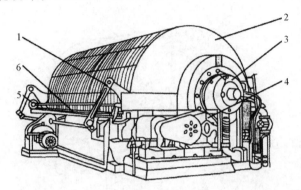

图 4.3.13 真空转鼓过滤机外形图 of vacuum drum filter

1—搅拌装置 Stirring device；2—转鼓 Drum；3—分配头 Distribution head；
4—传动系统 Drive system；5—钢丝缠绕装置 Wire winding device；
6—料浆储槽 Slurry storage tank

2. 真空转鼓过滤机结构 Vacuum drum filter

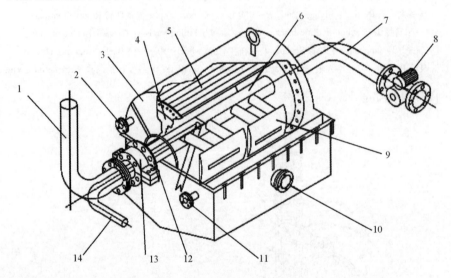

图 4.3.14 真空转鼓过滤机结构图 Structure drawing of vacuum drum filter

1—抽真空管 Pumping vacuum line；2—洗涤水入口 Wash water inlet；3—上盖 Upper cover；
4—转鼓 Drum；5—滤板 Filter plate；6—空心轴 Hollow shaft；7—反吹气管 Back flushing pipe；
8—脉冲阀 Pulse valve；9—反吹靴 Cleaning boots；10—视镜 Endoscopy；11—氮气入口 Nitrogen fill inlet；
12—链轮 Sprocket；13—轴承箱 Bearing box；14—滤液进口管 filter liquor inlet pipe

3. 真空转鼓过滤机主要零部件 Vacuum drum filter components

1) 反吹靴 Blowback boots

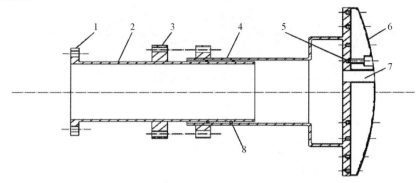

图 4.3.15 反吹靴结构图 Structure drawing of blowback boot

1—法兰 Flange；2—固定靴 Fixed boots；3—调整丝孔 Adjustment wire hole；4—活动靴 Activity boots；
5—内六角螺栓 Inner hexagonal bolts；6—聚四氟乙烯密封板 PTFE sealing plate；
7—反吹气体出口 Blowback gas outlet；8—O 形密封圈 O sealing ring

2) 转鼓过滤面 Drum filtering surface

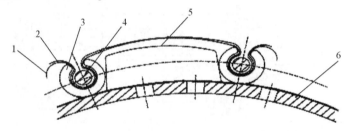

图 4.3.16 转鼓过滤面结构图 Structure drawing of drum filtering surface

1—衬布 Lining cloth；2—滤布 Filter cloth；3—滤布压管 Filter cloth pressure pipe；
4—滤布压条 Filter cloth mound layer；5—滤板 Filter plate；6—转鼓 Drum

4.3.1.9 三级蒸汽喷射泵 Level 3 steam-jet pump

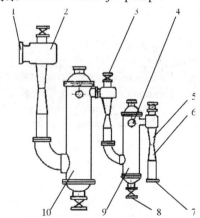

图 4.3.17 三级蒸汽喷射泵外形图 Structure drawing of level 3 steam-jet pump

1—真空接口 Vacuum interface；2—喷射器 Ejector；3—蒸汽进口 Steam inlet；4—冷却水进口 Cooling water inlet；
5—喷嘴 Nozzle；6—喇叭管 Horn pipe；7—尾气出口 Exhaust outlet；8—冷却水出口 Cooling water outlet；
9—二级蒸汽冷凝器 Level 2 steam condenser；10——级蒸汽冷凝器 Level 1 steam condenser

4.3.1.10 管道混合器 Pipeline mixer

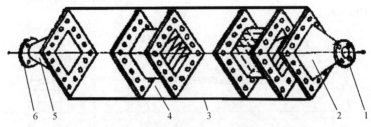

图 4.3.18 管道混合器结构图 Structure drawing of pipeline mixer

1、6—法兰 Flange；2、5—管接头 joint；3—筒体 Barrel；4—混合器 Mixer

4.3.1.11 溶解机 Dissolution machine

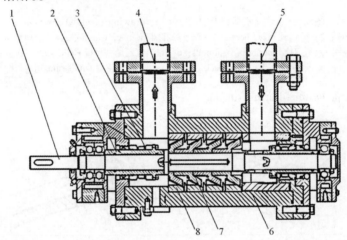

图 4.3.19 溶解机结构图 Structure drawing of dissolution machine

1—传动轴 Drive shaft；2—轴套 Sleeve；3—波纹管密封 Bellows seal；4—原液出口管 Liquid outlet pipe；5—淤浆进口管 Slurry inlet tube；6—机体 Compressor body；7—叶轮 Wheel；8—搅拌帽 Stir cap

4.3.1.12 干燥机 Drying machine

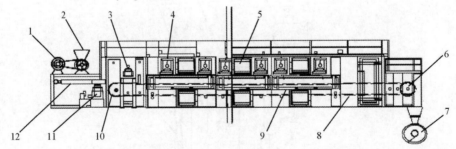

图 4.3.20 干燥机结构图 Structure drawing of drying machine

1、11—电机 Motor；2—挤条机 Extruding machine；3—热风机 Warm air machine；4—循环风机 Circulating fan；5—机门 Door；6—主动链轮 Drive sprocket；7—螺旋出料器 Spiral discharger；8—链条 Chain；9—加热器 Heater；10—从动链轮 Driven sprocket；12—铺匀机 Paving machine

4.3.1.13 挤条机 Banded extruder

1. 挤条机结构 Banded extruder Structure

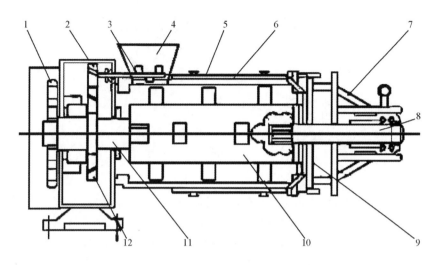

图 4.3.21 挤条机结构图 Structure drawing of banded extruder

1—链轮 Sprocket；2—从动齿轮 Driven gear；3—拨料轴 Dial feed axis；4—物料入口 Material inlet；
5—冷却夹套 Cooling jacket；6—机壳 Housing；7—挤出头 Extrusion head；8—刀片轴 Blade shaft；
9—面板 Panel；10—挤料轴 Material squeezing shaft；11—主动轴 Driving shaft；12—主动齿轮 Driving gear

2. 挤条机主要零部件 Banded extruder components

1）挤料轴 Crowded material shaft

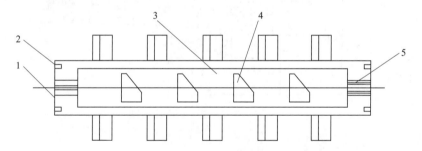

图 4.3.22 挤料轴结构图 Structure crowded material shaft

1—传动轴套 Drive shaft sleeve；2—连接螺栓 Connecting bolts；
3—空心管 Hollow pipe；4—推料翅片 Pushing fins；5—花键轴套 Splined sleeve

2）循环风机 Circulating fan

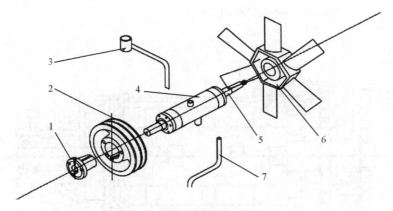

图 4.3.23　循环风机结构图 Structure drawing of circulating fan
1—锥形套 Cone set；2—皮带轮 Belt pulley；3—注油杯 Oil filling cup；
4—轴承箱 Bearing box；5—轴 Shaft；6—叶轮 Wheel；7—溢油管 Spill tube

4.3.1.14　马克混合机 Mark mixing machine

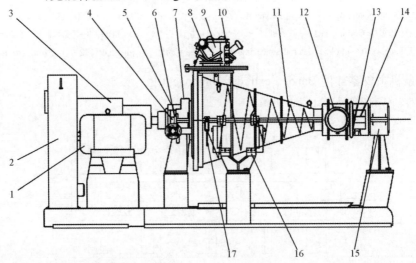

图 4.3.24　马克混合机结构图 Structure drawing of mark mixing machine
1—电机 Motor；2—皮带轮罩 Pulley cover；3—轴承箱 Bearing box；4—大端轴承 Big end bearings；
5—大端轴承座 Big end bearing seat；6—填料压盖 Stuffing gland；7、13—填料函 Stuffing box；
8—喷淋室放空管 Spray chamber vent pipe；9—喷淋室 Spray chamber；10—喷嘴 Nozzle；
11—锥形转子 Conical rotor；12—出料口 Discharge port；14—转子轴 Rotor shaft；
15—小端轴承座 Small end bearings base；16—支架 Support；17—锁紧压把 Locking pressure handle

4.3.2 纺丝设备 Spinning-equipment

4.3.2.1 纺丝机 Spinning machine

1. 纺丝机结构 Structure spinning machine

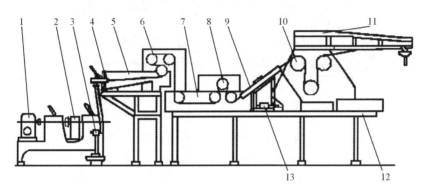

图 4.3.25 纺丝机结构图 Structure drawing of spinning machine

1—传动装置 Transmission；2—纺丝泵 Spinning pump；3—烛形过滤器 Candle filter；
4—喷丝头 Spinning nozzle；5—凝固浴槽 Coagulating tank；6—卷曲罗拉传动箱 Crimp roller gear box；
7—预热浴槽 Preheat bath；8—预拉伸罗拉传动箱 Pre-tension roller transmission box；
9—蒸汽拉伸盒 Steam stretching box；10—热拉伸罗拉传动箱 Hot stretch roller gearbox；
11—导丝架 Guide frame；12—机架 Rack；13—动力箱 Power control box

2. 纺丝计量泵 Spinning metering pump

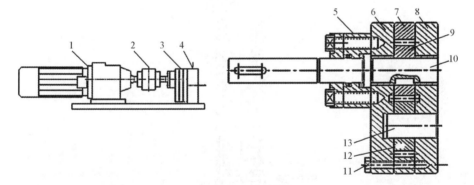

图 4.3.26 纺丝计量泵结构图 Structure drawing of spinning metering pump

1—电机及减速箱 Motor and gear box；2—联轴器 Coupling；3—泵体 Pump body；
4、8—后盖板 Rear cover；5—轴支座 Shaft bearing seat；6—前盖板 Front cover；
7—中间板 Intermediate plate；9—主动齿轮 Driving gear；10—主动轴 Driving shaft；
11—连接螺栓 Connecting bolts；12—从动齿轮 Driven gear；13—从动轴 Driven shaft

3. 烛形过滤器 Candle filter

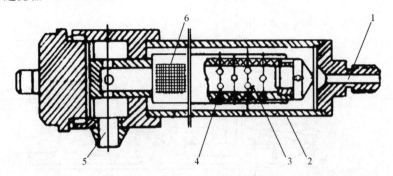

图 4.3.27　烛形过滤器结构图 Structure drawing of candle filter

1—原液进口 Stock solution import；2—壳体 Shell；3—滤芯 Filter element；
4—通液小孔 Liquid orifice；5—滤液出口 Filtered solution outlet；6—滤布 Filter cloth

4. 滤芯 Filter element

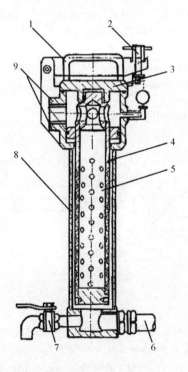

图 4.3.28　滤芯结构图 Structure filter element

1—压块 Briquetting；2—固定螺栓 Fixing screws；3—压盖 Gland；4—滤芯 Filter element；
5—滤材 Filter；6—接管 Connecting pipe；7—排放阀 Drain valve；8—壳体 Shell；9—密封环 Seal ring

4 合成纤维 Synthetic fiber

5. 凝固浴槽 Coagulating bath groove

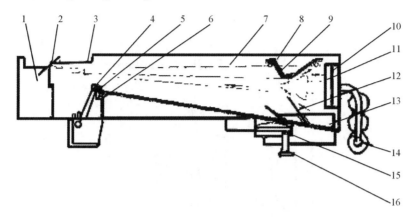

图 4.3.29 凝固浴槽结构图 Structure drawing of coagulating bath groove
1—回流槽 Reflux tank；2—液位调节板 Level adjustment plate；3—排液螺塞 Drain plug；
4—密封压板 Sealing plate；5—插座板 Socket board；6—插板 Flapper；7—浴槽 Bath；
8—短销 Short pin；9—导流盖 Diversion cap；10—后匀流板 After the uniform flow plate；
11—前匀流板 Uniform flow front plate；12—扰流板 Spoiler；13—接液盘 Wetted plate；
14—浴液进口管 Bath inlet tube；15—过滤网 Filter；16—放液管 Discharge tube

6. 湿法喷丝头 Wet process spinneret

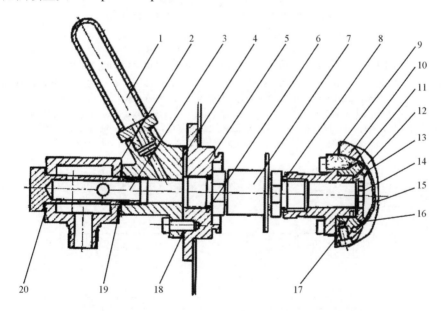

图 4.3.30 湿法喷丝头结构图 Structure drawing of wet process spinneret
1—稳流器 Current regulator；2—缓冲座 Buffering seat；3—缓冲管 Separator tube；4—稳流器座 Current regulator seat；
5—接头座 Joint seat；6、8、16、19、20—O 形密封圈 O sealing ring；7—喷丝头接头 Spinning nozzle joint；
9—喷丝头座 Spinning nozzle seat；10—半圆座 Semicircle seat；11—压紧棒 Pressing bar；12—压盖 Pressing cover；
13—分配板 Distributing plate；14—过滤板 Filter plate；15—喷丝头 Spinning nozzle；17、18—垫片 Gasker

7. 卷取罗拉箱 Crimping roller box

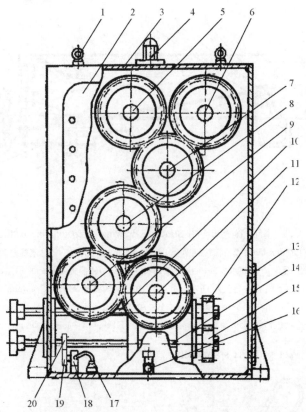

图 4.3.31　卷取罗拉箱结构图 Structure drawing of crimping roller box

1—吊环 Rings；2—箱盖 case cover；3—箱体 Box；4—示油器 Oil indicator；5、6、7—卷曲罗拉轴 Crimping roller shaft；8—过桥齿轮轴 Carrier gear shaft；9—导向罗拉轴 Guiding roller shaft；10—蜗杆 Worm rod；11—蜗轮轴 Worm-wheel shaft；12、15—总拉伸倍数变换齿轮 Total draw ratio change gear；13—侧盖 Side cover；14—传动轴 Transmission shaft；16—油位器 Oil level indicator；17—过滤器 Filter；18—油泵 Oil pump；19、20—油泵传动齿轮 Oil pump driving gear

8. 预热浴槽 Pre-heating bath groove

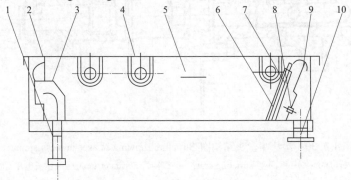

图 4.3.32　预热浴槽结构图 Structure drawing of pre-heating bath groove

1—进液管 Liquid inlet；2—档液板 Fluid baffle；3—匀流器 Flow homogenizer；4—上盖板 Upper cover；5—槽体 Groove；6—滤网架 Screen frame；7—滤网 Filter screen；8—排液螺塞 Drain plug；9—液位调节板 Liquid level adjusting plate；10—排液管 Bleeder tube

9. 预拉伸罗拉箱 Pre-drafting roller case

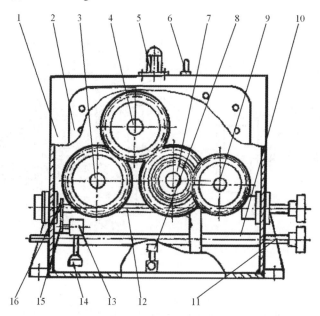

图 4.3.33 预拉伸罗拉箱结构图 Structure drawing of pre-drafting roller case
1—箱体 Box body；2—箱盖 Box cover；3、4、7—预拉伸罗拉轴 Pre-drafting roller shaft；
5—示油器 Oil indicator；6—吊环 Rings；8—油位器 Oil level indicator；9—蜗轮轴 Worm wheel shaft；
10—套管 Casing；11—传动轴 The shaft；12—蜗杆 Worm rod；13—油泵 Oil pump；
14—过滤器 Filter；15、16—油泵传动齿轮 Oil pump drive sprocket

10. 热拉伸罗拉箱 Thermal drafting roller case

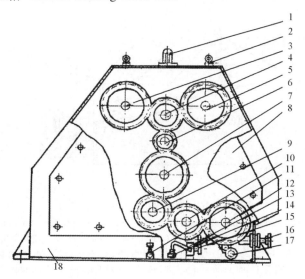

图 4.3.34 热拉伸罗拉箱结构图 Structure drawing of thermal drafting roller case
1—示油器 Oil indicator；2—吊环 Ring；3、5、7—罗拉轴 Roller shaft；4、6、9、10—齿轮轴 Carrier gear shaft；
8—箱盖 Case cover；11—油位器 Oil level indicator；12—蜗轮轴 Worm shaft；13—过滤器 Filter；
14—油泵 Oil pump；15—油泵传动齿轮 Oil pump driving gear；16—蜗杆 Worm rod；
17—自整角机 Selsyn；18—箱体 Box

11. 动力箱 Power supply box

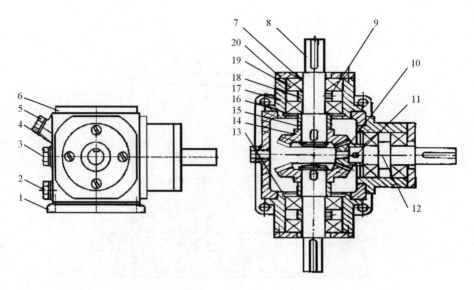

图 4.3.35 动力箱结构图 Structure drawing of power supply box
1—底座 Pedestal；2—排油螺塞 Oil extraction plug screw；3—油位螺塞 Oil level plug screw；4—箱体 Box；
5—加油螺塞 Oil-filler plug；6—箱盖 Case cover；7—轴封 Shaft sealing；8—输出轴 Output shaft；
9—紧固螺钉 Holding screw；10—主动齿轮 Driving gear；11—销子 Pin；12—主动轴 Driving shaft；
13、16—弹性挡圈 Collar；14—从动齿轮 Driven gear；15、18—定位环 Locating ring；17—轴承座 Bearing seat；
19—轴承 Bearing；20—压盖 Gland

4.3.2.2 斜底水平式纺丝机
Inclined bottom horizontal spinning machine

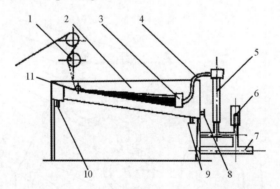

图 4.3.36 斜底水平式纺丝机结构图
Structure drawing of inclined bottom horizontal spinning machine
1—第一导辊 First guide roller；2—凝固浴槽 Coagulating bath groove；3—喷丝头 Spinning nozzle；
4—鹤颈管 Gooseneck；5—烛形过滤器 Candie filter；6—计量泵 Metering pump；
7—进浆管 Grout feeding pipe；8—凝固浴进口 Coagulation bath import；
9—液体排放口 Liquid discharge outlet；10—凝固浴出口 Coagulation bath outlet；
11—导丝辊 Godet

4.3.2.3 立管式纺丝机 Stand pipe spinning machine

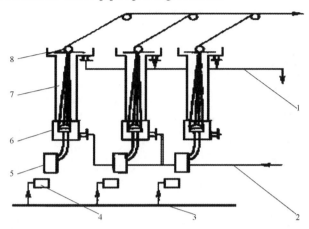

图 4.3.37 立管式纺丝机结构图 Structure drawing of stand pipe spinning machine
1—凝固浴出口 Coagulating bath outlet；2—凝固浴进口 Coagulating bath inlet；
3—纺丝原液管 Spinning solution；4—纺丝泵 Spinning pump；5—过滤器 Filter；
6—喷丝头 Spinning nozzle；7—凝固浴管 Coagulating bath pipe；
8—导丝辊 Thread guide roller

4.3.2.4 复合纺丝组件 Composite spinning components

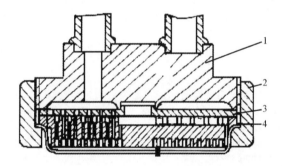

图 4.3.38 复合纺丝组件结构图 Structure drawing of composite spinning components
1—螺母 Nut；2—喷丝头座 Spinning nozzle seat；
3—分配板 Distributing plate；4—喷丝帽 Spinning cap

4.3.2.5 古典式纺丝机 Classic style spinning machine
4.3.2.6 干法纺丝机 Dry spinning machine

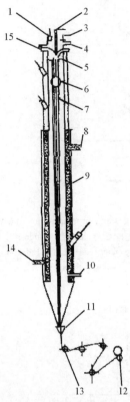

图 4.3.39 古典式纺丝机结构图
Structure drawing of classic style spinning machine
1—温度计 Thermometer；
2—原液导管 Stock soludtion guide pipe；
3—喷丝头 Spinning nozzle；
4—喷丝头载热体出口
Spinning nozzle heating medium outlet；
5—喷丝板 Spinning jet；6—加热气进口 Hot gas inlet；
7—丝束 Tow；8—加热载体出口 Heating carrier outlet；
9—纺丝甬道套管 Spinning shaft sleeve；
10—加热气出口 Heating outlet；
11—丝条出口 Stand outlet；
12—筒管 Boobin；13—导盘 Guide disc；
14—加热载体进口 Heating carrier inlet；
15—喷丝头载热体进口
Spinning nozzle heating medium inlet

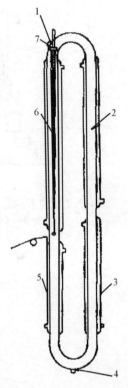

图 4.3.40 干法纺丝机结构图
Structure drawing of dry spinning machine
1—观察孔 Sight glass；
2—加热套管 Heating casing pipe；
3、5—低温夹套 Low temperature jacket；
4—冷凝后溶剂出口 Condensation solvent outlet；
6—纺丝甬道 Spinning channel；
7—喷丝头 Spinning nozzle

4 合成纤维 Synthetic fiber

4.3.2.7 牵引设备 Traction equipment
1. 牵引机 Traction machine

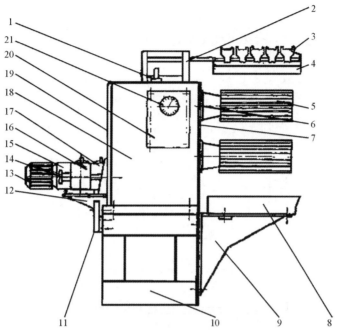

图 4.3.41　牵引机结构图 Structure drawing of traction machine
1—吊环 Ring；2—导丝架 Thread guide frame；3—导丝器 Thread guide；
4—小接液盘 Small liquid holding plate；5—牵引罗拉 Traction roller；
6—机罩 Hook cover；7—护板 Guard board；8—大接液盘 Big liquid holding plate；
9—接液盘托架 Liquid holding plate frame；10—机架 Rack；
11—调整螺栓座 Adjusting bolt seat；12—无级变速器支座 Infinitely variable transmission seat；
13—电机 Motor；14—调速手轮 Speed governing hand wheel；
15、17、21—减速器 Speed reducer；16—无级变速器 Continously ariable transmission；
18—箱体 Barrel；19—箱盖 Box cover；20—侧盖 Side cover；

2. 牵引罗拉座 Traction roller seat

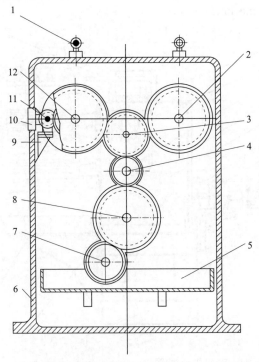

图 4.3.42 牵引罗拉座结构图 Structure drawing of traction roller seat
1—吊环 Ring；2—罗拉轴 Roller shaft；3、4、11—传动齿轮轴 Driving gear；
5—油槽 Oil groove；6—箱体 Box body；7—输入齿轮 Input gear；
8、12—牵引罗拉轴 Traction roller shaft；
9—速度指示器座 Speed indicator seat；10—速度指示器 Speed indicator

4.3.3 后加工设备 Post-processing equipment

4.3.3.1 长槽水洗机 Elongated slot washer

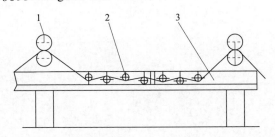

图 4.3.43 长槽水洗机结构图 Structure drawing of elongated slot washer
1—压辊 roller；2—水楔 wedge；3—水洗长槽 washing elongated slot

4.3.3.2 U形水洗机 U washer

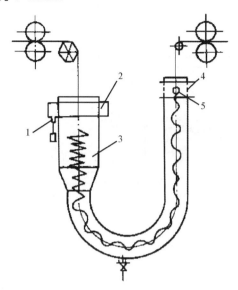

图 4.3.44 U形水洗机结构图 Structure drawing of U type washer
1—洗涤水出口 Wash water outlet；2、4—水孔 Water hole；
3—扩大箱 Enlarge case；5—洗涤水进口 Wash water inlet

4.3.3.3 圆网式干燥机 Circular screen dryer

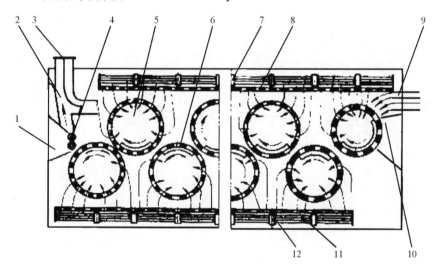

图 4.3.45 圆网式干燥机结构图 Structure drawing of circular screen dryer
1—纤维入口 Fiber inlet；2—排风道 Exhaust airway；3—排风口 Air outlet；
4—喂入罗拉 Feed roller；5—转鼓 Drum；6—挡风板 Wind shield；
7—上导流板 Upper guide plate；8—上匀流板 Upper uniform flow plate；
9—进风口 Air inlet；10—纤维出口 Fiber outlet；
11—下匀流板 Lower uniform flow plate；12—下导流板 Lower guide plate

4.3.3.4 汽蒸定形锅 Steaming sizing boiler

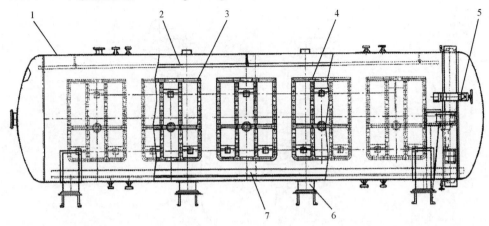

图 4.3.46 汽蒸定形锅结构图 Structure drawing of steaming sizing boiler
1—筒体 Shell；2—蒸汽管支承及受液槽 Steam pipe support & liquid groove；
3—蒸汽水汀排 Steam beside the row；4、7—加热蒸汽管 Heating steam pipe；
5—快开门装置 Quick door opener；6—支座 Support

4.3.3.5 集束架 Creel

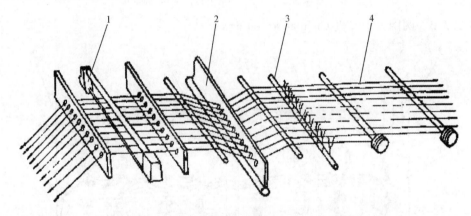

图 4.3.47 集束架结构图 Structure drawing of creel
1—飘丝检出装置 Filament detection device；2—打结检出装置 Kont detection device；
3—张力均匀装置 Tension uniform device；4—丝束 Tow

4.3.3.6 浸油槽 Oil-soaking groove

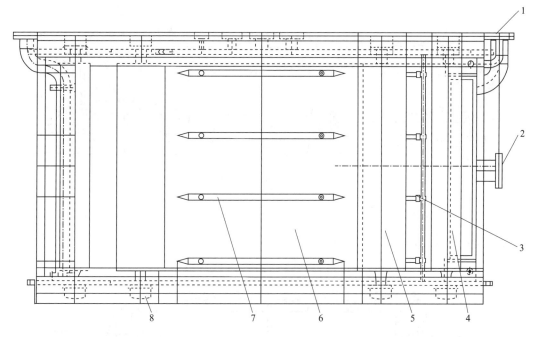

图 4.3.48 浸油槽结构图 Structure drawing of oil-soaking groove
1—油剂入口 Oil inlet；2—油剂出口 Oil outlet；3—分丝棒 Devillicate roll；
4—轴密封 Shaft sealing；5—导辊 Guide roll；6—槽头 Through；7—导流板 Guide plate；8—轴承 Bearing

4.3.3.7 牵伸机 Drafting machine

1. 牵引机结构 Structure of drafting machine

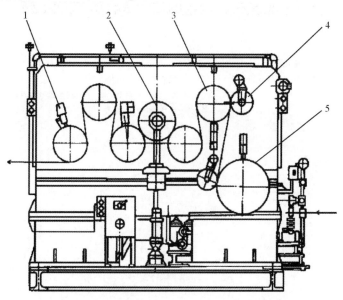

图 4.3.49 牵伸机结构图 Structure drawing of drafting machine
1—绕辊检测装置 detector；2—压紧辊 Press roll；
3—牵伸辊 Draft roll；4—压辊 Hydraulic roll；5—浸渍辊 Dip roll

2. 牵伸辊 Draft roll

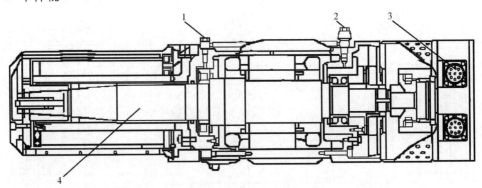

图 4.3.50　牵伸辊结构图 Structure drawing of draft roll
1—加热器 Heater；2—注油孔 Oil filling hole；3—连线插座 Attachment plug；4—轴 Shaft

3. 压辊 Press roll

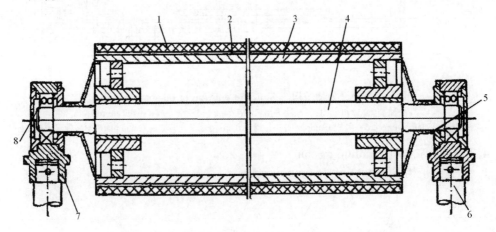

图 4.3.51　压辊结构图 Structure drawing of press roll
1—硬橡胶层 Hard rubber layer；2—软橡胶层 Soft rubber layer；3—辊体 Roller；
4—辊轴 Roller shaft；5—轴封 Roller sealing；6—机架 Rack；7—轴承座 Bearing block；
8—轴承 Bearing

4.3.3.8 蒸汽加热器 Stream heater

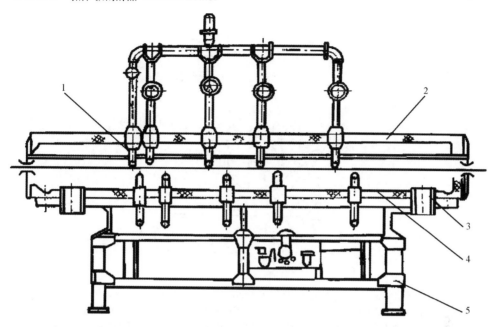

图 4.3.52 蒸汽加热器结构图 Structure drawing of steam heater
1—蒸汽管 pipe；2—上箱体 Upper box；3—冷凝水出口 outlet；4—下箱体 Lower outlet；5—机架 rack

4.3.3.9 卷曲机 Crimping machine

卷曲机用于干燥后丝束的卷曲，通过加热和机械挤压，使纤维获得一定的卷曲度和卷曲数，增加纤维间的抱合力，便于后续加工处理和满足纺织的使用性能。

The crimping machine is used to make the tow crimp. Through heating and mechanical squeeze, it make the fiber to get the crimpness and the number of the crimpness, which add the cohesive force between the fibers in order to be convenient to subsequent processing and meet the uing performance of textile.

1. 卷曲机 Crimping machine

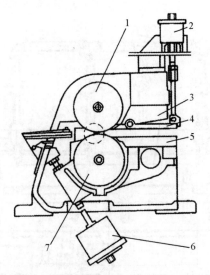

图 4.3.53　卷曲机结构图 Structure drawing of crimping machine
1—上卷曲轮 Upper crime wheel；2、6—汽缸 Cylinder；3—上卷曲刀 Upper Crimp knife；
4—活动板 Activite board；5—下卷曲刀 Lower Crimp knife；7—下卷曲轮 Lower shaft wheel

2. 卷曲轮 Crimping wheel

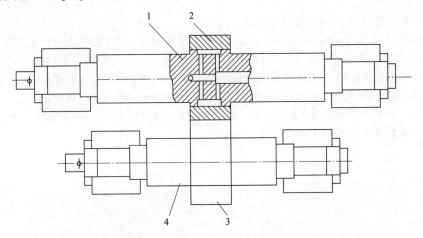

图 4.3.54　卷曲轮结构图 Structure drawing of crimping wheel
1—上卷曲轮轴 Upper crimping wheel shaft；2—上卷曲轮 Upper crimping wheel；
3—下卷曲轮 Lower crimping wheel；4—下卷曲轮轴 Lower crimping wheel shaft

4.3.3.10 热定形机 Heat-setting machine

1. 松弛热定形机 Loose heat-setting machine

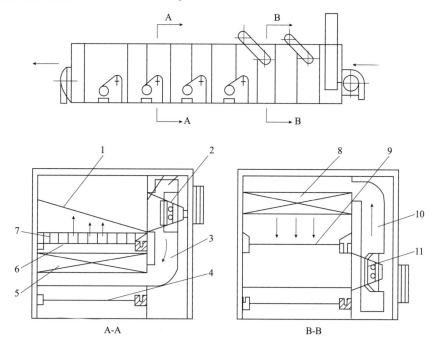

图 4.3.55 松弛热定形机结构图 Structure drawing of heat-setting machine
1—过滤网 Filter net；2、11—风机 Draught fan；3、10—风管 Air hole；4—下隔板 Lower clapboard；
5、8—加热器 warmer；6—链板 Scaping belt；7—铺丝漏斗 Wire laying funnel；9—上隔板 Upper clapboard

2. 张紧热定形机 Tension heat-setting machine

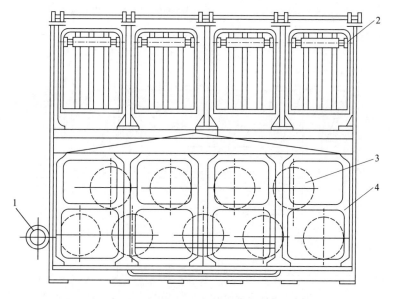

图 4.3.56 张紧热定形机结构图 Structure drawing of tension heat-setting machine
1—导丝辊 Godet；2—门罩平衡件 Unit of the door cover balance；
3—定形辊 Setting roller；4—门罩 Door cover

4.3.3.11 铺丝机 Fiber-spreading machine

铺丝机是将丝束均匀地铺放在链板上，使纤维得到均匀的热定形。

The fiber-spreading machine is puts the tows on the chain board uniformly, which make the fiber get the effort of the uniform heat setting.

1. 铺丝机结构 Fiber-spreading machine

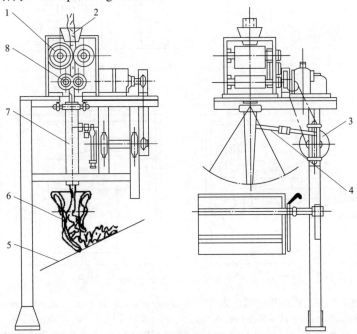

图 4.3.57 铺丝机结构图 Structure drawing of fiber-spreading machine
1—上辊 Upper roll；2—丝束 Tows；3、4—摆杆机构 bar organization；5—链板 Chain box；
6—翻斗 Tipping bucket；7—摆丝器 Reciprocator；8—下辊 Lower roll

2. 定形辊 Setting roller

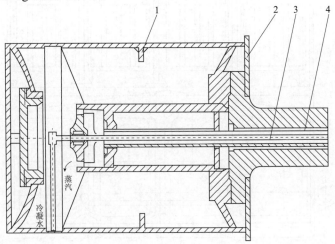

图 4.3.58 定形辊结构图 Structure drawing of setting roller
1—辊体 Roller；2—法兰 Flange；3—虹吸管 Siphon；4—进汽管 Steam inlet plate

4.3.3.12 绕辊检出装置 Around the roller detection device

为了防止丝束绕辊，影响定形效果，张紧热定形机上常装有绕辊检出装置。

In order to prevent the tows coil the roller and finalize the design effect, it is necessary to equip around the roller detection device on the tension heat-setting machine

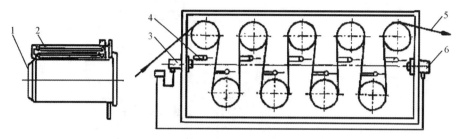

图 4.3.59 绕辊检出装置结构图 Structure drawing of around the roller detection device
1—定形辊 Setting roller; 2—自由辊 Loose roller; 3—遮光板 Visor;
4—发光器 Photophore; 5—丝束 Tows; 6—受光器 photic equipment

4.3.3.13 切断机 Cut-off machine

切断机用于腈纶纤维短纤加工，按照后道加工需要的纤维长度，纤维在刀盘上切断成需要的长度。

The cut-off is used to acrylic fiber staple fiber processing, which meet the needs of processing after the fiber length, then put the fiber into the needed length.

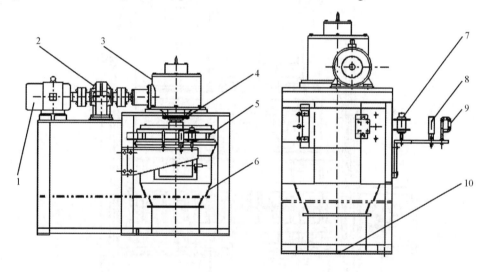

图 4.3.60 切断机结构图 Structure drawing of cut-off machine
1—电机 Motor; 2—减速箱 Reducer box; 3—齿轮箱 Gear box;
4—传动轴 Transmission shaft; 5—切断刀盘 Cutter; 6—落丝斗 Fall wire bucket;
7—压轮 Pinch roller; 8、9—导丝部件 Guide wire conponent; 10—落丝口 Fall wire port

4.3.3.14 打包机 Baler
1. 打包机结构 Structure of Baler

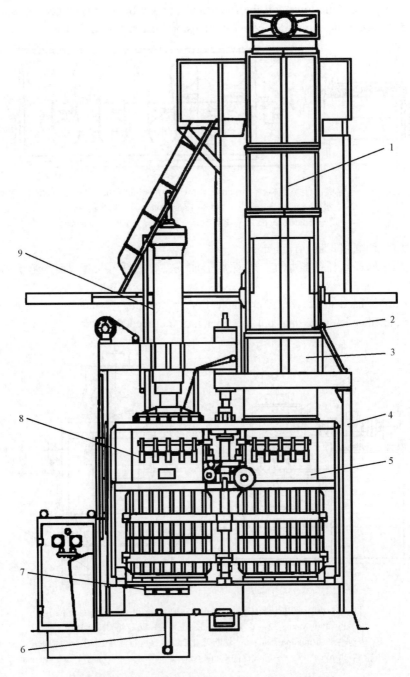

图 4.3.61 打包机结构图 Structure drawing of baler
1—预压汽缸 Prepressing cylinder；2—提箱机构 Lift box equipment；3—料仓 Stock bin；
4—机架 Rack；5—棉箱 Cotton box；6—推包机构 Bag-pushing mechanism；
7—转箱机构 Box mechanism；8—防回弹装置 Resilience-proof device；9—主压油缸 Main pressure cylinder

2. 主压油缸 Adovate cylinder

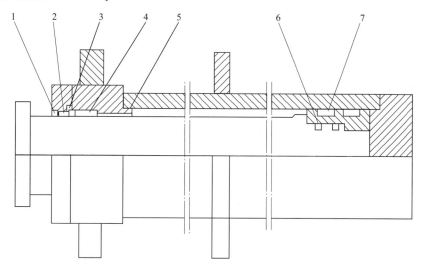

图 4.3.62 主压油缸结构图 Structure drawing of adovate cylinder
1—防尘圈 Scraper seal；2、4—轴封 Sealing；3、5、6—O 形密封圈 O sealing ring；7—孔圈 Hole ring

4.3.4 溶剂回收设备 Solvent-recovery equipment

4.3.4.1 降膜式蒸发器 Falling film evaporator

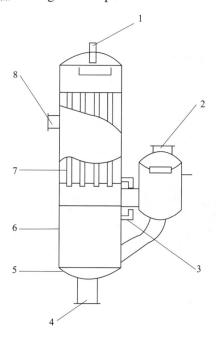

图 4.3.63 降膜式蒸发器结构图 Structure drawing of falling film evaporator
1—料液进口 Feed liquid inlet；2—蒸汽出口 Steam outlet；
3—冷凝水出口 Condensed water；4—料液出口 Feed liquid outlet；5—封头 Head；6—筒体 Barrel；
7—管束 Tube bundle；8—蒸汽入口 Steam inlet

4.3.4.2 外加热式换热器 External heating exchanger

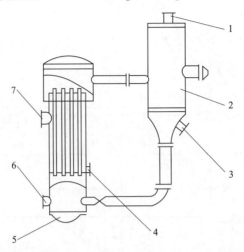

图 4.3.64 外加热式换热器结构图 Structure drawing of external heating exchanger
1—蒸汽出口 Steam outlet；2—蒸发室 Evaporation chamber；3—原料出口 Raw material outlet；
4—冷凝水 Condensed water；5—加热室 Heating chamber；6—原料入口 Raw material inlet；
7—蒸汽进口 Steam inlet

4.4 锦纶 Nylon

锦纶是聚酰胺纤维的商品名称，又称耐纶(Nylon)。

4.4.1 熔融聚合设备 Melt polymerization equipment

4.4.1.1 熔融设备 Melting equipment

1. 熔融釜 Melting kettle

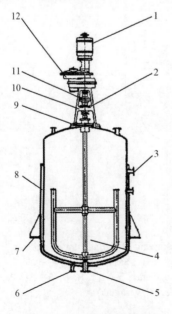

图 4.4.1 熔融釜结构图
Structure drawing of melting kettle

1—电机 Motor；2—轴承 Bearing；3—蒸汽进口 Steam inlet；4—搅拌器 Agitator；5—出料口 Discharge outlet；6—冷凝水出口 Condensate water outlet；7—支座 Support；8—蒸汽夹套 Steam jacket；9—填料函 Stuffing box；10—支架 Support；11—联轴器 Coupling；12—减速器 Reducer

4 合成纤维 Synthetic fiber

2. 螺旋进料器 Spiral feeder

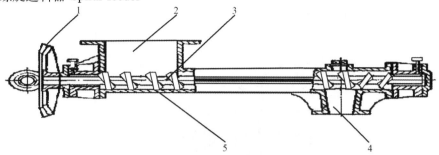

图 4.4.2 螺旋进料器结构图 Structure drawing of spiral feeder
1—传动装置 Transmission device；2—进料漏斗 Feed funnel；
3—螺杆 Screw；4—出料管 Discharge pipe；5—外壳 Shell

3. 盘式过滤器 Disc filter

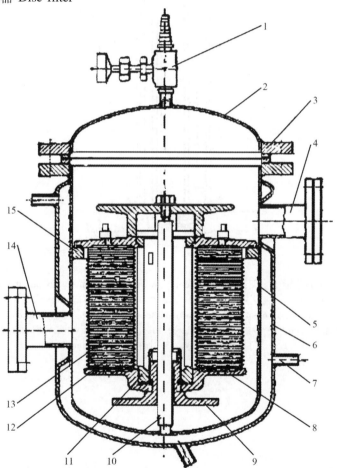

图 4.4.3 盘式过滤器结构图 Structure drawing of disc filter
1—针形阀 Needle valve；2—封头 Head；3—法兰 Flange；4—出料口 Discharge interface；
5—筒体 Barrel；6—加热夹套 Heating jacket；7—管接口 Pipe interface；8—圆盘 Disc；
9—螺母 Nut；10—轴 Shaft；11—套筒 Sleeve；12—过滤盘 Filter disc；13—过滤网 Filter screen；
14—进料口 Feeding inlet；15—支承环 Support ring

4.4.1.2 聚合设备 Polymerization device
1. U 形连续聚合管 U continuous polymerization pipe

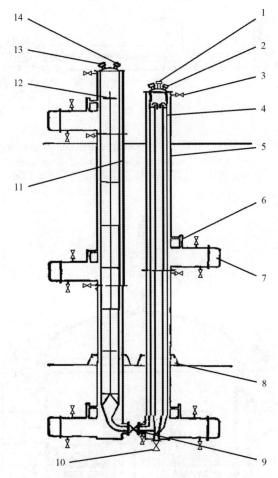

图 4.4.4 U 形连续聚合管结构图 Structure drawing of U continuous polymerization pipe
1、14—接管 Connectin pipe；2—视窗 Window；3—夹套放空口 Jacket unloading interface；4—中心管 Central tube；5—纺丝管 Spinning pipe；6—温度计接口 Thermometer interface；7—电加热棒 Electric heating rod；8—支座 Support；9—控制阀 Control valve；10—出料阀 Outlet valve；11—夹套 Jacket；12—有孔托盘 Hole tray；13—加料口 Feeding interface；

2. 直线连续聚合管 Linear continuous polymerization tube

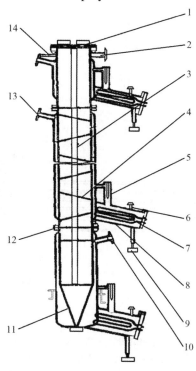

图 4.4.5 直线连续聚合管结构图 Structure drawing of linear continuous polymerization tube
1—顶盖 Top cover；2—加料口 Feed port；3—挡板 Baffle；4—多孔板 Perforated plate；
5—温度计套管 Thevmowell；6—联苯进液口 Biphenyl liquid inlet；7—电加热棒 Electric bar；
8—加热室 Heating chamber；9—联苯排液口 Biphenyl liquid outlet；10—联苯蒸汽入口
Biphenyl steam inlet；11—锥形底 Cone bottom；12—连接法兰 Connecting flange；
13—联苯蒸汽出口 Biphenyl steam outlet；14—氮气接口 Nitrogen supply port

3. 聚合管主要附件 Polymerization pipe main accessory
1）顶盖 Cap

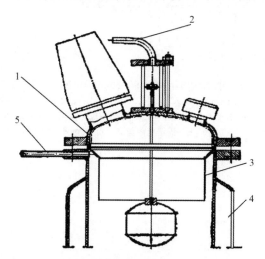

图 4.4.6 顶盖结构图 Structure drawing of top cap
1—顶盖 Top cover；2—排气管 Vent pipe；3—氮气罩 Nitrogen cover；
4—夹套 Jacket；5—氮气进口 Nitrogen inlet

2)加热器 Heater

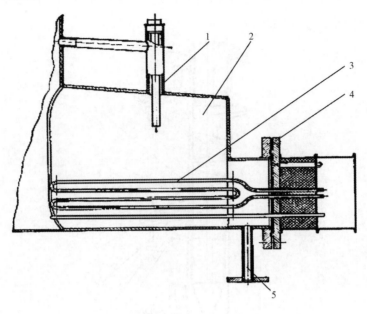

图 4.4.7 加热器结构图 Structure drawing of heater
1—进液口 Liquid inlet；2—加热室 Heating chamber；
3—电加热棒 Electric bar；4—法兰 Flange；5—排液口 Liquid outlet

3)电加热棒 Electric bar

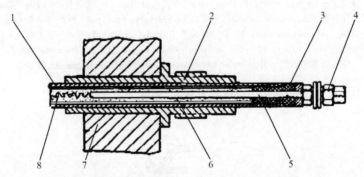

图 4.4.8 电加热棒结构图 Structure drawing of electric bar
1—无缝钢管 Seamless steel pipe；2—套管；3—瓷套 Porcelain bushing；
4—螺母 Nut；5—引出铜杆 Drawn copper rod；6—氧化镁填料 Magnesium oxide filler；
7—法兰 Flange；8—镍铬电阻丝 Nickel chrome resistance wire

4) 联苯冷凝器 Biphenyl condenser

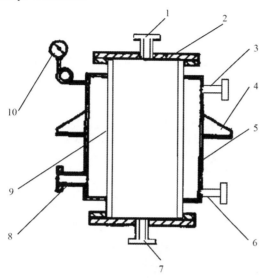

图 4.4.9 联苯冷凝器结构图 Structure drawing of biphenyl condenser
1—蒸汽出口 Steam outlet；2—顶盖 Top cover；3—排气管 Vent pipe；4—支座 Support；
5—夹套 Jacket；6—排液管 Bleeder pipe；7—蒸汽入口 Steam inlet；8—进液管 Liquid inlet；
9—内管 Inner pipe；10—压力表 Pressure gauge

4.4.2 切片设备 Slice equipment

4.4.2.1 铸带设备 Casting equipment

1. 铸带头 Casting head

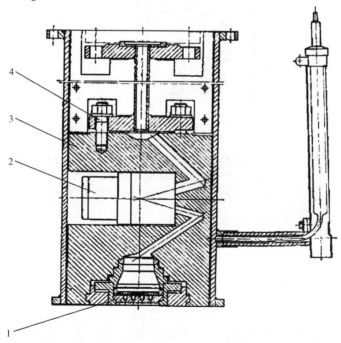

图 4.4.10 铸带头结构图 Structure drawing of casting head
1—注带板 Casting plate；2—齿轮泵 Gear pump；3—泵体 Pump body；4—螺栓 Bolt

2. 铸带槽 Casting groove

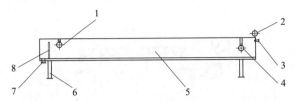

图 4.4.11 铸带槽结构图 Structure drawing of casting groove
1、4—滚筒 Roller；2—导辊 Guide roller；3—进水口 Water inlet；5—槽体 Cell body；
6—支脚 Support；7—溢流口 Overflow port；8—溢流板 Overflow plat

3. 牵引滚筒 Hauling drum

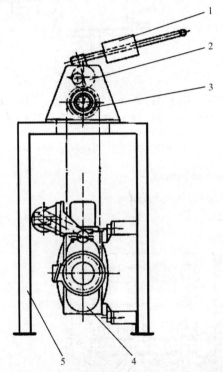

图 4.4.12 牵引滚筒结构图 Structure drawing of hauling drum
1—重锤 Hammer；2—压辊 Pressure roller；3—导辊 Guide roller；
4—无级变速器 Continuously variable transmission；5—机架 Rack

4.4.2.2 切片贮存桶 Slice storage barrel

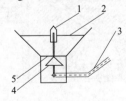

图 4.4.13 切片贮存桶结构图 Structure drawing of slice storage barrel
1—桶罩 Cover；2—固定棒 Fixed rod；3—手柄连杆 Handle connecting rod；
4—锥体 Cone；5—孔板 Orifice

4.4.2.3 萃取釜 Extraction kettle

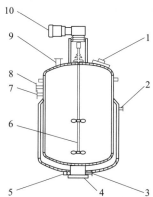

图 4.4.14 萃取釜结构图 Structure drawing of extraction kettle
1—加料口 Feeding port；2—蒸汽进口 Steam inlet；3—排水口 Drain port；
4—出料口 Discharge port；5—冷凝水出口 Condensate outlet；6—搅拌器 Agitator；
7—溢流口 Overflow pipe；8—进水口 Water inlet；9—排气口 Exhaust port；10—电机 Motor

4.4.2.4 干燥设备 Drying equipment

1. 水平式转鼓真空干燥机 Horizontal drum vacuum drying machine

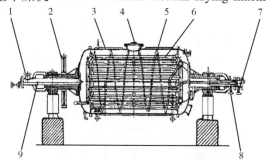

图 4.4.15 水平式转鼓真空干燥机结构图
Structure drawing of horizontal drum vacuum drying machine
1—接管 Connecting pipe；2—齿轮 Gear；3—夹套 Jacket；4—进料口 Feeding port；5—螺旋带 Spiral belt；
6—加热器 Heater；7—冷凝水出口 Condensate water outlet；8—蒸汽进口 Steam inlet；
9—主轴 Shaft

2. 倾斜式转鼓真空干燥机 Tilting drum vacuum drying machine

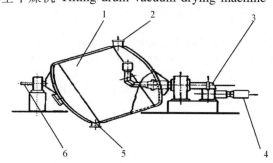

图 4.4.16 倾斜式转鼓真空干燥机结构图 Structure drawing of tilting drum vacuum drying machine
1—转鼓 Rotary drum；2—进料口 Feeding port；3—抽真空管道 Vacuum piping；
4—电机 Electric motor；5—出料口 Discharge port；6—蒸汽进口 Steam inlet

3. 双锥式转鼓真空干燥机 Double cone drum type vacuum drying machine
4. 感应加热式转鼓真空干燥机 Induction heating type drum vacuum drying machine

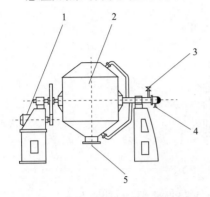
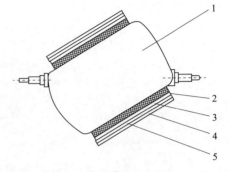

图 4.4.17 双锥式转鼓真空干燥机结构图
Structure drawing of double cone drum type vacuum drying machine
1—电机 Electric motor；2—转鼓 Rotary drum；
3—蒸汽进口 Steam inlet；
4—冷凝水出口 Condensate water outlet；
5—进/出料口 Discharge port

图 4.4.18 感应加热式转鼓真空干燥机结构图
Structure drawing of induction heating type vacuum drum drying machine
1—转鼓 Rotary drum；2—保温层 Insulation layer；
3—下夹板 Lower splint；4—上夹板 Upper splint；
5—感应线圈 Induction coil

5. 转鼓真空干燥机的主要零部件
The main components of rotary drum vacuum drying machine
1）加热轴头 Heating part shaft head

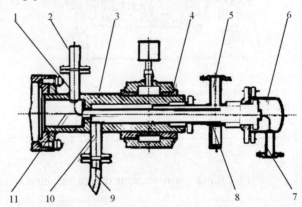

图 4.4.19 加热轴头结构图 Structure drawing of heating part shaft head
1、8—回流管 Return pipe；2、9—接管 Connecting pipe；3—轴头 Shaft head；
4—轴衬 Bushing；5—蒸汽进口 Steam inlet；6—集气室 Collection chamber；
7—蒸汽出口 Steam outlet；10—管接头 Pipe joint；11—回流室 Return chamber

2) 抽真空轴头 Vacuum part spindle nose

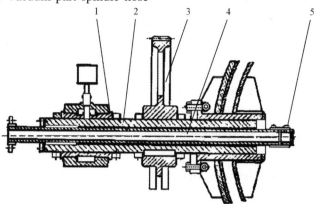

图 4.4.20 抽真空轴头结构图 Structure drawing of vacuum part spindle nose
1—轴衬 Bushing；2—轴头 Spindle nose；3—齿轮 Gear；4—真空管 Vaccum pipe；5—弯头 Bend

3) 除尘器 Dust separator

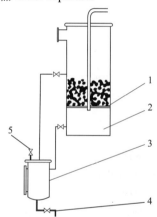

1—孔板 Bushing；2—除尘桶 Collecting dust tank；3—集水桶 Collecting water tank；4—进水口 Water inlet；5—出水口 Water outlet

图 4.4.21 除尘器结构图
Structure drawing of dust separator

4.4.2.5 输送振动器 Conveying vibrator

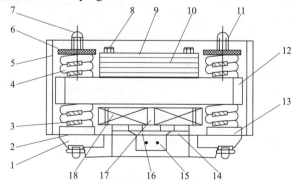

图 4.4.22 输送振动器结构图 Structure drawing of conveying vibrator
1—底座 Base；2—橡胶垫 Rubber mat；3—拉杆 Tie rod；4—弹簧 Spring；5—壳体 Shell；6—底盘 Chassis；7、11—压紧螺母 Compression nut；8—六角螺母 Hexagon nut；9—平衡板 Balance sheet；10—减振垫 Damping pad；12—振动板 Vibration plate；13—托盘 Tray；14—衔铁 Armature；15—接线盒 Junction box；16—接线板 Wiring board；17—线圈 Iron core；18—铁芯 Coil

4.4.3 纺丝设备 Spinning equipment

4.4.3.1 切片纺丝设备 Slice spinning equipment

将粒状或片状聚合物在纺丝机中纺制单丝、复丝、帘子线和短纤维等称为切片纺丝。切片纺丝设备的形式很多，常用的有炉栅纺丝和挤出纺丝两种。

Slice spinning is to use granular or flake polymer spinning monofilament and multifilament, cord and short fiber, etc in the spinning machine, Slice spinning equipment is usually divided into grate spinning and extrusion spinning machine.

1. 炉栅纺丝设备 Grid spinning equipment

1）炉栅纺丝器 Grid melt-spinning device

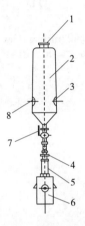

图 4.4.23　炉栅纺丝器结构图
Structure drawing of grid melt-spinning device
1—筒盖 Barrel cover；2—料筒 Charging barrel；
3、8—视窗 Window；4—补偿器 Compensator；
5—落料管 Feed drop tube；6—纺丝头 Spinning head；
7—加料阀 Feed valve

2）炉栅纺丝器主要零部件 Grid melt-spinning device main accessories

（1）炉栅 Grid melter

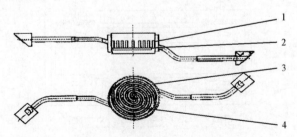

图 4.4.24　炉栅结构图 Structure drawing of grid melter
1—炉栅管 Grid melter tube；2—不锈钢丝 Stainless steel wire；
3—联苯排出环 Biphenyl discharge ring；4—联苯进入环 Biphenyls inlet ring

(2) 纺丝头 Spinning head

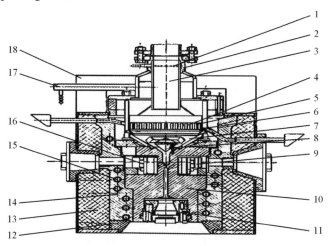

图 4.4.25 纺丝头结构图 Structure drawing of spinning head

1—充氮口 Nitrogen filling hole；2—钟型罩 Bell-shaped cover；3—落料管 Feed drop tube；
4—环形空腔 Annular cavity；5—炉栅管 Grid melter tube；6—不锈钢丝 Stainless steel wire；7—炉膛 Furnace chamber；
8—楔形块 Wedge block；9—计量泵 Metering pump；10—泵体 Pump body；11—喷丝头 Spinneret；
12—石棉水泥座 Asbestos cement block；13—玻璃纤维绝热层 Glass wool insulation；
14—加热夹套 Heating jacket；15—加热螺旋管 Heating coil；16—压力泵 Pressure pump；
17—排气管 Exhaust pipe；18—保温盖 Insulation cover

(3) 熔融纺丝头 Melt spinning head

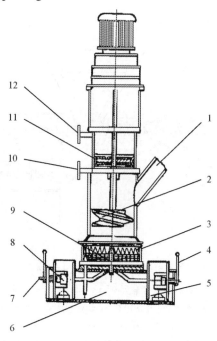

图 4.4.26 熔融纺丝头结构图 Structure drawing of melt spinning head

1—切片入口 Section inlet；2—挤压蜗杆 Extrusion worm；3—销钉式炉栅 Pin type grate；
4—张力器 Tension device；5—喷丝板 Spinneret；6—联苯加热区 Biphenyls heating zone；
7—泵轴 Pump shaft；8—纺丝泵 Pump；9—氮气管 Nitrogen pipe；10—空气出口 Air outlet；
11—压缩空气活塞 Compressed air piston；12—压缩空气入口 Compressed air inlet

(4) 喷丝头 Spinning nozzle

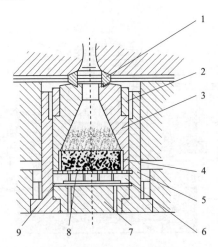

图 4.4.27 喷丝头结构图 Structure drawing of spinning nozzle
1、9—垫圈 Gasket；2—压紧螺旋 Pressed screw；3—内圈 Inner ring；
4—石英砂 Quartz sand；5—外圈 Outer ring；6—分配板 Distribution board；
7—喷丝板 Spinneret；8—过滤网 Filter screen；

2. 挤出纺丝设备 Extrusion spinning equipment
1) 螺杆挤出机 Screw extruder

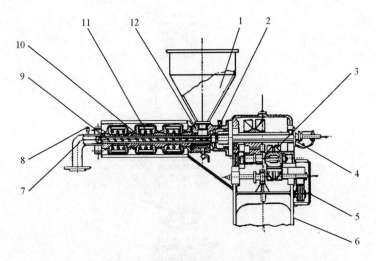

图 4.4.28 螺杆挤出机结构图 Structure drawing of screw extruder
1—料筒 Charging barrel；2—止推轴承 Thrust bearing；3—传动系统 Transmission；
4—冷却系统 Cooling system；5—整流电机 Rectifying motor；6—机体 Compressor body；
7—弯管 Elbow pipe；8—压力传感器 Pressure sensor；9—螺杆 Screw；10—套筒 Sleeve；
11—加热器 Heater；12—料筒支座 Charging barrel bearing

4 合成纤维 Synthetic fiber

2) 螺杆挤出机主要零部件 Screw extruder main accessories

(1) 喂入装置 Feeding equipment

(2) 止推轴承 Thrust bearing

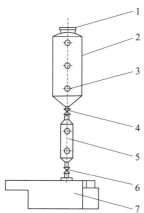

图4.4.29 喂入装置结构图
Structure drawing of feeding equipment

1—盖板 Cover；2—料筒 Charging barrel；
3—视窗 Window；4—加料阀 Charging vavle；
5—落料管 Drop feed pipe；6—进料阀 Feed vavle；
7—挤压机 Extruding machine

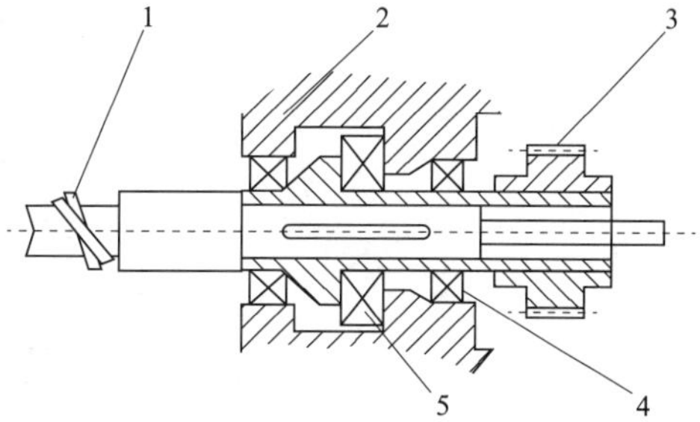

图4.4.30 止推轴承结构图
Structure drawing of thrust bearing

1—螺杆 Screw；2—箱体 Box；3—齿轮 Gear；
4—向心圆柱滚子轴承
Centripetal cylindrical roller bearing；
5—止推轴承 Thrust bearing

(3) 纺丝保温箱 Spinning incubator

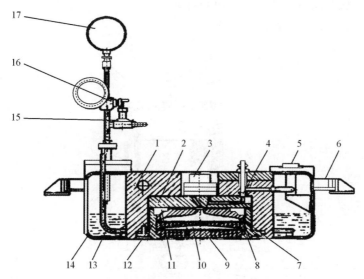

图4.4.31 纺丝保温箱结构图 Structure drawing of spinning incubator

1—纺丝头 Spinning nozzle；2—泵体 Pump body；3—计量泵 Metering pump；4—垫圈 Gasket；
5—进料口 Feeding inlet；6—支座 Support；7—喷丝头内圈 Spinneret inner ring；8—过滤网 Filter；
9—喷丝板 Spinneret；10—分配板 Allocation plate；11—压紧螺旋盖 Pressed screw cap；
12—喷丝头外圈 Spinnerets outer ring；13—电热棒 Electric rod；14—箱体 Box；15—针形阀 Needle valve；
16—铂电阻温度计 Platinum resistance thermometer；17—真空压力表 Vacuum gauge

4.4.3.2 直接纺丝设备 Direct spinning equipment

1. 辊子式薄膜蒸发器 Rolling thin film evaporator

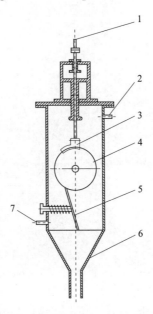

图 4.4.32 辊子式薄膜蒸发器结构图 Structure drawing of rolling thin film evaporator

1—进液管 Liquid inlet；2—氮气出口 Nitrogen outlet；3—分配器 Distributor；
4—辊子 Roller；5—刮板 Scarper blade；6—锥体 Cone；7—氮气进口 Nitrogen inlet

2. 振动式薄膜蒸发器 Vibrating film evaporator

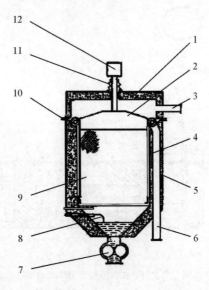

图 4.4.33 振动式薄膜蒸发器结构图 Structure drawing of vibrating film evaporator

1—壳体 Shell；2—顶罩 Top cap；3—排气口 Exhaust port；4—筒体 Barrel；5—夹套 Jacket；
6—进料管 Feeding pipe；7—齿轮泵 Gear pump；8—电接触丝 Electrical contact wire；
9—振动体 Vibrating body；10—环管 Ring pipe；11—密封装置 Sealing device；
12—振动器 Vibrator

3. 真空萃取器 Vacuum extractor

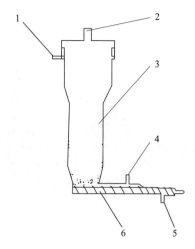

图 4.4.34 真空萃取器结构图 Structure drawing of vacuum extractor
1—熔体进口 Melt inlet；2—抽真空口 Vaccum port；3—萃取器 Extractor；
4—蒸汽入口 Steam inlet；5—出料口 Discharge port；6—输出螺杆 Output screw

4.4.3.3 纺丝泵 Spinning pump

1. 纺丝泵结构 Structure of spinning pump

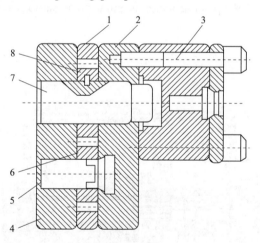

图 4.4.35 纺丝泵结构图 Structure drawing of spinning pump
1—中间板 Middle plate；2—后盖板 Back cover；3—紧固螺栓 Fastening bolt；
4—前盖板 Front cover；5—从动轴 Driven shaft；6—从动齿轮 Driven gear；
7—主动轴 Drive shaft；8—主动齿轮 Driving gear

2. 纺丝泵传动装置 Spinning pump transmission equipment

1）纺丝泵传动装置 Spinning pump transmission equipment

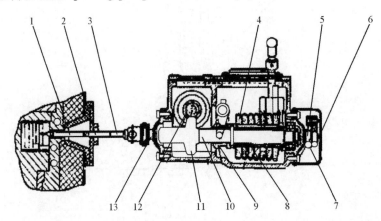

图 4.4.36　纺丝泵传动装置结构图 Structure drawing of spinning pump transmission equipment

1—泵轴接头 Pump shaft joint；2—十字头 Cruciform joint；3—调节轴 Adjustment shaft；
4—变换齿轮 Change gear；5—传动臂 Drive arm；6—泵轴 Pump shaft；7—保险销 Safety pin；
8—滑动套筒 Slide sleeve；9—滑动离合器 Slip clutch；10—蜗轮轴 Worm shaft；
11—蜗轮 Worm wheel；12—蜗杆 Worm rod；13—主轴 Shaft

2）纺丝计量泵传动装置 Spinning metering pump transmission equipment

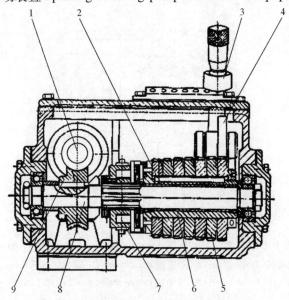

图 4.4.37　纺丝计量泵传动装置结构图
Structure drawing of spinning metering pump transmission equipment

1—蜗杆 Worm rod；2、6—滑动套筒 Slide supporting sleeve；3—调节手柄 Regulating handle；
4—过界齿轮 Overspill gear；5—变换齿轮 Change gear；7—电磁离合器 Electromagnetic clutch；
8—蜗轮轴 Worm shaft；9—蜗轮 Worm wheel

3）压力泵传动装置 Force pump transmission equipment

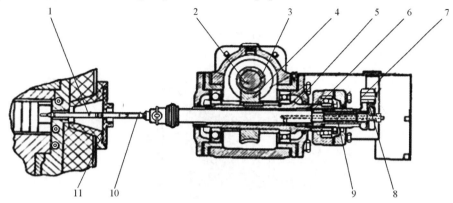

图 4.4.38 压力泵传动装置结构图
Structure drawing of force pump transmission equipment
1—泵轴接头 Pump shaft joint；2—主轴 Shaft；3—蜗杆 Worm rod；4—蜗轮 Worm wheel；
5—空心蜗轮轴 Hollow worm gear shaft；6—离合器 Clutch；7—传动臂 Drive arm；
8—保险销 Safety pin；9—泵轴 Pump shaft；10—调节轴 Regulating spindle；
11—十字头 Cross joint

4.4.3.4 冷却吹风装置 Air cooling and monomer suction device

1. 冷却吹风机 Cooling blower

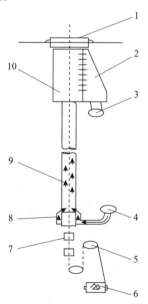

图 4.4.39 冷却吹风机结构图 Structure drawing of cooling blower
1—纺丝保温箱 Spinning incubator；2—吹风窗 Aeration window；
3—横吹风进风管 Transverse aeration inlet pipe；
4—下吹风进风管 Lower aeration inlet pipe；
5—纺丝盘 Spinning disc；6—卷丝筒子 Curl bobbin；
7—给丝上油盘 Wet oiling tray；8—吹风头 Aeration head；
9—冷却套管 Cooling sleeve；10—丝室 Wire room

2. 吹风头 Aeration head

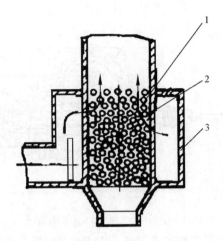

图 4.4.40 吹风头结构图 Structure drawing of aeration head
1—过滤网 Filter screen；2—冷却套管 Cooling casing pipe；3—套筒 Sleeve

4.4.3.5 单体抽吸装置 Monomer suction device

1. 带有喷水缝隙的抽吸装置
Suction device with hydraulic aperture

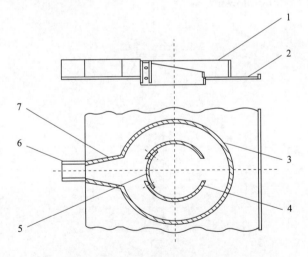

图 4.4.41 带有喷水缝隙的抽吸装置结构图
Structure drawing of suction device with hydraulic aperture
1—顶板 Roof；2—底板 Baseplate；3—外圈 Outer ring；4—内圈 Inner ring；
5—调节板 Adjusting plate；6—吸风管 Aspiration channel；7—扩散口 Diffusion port

2. 单体抽吸回收设备 Monomer suction recycling equipment

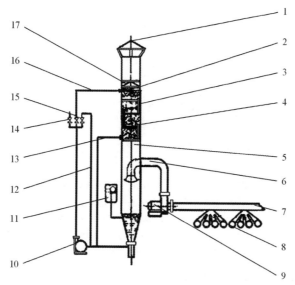

图4.4.42 单体抽吸回收设备结构图 Structure drawing of monomer suction recycling equipment
1—锥顶 Cone top；2—喷头 Nozzle；3—溢流管 Overflow pipe；4—筛板 Sieve plate；5—泡沫吸收塔 Foam absorber；
6—进风管 Inlet duct；7—抽吸总管 Suction manifold；8—抽吸装置 Suction device；9—鼓风机 Blowers；
10—循环水泵 Circulating pump；11—补充水槽 Supplement sink；12—分流器 Shunt；13—溢流管 Overflow pipe；
14—放液阀 Bleeder valve；15—分流阀 Shunt valve；16—循环水管 Circulating water pipe；
17—除沫装置 Demister device

4.4.3.6 卷绕设备 Rolling equipment
1. 移动式摩擦滚筒 Mobile friction drum

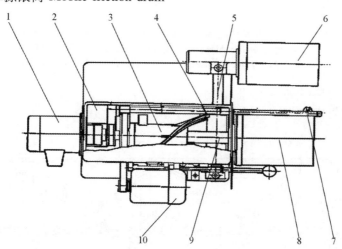

图4.4.43 移动式摩擦滚筒结构图 Structure drawing of mobile friction drum
1、10—电机 Motor；2—凸轮箱 Cam box；3—凸轮 Cam；4—往复导丝杆滑块 Reciprocating guiding screw slider；
5—滑动轴 Sliding shaft；6—筒管座 Bobbin holder；7—导丝器 Thread guide；
8—摩擦滚筒 Friction drum；9—摩擦滚筒轴 Friction drum shaft

2. 电动调频式导丝机构 Electric motor frequency modulation type thread guide

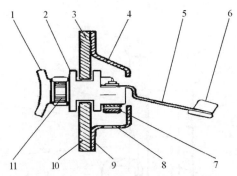

图 4.4.44　电动调频式导丝机构结构图
Structure drawing of electric motor frequency modulation type thread guide
1—滑梭 Shuttle；2—滑块 Slider；3—上部横动滑板 Upper traverse slider；4—上盖 Cover；
5—托板 Pallet；6—导丝器 Yarn guide；7—螺钉 Screw；8—视窗 Window；9—下盖 Lower cover；
10—下部横动滑板 Lower traverse slider；11—滚子 Roller

3. 导丝槽筒箱 Thread guide groove drum box

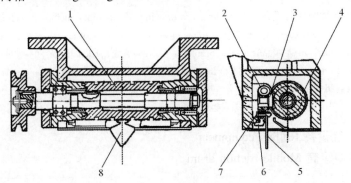

图 4.4.45　导丝槽筒箱结构图 Structure drawing of thread guide groove drum box
1、4—槽筒 Groove drum；2—导轴 Guide shaft；3—梭子 Shuttle；
5、8—导丝器 Thread guide；6—导板 Guide plate；7—滑块 Slider

4. 筒管座 Bobbin holder
1) 叶片式筒管座 Blade bobbin holder

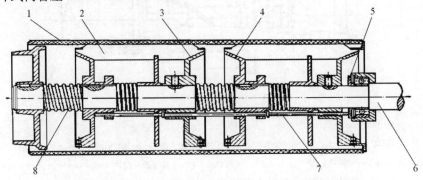

图 4.4.46　叶片式筒管座结构图 Structure blade bobbin holder
1—筒管 Bobbin；2—叶片 Vane；3—固定圆盘 Fixing disc；4—滑动圆盘 Sliding disc；
5—挡圈 Check ring；6—筒管轴 Bobbin shaft；7—推杆 Push rod；8—弹簧 Spring

2) 气动式筒管座 Gas operated bobbin holder

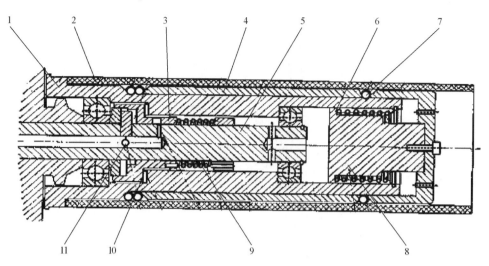

图 4.4.47 气动式筒管座结构图 Structure drawing of gas operated bobbin holder
1—筒管座 Bobbin holder；2—筒管 Bobbin；3、8—活塞 Piston；4—套筒 Sleeve；5—芯轴 Spindle；
6、9—弹簧 Spring；7—O 形密封圈 O sealing ring；10—刹车片 Brake pad；11—气孔 Stomata

3) 碟形弹簧式筒管座 Disc spring type bobbin holder

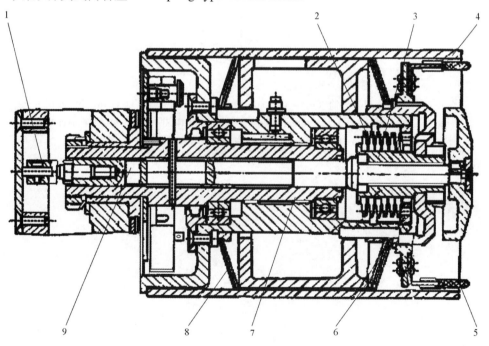

图 4.4.48 碟形弹簧式筒管座结构图
Structure drawing of disc spring type bobbin holder
1—滚动轴承 Rolling bearing；2—锥形弹簧 Conical spring；3—小套筒 Small sleeve；
4—筒管 Bobbin；5—钩丝器 Hook yarn；6—调节螺母 Adjusting nut；7—大套筒 Large sleeve；
8—碟形弹簧 Disc spring；9—芯轴 Spindle

4)弹簧圈握持筒管座 Spring coil grip bobbin holder

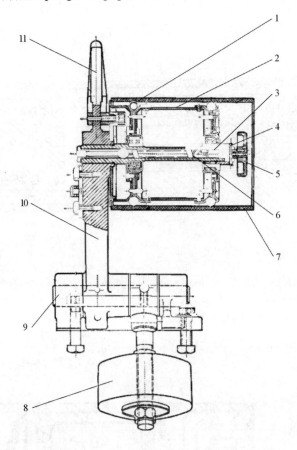

图 4.4.49 弹簧圈握持筒管座结构图 Structure drawing of spring coil grip bobbin holder
1、6—弹簧 Spring；2—滚筒 Roller；3—轴承 Bearing；4—销钉 Pin；5、11—手柄 Handle；
7—筒管 Bobbin；8—重锤 Hammer；9—托脚 Support；10—筒管架 Bobbin holder

4.4.4 牵伸加捻设备 Draft twisting device

4.4.4.1 牵伸加捻机 Draft twisting device

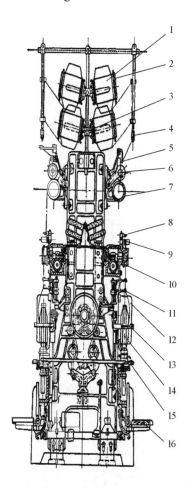

图 4.4.50 牵伸加捻机结构图 Structure drawing of draft twisting device

1—卷绕筒管 Take-up bobbin；2—丝筒盖 Wire cylinder cover；3—筒子架 Bobbin creel；
4—导丝棒 Guide wire rod；5—分丝棒 Devillicate rod；6—上压辊 Upper compression roller；
7—给丝罗拉 Wire roller；8—拉伸棒 Strech rod；9—小转子 Lesser trochanter；
10—牵丝盘 Lead wire plate；11—导丝钩 Guide wire hook；12—隔丝板 Wire insulation board；
13—钢丝钩 Steel wire hook；14—筒管 Tube；15—锭带 Spindle tape；16—踏脚板 Footboard

4.4.4.2 喂入设备 Feeding equipment

1. 斜插固定式筒子架 Diagonal fixed bobbin creel

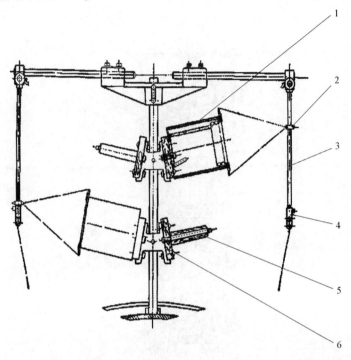

图 4.4.51 斜插固定式筒子架结构图 Structure drawing of diagonal fixed bobbin creel

1—铁丝架 Iron wire rack；2—导丝钩 Guide wire hook；3—圆棒 Rod；
4—导丝棒 Thread guide rod；5—插座 Socket；6—筒子架木板 Creel board

2. 双蜗轮式横动装置 Double worm gear type traverse device

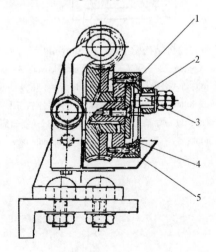

图 4.4.52 双蜗轮式横动装置结构图
Structure drawing of double worm gear type traverse device

1—固定心轴 Stationary shaft；2—轴套 Sleeve；3—短轴 Minor shaft；
4—盖板 Cover plate；5—销钉 Pin

4.4.4.3 牵伸上压辊 Drafting equipment

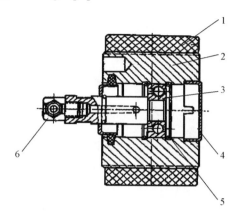

图 4.4.53 牵伸上压辊结构图

Structure drawing of drafting compression roller

1—丁腈橡胶套 Sleeve；2—铁芯 Iron core；3—轴承 Bearing；4—端盖 Cover；
5—弹簧挡圈 Spring ring；6—销轴 pin

4.4.4.4 加捻设备 Twisting device

1. 升降式导丝钩 Elevating thread guide hook

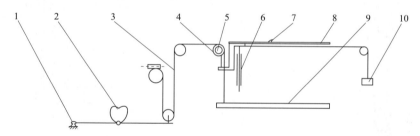

图 4.4.54 升降式导丝钩结构图

Structure drawing of elevating thread guide hook

1—支点 Fulcrum；2—凸轮 Cam；3—皮带 Belt；4—皮带轮 Belt wheel；
5—链轮 Chain wheel；6—轴 Shaft；7—导丝钩 Thread guide hook；
8—扁铁板 Flat iron；9—钢领板 Steel collar plate；
10—平衡重锤 Counterweight hammer

2. 滚柱轴承式锭子 Roller bearing spindle

4.4.4.5　卷绕机 Roller

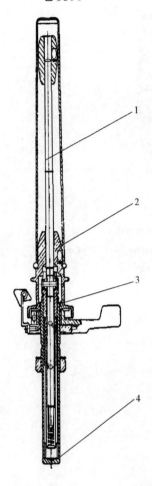

图 4.4.55　滚柱轴承式锭子结构图 Structure drawing of roller bearing spindle

1—锭杆 Spindle blade;
2—锭盘 Wharve;
3—锭胆 Bolster;
4—锭脚 Spindle foot

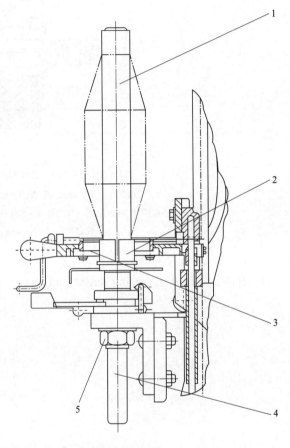

图 4.4.56　卷绕机结构图 Structure drawing of roller

1—筒管 Bobbin;
2—废丝盘 Silk waste plate;
3—钢领 Steel collar;
4—锭子 Spindle; 5—螺母 bolt

4.4.5 后加工设备 Post-processing equipment

4.4.5.1 长丝后加工设备 Filament post-processing equipment

1. 加捻机 Twisting machine

1) 层式加捻机（立式）Layer type twisting machine (vertical)

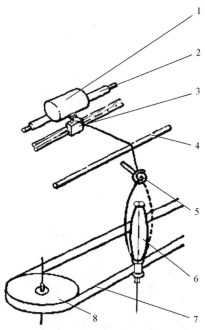

图 4.4.57 层式加捻机（立式）结构图
Structure drawing of layer type twisting machine (vertical)

1—有孔筒管 Hole bobbin；2—摩擦辊筒 Friction roller；3—导丝器 Thread guide；
4—导丝棒 Thread guide bar；5—导丝钩 Thread guide hook；6—管丝 Tube filament；
7—皮带 Belt；8—皮带轮 Belt wheel

2) 层式加捻机（卧式）Layer-type twisting machine (horizontal)

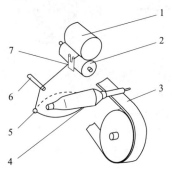

图 4.4.58 层式加捻机（卧式）结构图 Structure drawing of layer-type twisting machine (horizontal)

1—有孔筒管 Hole bobbin；2—摩擦辊筒 Friction roller；3—皮带 Belt；
4—管丝 Tube filament；5—导丝钩 Thread guide hook；6—导丝棒 Thread guide bar；
7—导丝器 Thread guide

2. 加捻锭子 Twisting spindle

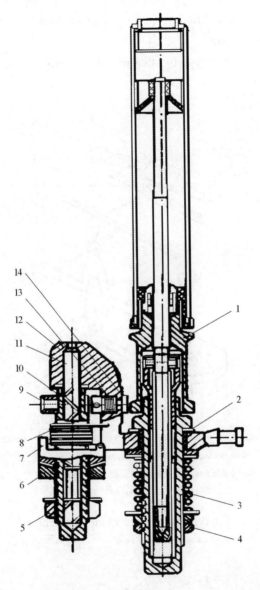

图 4.4.59 加捻锭子结构图 Structure drawing of twisting spindle
1—锭子 Spindle；2—衬垫 Liner；3—压簧 Spring；4—螺帽 Nut；5—螺母 Screw nut；
6—垫块 Heel block；7—托脚 Holder；8—扭簧 Torsion spring；9—偏心块 Eccentric block；
10—套筒 Sleeve；11—轴 Shaft；12—摇臂 Rocker arm；13—弹簧 Spring；
14—刹停器 Braking device

4.4.5.2 调幅式往复渐缩机构
Amplitude modulation type reciprocating reducing mechanism

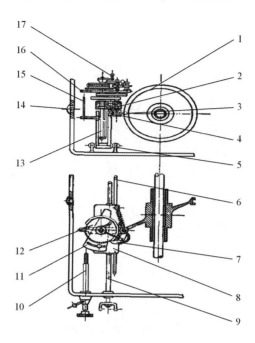

图 4.4.60 调幅式往复渐缩机构结构图
Structure drawing of amplitude modulation type reciprocating reducing mechanism

1—凸轮 Cam；2—座架 Seat frame；3—转子 Rotor；4—滑块 Slider；5—摇臂 Rocker arm；6—导轨 Rail；
7—撑头 Support head；8—托板 Pallet；9—往复导杆 Reciprocating rod guide；10—撞针 Striker；
11—扇形板 Fan-shaped plate；12—棘轮 Ratchet；13—螺杆 Screw；14—视窗 Window；
15—指示棒 Designating rod；16—托脚 Standoff；17—手柄 Handle

1. 调幅式往复机构 Amplitude modulation type reciprocating mechanism

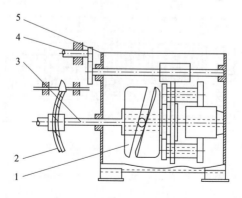

图 4.4.61 调幅式往复机构结构图
Structure drawing of amplitude modulation type reciprocating mechanism

1—辅凸轮 Auxiliary cam；2—主凸轮 Main cam；3—往复导丝轴 Traverse shaft；
4—传动轴 Drive shaft；5—凸轮箱 Cam box

2. 压洗设备 Extrusion washing equipment

1）压洗锅 Extrusion washing boiler

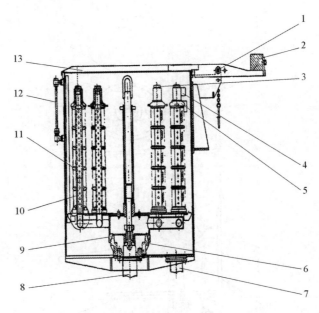

图 4.4.62 压洗锅结构图 Structure drawing of extrusion washing boiler
1—销轴 Shaft；2—重锤 Hammer；3—螺母 Nut；4—溢流口 Overflow port；5—压盖 Gland；
6—锥底 Conical bottom；7—排水口 Outlet；8—进水口 Intake；9—锥体 Cone；
10—洗柱 Wash column；11—垫圈 Washer；12—液位指示器 Level indicator；
13—盖板 Cover

2）压洗槽 Extrusion washing groove

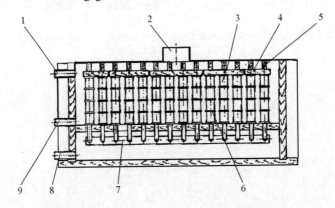

图 4.4.63 压洗槽结构图 Structure drawing of extrusion washing groove
1—溢流口 Overflow pipe；2—排气口 Exhaust port；3—洗筒 Cheese；4—压板 Plate；
5—横系 Horizontal lines；6—垫圈 Washer；7—进水管 Water inlet；
8—排放口 Discharge outlet；9—出水管 Water outlet

4 合成纤维 Synthetic fiber

3) 过滤器 Filter

4) 电磁阀 Solenoid valve

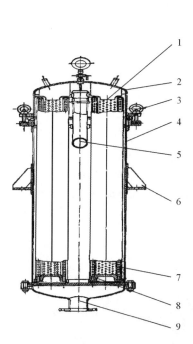

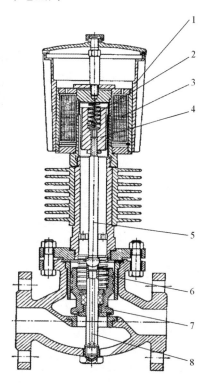

图4.4.64 过滤器结构图 Structure drawing of filter

1—定位板 Positioning plate；2—封头 Head；
3—吊耳 Lifting lug；4—筒体 Barrel；
5—进水口 Water inlet；6—支座 Support；
7—过滤管 Filter tube；8—链孔 Chain hole；
9—排放口 Discharge outlet

图4.4.65 电磁阀结构图
Structure drawing of solenoid valve

1—外壳 Shell；2—弹簧 Spring；3—线圈 Coil；
4—铁芯 Iron core；5—阀杆 Valve stem；
6—压力弹簧 Pressure spring；
7—阀芯 Spool；8—套管 Casing pipe

4.4.5.3 纺丝及定形设备 Spinning and sizing equipment

1. 络丝机 Thread winder

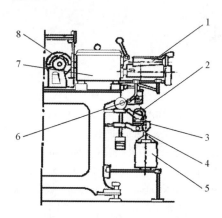

图4.4.66 络丝机结构图
Structure drawing of thread winder

1—锥形筒管 Cone bobbin；2—给油盘 Oiling pan；3—清洁器 Cleaner；4—导丝钩 Thread guide hook；5—筒管 Bobbin；6—张力器 Tensioner；7—锭箱 Ingot box；8—传动装置 Transmission equipement

2. 张力器 Tensioner

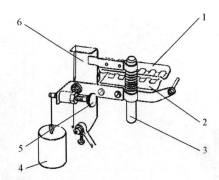

图4.4.67 张力器结构图 Structure drawing of tensioner

1—上梳片 Upper comb slice；2—下梳片 Lower comb slice；3—阻尼器 Damper；
4—重锤 Hammer；5—调节螺栓 Adjustment bolt；6—支架 Support

3. 锭轴变速机构 Spindle speed change mechanism

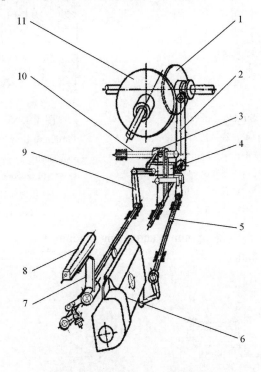

图4.4.68 锭轴变速机构结构图
Structure drawing of spindle speed change mechanism

1—主动盘 Drive plate；2、9—连接件 Connector；3—螺母 Nut；4—滑块 Sliding block；
5—轴 Shaft；6—导丝盒 Thread guide box；7—杠杆 Lever；8—小转子 Small rotor；
10—螺杆 Screw；11—从动盘 Driven plate

4. 往复导丝器 Reciprocatingthread guide

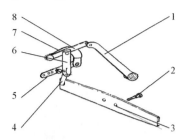

图 4.4.69　往复导丝器结构图 Structure drawing of reciprocating thread guide
1—牵引杆 Tow bar；2—螺栓 Screw；3—螺栓孔 Screw hole；4、7—连杆 Linkage；
5—调节板 Adjustment board；6—升降杆 Lift rod；8—杠杆 Lever

5. 断头满管自停机构 Decollation full pipe automatic stop mechanism

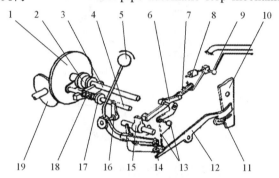

图 4.4.70　断头满管自停机构结构图
Structure drawing of decollation full pipe automatic stop mechanism
1—摩擦盘 Friction disc；2—套筒 Sleeve；3—锭轴 Spindle；4—小轴 Small shaft；5、13—手柄 Handle；
6、9、12—杠杆 Lever；7—连杆 Link；8—重锤 Hammer；10、16—摇臂 Arm；11—销钉 Pin；
14—调节螺栓 Adjustment screw；15—凸块 Bumps；17—挡圈 Ring；18—弹簧 Spring；19—传动盘 Disk drive

6. 断头满管电气自停机构 Decollation full pipe electric automatic stop mechanism

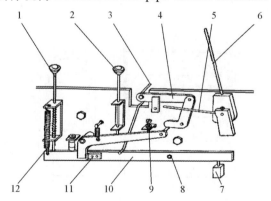

图 4.4.71　断头满管电气自停机构结构图
Structure drawing of decollation full pipe electric automatic stop mechanism
1、2—按钮 Button；3、6—停车杆 Parking rod；4—连杆 Linkage；5—压杆 Compression bar；
7—电源开关 Power switch；8—销钉 Pin；9—调节螺栓 Adjustment screw；10—杠杆 Lever；
11—连接板 Connection plate；12—拉簧 Tension spring

4.4.5.4 短丝后处理设备 Short silk post-processing equipment

1. 五辊牵伸机 Five-roller drafting machine

1) 缠辊自停装置 Roll automatic stop device

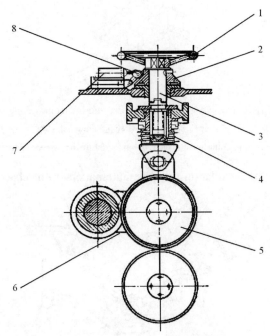

图 4.4.72 缠辊自停装置结构图
Structure drawing of roll automatic stop device

1—手轮 Hand wheel；2—手轮座 Hand wheel seat；3—手轮轴 Hand wheel；4—弹簧 Spring；
5—牵伸辊 Draft roll；6—摆臂 Arm；7—限位开关 Limit switch；8—转子 Rotor

2) 牵伸辊防护罩开启自停装置
Drafting roller cover open automatic stop device

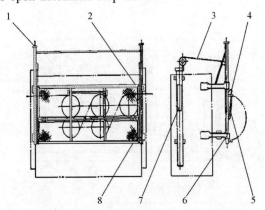

图 4.4.73 牵伸辊防护罩开启自停装置结构图
Structure drawing of drafting roller cover open automatic stop device

1—导轨 Guide rail；2—上护罩 Upper shield；3—钢丝绳 Steel wire rope；4—撑脚 Rack；
5—斜面板 Skewback；6—托脚 Holder；7—重锤 Hammer；8—下护罩 Lower shield

3) 五辊牵伸机润滑装置 Five-roller drafting machine lubrication device

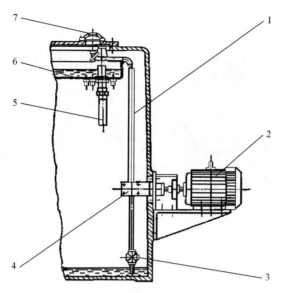

图 4.4.74　五辊牵伸机润滑装置结构图
Structure drawing of five-roller drafting machine lubrication device
1—油管 Oil tube；2—电机 Motor；3—磁性过滤器 Magnetic filter；4—齿轮泵 Gear pump；
5—溢流管 Overflow pipe；6—油盘 Oil tray；7—玻璃视窗 Glass window

4) 五辊牵伸机润湿装置 Five-roller drafting machine wetting device

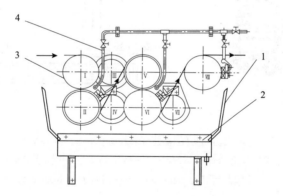

图 4.4.75　五辊牵伸机润湿装置结构图
Structure drawing of five-roller drafting machine wetting device
1—挡水板 Water fender；2—盛水槽 Water filling sink；
3—海绵 Sponge；4—冷水管 Cold water pipe

2. 卷曲装置 Crimping device

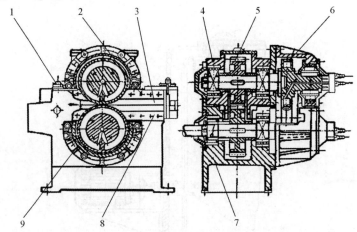

图 4.4.76　卷曲装置结构图 Structure drawing of crimping device

1—进口导板 Inlet guide plate；2—上刮板 The blade；3—导板 Guide plate；4—上箱体 Upper box；
5—齿轮 Gear；6—夹套 Jacket；7—下箱体 Lower cabinet；8—下刮板 Lower scraper；9—卷曲轮 Curled wheel

3. 沟轮式切断机 Furrow wheel cutter

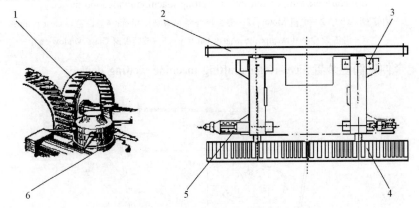

图 4.4.77　沟轮式切断机结构图 Structure drawing of furrow wheel cutter

1、4—沟轮 Furrow wheel；2—齿轮 Gear；3—轴承 Bearing；
5—顶紧装置 Top tight unit；6—回转刀盘 Rotary cutter

4. 履带式干燥机 Crawler drier

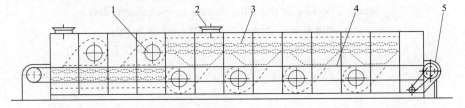

图 4.4.78　履带式干燥机结构图 Structure drawing of crawler drier

1—鼓风机 Blower；2—排气口 Exhaust port；3—散热管 Radiating pipe；
4—履带 Crawler；5—链轮 Sprocket

4 合成纤维 Synthetic fiber

5. 多带式干燥机 Multi-belt drier

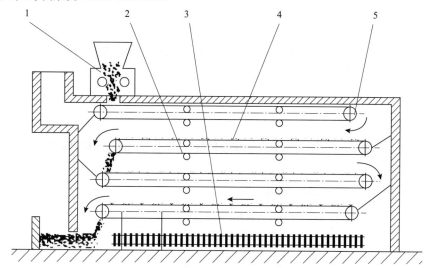

图 4.4.79 多带式干燥机结构图 Structure drawing of multi-belt drier
1—进料口 Feeder；2—支承轮 Support wheel；3—预热器 Preheater；
4—运输带 Conveyor belt；5—转辊 Roller

6. 切断机刀头箱 Cutting machine tool bit box

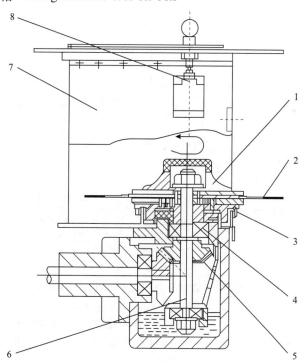

图 4.4.80 切断机刀头箱结构图 Structure drawing of cutting machine tool bit box
1—上刀盘 Upper cutter；2—切断刀 Cut-off tool；3—下刀盘 Lower cutter；4—轴承 Bearing；
5—伞齿轮 Bevel gear；6—刀盘轴 Cutter shaft；7—刀头箱 Tool bit box；8—电气开关 Electric switch

5 合成原料 Synthetic materials

5.1 精对苯二甲酸(PTA) Purified terephthalic acid(PTA)

5.1.1 反应设备 Reaction equipment

5.1.1.1 氧化反应器 Oxidation reactor

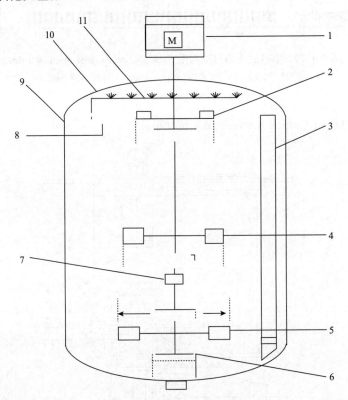

图 5.1.1 氧化反应器结构图 Structure drawing of oxidation reactor
1—搅拌器 Stirrer；2—回流甩液盘 Vaned reflux slinger；3—挡板 Baffles；
4—升液搅拌桨 Bladed upflow impeller；5—径向搅拌桨 Curved blade radial impeller；
6—小型升液搅拌桨 Small bladed upflow impeller；7—轴承 Bearing；
8—溶剂喷射口 Solvent sprays；9—筒体 Barrel；10—封头 Head；11—喷嘴 Nozzle

5 合成原料 Synthetic materials

5.1.1.2 结晶器 Crystallizer
1. 第一结晶器 Firs tcrystallizer

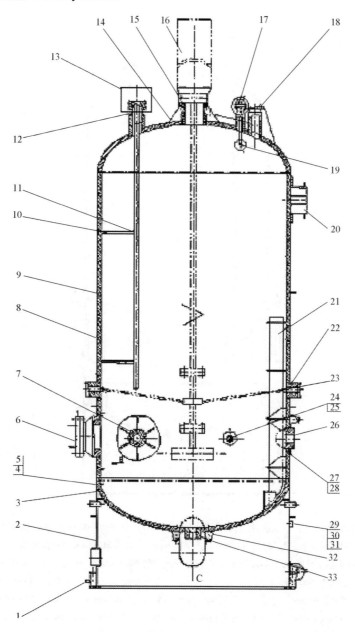

图 5.1.2 第一结晶器结构图 Structure drawing of first crystallizer

1—接地板 Earth plate；2—裙座 Skirt；3—封头 Head；4—梯子 Ladder；5—把手 Handle；6—人孔 Manhole；
7、18、25—法兰 Flange；8—筒体 Barrel；9—保温支承圈 Insulation bearing ring；10—管支承 Tubular brace；
11—支承垫板 Support plate；12—液位计接口 Liquid level gauge interface；13—液位计 Level gauge；
14—搅拌器支座 Stirrer support；15—机架 Rack；16—搅拌器 Stirrer；17—喷淋接口 Spray interface；19—喷嘴 Nozzle；
20、32—吊耳 Lifting lug；21—折流板 Baffle plate；22、26、33—凸缘 Flange；23—轴支承固定件 Shaft support fixed part；
24—厚壁接管 Thick wall adaper；27—补强板 Base board；28—防冲挡板 Inpingement baffle；29—铭牌 Nameplate；
30—铭牌架 Nameplate support；31—铆钉 Rivet

2. 第二结晶器 Second crystallizer

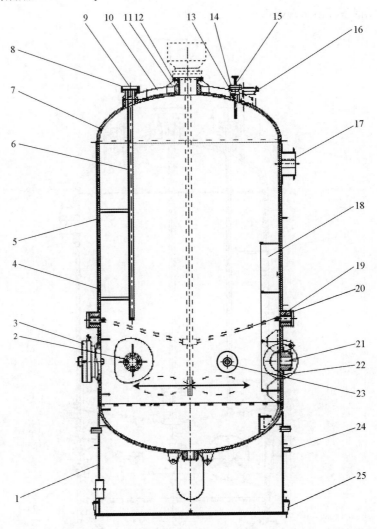

图 5.1.3　第二结晶器结构图结构图 Structure drawing of second crystallizer
1—裙座 Skirt；2、12、13、16—法兰 Flange；3—人孔 Manhole；4—筒体 Barrel；
5—保温支承圈 Insulation bearing ring；6—内伸管 Inner stretch pipe；7—封头 Head；
8—内伸管接口 Inner stretch pipe interface；9—法兰盖 Flange cover；10—加强筋 Reinforcing rib；
11、22—补强圈 Reinforcing ring；14—双头螺柱 Stud；15—内插管 Inner intubation；17—吊耳 Lifting lug；
18—挡板 Baffle；19、21—凸缘 Flange；20—轴支承固定件 Shaft support fixed part；
23—厚壁接管 Thick wall adaper；24—铭牌 Nameplate；25—接地板 Earth plate

5 合成原料 Synthetic materials

3. 第三结晶器 Third crystallizer

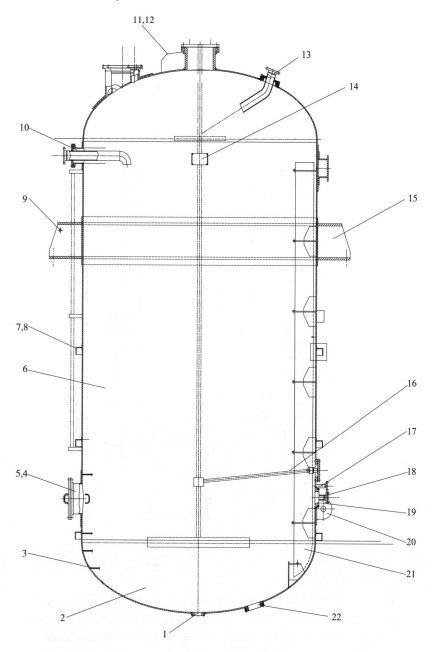

图 5.1.4　第三结晶器结构图 Structure drawing of third crystallizer

1、19、22—凸缘 Flange；2—封头 Head；3—梯子 Ladder；4—人孔 Manhole；5、10—补强圈 Stiffening ring；
6—筒体 Barrel；7—保温支承圈 Insulation bearing ring；8—补强板 Stiffening plate；9—接地柱 Earth plate；
11—搅拌器支座 Stirrer support；12—筋板 Rib plate；
13—清扫管接口 Clean pipe interface；14—铭牌 Nameplate；15—支座 Support；
16—轴支承固定件 Shaft support fixed part；17—温度计接口 Thermometer；
18—液位计接口 Liquid level gauge interface；20—吊耳 Lifting lug；21—折流板 Baffle plate

5.1.1.3 溶解反应罐 Dissolving reaction tank

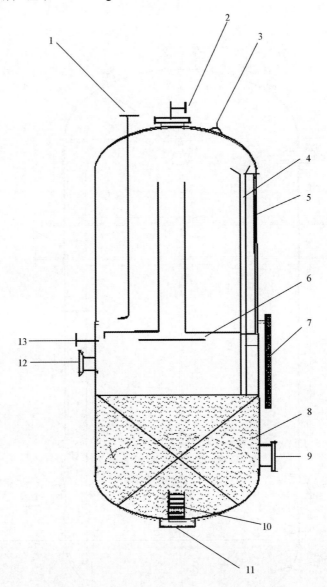

图 5.1.5 溶解反应罐结构图 Structure drawing of dissolving reaction tank
1—溶液入口 Solution inlet；2—氢气入口 Hydrogen inlet；3—减压口 Decompression mouth；
4—液位导管 Liquid level guide pipe；5—筒体 Barrel；6—分布器 Distributor；
7—液位计 Lquid level meter；8—催化剂床层 Catalyst bed；9、12—人孔 Manhole；
10—筛网 Screen cloth；11—溶液出口 Solution outlet；13—排放口 Discharge outlet

5.1.2 塔设备 Comlumn equipment

5.1.2.1 尾气洗涤塔 Tail gas washing tower

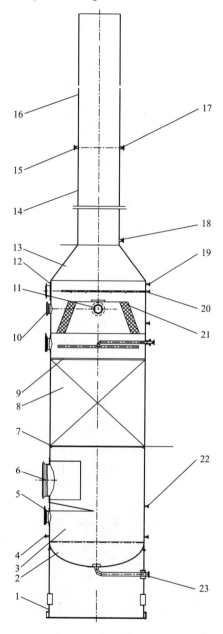

图 5.1.6 尾气洗涤塔结构图 Structure drawing of tail gas washing tower

1—裙座 Skrit；2—封头 Head；3、12、14、16—筒体 Barrel；4、19—温度计接口 Thermometer interface；
5—人孔 Manhole；6—尾气入口 Tail gas inlet；7—填料支承 Racking support；8—填料 Packing；
9—填料压环 Packing press ring；10、11—洗涤液入口 Washing liquid inlet；13—锥体 Cone；
15、17—吊耳 Lifting lug；18—蒸汽入口 Steam inlet；20—添加剂入口 Additive inlet；21—除沫器 Demister；
22—液位计接口 Liquid level gauge interface；23—排放口 Discharge outlet

5.1.2.2 高压吸收塔 High pressure absorption tower

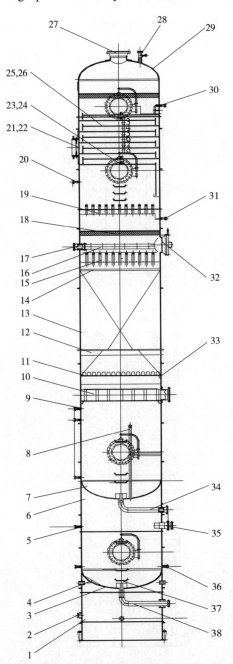

图 5.1.7 高压吸收塔结构图 Structure drawing of high pressure absorption tower

1—裙座 Skirt；2—铭牌 Nameplate；3、7、29—封头 Head；4—梯子 Ladder；5、31—温度计接口 Thermometer interface；6、13—筒体 Barrel；8—升气管 Standpipe；9、36—液位计接口 Liquid level gauge interface；10—气体分布器 Gas distributor；11—驼峰支承 Hump supporting；12—散堆填料 Random packing；14—填料压环 Stuffing ring；15—液体分布器 Liquid dristributor；16—进料分布器 Feed distributor；17—进料口 Feed inlet；18—丝网除沫器 Demister；19—集液箱 Collecting box；20、28、30—接管 Connecting pipe；21—补强板 Stiffening plate；22—轴式吊耳 Shaft lifting lug；23、25—塔盘 Tray；24、26—塔盘支承件 Tray support unit；27—气相出口 Gaseous phrase outlet；32—人孔 Manhole；33—筋板 Rib plate；34—出料口 Feed outlet；35—气相入口 Gaseous phrase inlet；37—防涡流器；38—液相出口 Liquid phrase outlet

5.1.2.3 溶剂脱水塔 Solvent dehydration tower

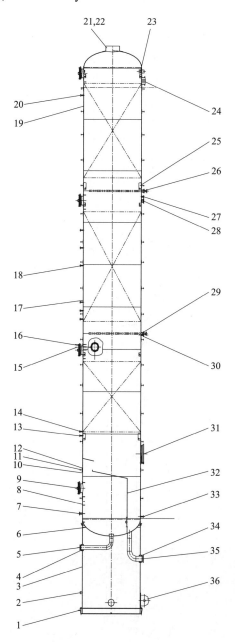

图 5.1.8 溶剂脱水塔结构图 Structure drawing of solvent dehydration tower

1—接地板 Earth plate；2—铭牌 Nameplate；3—裙座 Skirt；4、22、23、34—补强圈 Stiffening ring；
5—液相出口 Liquid phrase outlet；6—封头 Head；7—液位计接口 Liquid level gauge interface；
8、10、17、19—筒体 Barrel；9—人孔 Manhole；11—保温支承圈 Insulation bearing ring；
12—补强板 Stiffening plate；13、15、18、27、33—温度计接口 Thermometer connection；14、20—压力计接口 Pressure tap；16、30—衬套管 Bush pipe；21—气相出口 Gaseous outlet；24、36—吊耳 Lifting lug；
25—塔内件支承 Tower internals support；26—塔顶进料口 Tower top feed inlet；28—回流口 Reflux inlet；
29—中部进料口 Midale feed inlet；31—气相入口 Gaseous inlet；32—堰板 Weir plate；35—水出口 Water outlet

5.1.2.4 回收塔 Recovery tower

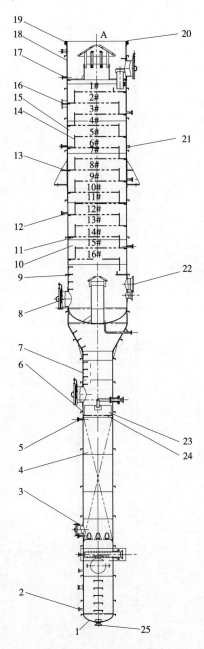

图 5.1.9 回收塔结构图 Structure drawing of recovery tower
1—封头 Head；2—排放口 Discharge outlet；3—手孔 Handhole；4—填料 Packing；
5、12、17—温度计接口 Thermometer interface；6、9—保温支承圈 Insulation bearing ring；
7、10、15、18—筒体 Barrel；8—人孔 Manhole；11—塔盘支承圈 Tray support ring；
13—支座 Support；14—塔盘 Tray；16—吊耳 Lifting lug；19、20—连接法兰 Connecting flange；
21—铭牌 Nameplate；22—气相入口 Gaseous inlet；23—喷淋分配器 Spray distributor；
24—除沫器 Demister；25—液相出口 Liquid phrase outlet

5.1.2.5 PX 采出塔 PX extraction tower

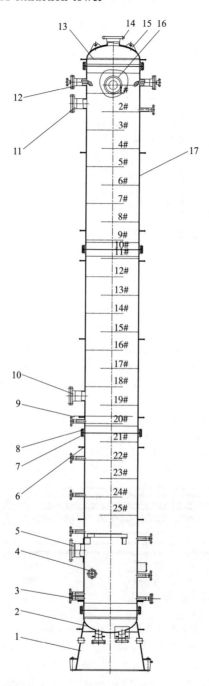

图 5.1.10 PX 采出塔结构图 Structure drawing of PX extraction tower
1—裙座 Skrit；2—封头 Head；3—出料口 Feed outlet；4、9—备用口 Backup mouth；
5—气相入口 Gaseous phrase inlet；6—保温支承圈 Insulation bearing ring；7—采出口 extraction tower；
8—补强圈 Stiffening ring；10、11—人孔 Manhole；12—进料口 Feed inlet；13—防护板 Preventer plate；
14—气相出口 Gaseous phrase outlet；15、16—吊耳 Lifting lug；17—筒体 Barrel

· 453 ·

5.1.3 换热设备 Heat-exchanger equipment

5.1.3.1 热媒排气冷凝器 HTM exhaust condenser

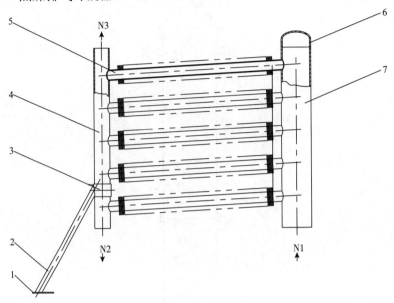

图 5.1.11 热媒排气冷凝器结构图 Structure drawing of HTM exhaust condenser

1—垫板 Baseboard；2—支承角钢 Supporting angle steel；3—垫环 Backing ring；4—出口管 Outlet pipe；
5—翅片管 Finned tube；6—管帽 Pipe cap；7—进口管 Inlet pipe

5.1.3.2 开车加热器 Driving heater

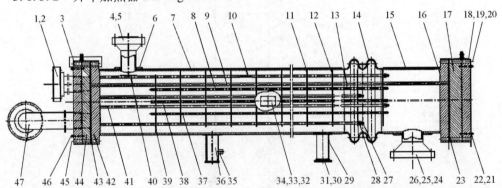

图 5.1.12 开车加热器结构图 Structure drawing of driving heater

1、4、24—接管 Connecting pipe；2、5、25、47—法兰 Flange；3—前端管箱压盖 Front tube box gland；
6、26—补强圈 Stiffening ring；7、15—筒体 Barrel；8、37、38—定距管 Spacer tube；9—换热管 Heat exchange tube；
10、27、41—拉杆 Tie rod；11、12、39—折流板 Baffle plate；13—挡管 Dummy tube；14—膨胀节 Expansion joint；
16—后端管板 Back-end tube sheet；17—后端管箱 Back-end tube box；18—后端管箱压盖 Back-end tube box cover；
19、45—双头螺柱 Stud；20、22、28—螺母 Nut；21、46—螺栓 Bolt；23—后管箱垫片 Back-end tube box gasket；
29、36—支座 Support；30—滑动底板 Sliding plate；31—滑动垫 Sliding pad；32—铭牌架 Nameplate support；
33—铭牌 Nameplate；34—铆钉 Rivet；35—接地板 Earth plate；40—防冲挡板 Impingement baffle；
42—前端管板 Front-end tube sheet；43—前端管箱垫片 Front-end tube box gasket；
44—前端管箱 Front-end tube box

5.1.3.3 CTA 第三结晶冷凝器 CTA third crystallization condenser

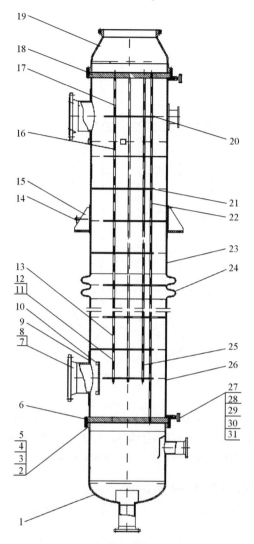

图 5.1.13 CTA 第三结晶冷凝器结构图
Structure drawing of CTA third crystallization condenser

1—下管箱 Lower channel；2、30—双头螺栓 Screw bolt；3、12、31—螺母 Nut；4、29—垫片 Gasket；
5、7、27—法兰 Flange；6—下管板 Lower tube plate；8—接管 Connecting pipe；9—补强圈 stiffening ring；
10—防冲挡板 Impingement baffle；11、13、16、17—拉杆 Tie rod；14—接地板 Ground plate；
15—支座 Support；18—上管板 Upper plate；19—上管箱 Upper channel；
20、21—折流板 Baffle plate；22—换热管 Heat exchange tube；23—上筒体 Upper barrel；
24—膨胀节 Expansion joint；25—定距管 Spacer tube；26—下筒体 Lower barrel；
28—厚壁管 Thick wall pipe

5.1.3.4 回收塔冷凝器 Recovery tower condenser

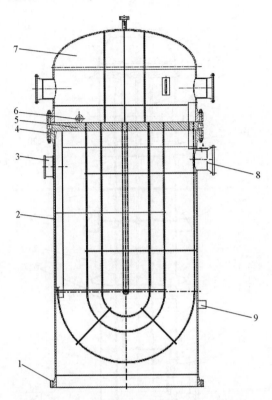

图 5.1.14 回收塔冷凝器结构图 Structure drawing of recovery tower condenser

1、4、8—法兰 Flange；2—筒体 Barrel；3、6—吊耳 Lifting lug；
5—管板 Tube sheet；7—管箱 Channel；9—铭牌 Nameplate

5.1.3.5 凝液再冷器 Condensate re-cooler

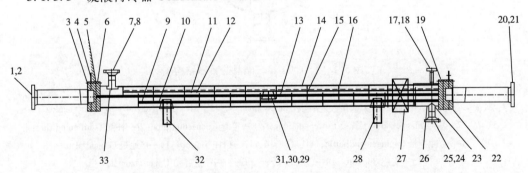

图 5.1.15 凝液再冷器结构图 Structure drawing of condensate re-cooler

1、7、17、20、24—接管 Connecting pipe；2、8、18、21、25—法兰 Flange；3、22—封头 Head；4—管箱 Channel；
5—垫片 Gasket；6—左管板 Left tube sheet；9、10—定距管 Spacer tube；11、12—拉杆 Tie rod；
13—换热管 Heat exchange tube；14、26—筒体 Barrel；15、16—折流板 Baffle plate；19—右管板 Right tube sheet；
23—补强圈 Stiffening ring；27—膨胀节 Expansion joint；28、32—支座 Support；29—铭牌架 Nameplate support；
30—铭牌 Nameplate；31—铆钉 Rivet；33—防冲挡板 Impingement baffle

5.1.3.6 精制第一预热器 Refined first preheater

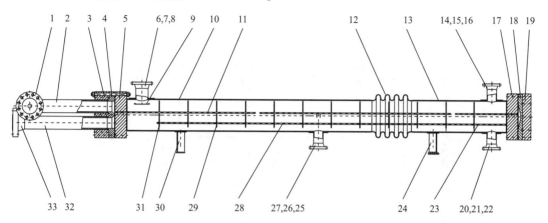

图 5.1.16 精制第一预热器结构图 Structure drawing of refined first preheater

1、6、14、20、25、33—法兰 Flange；2、7、15、21、26、32—接管 Connecting pipe；
3—前端管箱盖 Front-end tube box cover；4—前端管箱 Front-end tube box；5—前端管板 Front-end tube plate；
8、16、22、27—补强圈 Stiffening ring；9—防冲板 Impingement plate；10、13—筒体 Shell；
11—换热管 Heat exchange tube；12—膨胀节 Expansion joint；17—后端管板 Back-end tube plate；
18—后端管箱 Back-end tube box；19—后端管箱盖 Back-end tube box cover；
23、28—拉杆 Tie rod；24—滑动支座 Sliding support；29、31—折流板 Baffle plate；
30—固定支座 Fixed support

5.1.3.7 再打浆水加热器 Re-pulping water heater

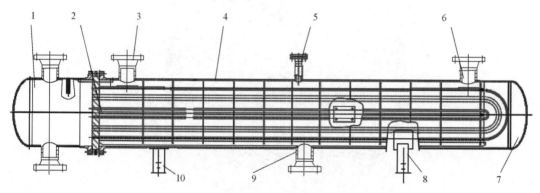

图 5.1.17 再打浆水加热器结构图 Structure drawing of re-pulping water heater

1—管箱 Channel；2—管束 Tube bundles；3、5、6、9—接管 Connecting pipe；4—筒体 Barrel；
7—封头 Head；8、10—支座 Support

5.1.3.8 放空洗涤塔冷凝器 Emptying washing tower condenser

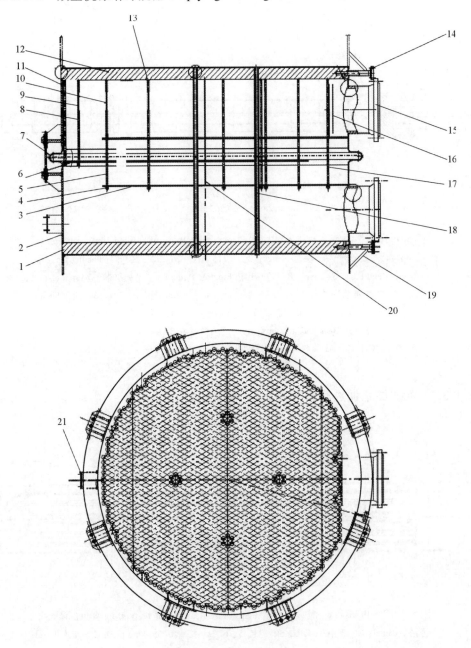

图 5.1.18 放空洗涤塔冷凝器结构图 Structure drawing of emptying washing tower condenser
1—下管板 Lower tube plate；2、11—筒体 Barrel；3、6—折流板 Baffle plate；4—螺母 Nut；
5、10、13、17—拉杆 Tie rod；7—膨胀节 Expansion joint；8、9—定距管 Spacer tube；
12—上管板 Upper tube plate；14—螺栓 Bolt；15—法兰 Flange；16—防冲挡板 Impingement baffle；
18—换热管 Heat exchanger tube；19—加强筋 Reinforcing rib；20—防短路板 Prevent short circuit board；
21—铭牌 Nameplate

· 458 ·

5 合成原料 Synthetic materials

5.1.3.9 回收塔汽凝器 Recovery Column stream Condenser

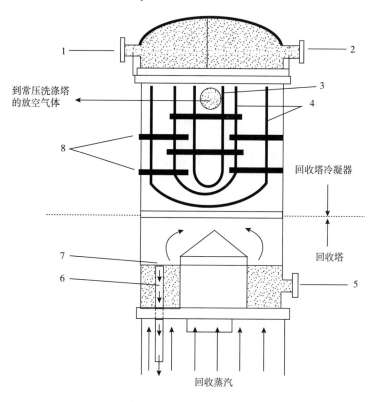

图 5.1.19 回收塔汽凝器结构图 Structure drawing of recovery Column stream Condenser

1—冷却水出口 Cooling water outlet；2—冷却水进口 Cooling water inlet；
3—蒸汽出口 Vapour outlet；4—U形管 U Tubes；5—冷凝液出口 Condensate outlet；
6—回流立管 Reflux standpipe；7—液体液位线 Liquid level；8—缓冲板 Baffles

5.1.3.10 空气预热器 Air preheater

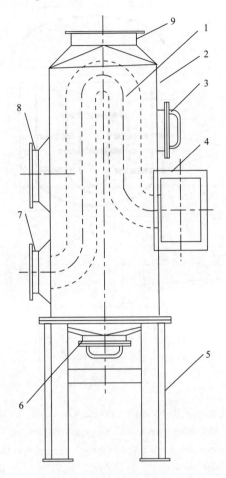

图 5.1.20 空气预热器结构图 Structure drawing of air preheater
1—换热箱 Heatexchange plate box；2—筒体 Outer shell；3—侧清灰口 Side cleaning door；
4—冷风进口 Cold wind inlet；5—支架 Support；6—底清灰口 Base cleaning door；
7—热风出口 Hot wind inlet；8—烟气进口 Smoke inlet；9—烟气出口 Smoke outlet

5 合成原料 Synthetic materials

5.1.4 加热炉 Heating furnace

5.1.4.1 热油炉 Heating furnace

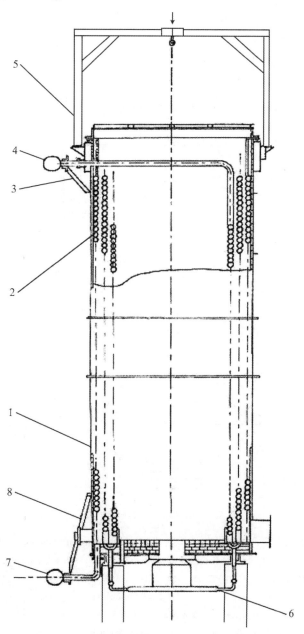

图 5.1.21 热油炉结构图 Structure drawing of heating furnace

1—炉体 Furnace body；2—炉管 Furnace pipe；3—上连箱支承 Upper header support；4—上连箱 Upper header；
5—吊架 Hanging bracket；6—排液管 Fluid-discharge tube；7—下连箱 Lower header；
8—下连箱支承 Lower header support

5.1.4.2 催化焚烧炉 Catalytic oxidizer

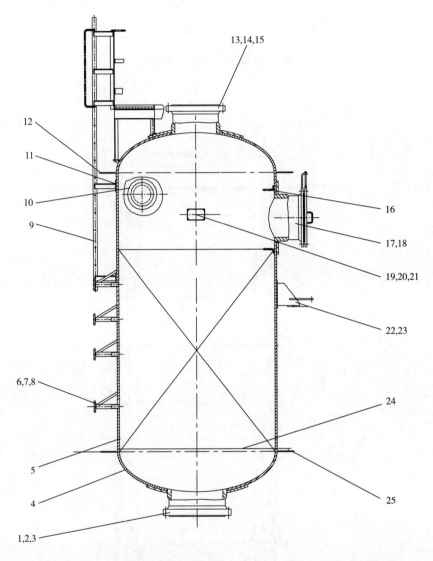

图 5.1.22 催化焚烧炉结构图 Structure drawing of catalytic oxidizer

1、6、13—法兰 Flange；2、7、14—接管 Connecting pipe；3、15、17—补强圈 Stiffening ring；4—封头 Head；
5—筒体 Barrel；8—拉筋 Bracing flat steel；9、16—梯子 Ladder；10—吊耳 Lug；
11—补强板 Stiffening plate；12、25—保温支承圈 Insulation bearing ring；18—人孔 Manhole；
19—铭牌 Nameplate；20—铭牌支架 Nameplate support；21—铆钉 Rivet；22—接地板 Earth plate；
23—支座 Support；24—催化床支承 Catalytic bed support

5.1.5 泵 Pump

5.1.5.1 高速离心泵 High speed centrifugal pump

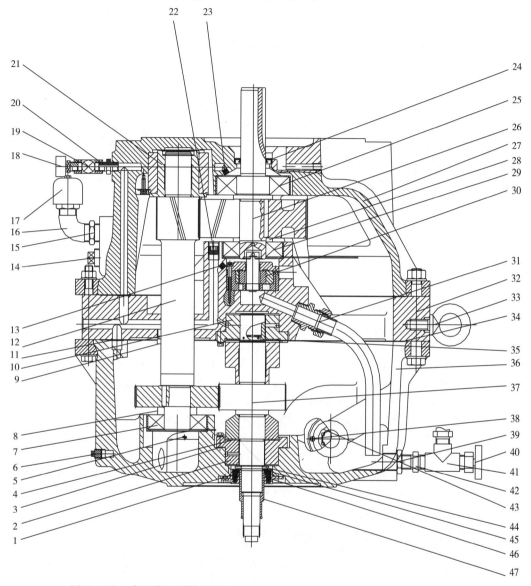

图 5.1.23 高速离心泵结构图 Structure drawing of high speed centrifugal pump

1、11、45—O 形密封圈 O sealing ring；2、10、21—滑动轴承 Sliding bearing；3—调整垫 Adjusting pad；
4、35—止推垫片 Thrust gasket；5、13、23—喷油嘴 Fuel spray nozzle；6、14、22—丝堵 Plug；
7、25、29—球轴承 Angular contact ball bearing；8、28—隔套 Spacer bush；9—卸压阀 Pressure relief valve；
12—中速轴 Medium speed shaft；15、43—视窗 Oil windows；16—弯头 Elbow；17—管帽 Pipe sleeve；
18—压力表 Pressure gauge；19—仪表针阀 Globe Valves；20、42—对丝 Double nipple；24—骨架密封 Reinforced seal；
26—低速轴 Low speed shaft；27—上箱体 Upper box；30—内油泵 Inner oil pump；31—吸油管 Oil sucking pipe；
32—中箱体 Middle gear box；33—吊环 Lifting ring；34—橡胶垫 Gear box rubber blanket；36—下箱体 Lower box；
37—高速轴 High speed shaft；38—圆柱头螺钉 Cross recessed cheese head screws；39—接头管 Niple；
40—双金属温度计 Bimetal thermometer；41—三通 Tee；44—机械密封动环 Mechanical rotating seal ring；
46—机械密封静环 Mechanical stationary seal ring；47—轴套 Sleeve

5.1.5.2 液环真空泵 Liquid ring vacuum pump

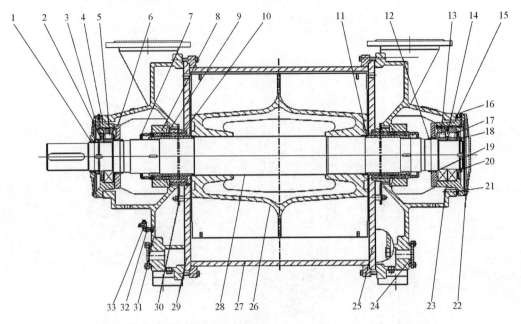

图5.1.24 液环真空泵结构图 Structure drawing of liquid ring vacuum pump

1、12—轴封 Shaft seal ring；2—前挡油环 Front oil retainer；3—前轴承外盖 Front bearing outer cover；
4—前轴承座 Front bearing support；5—圆柱滚子轴承 Cylindrical roller bearing；6—轴承内盖 Bearing inner cover；
7—填料压盖 Stuffing gland；8—填料函 Stuffing box；9—填料环 Padding ring；10—填料 Padding；
11—O形密封圈 O sealing ring；13—圆锥滚子轴承 Cylindrical roller bearing；14—调整垫圈 Ajusting washer；
15—补偿垫圈 Compensation washer；16—油杯 Oil cup；17—后挡油环 Rear oil retainer；
18—止动垫圈 Lock washer；19—油毡 Linoleum；20—螺母 Nut；21—螺栓 Bolt；
22—后轴承外盖 Rear bearing outer cover；23—后轴承座 Rear bearing support；
24—后侧盖 Rear and side cover；25—后分配板 Rear distribution board；26—叶轮 Wheel；
27—泵体 Pump body；28—泵轴 Shaft；29—前分配板 Front distribution board；30—柔性阀板 Flexible valve plate；
31—前侧盖 front side cover；32—自动排水阀 Automatic drain valve；33—阀球 Valve ball

5　合成原料 Synthetic materials

5.1.5.3　屏蔽泵 Shield pump

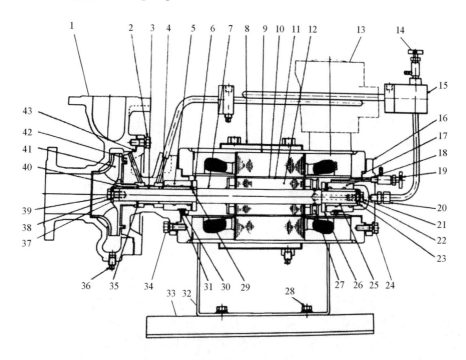

图 5.1.25　屏蔽泵结构图 Structure drawing of shield pump

1—泵体 Pump body；2、21、24、28、34、39—螺栓 Bolt；3、25—轴套 Sleeve；4—衬套 Shaft sleeve；
5、18—轴承 Bearing；6、26—推力盘 Thrust plate；7—轴 Shaft；8—冷却水套 Cooling water jacket；9—定子 Stator；
10—定子屏蔽套 Stator shield sleeve；11—转子屏蔽套 Rotor shield sleeve；12—转子 Rotor；13—接线盒 Junction box；
14、19—排气阀 Vent valve；15—热交换器 Heat exchanger；16—后端盖 Back-end cover；
17、43—密封垫圈 Sealing washer；20—活接接头 Union joint；22、38—止动垫圈 Locking washer；
23、31、37—垫圈 Washer；27—副叶轮 Vice-impeller；29、40—键 Key；30—紧固螺钉 Set screw；
32—机架 Rack；33—底座 Base；35—调整垫圈 Ajusting washer；36—丝堵 Plug；
41—叶轮 Wheel；42—连接体 Connecting body

5.1.6 压缩机 Compressor

5.1.6.1 空压机 Air compressor

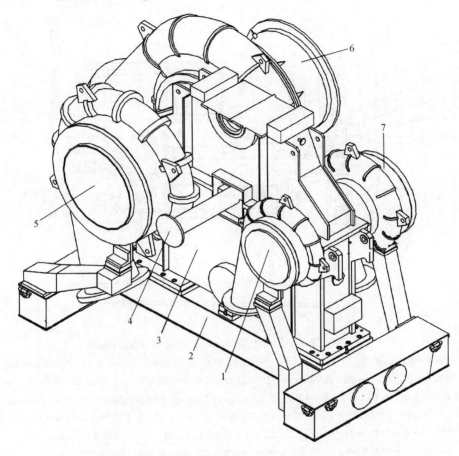

图 5.1.26 空压机外形图 Structure drawing of air compressor
1—压缩机第 4 级 Compressor stage 4；2—底座 Base frame；
3—齿轮箱 Gear case；4—轴 Shaft；5—压缩机第 2 级 Compressor stage 2；
6—压缩机第 1 级 Compressor stage 1；7—压缩机第 3 级 Compressor stage 3

5.1.6.2 尾气膨胀机 Tail-gas expansion turbine

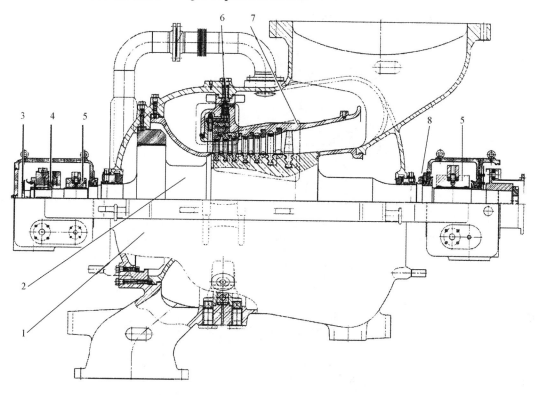

图 5.1.27　尾气膨胀机结构图 Structure drawing of tail-gas expansion turbine

1—机体 Compressor body；2—转子 Rotor；3—齿轮 Gear；4—双推力轴承 Double thrust bearing；
5—径向轴承 Radial bearing；6—定子叶片制动器 Stator blade actuator；
7—定子叶片承载架 Stator blade carrier with stator blades；
8—防泄漏装置 Leakage current arrestor

5.1.6.3 氮气压缩机 Nitrogen compressor

1. 氮气压缩机外形 Nitrogen compressor shape

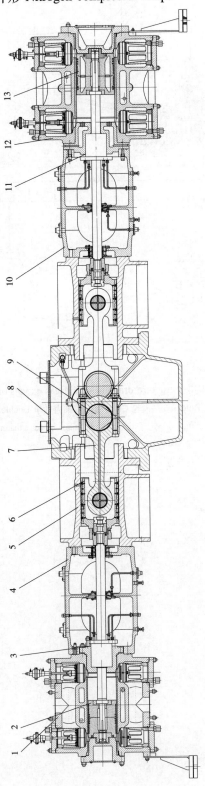

图 5.1.28 氮气压缩机外形图 Shape of compressor

1—二级气缸 Second cylinder；2—二级活塞 Second piston；3—二二级填料 Second feed；4—二级接筒 Second connecting tube；5—中体 Midbody；6—十字头 Crosshead；7—连杆 Connecting rod；8—机身 Body；9—曲轴 Crankshaft；10—一级接筒 First connect tube；11—一级填料 First feed；12—一级气缸 First cylinder；13—一级活塞 First piston

2. 氮气压缩机结构 Nitrogen compressor structure

1）一级气缸 First level cylinder

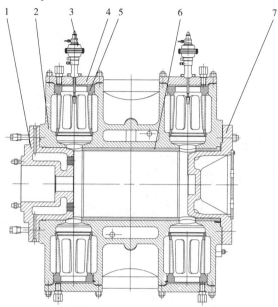

图 5.1.29　一级气缸结构图 Structure drawing of first level cylinder

1—缸盖 Cover；2—缸体 Body；3—压阀罩 Valve cover；4—阀孔盖 Valve hole cover；
5—卸荷器 Unloading device；6—缸套 Liner；7—缸座 Block

2. 二级气缸 Second level cylinder

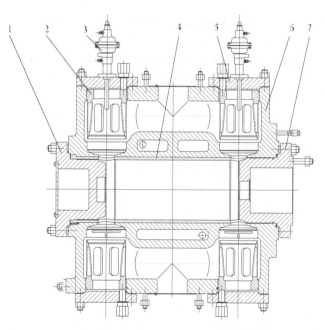

图 5.1.30　二级气缸结构图 Structure drawing of second level cylinder

1—缸盖 Cover；2—卸荷器 Unloading device；3—压阀罩 Valve cover；4—缸套 Liner；
5—阀孔盖 Valve hole cover；6—缸体 Body；7—缸座 Block

5.1.7　其它设备 Other equipments

5.1.7.1　蒸汽轮机 Seam turbine

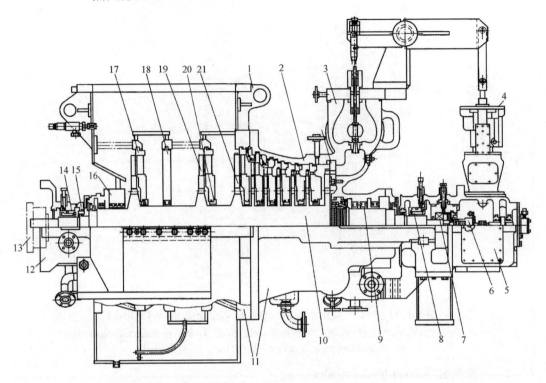

图 5.1.31　蒸汽轮机结构图 Condensing multi-stage steam turbine

1—排气缸 Row cylinder；2—上部汽缸 Upper cylinder；3—调节阀 Regulating valve；4—调速机构 Speed governor；5—前轴承箱 Front bearing casing；6—危急保安器 Crisis protector；7—推力轴承 Thrust bearing；8—前轴承 Front bearing；9—高压端气封 High pressure end gas seal；10—转子 Rotor；11—下部汽缸 Cylinder；12—后轴承箱 Bearing housing；13—联轴器 Coupling；14—后轴承 Rear bearing；15—油封 Oil seal；16—低压端气封 Low pressure end gas seal；17、18——隔板 diaphragm；19——动叶片 Moving blade；20——隔板气封 diaphragm gas seal；21—叶轮 Wheel

5.1.7.2 旋转真空过滤机 Rotary vacuum filter

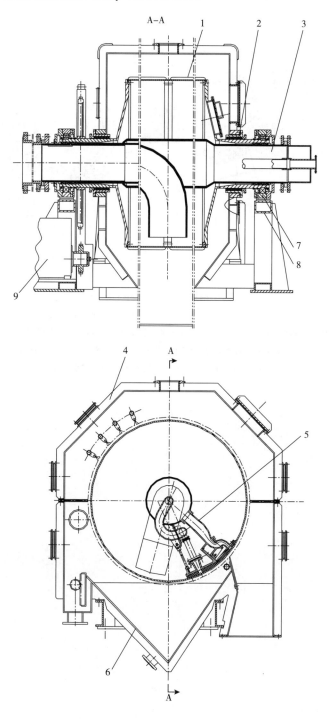

图 5.1.32 旋转真空过滤机结构图 Structure drawing of rotary vacuum filter
1—转鼓 Rotor drum；2—密封 Seal；3—空心轴 Hollow shaft；4—上壳体 Upper housing；
5—反吹系统 Blowback system；6—下壳体 Lower shell；7—轴承 Left and right support；
8—轴承座 Basis；9—驱动装置 Transmission

5.1.7.3 干燥机 Drying machine

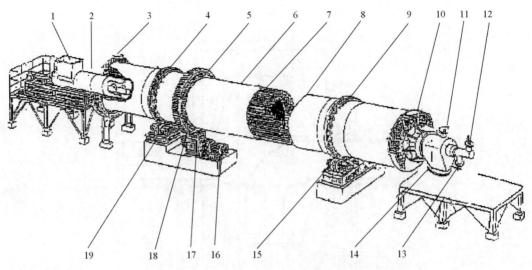

图 5.1.33　干燥机结构图 Structure drawing of drying machine

1—进料口 Feed inlet；2—进料螺杆 Feed inlet screw；3—集汽管 Gas collecting tube；4、9—轮箍 Tyre；
5—圆周齿轮 Circular gear；6—壳体 Shell；7—加热管 Heating pipe；8—干燥器 Drying machine；
10—歧管 Manifold；11—载气进口 Carrier gas inlet；12—蒸汽进口 Steam inlet；13—排水口 Water outfall；
14—产品出口 Products outlet；15、19—辊子 Roller；16—电机 Motor；17—减速器 Reducer；18—小齿轮 Small gear

5.1.7.4 离心机 Centrifuge

1. 草酸离心机 Oxalate centrifuge

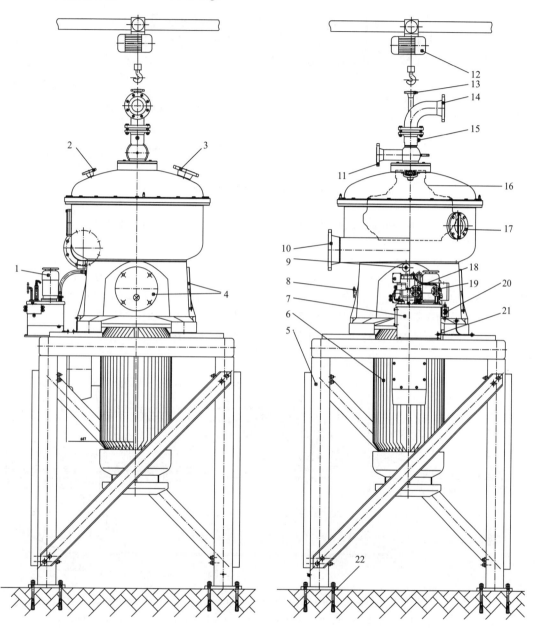

图 5.1.34 草酸离心机结构图 Structure drawing of oxalate centrifuge

1—润滑油电动机 Lubricating oil motor；2—氮气口 Nitrogen mouth；3—通风口 Ventilation opening；
4、17—安装口 Installation mouth；5—框架 Frame；6—主电机 Main motor；
7—润滑油装置 Outer lubricating oil equipment；8—泄漏液排放口 Leaking fluid discharge outlet；9—振动监测器 Vibration monitor；10—排放口 Discharge outlet；11—卸料口 Discharge opening；12—电动单轨吊车 Telpher；13—凝缩循环口 Condensed circulation port；14—进料口 Feed inlet；15—排气口 Exhaust port；
16—转鼓 Rotor drum；18—冷却水出口 Cooling water outlet；19—冷却水进口 Cooling water inlet；
20—接线盒 Terminal box；21—接地板 Earth plate；22—地脚螺栓 Anchor bolt

2. 常压离心机 Normal pressure centrifuge

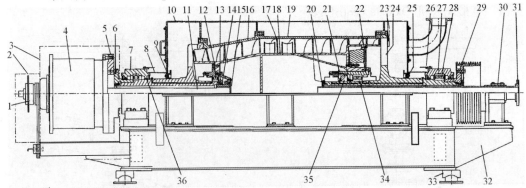

图5.1.35 常压离心机结构图 Structure drawing of normal pressure centrifuge
1—驱动装置 Torquer；2、3、8—防护罩 Shield；4—齿轮箱 Gear box；5、6—法兰 Flange；
7、21、27—轴承箱 Bearing box；9—密封环 Sealing ring；10—机壳 Housing；11—驱动轴 Drive shaft；
12—内轴承座 Inner bearing support；13、15—端盖 End cover；14—轴承压盖 Bearing end cover；
16—锥形转筒 Conical drum；17—叶片 Vane；18—传送装置 Conveyer；19—圆柱转筒 Cylinder drum；
20—盖板 Cover plate；22—转筒端盖 Drum end cover；23—喷嘴 Nozzle；24—转筒头 Drum head；
25—密封环 Sealing ring；26—蒸汽屏 Steam screen；28、34—球面滚柱轴承 Spherical roller bearing；
29—从动带轮 Driven pulley；30—管座 Tube socket；31—管接口 Tube support；32—底座 Base；
33—减震器 Shock absorber；35—球形止推轴承 Special spherical thrust bearings；36—油封 Oil seal

3. 压力离心机 Pressure centrifuge

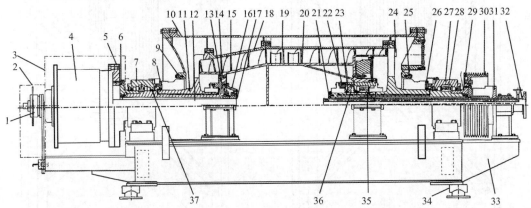

图5.1.36 压力离心机结构图 Structure drawing of pressure centrifuge
1—驱动装置 Torquer；2、3、8、26—防护罩 Shield；4—齿轮箱 Gear box；5、6—法兰 Flange；
7、27—轴承箱 Bearing box；9、25—密封环 Sealing ring；10—壳体 Shell；11—转筒头 Drum head；
12—传动轴 Transmission shaft；13—内部轴承箱 Inner bearing box；14—滚柱轴承 Roller bearing；
15—输送机叶片 Conveyor vane；16—内部轴承箱半封闭端盖 Inner bearing box semi-closed end cover；
17—内部轴承端盖 Inner bearing end cover；18—锥形转筒 Conical centrifuge drum；
19—圆柱转筒 Parallel centrifuge drum；20—输送机 Conveyor；21、22—端盖 End cover；
23—止推轴承压紧器 Thrust bearings pressing device；24—节流孔板 Restriction orifice；
28、35—球面滚柱轴承 Spherical roller bearing；29—减震法兰 Decouple flange；
30—从动皮带轮 driven pulley；31—供料管托架 Feed tube carrier；32—冲洗管接口 Wash pipe interface；
33—底座 Base；34—减震器 Shock absorber；36—球形止推轴承 Spherical thrust bearings；37—油封 Oil seal

5.1.7.5 螺旋输送设备 Auger delivery equipment

1. 过滤机出料螺杆 Filter discharging screw rod

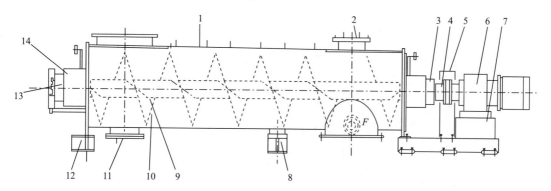

图 5.1.37 过滤机出料螺杆结构图 Structure drawing of filter discharging screw rod

1—上盖 Upper cover；2—上盖板 Upper board；3—驱动端轴承 Drive end bearing；
4—联轴器 Coupling；5—联轴器罩 Coupling cover；6—减速机 Reduction box；
7—驱动底座 Drive end base；8—中间底座 Middle base；9—螺杆 Screw shaft；
10—壳体 Shell；11—下盖板 Lower board；12—压板 Pressing plate；
13—测速盘 Tachometer disc；14—从动端轴承 Driven end bearing；

2. 干燥机进料螺旋 Dryer feed screw

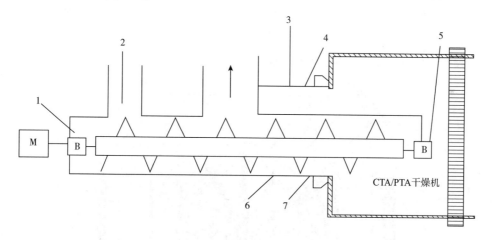

图 5.1.38 干燥机进料螺旋结构图 Structure drawing of dryer feed screw

1—至轴承的输送气体 bearing conveying gas；
2—粗对苯二甲酸潮湿粉末 Coarse formic acid wet powder；
3、6—输送气体 Transmission gas；
4、7—至密封填料的冷却水 Sealing packing cooler water；
5—悬挂轴承 Hanger bearing

5.1.7.6 旋转下料阀 Rotating discharge valve

1. 旋转下料阀外形 Rotating discharge valve shape

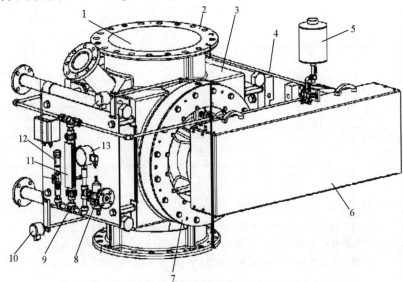

图 5.1.39　旋转下料阀外形图 Structure drawing of rotating discharge valve shape

1—转子 Rotor；2—进料口 Feed inlet；3—旋转下料阀 feed inlet valve；4—电机支架 Motor support；5—油壶 Oiler；6—链条防护罩 Chain shield；7—连接法兰 Connecting flange；8—降压阀 Valve；9—净化气测量系统 Purify air measurement system；10—测温器 Temperature detector；11—流程指示管 Process indicator tube；12—减压阀 Reducing valve；13—压力计 Pressure indicator

2. 旋转下料阀结构 Rotating discharge valve structure

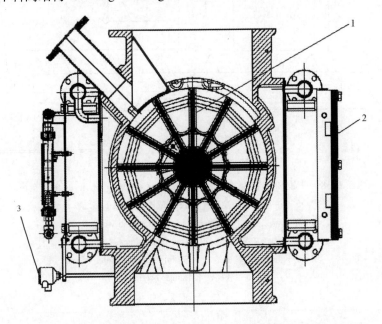

5.1.41　旋转下料阀结构图 Structure drawing of rotating discharge valve structure

1—转子 Rotor；2—电机支架 Motor support；3—测温器 Temperature detector

5.1.7.7 薄膜蒸发器 Film evaporator

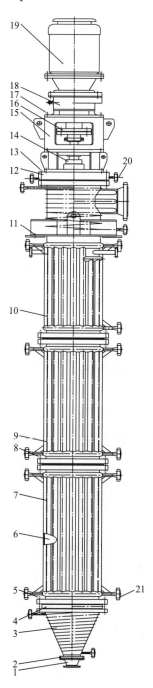

图 5.1.41 薄膜蒸发器结构图 Structure drawing of film evaporator

1、3—下料锥 Feed discharge cone；2、5、8—法兰 Flange；4—底轴支架 Base shaft support；
6—转子 Rotor；7—下部筒体 Lower barrel；9—中部筒体 Middle barrel；10—上部筒体 Upper barrel；
11、13、17—螺栓 Bolt；12—上盖 Upper cover；14—机械密封 Mechanical seal；15—支架 Support；
16—联轴器 Coupling；18—减速机 Reducer；19—电机 Motor；20、21—接管 Connecting pipe

5.1.7.8 PTA 母液过滤器 PTA mother liquid filter

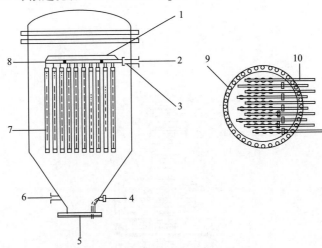

图 5.1.42 PTA 母液过滤器结构图 Structure drawing of PTA mother liquid filter
1—分配管吊耳 Register lifting lugs；2、10—滤液出口 Discharge outlet；3—分配管接头 Distributor coupling；4—排放口 Discharge opening；5—物料入口 Material inlet；6—滤液进口 Feed nozzle；7—过滤棒 Candles；8、9—分配器 Distributor

5.2 己内酰胺 Caprolactam

5.2.1 反应设备 Conversion equipment

5.2.1.1 加氢反应器 Hydrogenation reactor

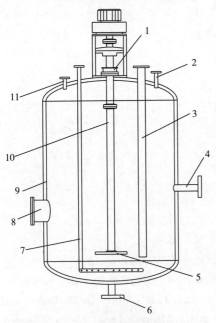

1—轴封 Shaft seal；2—溢流液回流管 Overflow fluid return pipe；3—进料管 Feed pipe；4—出料口 Discharge hole；5—搅拌桨叶 Paddle；6—放料口 Feed pipe；7—氢气进料管 Hydrogen feed pipe；8—人孔 Manhole；9—筒体 Shell；10—搅拌器 Stirrer；11—安全阀接口 Safety valve interface

图 5.2.1 加氢反应器结构图
Structure drawing of hydrogenation reactor

5.2.1.2 磁稳定床 Magnetic stabilized bed

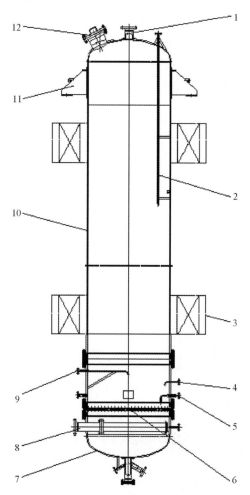

图 5.2.2 磁稳定床结构图 Structure drawing of magnetic stabilized bed
1—出料口 Discharge hole；2—催化剂循环管 Catalyst circulation pipe；
3—电磁线圈 Eectromagnetic coil；4—催化剂床层整理口 Catalyst bed tidy mouth；
5—催化剂卸出口 Catalyst discharge outlet；6—分布器 Distributor；7—封头 Head；
8—进料口 Feed inlet；9—催化剂膨胀口 Catalyst expansion mouth；
10—筒体 Barrel；11—支座 Support；12—人孔 Manhole

5.2.1.3 蒸发结晶罐 Evaporation crystallization cans

5.2.1.4 重排混合器 Rearrangement mixer

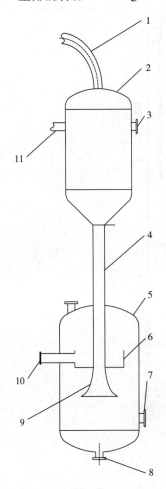

图 5.2.3　蒸发结晶罐结构图
Structure drawing of evaporation crystallization cans
1—气体出口 Gas outlet；2—蒸发罐 Vaporizing tank；
3—人孔 Manhole；4—导流筒 Draft tube；
5—结晶罐 Crystallizer；6—溢流盘 Over-flow plate；
7—出料口 Feed outlet；8—排放口 Discharge outlet；
9—扩散口 Horn mouth；
10—循环液出口 Circulating fluid outlet；
11—循环液进口 Circulating fluid inlet

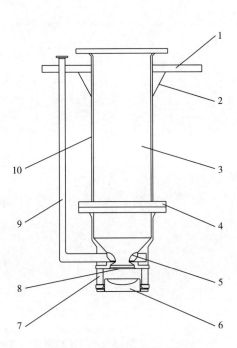

图 5.2.4　重排混合器结构图
Structure drawing of rearrangement mixer
1—顶板 Crown sheet；2、7—筋板 Rib plate；
3—循环管 Circulating tube；4—法兰 Flange；
5—喉管 Throat pipe；6—底板 Baseboard；
8—进料口 Feed inlet；9—环己酮肟管
Cyclohexanone oxime tube；10—管壁 Tube wall

5 合成原料 Synthetic materials

5.2.1.5 硫铵结晶反应器 Ammonium sulphate crystallization reactor

5.2.2 塔设备 Column equipment

5.2.2.1 转盘塔 Rotating disc column

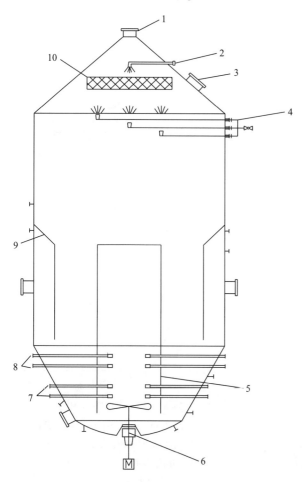

图 5.2.5 硫铵结晶反应器结构图
Structure drawing of ammonium sulphate crystallization reactor
1—不凝气出口 Noncondensable gas outlet；
2—喷淋水入口 Spray water inlet；3—人孔 Manhole；
4—冲洗水入口 Wash water inlet；
5—折流板 Baffle plate；6—搅拌器 Stirrer；
7—重排液进口 Rearrangement liquid inlet；
8—氨水进水 Ammonium hydroxide inlet；
9—混流筒 Mixed flow barrel；
10—除沫网 Despumation net

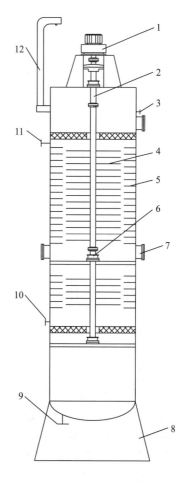

图 5.2.6 转盘塔结构图
Structure drawing of rotating disc column
1—减速机 Reducer；2—主轴 Shaft；
3—出料口 Feed outlet；4—动盘 Dynamic plate；
5—定盘 Fixed plate；6—轴承 Shaft；
7—人孔 Manhole；8—裙座 Skirt；
9—残液出口 Raffinate outlet；
10—苯进料口 Benzene feed inlet；
11—进料口 Feed inlet；12—塔顶吊柱 Top davit

5.2.2.2 反萃塔 Reverse extraction tower

5.2.2.3 蒸发塔 Evaporation tower

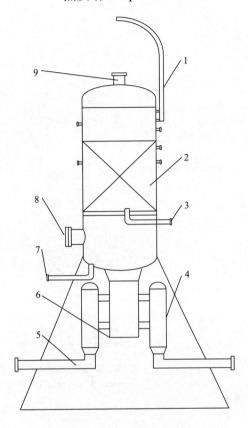

图 5.2.7 反萃塔结构图
Structure drawing of reverse extraction column
1—塔顶吊柱 Top davit；2—填料 Packing；3—进料口 Feed inlet；
4—脉冲罐 Pulse tank；5—循环管 Circulating pipe；
6—旋转阀 Rotary valve；7—出料口 Feed outlet；
8—人孔 Manhole；9—工艺水进口 Process water inlet

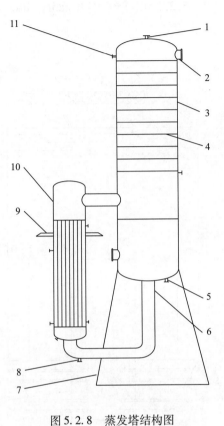

图 5.2.8 蒸发塔结构图
Structure drawing of evaporation column
1—气体出口 Gas outlet；2—人孔 Manhole；
3—筒体 Barrel；4—塔板 Tray；
5—物料出口 Material outlet；
6—循环管 Circulating pipe；
7—裙座 Skirt；
8—物料进口 Material inlet；
9—支座 Support；
10—再沸器 Reboiler；
11—工艺水进口 Pocess water inlet

5.2.3 离子交换器 Ion exchanger

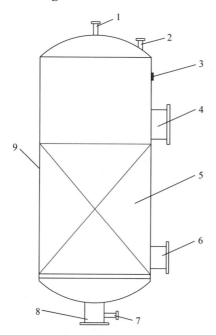

图 5.2.9 离子交换器结构图 Structure drawing of ion exchanger

1—己水进料口 Water supply feed inlet；2—排气管 Exhaust pipe；3—视镜 Sight glasses；
4—人孔 Manhole；5—树脂 Resin；6—手孔 Handhole；7—己水出料口 Water supply feed outlet；
8—釜液出口 Residue outlet；9—筒体 Barrel

5.2.4 换热设备 Heat-exchange equipment

5.2.4.1 加氢进料加热器 Hydrogenation feed heater

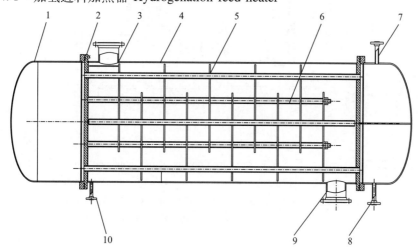

图 5.2.10 加氢进料加热器结构图 Structure drawing of hydrogenation feed heater

1—管箱 Tube box；2—管板 Tube sheet；3—物料进口 Material inlet；4—筒体 Barrel；5—折流板 Baffle plate；
6—拉杆 Tie rod；7—蒸汽进口 Steam inlet；8—蒸汽出口 Steam outlet；9—物料出口 Material outlet；
10—排放口 Drain valve

5.2.4.2 重排冷却器 Rearrangement cooler

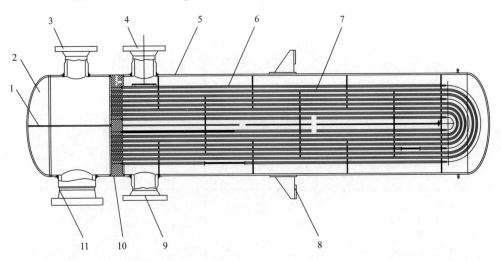

图 5.2.11 重排冷却器结构图 Structure drawing of rearrangement cooler
1—隔板 Clapboard；2—管箱 Tube box；3—冷却水出口 Cooling water outlet；4—物料出口 Transposition ester outlet；
5—筒体 Shell；6—拉杆 Tie rod；7—换热管 Heat exchange tube；8—支座 Support；
9—物料进口 Transposition ester inlet；10—管板 Tube sheet；11—冷却水进口 Cooling water inlet

5.2.5 泵 Pump

5.2.5.1 重排循环泵 Rearrangement circulating pump

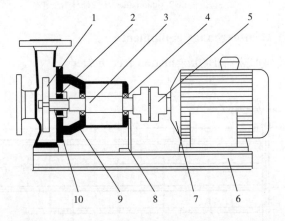

图 5.2.12 重排循环泵结构图 Structure drawing of rearrangement circulating pump
1—叶轮 Wheel；2—机械密封 Mechanical seal；3—轴 Shaft；4—轴承 Shaft bearing；
5—联轴器 Coupling；6—底座 Base；7—电机 Motor；8—轴承支架 Support；
9—轴承箱 Bearing box；10—轴套 Sleeve

5.2.5.2 磁力泵 Magnetic drive pump

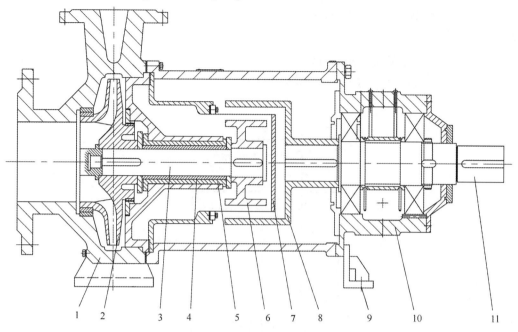

图 5.2.13 磁力泵结构图 Structure drawing of magnetic drive pump

1—泵体 Pump body；2—叶轮 Wheel；3—泵轴 Pump Shaft；4—轴套 Sleeve；5—滑动轴承 Silding Bearing；
6—内磁转子 Inner magnetic Rotor；7—隔离套 Can；8—外磁转子 Outer Magnetic；
9—辅助支架 Aided Suppo；10—轴承箱 Bearing Box；11—驱动轴 Driving Shaft；

5.2.6 离心机 Centrifugal machine

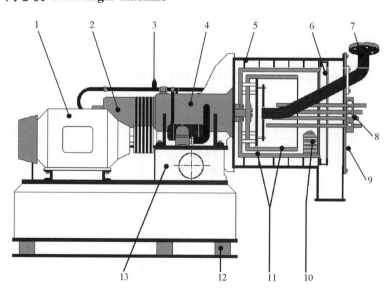

图 5.2.14 离心机结构图 Structure drawing of centrifugal machine

1—电机 Motor；2—传动机械 Gearing；3—油管 Oil pipe；4—轴承箱 Bearing box；5—转鼓外壳 Drum shell；
6—收集槽 Collecting tank；7—下料管 Discharge pipe；8—冲洗装置 Flusher；9—前板 Foreplate；
10—筛网 Screen cloth；11—转鼓 Drum；12—防振支座 Antivibration supports；13—底座 Base

5.3 环己酮 Cyclohexanone

5.3.1 反应设备 Reaction equipment

5.3.1.1 氧化反应器 Oxidation reactor

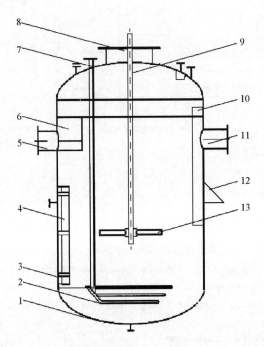

图 5.3.1 氧化反应器结构图 Structure drawing of oxidation reactor
1—封头 Head；2—空气分布器 Air distributor；3—筒体 Shell；4—支承板 Support plate；
5—物料入口 Material inlet；6、10—挡板 Baffle；7—空气入口管 Air inlet pipe；
8—传动支座 Transmission bearing；9—搅拌轴 Agitating shaft；11—物料出口 Material outlet；
12—支座 Support；13—搅拌桨叶 Stirring blade

5.3.1.2 搅拌器 Stirrer

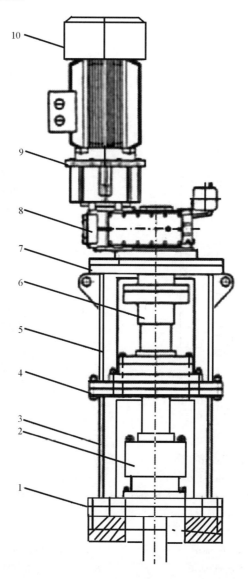

图 5.3.2 搅拌器结构图 Structure drawing of stirrer
1—底板 Base board；2—机械密封 Mecanical seal；3—下机架 Lower bracket；
4、7—机架过渡板 Rack transition plate；5—上机架 Upper bracket；6—联轴器 Coupling；
8—减速机 Reducer；9—弹簧垫圈 Spring gasket；10—电机 Motor

5.3.1.3 分解反应器 Decomposition reactor
5.3.1.4 脱氢反应器 Dehydrogenation reactor

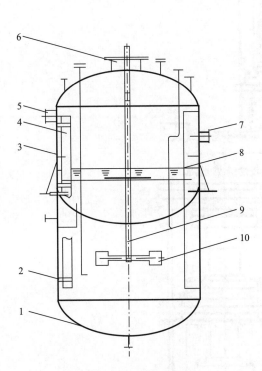

图5.3.3 分解反应器结构图
Structure drawing of decomposition reactor

1—封头 Head；2—支承板 Bearing plate；3—筒体 Shell；
4、8—挡板 Baffle；5—物料入口 Material inlet；
6—传动支座 Transmission bearing；7—物料出口 Material outlet；
9—搅拌轴 Agitating shaft；
10—搅拌桨叶 Stirring blade

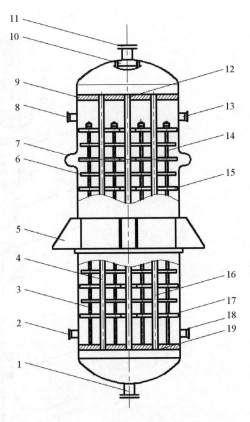

图5.3.4 脱氢反应器结构图
Structure drawing of dehydrogenation reactor

1—气相出口 Gaseous phase outlet；
2、18—导热油出口 Heat-transfer oil outlet；
3、14—拉杆 Tie rod；
4、12、16—换热管 Heat exchange tube；5—支座 Support；
6、15—折流板 Baffle plate；7—膨胀节 Expansion joint；
8、13—导热油入口 Heat-transfer oil inlet；
9、19—管板 Tube sheet；10—气体分布器 Distributor cap；
11—气相入口 Gaseous phase inlet；17—筒体 Barrel

5.3.1.5 主反应器 Main reactor

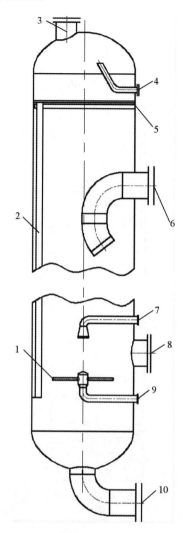

图 5.3.5 主反应器结构图 Structure drawing of main reactor

1—气体分布器 distributor；2—支承板 Support plate；3—人孔 Manhole；4—气体出口 Gas outlet；
5—筛板 Sleeve plate；6—循环液入口 Circulating fluid inlet；7—苯入口 Benzene inlet；
8—人孔 Manhole；9—氢气入口 Hydrogen inlet；10—循环液出口 Circulating fluid outlet

5.3.1.6 后反应器 Post reactor

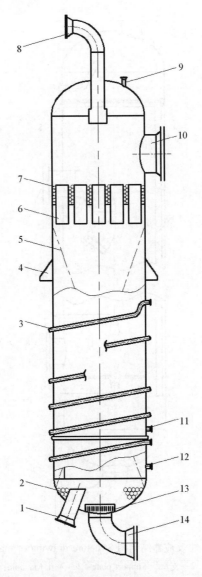

图 5.3.6 后反应器结构图 Structure drawing of post reactor
1—催化剂卸料口 Catalyst discharge outlet；2—氧化铝球 Alumina balls；3—盘管 Coiler；
4—支座 Support；5—触媒 Catalyzer；6—气体分布盘 Gas distribution plate；
7—筒体 Barrel；8—反应物入口 Material inlet；9—安全阀接口 Relief valve connector；
10—人孔 Manhole；11—蒸汽出口 Steam outlet；12—冷凝液出口 Condensate outlet；
13—气体过滤器 Gas filter；14—反应物出口 Reactant outlet

5.3.2 塔设备 Column equipment

5.3.2.1 冷却洗涤塔 Cooling scrubbing cloumn

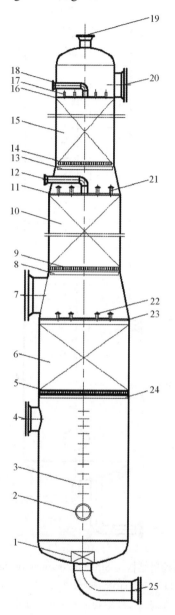

图 5.3.7 冷却洗涤塔结构图 Structure drawing of cooling scrubbing column

1—防涡流器 Vortex breaker；2、7、20—人孔 Manhole；3—梯子 Ladder；4—气体入口 Gas inlet；
5、9、14—填料支承格栅 Packing support grille；6、10、15—规整填料 Structured packing；
8、13、24—填料支承 Packing support plate；11、16、23—分配器支承圈 Distributor support ring；
12—反应物出口 Reactant outlet；17、21、22—液体分配器 Liquid distributor；18—冷烷入口 Cold alkanes inlet；
19—气体出口 Gas inlet；25—液体出口 Liquid outlet

5.3.2.2 蒸馏塔 Distillation tower

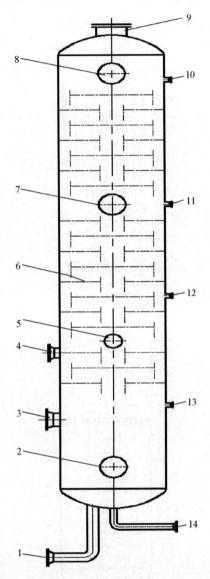

图 5.3.8 蒸馏塔结构图 Structure drawing of distillation column
1—液相出口 Liquid phrase outlet；2、5、7、8—人孔 Manhole；
3—气相入口 Gaseous phase inlet；
4—进料口 Feed inlet；6—塔盘 Tray；9—气相出口 Gaseous phase outlet；
10—温度计接口 Thermometer connection；
11、12、13—压力计接口 Pressure gauge connection；14—釜液出口 Residue outlet

5.3.2.3 酮塔 Ketone column

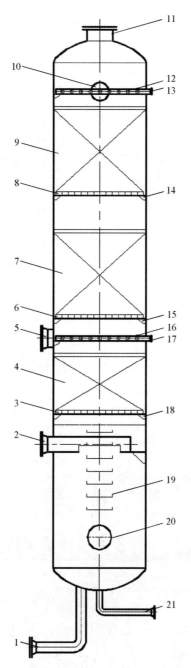

图 5.3.9 酮塔结构图 Structure drawing of ketone tower

1—液相出口 Liquid phrase outlet；2—气相入口 Gaseous phase inlet；3、6、8—填料支承栅板 Packing support grille；
4、7、9—填料 Packing；5、10、20—人孔 Manhole；11—气相出口 Gaseous phase outlet；
12、16—液体分布器 Liquid distributor；13—物料入口 Material inlet；14、15、18—栅板支承圈 Grid support ring；
17—物料出口；19—梯子 Ladder；21—釜液出口 Residue outlet

5.3.3 泵 Pump

5.3.3.1 水环真空泵 Water ring vacuum pump

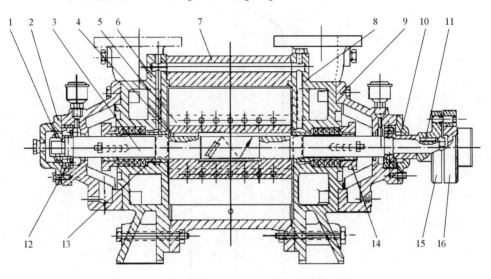

图 5.3.10 水环真空泵结构图 Structure drawing of water ring vacuum pump

1—锁紧螺母 Round nut；2、10—轴承压盖 Bearing cover；3—泵轴 Bearing shaft；4、11—键 Flat key；
5—叶轮 Wheel；6—后盖 Rear cover；7—泵体 Pump body；8—前盖 Front cover；9—轴承箱 Bearing box；
12—滚珠轴承 Ball bearing；13—轴套 Sleeve；14—填料压盖 Stuffing gland；15、16—联轴器 Coupling

5.3.3.2 苯进料泵 Benzene charge pump

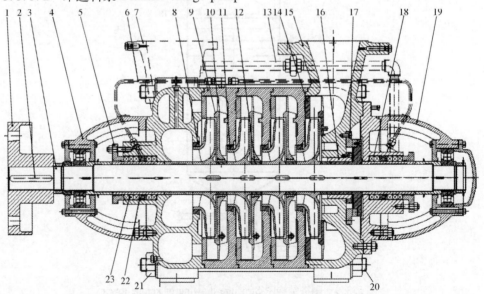

图 5.3.11 苯进料泵结构图 Structure drawing of Benzene charge pump

1—联轴器 Coupling；2—键 Key；3—轴 Shaft；4、19—轴承箱 Bearing box；
5—填料压盖 Stuffing gland；6—油封管 Sealing water pipe；7—入口法兰 Inlet flange；
8—叶轮 Wheel；9—隔板 Clapboard；10、14—导叶 Guide vane；11—密封环 Sealing ring；
12—级间轴套 Guide vane sleeve；13—平衡管 Balance pipe；15—出口法兰 Outlet flange；
16—平衡座 Balance sleeve；17—平衡盘 Balance ring；18—填料函 Stuffing box；
20—螺栓 Bolt；21—螺母 Nut；22—轴封 Packing ring；23—填料 Packing

5.3.4 压缩机 Compressor

5.3.4.1 空压机 Air compressor

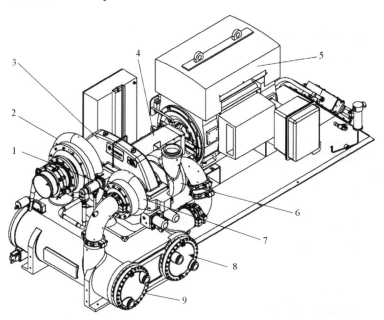

图 5.3.12 空压机外形图 Structure drawing of air compressor

1—主油泵 Main oil pump；2——级蜗壳 First level volute；3—齿轮箱 Gear box；4—联轴器 Coupling；
5—电机 Motor；6—三级蜗壳 Third level volute；7—二级蜗壳 Second level volute；
8—二级冷却器 Second level cooler；9——级冷却器 First level cooler

5.3.4.2 氢气压缩机 Hydrogen compressor

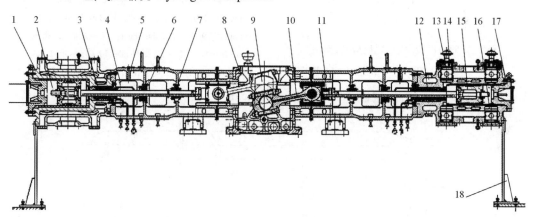

图 5.3.13 氢气压缩机结构图 Structure drawing of hydrogen compressor

1、17—气缸 Cylinder；2——级活塞 First level piston；3、12—填料 Packing；4—保护系统 Protecting system；
5—中体 Midbody；6—填料压盖 Stuffing gland；7—刮油环 Scraper；8—机体 Compressor body；
9—曲轴 Crankshaf；10—连杆 Connecting rod；11—十字头 Crosshead；13—二级进气阀 Second level inlet valve；
14—顶开阀 Open valve；15—二级活塞 Second level piston；
16—二级排气阀 Second level vent valve；18—支座 Support

5.4 乙二醇 Ethylene glycol

5.4.1 反应设备 Reaction equipment

5.4.1.1 环氧乙烷反应器 Ethylene oxide reactor

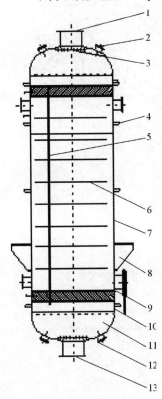

图 5.4.1 环氧乙烷反应器结构图
Structure drawing of ethylene oxide reactor

1—反应器入口 Reactor inlet；2—上人孔 Upper manhole；3—分布器 Distributor；4—保温支承圈 Insulation supports；5—反应管 Reaction tube；6—折流板 Baffle plate；7—筒体 Barrel；8—支座 Support；9—管板 Tube sheet；10—短筒节 Short barrel section；11—封头 Head；12—下人孔 Lower manhole；13—反应器出口 Reactor outlet

5.4.1.2 乙二醇反应器 Ethylene glycol reactor

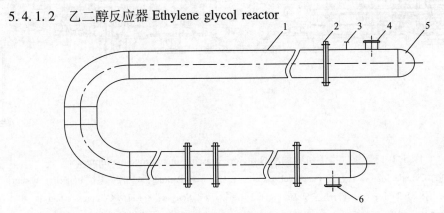

图 5.4.2 乙二醇反应器结构图 Structure drawing of ethylene glycol reactor
1—筒体 Barrel；2—法兰 Flange；3—安全阀接口 Safety valve interface；4—物料出口 Material outler；
5—封头 Head；6—物料入口 Material inlet

5.4.2 塔设备 Column equipment

5.4.2.1 环氧乙烷吸收塔 Ethylene oxide absorption column

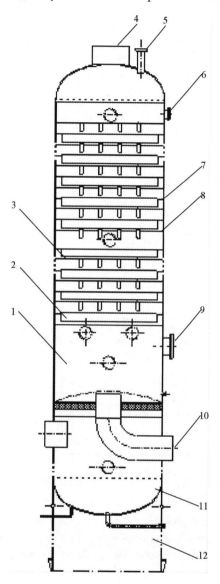

图 5.4.3 环氧乙烷吸收塔结构图 Structure drawing of ethylene oxide absorption tower

1—筒体 Barrel；2—塔盘 Tray；3—塔盘支承 Supporter；4—气相出口 Gaseous phase outlet；
5—安全阀接口 Safety valve interface；6—物料入口 Material inlet；7、8—密封压板 Sealing press plate；
9—人孔 Manhole；10—物料出口 Material outlet；11—封头 Head；12—裙座 Skirt

5.4.2.2 环氧乙烷解析塔 Ethylene oxide analytic column

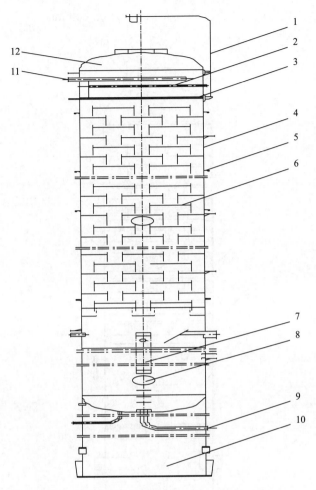

图 5.4.4 环氧乙烷解析塔结构图
Structure drawing of ethylene oxide analytic tower
1—塔顶吊柱 Top davit；2—回流管 Return pipe；3—支承架 Support frame；4—筒体 Barrel；
5—保温支承圈 Insulation support ring；6—塔盘 Tray；7—梯子 Ladder；8—人孔 Manhole；
9—物料出口 Material outlet；10—裙座 Skirt；11—物料入口 Material inlet；12—封头 Head

5.4.2.3 乙二醇精制塔 Ethylene glycol treating column

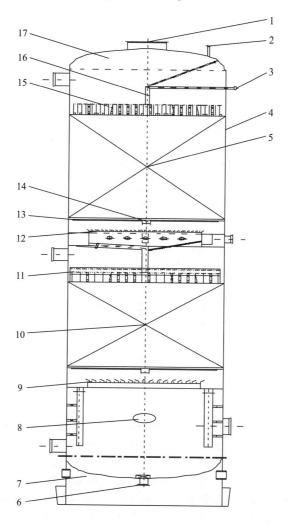

图 5.4.5　乙二醇精制塔结构图 Structure drawing of ethylene glycol treating column

1—气相出口 Gaseous phase outlet；2—安全阀接口 Safety valve interface；
3—物料入口 Material inlet；4—筒体 Barrel；5、10—规整填料 Structured packing；
6—物料出口 Material outlet；7—裙座 Skirt；8—人孔 Manhole；9、12—收集器 Accumulator；
11、15—分布器/定位格栅 Distributor/ location grid；13—支承格栅 Supporting grid；
14—支承架 Support frame；16—进料管 Feed pipe/Down comer；17—封头 Head

5.4.3 换热设备 Heat-exchange equipment

5.4.3.1 乙二醇一段高效再沸器 Ethylene glycol one passage efficient reboiler

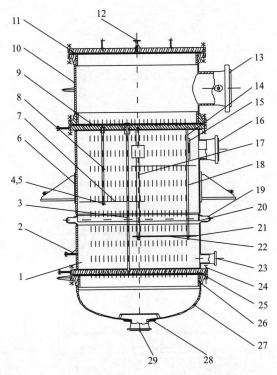

图 5.4.6 乙二醇一段高效再沸器结构图 Structure drawing of ethylene glycol one passage efficient reboiler

1、10—筒体 Barrel；2—排放口 Discharge outlet；3—换热管 Heat exchange tube；4、8—定距管 Distance sink tube；
5—螺母 Nut；6—支座 Support；7、22—折流板 Baffle plate；9—上管板 Upper tube plate；
11—平盖 Flat cap；12—放空口 Vent；13—物料出口 Material outlet；14、17、18、21—拉杆 Tie rod；
15—防冲挡板 Impingement baffle；16—蒸汽入口 Steam inlet；19—膨胀节 Expansion joint；20—衬筒 Bushing；
23—蒸汽出口 Steam outlet；24—螺栓 bolt；25—下管板 Lower tube plate；26—衬板 Scaleboard；
27—封头 Head；28—补强圈 Stiffening ring；29—物料入口 Material inlet

5.4.3.2 环氧乙烷精制塔冷凝器 Ethylene glycol treating column condenser

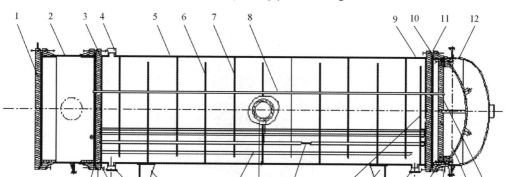

图 5.4.7 环氧乙烷精制塔冷凝器结构图 Structure drawing of ethylene glycol treating tower condenser

1—管箱盲板 Flange cover；2—管箱 Left channel；3—管板 Left tube plate；4、22—物料入口 Material inlet；
5—筒体 Barrel；6、7—折流板 Baffle plate；8—换热管 Heat exchange tube；9—衬筒 Bushing；
10—凸面法兰 Male flange；11、15、27—垫片 Gasket；12—外头盖 Outer cover；
13—浮动管板 Floating head tube plate；14—浮头盖 Floating head cover；16—钩圈 Hook ring；
17、26—物料出口 Material outlet；18、25—支座 Support；19—支承板 Support plate；20—拉杆 Tie rod；
21—接管 Connecting pipe；23—防冲挡板 Inpingement baffle；24—滚轮滑道 Wheel slide；
28—螺栓 Bolt；29—法兰 Flange

5.4.3.3 吸收水冷却器 Absorbed water cooler

1. 吸收水冷却器外形 Absorbed water cooler outline

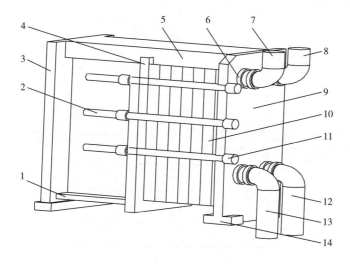

图 5.4.8 吸收水冷却器外形图 Outline of absorbed water cooler

1—下导杆 Guide rod；2—夹紧螺栓 Thrust bolt；3—支柱 Brace；4—中间隔板 Active press plate；
5—上导杆 Upper guide rod；6—接管法兰 Pipe connecting flange；7—物料入口 Material inlet；
8—冷却水出口 Cooling water outlet；9—固定压紧板 Dead plate；10—换热板 Heat exchange plate；
11—夹紧螺母 Clamping nut；12—物料出口 Material outlet；13—冷却水入口 Cooling water inlet；
14—支座 Support

2. 吸收水冷却器结构 Absorbed water cooler structure

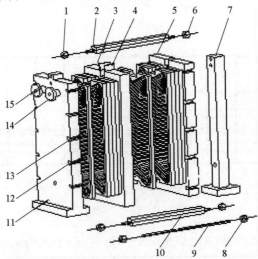

图 5.4.9 吸收水冷却器结构图 Structure drawing of absorbed water cooler

1、6、8—螺母 Nut；2—上导杆 Upper guide rod；3—中间隔板 Intermediate bulkhead；
4—滚动机构 Rolling mechanism；5—活动压紧板 Activity press plate；7—立柱 Upright；
9—夹紧螺柱 Clamping screw；10—下导杆 Lower guide rod；11—固定压紧板 Dead press plate；
12—板片 Plate；13—垫片 Gasket；14—法兰 Flange；15—接管 Connecting pipe

5.4.4 容器与储罐 Container and tank

5.4.4.1 脱氧槽及水箱 Deoxygenation tank and water tank

1. 脱氧槽 Deoxygenation tank

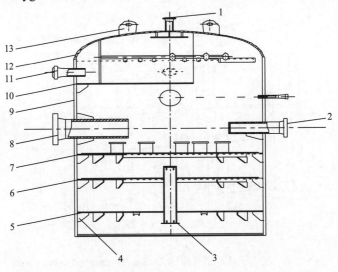

图 5.4.10 脱氧槽结构图 Structure drawing of deoxygenation tank

1—放空口 Vent；2、8、11—物料入口 Material inlet；3—进气管 Intake pipe；
4—筋板 Rib plate；5—孔板 Hole plate；6—淋水盘 Water spraying tray；
7—淋水装置 Liquid distribution equipment；9—筒体 Barrel；10—托板 Layer board；
12—封头 Head；13—吊耳 Lifting lug

2. 水箱 Water tank

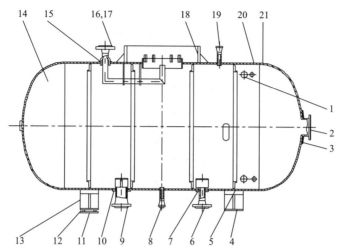

图 5.4.11 水箱结构图 Structure drawing of water tank

1—仪表接口 Instrumentation tap；2—人孔 Manhole；3、7、10、15—补强圈 Stiffening ring；4—固定支座 Fixed support；5—补强板 Stiffening plate；6—出水管 Discharging tube；8—排放口 Discharge outlet；9—物料出口 Material outlet；11—接地板 Floor；12—滚子 Roller；13—活动支座 slide support；14、21—封头 Head；16—接管 Connecting pipe；17—法兰 Flange；18—接口圈 Connecting ring；19—安全阀接口 Safety valve interface；20—筒体 Barrel

5.4.4.2 乙二醇产品储罐 Ethylene glycol storage tank

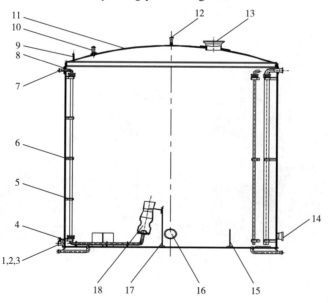

图 5.4.12 乙二醇产品储罐结构图 Structure drawing of ethylene glycol storage tank

1—接管 Connecting pipe；2—排放口 Discharge outlet；3、8—补强圈 Stiffening ring；4、9—仪表接口 Instrument tap；5—筒体 Barrel；6—卡子 Clip；7—物料入口 Material inlet；10—安全阀接口 Safety valve interface；11—顶盖 Top cover；12—氮气入口 Nitrogen inlet；13—呼吸阀接口 Breather valve connector；14—物料出口 Material outlet；15—隔板 Clapboard；16—人孔 Manhole；17—底板 Baseboard；18—喷射器 Ejector

5.4.4.3 环氧乙烷球罐 Epoxy ethane spherical tank

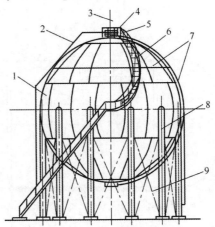

图 5.4.13　环氧乙烷球罐结构图 Structure of epoxythane spherical tank

1—球壳 Spherial shell；2—液位计导管 Level gauge duct；3—避雷针 Lightening rod；4—安全泄放阀 Relief valve；
5—操作平台 Operating platform；6—盘梯 Stairway；7—喷淋水管 Spraying water pipe；8—支柱 Support；9—拉杆 Tie rod

5.4.5 泵 Pump

5.4.5.1 乙二醇产品输送泵 Ethylene glycol delivery pump

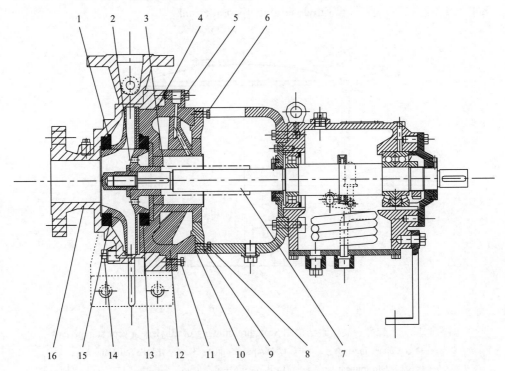

图 5.4.14　乙二醇产品输送泵结构图 Structure drawing of ethylene glycol delivery pump

1、15—泵体密封环 Impeller sealing ring；2—叶轮 Wheel；3—密封环 Sealing ring；4、9—密封垫 Sealing gasket；
5、6、8、11—六角螺栓 Hexagon bolt；7—轴 Shaft；10—O 形密封圈 O sealing ring；12—泵盖 Pump cover；
13—叶轮密封环 Impeller labyrinth seal ring；14—丝堵 Plug；16—泵体 Pump body

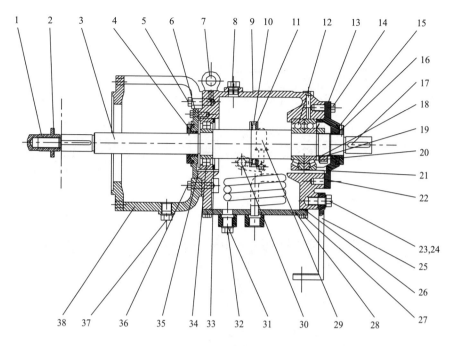

图5.4.15 乙二醇产品输送泵轴承箱结构图
Structure drawing of ethylene glycol delivery pump bearing box

1—叶轮螺母 Impeller nut；2、22、26、32、35—密封垫 Sealing gasket；3—轴 Shaft；4、10—紧固螺栓 Fastening screw；
5、14、23、27、36—六角螺栓 Hexagon bolt；6、13—轴承箱压盖 Bearing box gland；7—吊环 Lifting ring；
8—放气塞 Relief plug；9—轴承箱 Bearing box；11—溅油盘 Splashing oil pan；12、34—滚动轴承 Rolling bearing；
15—O形密封圈 O sealing ring；16—防尘盘 Dust proof plate；17—键 Key；
18—轴承锁紧螺母 Bearing lock nut；19—止动垫圈 Locking gasket；20—调整垫 Adjusting gasket；
21—轴承隔环 Bearing spacer ring；24—平垫片 Flat gasket；25—轴承箱支架 Bearing box bracket；
28—油室压盖 Oil chamber gland；29—油杯 Oil cup；30—视油窗 Oil sight window；
31—放油塞 Oil drain plug；33—挡圈 Retainer ring；37—丝堵 Plug；
38—中间连接体 Intermediate junction segment

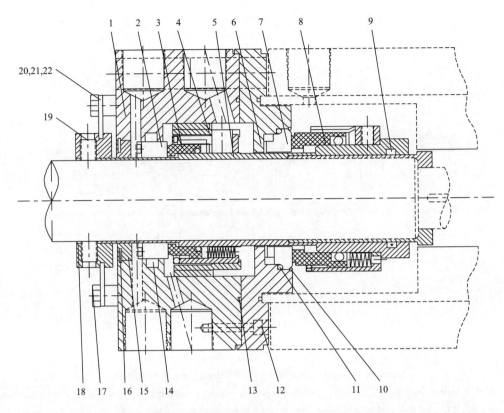

图 5.4.16　乙二醇产品输送泵密封结构图 Structure drawing of ethylene glycol delivery pump seal
1—密封压盖 Outer flange；2、7—静环 Stationary seal ring；3、8—动环 Rotating ring；
4—泵效环 Pump efficiency ring；5—轴套 Sleeve；6—内法兰 Inner flange；
9、11、13、14—O 形密封圈 O sealing ring；10—开口挡圈 Split washer；12、17、19、21—螺栓 Bolt；
15—节流环 Restrictor ring；16—卡簧 Snap spring；18—传动环 Drive lock ring；
20—偏心垫圈 Eccentricity gasket；22—限位套 Stop collar

5.4.5.2 碱计量泵 Alkali dosing pump

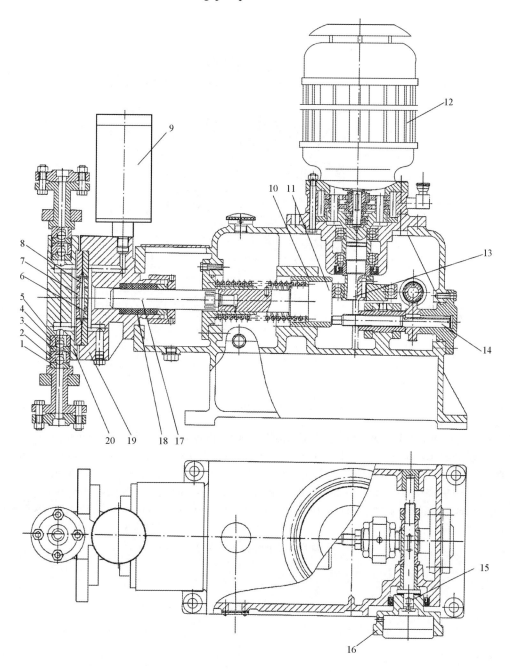

图 5.4.17 碱计量泵结构图 Structure drawing of alkali dosing pump

1—下阀座 Lower valve seat；2—下导向套 Lower guide sleeve；3—上阀座 Upper valve seat；
4—上导向套 Upper guide sleeve；5—阀球 Vavle ball；6—前限制板 Front limiting plate；
7—后限制板 Rear limiting plate；8—隔膜 Membrane；9—安全补油阀 Safe oil supplementary valve；
10—导向套 Guide sleeve；11—十字头 Crosshead；12—电机 Motor；13—轴 Output shaft；
14—调节杆 Adjusting rod；15—调节轴 Adjusting shaft；16—调节转盘 Adjusting rotary table；
17—柱塞 Plunger；18—密封环 Sealing ring；19—前缸头 Front cylinder end；20—后缸头 Rear cylinder end

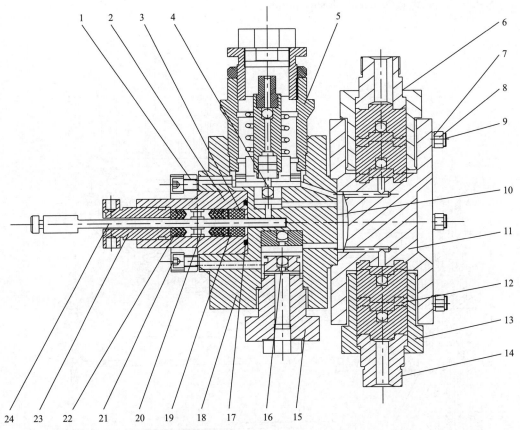

图 5.4.18 碱计量泵液力端结构图 Structure drawing of alkali dosing pump
1—压缸法兰 Cylinder flange；2—液缸 Hydraulic cylinder；3—衬套 Bush；4—内排阀 Inner exhausting valve；
5—放气阀 Vent valve；6—出口管 Outlet pipe；7—垫圈 Gasket；8—螺母 Nut；9—螺栓 Bolt；
10—隔膜 Membrane；11—缸盖 Cylinder head；12—球阀 Ball valve；13—阀螺母 Valve nut；
14—入口管 Inlet pipe；15—补油阀 Supplementary valve；16—内吸阀 Inner suction valve；17—O 形密封圈 O sealing ring；
18—缸体 Cylinder；19—密封环 Sealing ring；20—隔环 Distance ring；21—沉头螺栓 Bolt；22—填料 Support ring；
23—填料压盖 Packing pressure nut；24—柱塞 Plunger

5.4.5.3 洗涤水高速泵 Washing water high speed pump

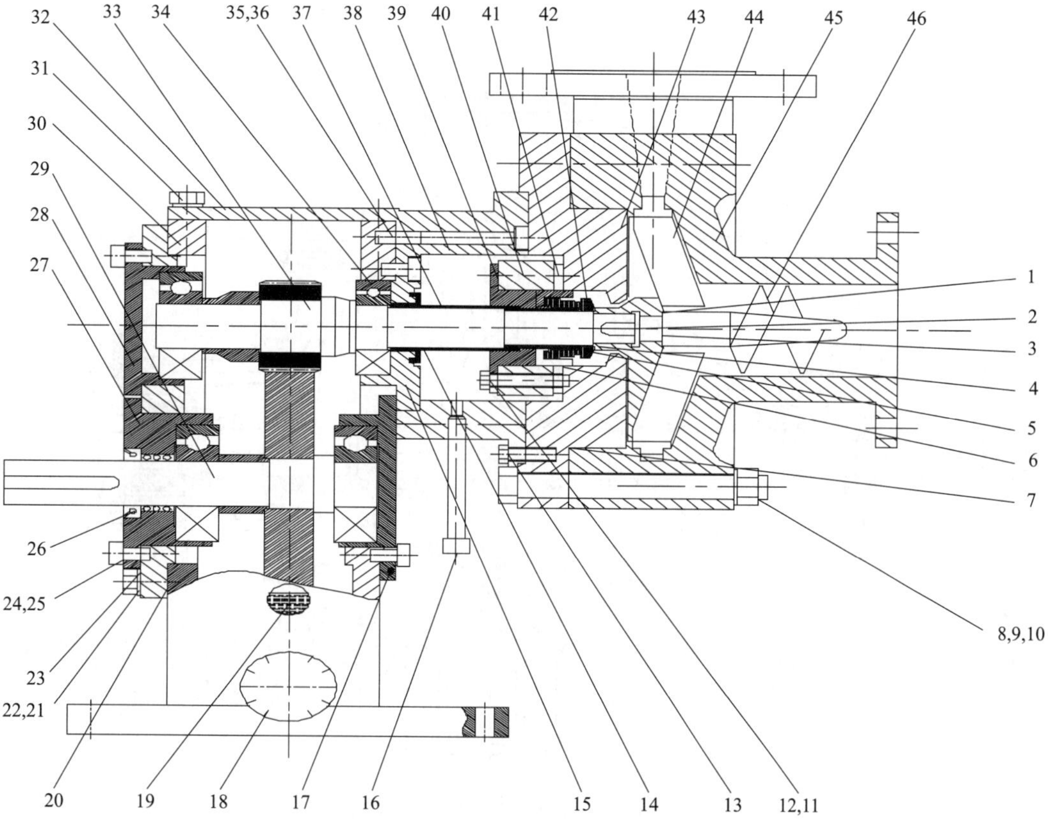

图 5.4.19 洗涤水高速泵结构图
Structure drawing of washing water high speed pump

1—止动垫片 Locking gasket；2—诱导轮连接螺栓 Inducer connection bolt；3、4、5、6、7、14—O 形密封圈 O sealing ring；
8、11、21、24、35—垫圈 Gasket ring；9—螺母 Nut；10、22、25、31—螺栓 Bolt；
12、13、36—内六角螺栓 Inner hexagon bolt；
15—高速轴前轴承座 High speed front bearing housing；16—排液接管 Discharge adapter；
17—低速轴前轴承座 Lower speed front bearing housing；18—温度计 Thermometer；19—油标 Oil pointer；
20、23—密封垫片 Sealing gasket；26—唇形密封圈 Lip sealing ring；
27—低速轴后轴承座 Lower speed rear bearing housing；28—高速轴后轴承座 High speed rear bearing housing；
29—低速轴 Lower speed shaft；30—箱体 Tank；32—箱盖 Tank cover；33—高速轴 High speed shaft；
34—油封 Oil seal；37—轴套 Sleeve；38—连接支架 Attachment bracket；
39—节流圈 Throttle；40—密封腔 Annular seal space；41—调整垫 Adjusting gasket；42—机械密封 Mechanical seal；
43—后盖 Rear cover；44—叶轮 Wheel；45—泵体 Pump body；46—诱导轮 Inducer

5.4.5.4 环氧乙烷吸塔水泵 Epoxy ethane absorbing water pump

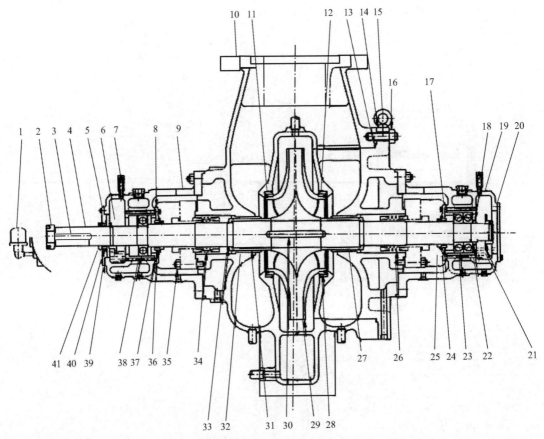

图 5.4.20 环氧乙烷吸塔水泵结构图 Structure drawing of epoxy ethane absorbing colum water pump

1—油杯 Oil cup；2、35—螺母 Nut；3、30—键 Key；4—轴 Shaft；5、8、17—轴承盖 Bearing cover；
6、21—甩油板 Flinger；7—通气孔 Air vent；9—轴承座 Bearing housing；10—壳体 Shell；
11—壳体环 Shell ring；12—吸入端口环 Suction port ring；13—垫片 Gasket；14—吸入端泵盖 Suction port pump gland；
15—吊环 Eye bolt；16—泵盖螺母 Stud and nut；18—轴承螺母 Bearing nut；19—止推轴承座 Thrust bearing housing；
20—止推轴承盖 Thrust bearing cover；22—轴承锁紧垫 Bearing locking gasket；23—止推轴承 Thrust bearing；
24、36、40—轴封 Shaft seal；25、41—挡水环 Deflector；26、34—机械密封 Mechanical seal；
27、31—定距套 Fixed pitch sleeve；28—叶轮环 Impeller Ring；29—叶轮 Wheel；32—节流环 Throat bushing；
33—开口环 Snap ring；37、39—轴承盖垫片 Bearing cover gasket；38—径向轴承 Radial Bearing

5.4.5.5 热水收集槽泵 Hot water collecting tank pump

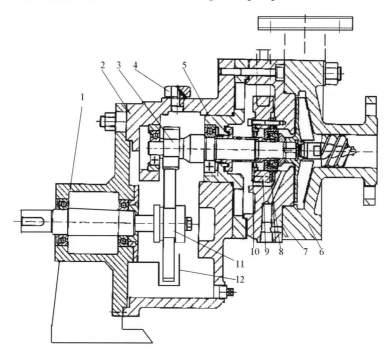

图 5.4.21　热水收集槽泵结构图
Structure drawing of hot water collecting tank pump
1—轴承箱 Bearing box；2—齿轮箱 Gear box；3—高速转子 High speed rotor；4—排气螺塞 Bleed screw；
5—齿轮箱盖 Gear casing cover；6—泵体 Pump body；7—密封箱 Sealing box；
8—机械密封 High speed mechanical seal；9—冲液腔 Liquid cavity；
10—节流环 Restrictor ring；11—低速转子 Lower speed rotor；12—隔油套 Oil proof sleeve

5.4.5.6 锅炉给水泵 Boiler feed pump

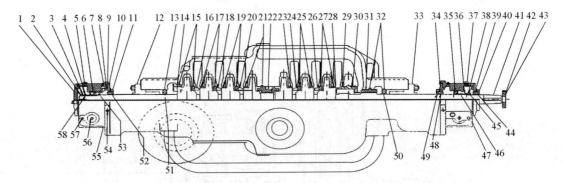

图 5.4.22 锅炉给水泵结构图 Structure drawing of boiler feed pump

1—轴承箱盖 Bearing box cover；2、38—锁紧螺母 Locknut；3、45—垫片 Locking gasket；
4、36—甩油环 Oil flinger；5、37—呼吸帽 Breath cap；6—止推轴承 Thrust bearing；
7—水夹套丝堵 Water jacketed plug；8—O 形密封圈 O sealing ring；9—轴承盖垫片 Bearing cover gasket；
10、34、39—轴承盖 Bearing cover；11、40、48—油封 Oil seal；12、33—机械密封 Mechanical seal；
13、54—定位销 Locating pin；14—壳体环 Shell ring；15—叶轮环 Impeller ring；
16、18、20、24、26、28—隔板 Clapboard；17、19、21、23、25、27—叶轮环 impeller ting；
22、50、51—节流环 Restrictor ring；29、32—定位环 Locating ring；30、58—定距套 Fixed pitch sleeve；
31—平衡鼓 Balancing drum；35—径向轴承 Radial bearing；
41—轴 Shaft；42—联轴器柱销 Coupling pin；43—联轴器 Coupling；
44—油封定位销 Oil seal alignment pin；46—轴承定位套 Bearing locating sleeve；
47、57—油杯接口 Oil cup interface；49、52—锁紧螺栓 Lock screw；53—轴承垫 Bearing gasket；
55—轴承箱 Bearing box；56—油视镜 Oil level glass

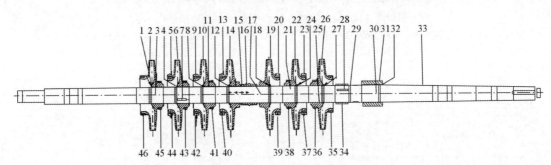

图 5.4.23 锅炉给水泵轴结构图 Structure drawing of boiler feed pump shaft

1、4、8、12、18、24、27—叶轮固定环 Impeller fixing ring；2、6、10、14、19、22、26—叶轮 Wheel；
3、7、11、15、17、21、25、28、30—键 Key；5、9、13、20、23—壳体口环 Shell choma；
16—中间轴套 Intermediate shaft sleeve；29—轴套固定环 Shaft sleeve fixing ring；
31—平衡鼓 Balancing drum；32—平衡鼓固定环 Balancing drum fixing ring；
33—轴 Shaft；34—轴套 Sleeve；
35、36、37、38、39、40、41、42、43、44、45、46—叶轮环 Impeller

5.4.6 压缩机 Compressor
5.4.6.1 循环气压缩机 Recycle gas compressor

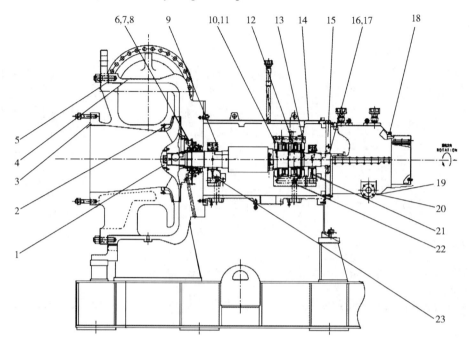

图 5.4.24 循环气压缩机结构图 Structure drawing of recycle gas compressor

1—转子 Rotor；2—迷宫密封 Seal；3—进气缸 Inlet casing；4—排气缸 Discharge casing；
5、17、18—O 形密封圈 O sealing ring；6—干气密封 Dry gas seal；7—沉头螺栓 Socket head cap screw；
8—锁紧螺钉 Lock bolt；9—径向轴承 Radial bearing；10、13—护圈板 Retainer plate；11—密封圈 Seal ring；
12—推力轴承 Thrust bearing；14—轴承箱 Bearing box；15—端盖 End cover；16—轴封 Shaft seal；
19—侧盖 Cover；20—双头螺栓/螺母 Stud bolt/nut；21、22、23—节流塞 Orifice plug

5.4.6.2 尾气压缩机 Tail gas compressor

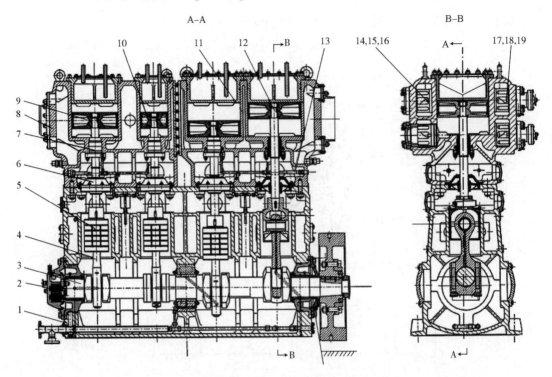

图 5.4.25 尾气压缩机结构图 Structure drawing of tail gas compressor

1—曲轴箱 Crankshaft box；2—油泵 Oil pump；3—曲轴 Crankshaft shaft；4—连杆 Connecting rod；
5—十字头 Crosshead；6—导向轴承 Guide bearing；7—填料 Padding；
8—二级气缸 Second level cylinder；9—二级活塞 Second level piston；
10—三级活塞 Third level piston；11——级气缸 First level cylinder component；
12——级活塞 First level piston；13—填料漏气回收管 Packing leakage return pipe；
14——级排气阀 First level vent valve；15—二级进气阀 Second level gas inlet valve；
16—三级排气阀 Third level gas exhaust；17——级进气阀 First level inlet valve；
18—二级排气阀 Second level gas exhaust；19—三级进气阀 Third level gas inlet valve

5.5 丙烯腈 Acrylonitrile

5.5.1 丙烯腈反应器 Acrylonitnle reactor

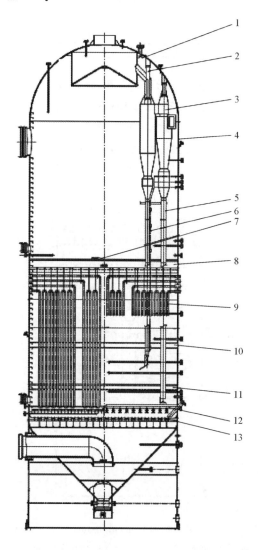

图 5.5.1 丙烯腈反应器结构图 Structure drawing of Acrylonitnle reactor

1—内集气室 Internal plenum；2—旋风分离器吊架 Cyclone separator hanger；3—旋风分离器 Cyclone separator；
4—筒体 Shell；5—料腿 Dipleg；6—料腿吹扫管 Dipleg venting pipe；7—隔板 Beam cover；
8—顶梁 Top beam；9—冷却水管 Cooling water pipe；10—中梁 Middle beam；11—底梁 Bottom beam；
12—丙烯腈分布器 Acrylonitnle distributor；13—空气分布板 Air distribution plate

5.5.2 塔设备 Column equipment

5.5.2.1 急冷塔 Quench column

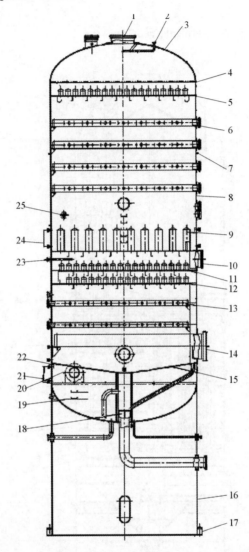

图 5.5.2 急冷塔结构图 Structure drawing of quench column

1—工艺气出口 Process gas outlet；2—上段补充水入口 Upper section supplementary water inlet；
3—封头 Head；4、7—C 形保温圈 C-type insulation ring；5—上段除沫塔盘 Upper section demister tray；
6—上段喷淋装置 Upper section spray assembly；8—筒体 Cylinder；9—升气塔盘 Ascending gas tray；
10—上段循环液出口 Circulating fluid outlet of upper segment；11、12—除沫塔盘 Under section demister tray；
13—下段喷淋装置 Lower section spray assembly；14—工艺气入口 Process gas inlet；
15—塔釜内件 Column internals；16—裙座 Skirt；17—接地板 Ground plate；
18—防涡器 Vortex breaker；19—梯子 Ladder；20—把手 Handle；
21、24—磁性浮子液位计 Magnetic float level gauge；22—补强圈 Reinforcing ring；
23—中段补充水入口 Middle section supplementary water inlet；25—加硫酸口 Sulfuric acid inlet

5.5.2.2 回收塔 Recovery column

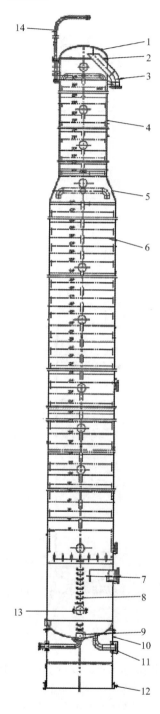

图 5.5.3 回收塔结构图 Structure drawing of recovery column
1—封头 Head；2—保温圈 Insulation ring；3—塔顶气相出口 Overhead gas outlet；4—筒体 Cylinder；
5—锥体 Conical shell；6—塔盘 Tray；7—再沸器物料返回口 Reboiler material returning port；
8—梯子 Ladder；9—防涡器 Vortex breaker；10—裙座 Skirt；11—筋板 Reinforcing plate；
12—接地板 Ground plate；13—人孔 Manhole；14—塔顶吊柱 Top davit

5.5.2.3 脱氢氰酸塔 Dehydrogenation cyanate column

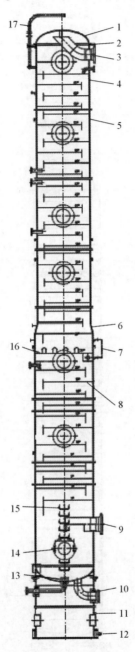

图 5.5.4 脱氢氰酸塔结构图 Structure drawing of dehydrogenation cyanate column
1—封头 Head；2—保温圈 C-type insulation ring；3—气相出口 Gas outlet；4—筒体 Cylinder；
5—保温支承圈 Insulation ring；6—锥体 Conical shell；7—磁性浮子液位计 Magnetic float level gauge；
8—塔盘 Tray；9—物料入口 Material inlet；10—筋板 Reinforcing plate；11—裙座 Skirt；
12—接地板 Ground plate；13—防涡器 Baffle breaker；14—人孔 Manhole；15—梯子 Ladder；
16—升气塔盘 Ascending gas tray；17—塔顶吊柱 Top davit

5.5.2.4 吸收塔 Absorption column

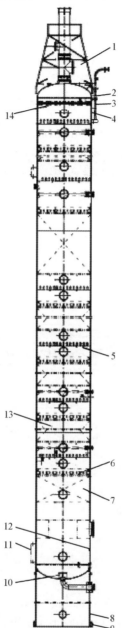

图 5.5.5　吸收塔结构图 Structure drawing of absorption column

1—排气装置 Exhaust device；2—封头 Head；3—塔顶吊柱 Top davit；4—筒体 Cylinder；
5—气体分配器 Gas distribution plate；6—液体分布器 Liquid distributor；7—散堆填料 Random packing；
8—裙座 Skirt；9—接地板 Ground plate；10—防涡器 Vortex breaker；
11—磁性浮子液位计 Magnetic float level gauge；12—缓冲板 Buffer board；
13—规整填料 Structured packing；14—丝网除沫器 Wire mesh demister

5.5.3 容器与储罐 Container and tank

5.5.3.1 丙烯/氨排液槽 Propylene drainage trough

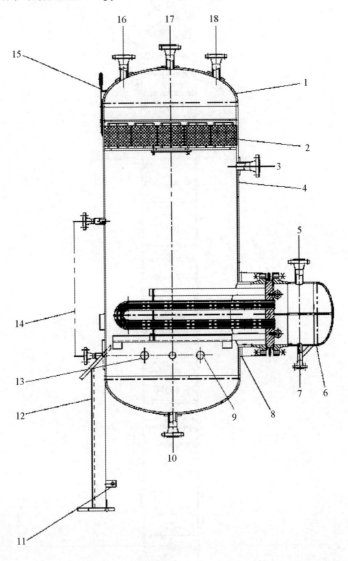

图 5.5.6 丙烯/氨排液槽结构图 Structure drawing of propylene drainage trough

1—封头 Head；2—丝网除沫器 Wire mesh demister；3—压力计接口 Manometer interface；4—筒体 Cylinder；5—水蒸气入口 Steam inlet；6—加热器 Heater；7—蒸汽凝液出口 Steam condensate outlet；8—补强圈 Reinforcing ring；9—液体丙烯入口 Liquid propylene inlet；10—液体丙烯出口 Liquid propylene outlet；11—接地板 Ground plate；12—支腿 Support；13—排气口 Exhaust port；14—磁性浮子液位计 Magnetic float level gauge；15—吊耳 Lug；16—排气口 Exhaust port；17—气体丙烯出口 Propylene gas outlet；18—安全阀接口 Safety valve interface

5.5.3.2 回收塔分层器 Recovery column delaminator

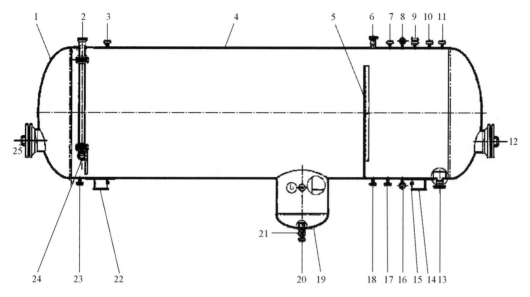

图 5.5.7 回收塔分层器结构图 Structure drawing of recovery column delaminator

1、19—封头 Head；2—液体入口 Liquid inlet；3、11—排气口 Exhaust port；4—筒体 Cylinder；5—隔板 Partition；6—安全阀接口 Safety valve interface；7—压力计接口 Manometer interface；8、16、20—液位计接口 Liquid level gauge interface；9—自控液位计口 Automatic level gauge opening；10—备用口 Spare opening；12、25—人孔 Manhole；13、21—液体出口 Liquid outlet；14、22—支座 Support；15—接地板 Ground plate；17、23 蒸汽吹扫口 Steam purge opening；18—排放口 Drain opening；24—导液管 Catheter

5.5.3.3 脱氰塔分层器 Decyanation column delaminator

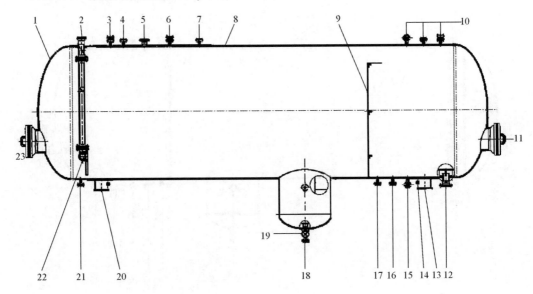

图 5.5.8 脱氰塔分层器结构图 Structure drawing of decyanation column delaminator
1—封头 Head；2—液体入口 Liquid inlet；3、6—备用口 Spare opening；4—不凝气出口 Non-condensable gas outlet；
5—安全阀接口 Safety valve interface；7—压力计接口 Manometer interface；8—筒体 Cylinder；9—隔板 Partition；
10—自控液位计接口 Automatic level gauge interface；11、23—人孔 Manhole；12、18—液体出口 Liquid outlet；
13、20—支座 Support；14—接地板 Ground plate；15、19—液位计接口 Liquid level gauge interface；
16、21—蒸汽吹扫口 Steam purge opening；17—排放口 Drain opening；22—导液管 Catheter

5.5.3.4 急冷塔顶除沫器 Quench column top demister

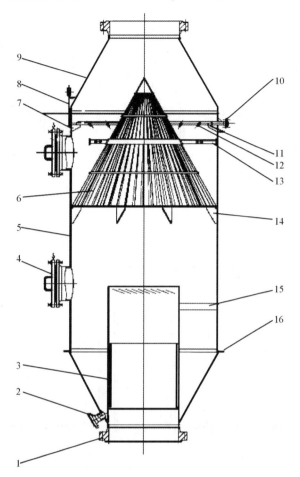

图 5.5.9 急冷塔顶除沫器结构图 Structure drawing of quench column top demister

1—法兰 Flange；2、10—接管 Connecting pipe；3—出口 Outlet；4—人孔 Manhole；
5—筒体 Cylinder；6—旋流板 Swirl plate；7、14—支承板 Support plate；8—吊耳 Lug；
9—锥壳 Conical shell；11—冲洗装置 Flushing device；12—喷嘴 Nozzle；
13—固定支承架 Fixed bracket；15—连接板 Connecting plate；
16—保温支承圈 Insulation bearing ring

5.5.3.6 催化剂沉降槽 Catalyst settlement trough

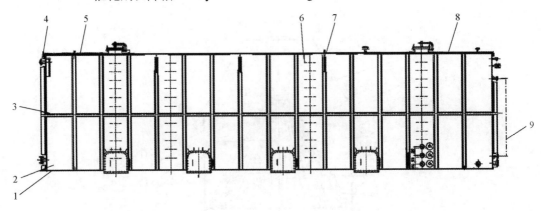

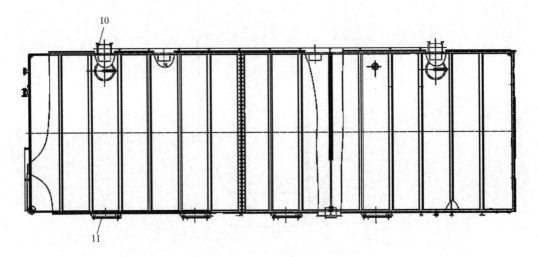

图 5.5.11 催化剂沉降槽结构图 Structure drawing of catalyst settlement trough
1—底板 Bottom panel；2—侧板 Side panel；3—垫板 Plate；4—法兰 Flange；5、8—顶盖 Top cover；
6—把手 Handlebar；7—隔板 Partition；9—磁性浮子液位计 Magnetic float level gauge；10—梯子 Ladder；
11—清理口 Cleanup opening

5.5.3.7 尾气洗涤器 Exhaust gas scrubber

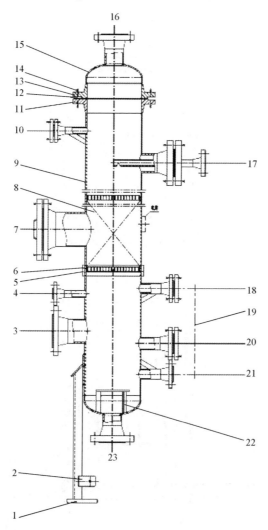

图 5.5.12 尾气洗涤器结构图 Structure drawing of exhaust gas scrubber

1—支腿 Support；2—接地板 Ground plate；3—气体入口 Gas inlet；4、21—压力计接口 Manometer interface；
5—支承圈 Support ring；6—钢圈 Steel ring；7、16—气体出口 Gas outlet；8—填料 Packing；9—筒体 Cylinder；
10、18—备用口 Spare opening；11、13—法兰 Flange；12—缠绕垫 Winding pad；
14—全螺纹螺柱、螺母 Fully threaded stud, Nut；15—封头 Head；17—液体进口 Liquid inlet；
19—液位计 Level gauge；20—吹扫口 Purging opening；22—防涡器 Vortex breaker；
23—液体出口 Liquid outlet

5.5.4 换热设备 Heat exchanger
5.5.4.1 反应气体冷却器 Reaction gas cooler

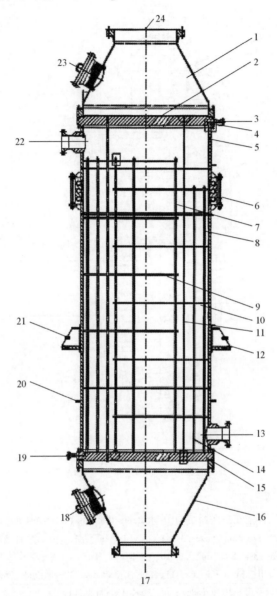

图 5.5.13 反应气体冷却器结构图 Structure drawing of reaction gas cooler

1—上管箱 Upper tube box；2—牺牲阳极板 Sacrificial anode plate；3—管板排气口 Upper vent of tube plate；4—上管板 Upper tube sheet；5—筒体 Cylinder；6—膨胀节 Expansion joint；7—拉杆 Tie rod；8、14—定距管 Spacer tube；9、10—折流板 Baffle；11—换热管 Heat exchange tube；12—支座 Support；13—冷却水入口 Boiler cooling water inlet；15—下管板 Lower tube sheet；16—下管箱 Lower tube box；17—反应气出口 Reaction gas outle；18、23—人孔 Manhole；19—管板排液口 Upper discharge port of tube plate；20—保温支承圈 Insulation bearing ring；21—接地板 Ground plate；22—冷却水出口 Boiler cooling water outlet；24—反应气入口 Reaction gas inlet

5.5.4.2 丙烯/氨蒸发器 Propylene / ammonia evaporator

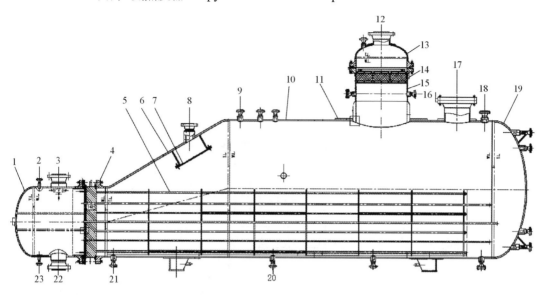

图 5.5.14　丙烯/氨蒸发器结构图 Structure drawing of propylene / ammonia evaporator

1—管箱 Tube box；2—管程放空口 Vent tube；3—贫水出口 Poor water outlet；4—壳程法兰 Shell flange；
5—管束 Tube bundle；6—斜锥壳 Slant conical shell；7—缓冲挡板 Buffer baffle；
8—丙烯/氨液体入口 Propylene / ammonia liquid inlet；9—吹扫口 Purging opening；
10、15—筒体 Cylinder；11—补强圈 Reinforcing ring；12—丙烯/氨气体出口 Propylene / ammonia gas outlet；
13、19—封头 Head；14—丝网除沫器 Wire mesh demister；16—气相返回口 Gas phase return port；
17—人孔 Manhole；18—安全阀接口 Safety valve interface；20—丙烯/氨液体出口 Propylene / ammonia liquid outlet；
21—壳程排放口 Shell drain opening；22—贫水入口 Poor water inlet；23—管程排液口 Tube discharge port

5.5.4.3 脱氰塔冷凝器 Decyanation column condenser

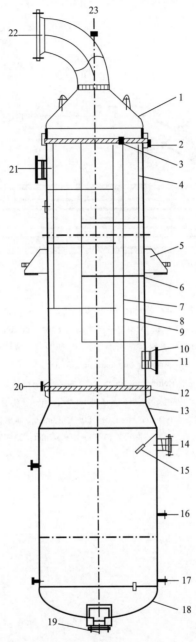

图 5.5.15　脱氰塔冷凝器结构图 Structure drawing of decyanation column condenser

1、13—锥形筒体 Conical cylinder；2—排气口 Exhaust port；3—上管板 Upper tube sheet；
4—定距管 Spacer tube；5—支座 Support；6—折流板 Baffle；7、9—换热管 Heat exchange tube；
8—筒体 Cylinder；10—缓冲挡板 Buffer board；11—冷冻盐水入口 Frozen brine inlet；
12—下管板 Lower tube sheet；14—不凝气体出口 Non-condensable gas outlet；
15—挡板 Baffle；16—备用口 Spare opening；17—蒸汽吹扫口 Steam purge opening；
18—封头 Head；19—氢氰酸液体出口 Hydrocyanic acid liquid outlet；20—排放口 Drain opening；
21—冷冻盐水出口 Frozen brine outlet；22—氢氰酸蒸气入口 Hydrocyanic acid vapor inlet；
23—阻聚剂入口 Polymerization inhibitor inlet

5.5.4.4 废水蒸发器 Waste water evaporator

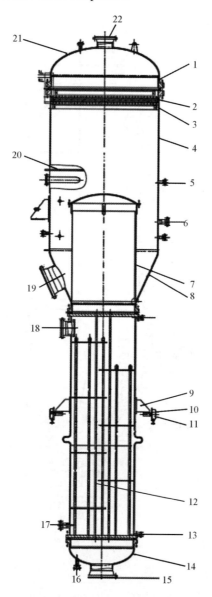

图 5.5.16 废水蒸发器结构图 Structure drawing of waste water evaporator

1—凸面法兰 Convex flange；2—凹面法兰 Concave flange；3—丝网除沫器 Wire mesh demister；
4—筒体 Cylinder；5—水冲洗口 Water rinse opening；6—排气口 Exhaust port；7—升气管 Riser pipe；8—锥体 Cone；
9—支座 Support；10—垫板 Plate；11—滚轮 Roller；12—换热器 Heat exchanger；13—排液口 Discharge port；
14—管箱 Tube box；15—循环物料入口 Circulation material inlet；16—蒸汽吹扫入口 Steam purge inlet；
17—蒸汽凝液出口 Steam condensate outlet；18—蒸汽入口 Steam inlet；19—循环物料出口 Circulation material outlet；
20—废水入口 Waste water inlet；21—封头 Head；22—二次出口 Secondary outlet

5.5.5 工业炉 Industrial furnace

5.5.5.1 开工加热炉 Start-up furnace

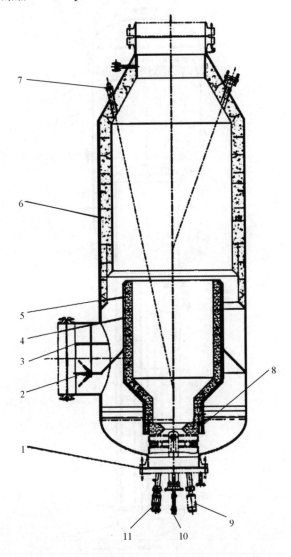

图 5.5.17 开工加热炉结构图 Structure drawing of start-up furnace

1—燃烧器 Burner；2、3—调节板 Adjustment plate；4—刚玉浇注衬里 Jade castable；
5—背衬板 Backing plate；6—轻质耐火浇注衬里 Lightweight refractory castable；
7、9—火焰监测器 Flame monitor；8—高铝陶瓷纤维毯衬里 Alumina ceramic fiber blanket；
10—点火器 Igniter；11—看火孔 Kiln eye

5.5.5.2 焚烧炉 Incinerator

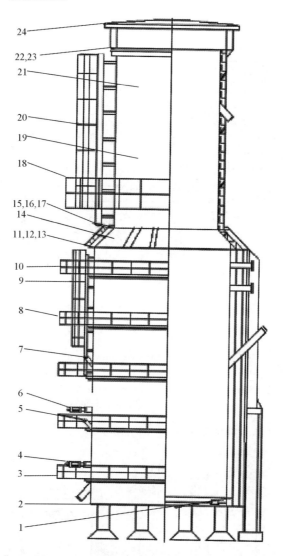

图 5.5.18 焚烧炉结构图 Structure drawing of incinerator

1—耐火衬里 Refractory liner；2—炉体钢结构 Furnace steel structure；3—下部平台 Lower platform ladder；
4—氢氰酸喷嘴 Hydrocyanic acid nozzle；5—废水喷嘴 Waste water nozzle；6—油-气联合燃烧器 Oil-gas burner；
7—硫铵液喷嘴 Ammonium sulfate solution nozzle；8、10、18—平台 Platform；9—下段直梯 Lower section ladders；
11、15、22—螺栓 Stud；12、16、23—螺母 Nut；13、17—垫片 Gasket；14—锥段 Cone section；
19—烟囱下段 Chimney lower section；20—上段直梯 Upper section ladder；21—烟囱上段 Chimney upper section；
24—防雨帽 Rainhat

5.5.6 泵 Pump

5.5.6.1 离心泵 Centrifugal pump

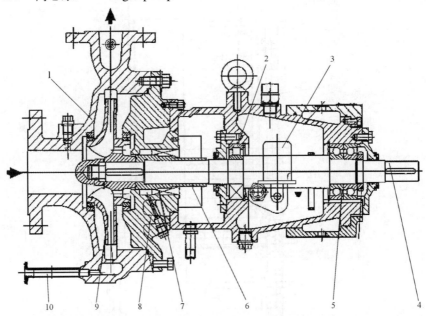

图 5.5.19 离心泵结构图 Structure drawing of centrifugal pump

1—泵体 Pump body；2、5—轴承 Bearing；3—恒位油杯 Constant level oiler；4—轴 Shaft；
6—轴套 Sleeve；7—机械密封 Mechanical seal；8—泵盖 Pump cover；
9—叶轮 Wheel；10—泵体排液管 Pump drain pipe

5.5.6.2 隔膜计量泵 Diaphragm metering pump

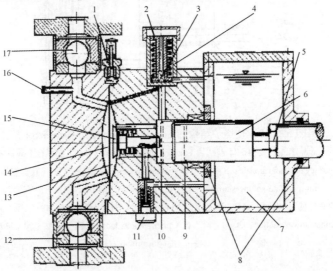

图 5.5.20 隔膜计量泵结构图 Structure drawing of diaphragm metering pump

1—排放接头 Discharge connection；2—泄压阀 Pressure relief valve；3—排气阀 Exhaust valve；
4—双功能液压阀 Dual-function hydraulic valve；5—锁紧螺母 Lock nut；6—柱塞 Plunger；
7—液压油箱 Hydraulic oil tank；8—填料密封 Packing seal；9—后止点 Rear check point；
10—前止点 Front check point；11—补油阀 Fill oil valve；12、17—单向阀 Check valve assembly；
13—液压室 Hydraulic chamber；14—隔膜 Diaphragm；15—推杆 Push rod；16—排空阀 Drain valve

5.5.6.3 齿轮泵 Gear pump

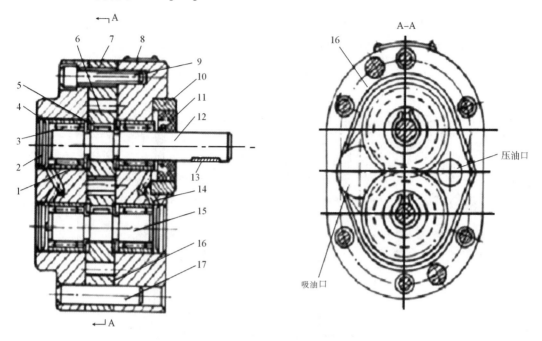

图 5.5.21 齿轮泵结构图 Structure drawing of gear pump

1—轴承外环 Bearing outer ring；2—堵头 Plug；3—滚子轴承 Roller bearing；4—后泵盖 Rear pump cover；
5、13—键 Key；6—齿轮 Gear；7—泵体 Pump body；8—前泵盖 Front pump cover；9—螺栓 Screw；
10—压环 Pressure ring；11—密封环 Sealing ring；12—主动轴 Drive shaft；14—泻油孔 Oil discharge hole；
15—从动轴 Driven shaft；16—泻油槽 Oil discharge slot；17—定位销 Positioning pin

5.5.6.4 液环式真空泵 Liquid ring vacuum pump

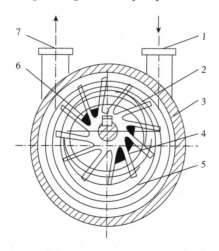

图 5.5.22 液环式真空泵结构图 Structure drawing of liquid ring vacuum pump

1—吸气口 Suction port；2—叶轮 Wheel；3—泵体 Pump body；4—吸气孔 Suction hole；
5—液环 Liquid ring；6—排气孔 Vent hole；7—排气口 Exhaust port

5.5.7 压缩机 Compressor

5.5.7.1 空压机 Air compressor

1. 空压机结构 Structure of air compressor

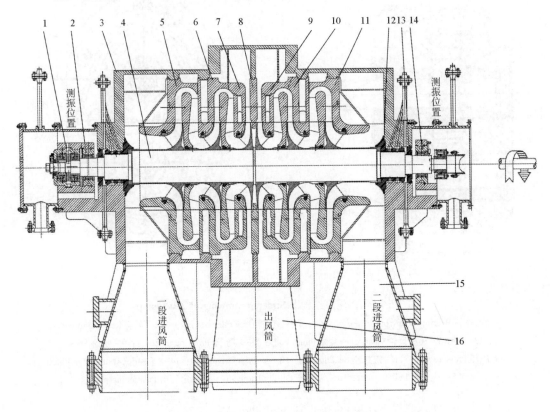

图 5.5.23 空压机结构图 Structure drawing of air compressor

1—推力轴承 Thrust bearing；2、14—可倾瓦轴承 Tilting tile bearing；3—壳体 Shell；4—主轴 Shaft；
5、11—进口隔板 Partition of inlet；6、7、9、10—隔板 Partition；8—中间隔板 Middle partition；
12—轴端密封 Shaft seal；13—油封 Oil seal；15—进气口 Gas inlet；16—排气口 Exhaust port

2. 转子 Rotor

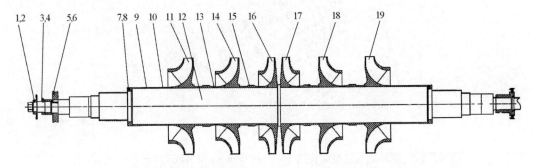

图 5.5.24 转子结构图 Structure drawing of rotor

1、4、7—螺栓 Screw；2—圆盘 Disc；3、8—螺母 Nut；5—推力盘 Thrust plate；6—键 Key；
9、10、13、15—隔套 Spacer；11、14、16、17、18、19—叶轮 Wheel；12—主轴 Shaft

5 合成原料 Synthetic materials

5.5.7.2 大冰机 Big ice machine

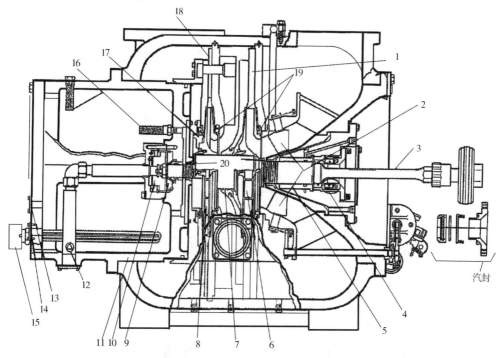

图 5.5.25 大冰机结构图 Structure drawing of big ice machine

1—一级扩压器 Diffuser level 1；2—预旋叶片 Prewhirl blade；3—主动轴 Drive shaft；4—汽封 Steam seal；5—吸入端轴承 Suction side bearing；6—一级叶轮 Impeller level 1；7—迷宫密封 Shaft labyrinth seal；8—二级叶轮 Impeller level 2；9—径向轴承和推力轴承 Radial and thrust bearings；10—油泵 Oil pump；11—油箱 Fuel tank；12—油加热器 Oil heater；13—上视镜 Upper endoscopic；14—下视镜 Lower sight glass；15—轴头泵 Pump of shaft head；16—油雾器 Lubricator；17—平衡塞 Balanced plug；18—二级扩压器 Diffuser level 2；19—叶轮进口密封 Impeller inlet seal；20—油封 Oil seal

5.5.7.3 螺杆制冷压缩机 Screw refrigeration compressor

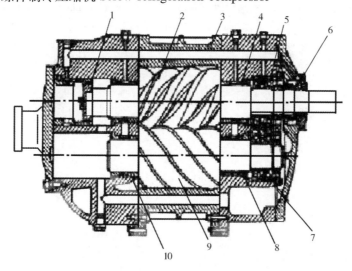

图 5.5.26 螺杆制冷压缩机结构图 Structure drawing of screw refrigeration compressor

1—平衡盘 Balance disc；2—阳转子 Male rotor；3—机体 Compressor body；4、8、10—滑动轴承 Sliding bearing；5、7—止推轴承 Thrust bearing；6—轴封 Shaft seal；9—阴转子 Female rotor

5.5.7.4 气氨压缩机 Ammonia gas compressor

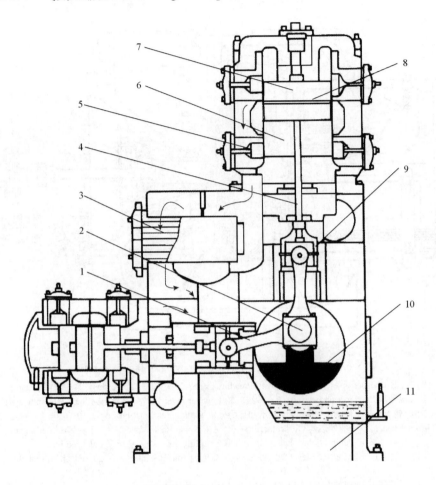

图 5.5.27 气氨压缩机结构图 Structure drawing of ammonia gas compressor
1—连杆 Connecting rod；2—曲轴 Crankshaft；3—中间冷却器 Intercooler；4—活塞杆 Piston rod；
5—气阀 Gas valve；6—气缸 Cylinder；7—活塞 Piston；8—活塞环 Piston ring；
9—十字头 Crosshead；10—平衡重 Balance weight；11—机身 Fuselage

5.6 聚酰胺 Polyamide

5.6.1 反应设备 Reaction equipment

5.6.1.1 加压聚合器 Pressurized aggregator

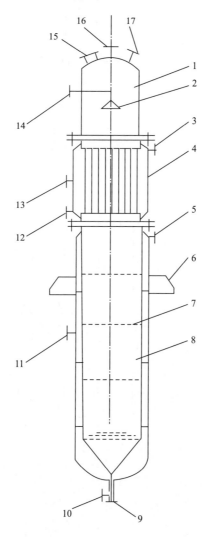

图 5.6.1 加压聚合器结构图 Structure drawing of pressurized aggregator

1—上段筒体 Upper section cylinder；2—分配器 Distributor；3、5—汽态联苯出口 Biphenyl vapor outlet；
4—换热器 Tube heat exchanger section；6—支座 Support；7—筛板 Sieve plate；8—下段筒体 Lower section cylinder；
9—己内酰胺出料口 Caprolactam outlet；10、12—液态联苯出口 Liquid biphenyl outlet；
11、13—联苯进口 Biphenyl inlet；14—己内酰胺进料口 Caprolactam inlet；15—人孔 Manhole；
16—排气口 Exhaust port；17—泄压口 Pressure relief port；

5.6.1.2 后聚合器 Rear aggregator

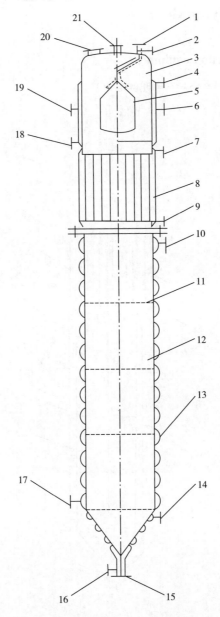

图 5.6.2 后聚合器结构图 Structure drawing of rear aggregator

1—聚酰胺进料口 Polyamide inlet；2、4、7—汽态联苯出口 Vaporous biphenyl outlet；
3—上段筒体 Upper section cylinder；5—进料分配器 Feed distributor；
6、9—汽态联苯进口 Vaporous biphenyl inlet；8—换热器 Heat exchanger；
10、14—液态联苯出口 Liquid biphenyl outlet；11—筛板 Sieve plate；
12—下段筒体 Lower section cylinder；13—夹套盘管 Jacketed coil；
15—聚酰胺出料口 Polyamide outlet；16、17—液态联苯进口 Liquid biphenyl inlet；
18—联苯出口 Diphenyl outlet；19—联苯进口 Diphenyl inlet；20—人孔 Manhole；
21—抽真空口 Evacuation port

5.6.1.3 氮气脱氧器 Nitrogen deaerator

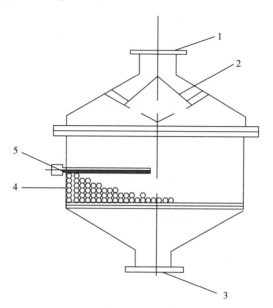

图 5.6.3 氮气脱氧器结构图 Structure drawing of nitrogen deaerator
1—氢气入口 Hydrogen inlet；2—气体分布器-Gas distributor；3—氢气出口 Hydrogen outlet；
4—催化剂 Catalyst；5—热电偶 Thermocouple

5.6.2 塔设备 Column equipment

5.6.2.1 萃取塔 Extraction column

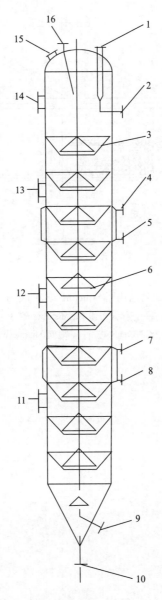

图 5.6.4 萃取塔结构图 Structure drawing of extraction column

1—溢流管接口 Overflow pipe interface；2—萃取水出口 Extract water outlet；3—分配环 Distribution ring；4、7—夹套蒸汽入口 Jacketed steam inlet；5、8—夹套蒸汽出口 Jacketed steam outlet；6—分配器 Distributor；9—萃取水进口 Extract water inlet；10—切片出料口 Slicing discharge port；11、12、13、14—人孔 Manhole；15—视镜 Endoscopy；16—切片进料口 Slicing feed port

5.6.2.2 干燥塔 Drying column

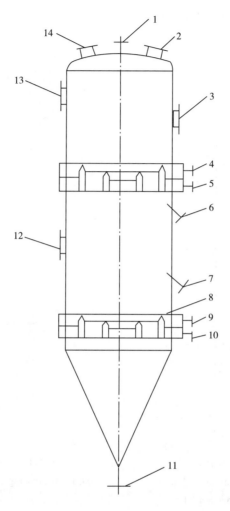

图 5.6.5 干燥塔结构图 Structure drawing of drying column

1—切片进料口 Slicing feed port；2—氮气出口 Nitrogen outlet；3、12—人孔 Manhole；
4—上段氮气入口 Upper section nitrogen inlet；5—上段氮气出口 Upper section nitrogen outlet；
6、7—取样口 Sampling port；8—气体分配器 Gas distributor；9—下段氮气入口 Lower section nitrogen inlet；
10—下段氮气出口 Lower section nitrogen outlet；11—切片出料口 Slicing discharge port；
13、14—视镜 Endoscopy

5.6.2.3 冷却塔 Cooling column
5.6.2.4 蒸发塔 Evaporation column

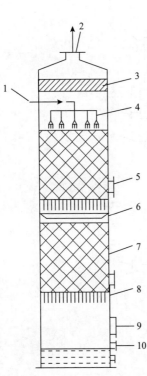

图 5.6.6 冷却塔结构图
Structure drawing of cooling column
1—冷却水入口 Cooling water inlet;
2—氮气出口 Nitrogen outlet;
3—丝网除沫器 Wire mesh demister;
4—分配管 Distribution pipe;
5—卸料口 Discharge outlet;
6—液体再分配器 Liquid redistributor;
7—填料 Packing;
8—支承格栅 Support grid;
9—氮气入口 Nitrogen inlet;
10—冷却水出口 Cooling water outlet

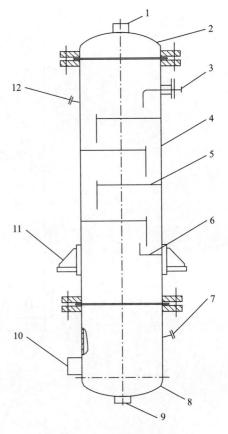

图 5.6.7 蒸发塔结构图
Structure drawing of evaporation column
1—气相出口 Gas phase outlet;
2—上封头 Upper head;
3—液相入口 Liquid inlet;
4—筒体 Cylinder;
5—浮阀塔盘 Valve tray;
6—受液盘 Liquid tray;
7、12—视镜 Endoscopy;
8—下封头 Lower head;
9—液相出口 Liquid outlet;
10—气相入口 Gas phase inlet;
11—支座 Support

5.6.3 换热设备 Heat exchange equipment

5.6.3.1 己内酰胺预热器 Caprolactam preheater

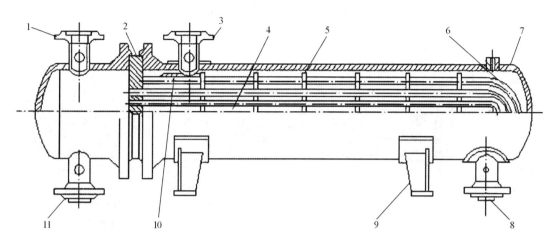

图 5.6.8 己内酰胺预热器结构图 Structure drawing of caprolactam preheater
1—己内酰胺入口 Caprolactam inlet；2—管板 Tube sheet；3—液态联苯入口 Liquid biphenyl inlet；
4—中间挡板 Middle baffle；5—折流板 Baffle；6—管束 Tube bundle；7—筒体 Cylinder；
8—液态联苯出口 Liquid biphenyl outlet；9—支座 Support；
10—防冲挡板 Impingement baffle；11—己内酰胺出口 Caprolactam outlet

5.6.3.2 氮气加热器 Nitrogen heater

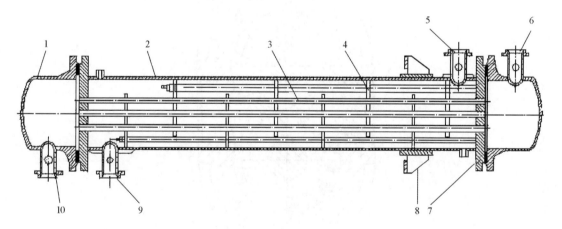

图 5.6.9 氮气加热器结构图 Structure drawing of nitrogen heater
1—管箱 Head cap；2—筒体 Shell；3—管束 Tube bundle；4—折流板 Baffle damper；
5—蒸汽入口 Steam inlet；6—氮气入口 Nitrogen inlet；7—管板 Tube sheet；
8—支座 Support；9—蒸汽出口 Steam outlet；10—氮气出口 Nitrogen outlet

5.6.4 泵 Pumps

5.6.4.1 出料齿轮泵 Discharge gear pump

1. 出料齿轮泵外形 Appearance of discharge gear pump

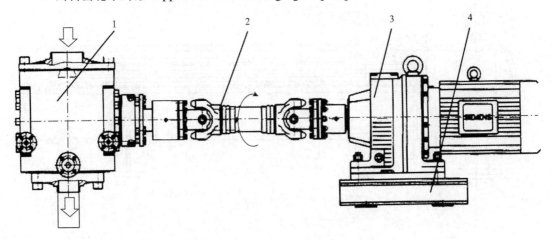

图 5.6.10 出料齿轮泵外形图 Outline drawing of discharge gear pump

1—泵体 Pump body；2—万向联轴器 Universal coupling；3—减速电机 Geared motor；4—底座 Base

2. 出料齿轮泵结构 Structure of discharge gear pump

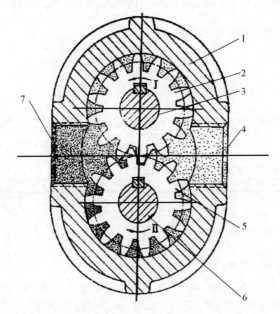

图 5.6.11 出料齿轮泵结构图 Structure drawing of discharge gear pump

1—泵体 Pump body；2—主动齿轮 Driving gear；3—主动轴 Drive shaft；4—进料口 Feed inlet；5—从动齿轮 Driven gear；6—从动轴 Driven shaft；7—出料口 Discharge opening

5 合成原料 Synthetic materials

3. 齿轮箱 Gearbox

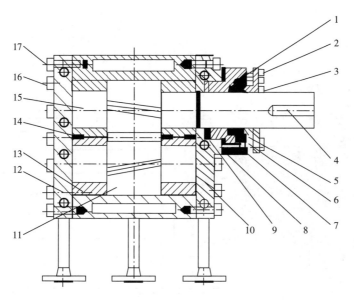

图 5.6.12 齿轮箱结构图 Structure drawing of gearbox

1—迷宫密封 Labyrinth seal；2—压盖 Gland；3、17—六角螺栓 Hexagon bolt；
4—键 Key；5—填料 Packing；6—隔圈 Spacer；7、16—内沉六角螺钉 Inner hexagon screw；
8—挡圈 Retaining ring；9—密封垫圈 Sealing washer；10—前侧板 Front side panel；
11—从动齿轮 Driven gear；12—箱体 Shell；13—轴承 Bearing；
14—定位键 Locating key；15—主动齿轮 Driving gear

5.6.4.2 热媒循环泵 Heat medium circulation pump

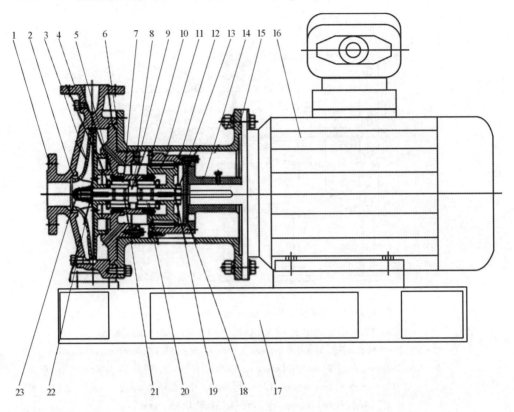

图 5.6.13 热媒循环泵结构图 Structure drawing of heat medium circulation pump
1—泵体 Pump body；2—密封环 Sealing ring；3—叶轮 Wheel；4—止推环 Thrust ring；
5—轴承座 Bearing seat；6—泵盖 Pump cover；7—滑动轴承 Sliding bearing；
8—隔离套压环 Isolated casing pressure ring；9—泵轴 Pump shaft；
10—外磁挡板 Outer magnetic damper；11—内磁总成 Internal magnetic assembly；
12—外磁总成 External magnetic assembly；13—锁紧螺母 Lock nut；
14—传动套 Transmission sleeve；15—联接架 Connecting rack；16—电机 Motor；
17—底座 Base；18—隔离套 Spacer sleeve；19—止动垫圈 Lock washer；
20—轴套 Sleeve；21—鼓型圈 Drum circle；22—叶轮垫圈 Impeller washer；
23—叶轮螺母 Impeller nut

5.6.4.3 回流泵 Reflux pump

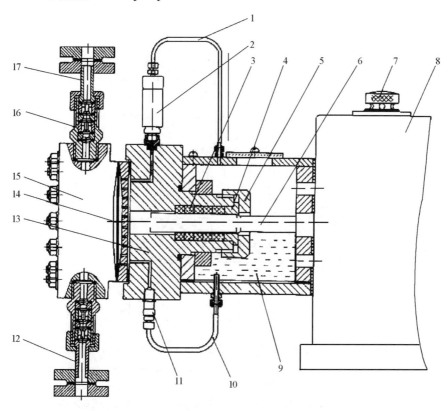

图 5.6.14 回流泵结构图 Structure drawing of reflux pump

1—泄压回流管 Pressure backflow pipe；2—二阀组 Two valve group；3—填料 Packing；
4—填料衬套 Packing liner；5—压紧螺母 Gland nut；6—柱塞 Plunger；7—透气螺塞 Vent plug；
8—动力箱 Power end；9—油池 Oil pool；10—补油管 Filling pipe；11—补油阀 Filling valve；
12—进液管 Inlet tube；13—缸体 Cylinder；14—膜片 Membrane；15—泵头 Pump head；
16—单向阀 Check valve；17—出液管 Outlet pipe

5.6.5 氮气压缩机 Nitrogen compressor

5.6.5.1 氮气压缩机外形 Nitrogen compressor appearance

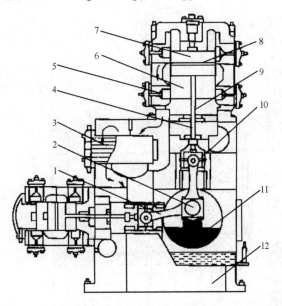

图 5.6.15 氮气压缩机外形图 Outline drawing of nitrogen compressor
1—连杆 Connecting rod；2—曲轴 Crankshaft；3—中间冷却器 Intercooler；4、9—活塞杆 Piston rod；
5—气阀 Gas valve；6—气缸 Cylinder；7—活塞 Piston；8—活塞环 Piston ring；
10—十字头 Crosshead；11—平衡重 Balance weight；12—机体 Compressor body

5.6.5.2 氮气压缩机主要零部件 Major parts of nitrogen compressor

1. 曲轴 Crankshaft

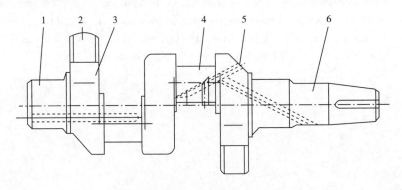

图 5.6.16 曲轴 Structure of crankshaft
1—主轴颈 Main journal；2—平衡块 Balance weight；3—曲柄 Crank；4—曲轴颈 Crank pin；
5—油孔 Oil hole；6—轴颈 Journal

2. 汽缸套和吸排气阀 Jacket and suction exhaust valve

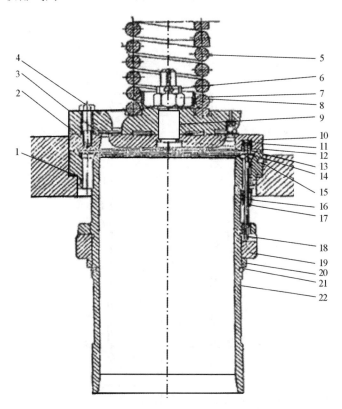

图 5.6.17　汽缸套和吸排气阀结构图 Structure drawing of jacket and suction exhaust valve
1—调整垫片 Adjusting washer；2—阀盖 Valve cover；3—排气阀片 Exhaust valve slice；4—螺栓 Bolt；
5—假盖弹簧 False cover spring；6、17—开口销 Cotter pin；7—螺母 Nut；8—钢碗 Steel bowl；
9—气阀螺栓 Valve bolts；10—外阀座 Outer seat；11—内阀座 Inner seat；12—垫片 Gasket；
13—气阀弹簧 Valve spring；14—吸气阀片 Air suction valve slice；15—圆柱销 Cylindrical pin；
16—顶柱弹簧 Top column spring；18—顶杆 Top bar；19—转动环 Rotating ring；20—垫圈 Washer；
21—弹性圈 Elastic ring；22—气缸套 Cylinder liner

5.6.6 预萃取水罐 Preliminary extraction tank

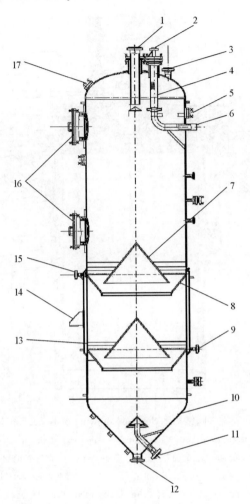

图 5.6.18 预萃取水罐结构图 Structure drawing of preliminary extraction tank
1—进料口 Feed inlet；2—氮气入口 Nitrogen inlet；3—放空口 Vent；4—过滤器 Filter；
5、17—视镜 Endoscopy；6—萃取水出口 Extraction water outlet；7—分配锥 Distribution cone；
8—分配环 Distribution ring；9—冷凝水出口 Condensed water outlet；10—锥体 Cone；
11—萃取水入口 Extraction water inlet；12—出料口 Discharge opening；
13—夹套 Jacketed cylinder；14—支座 Support；15—蒸汽入口 Steam inlet；
16—人孔 Manhole

5.6.7 其它设备 Other equipments

5.6.7.1 熔体过滤器 Melt filter

1. 熔体过滤器结构 Melt filter structure

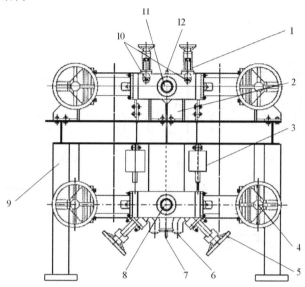

图 5.6.19 熔体过滤器结构图 Structure drawing of melt filter

1—排气放料阀 Exhaust discharge valve；2—过滤室 Filter chamber；3—弹簧吊架 Spring hanger；4—熔体三通阀 Melt three-way valve；5—排污放料阀 Sewage discharge valve；6—排污口 Discharge port；7—热媒夹套出口 Heat medium jacket outlet；8—熔体进口 Melt inlet；9—支架 Support；10—取样口 Sampling port；11—熔体出口 Melt outlet；12—热媒夹套进口 Heat medium jacket inlet

2. 过滤室 Filter chamber

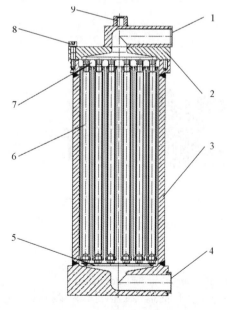

图 5.6.20 过滤室结构图 Structure drawing of filter chamber

1—熔体出料口 Melt discharge port；2—出料端盖 Discharge end cover；3—筒体 Cylinder；4—熔体进料口 Melt inlet；5—定位板 Positioning plate；6—过滤芯 Filter element；7—安装板 Mounting plate；8—紧固螺栓 Fastening screw；9—吊环 Lifting bolt

5.6.7.2 旋转锁定送料阀 Rotary lock feed valve

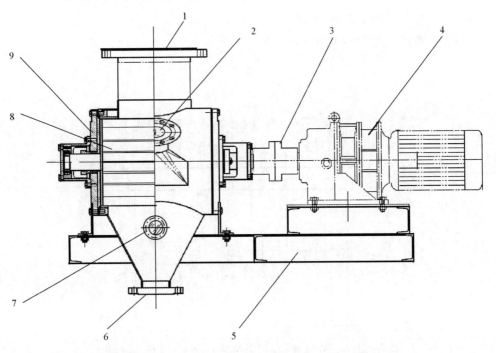

图 5.6.21 旋转锁定送料阀结构图 Structure drawing of rotary lock feed valve

1—切片进口 Slice inlet；2—进水口 Water inlet；3—联轴器 Coupling；4—减速电机 Gear motor；
5—底座 Base；6—切片出口 Slice outlet；7—视镜 Endoscopy port；8—叶轮 Wheel；9—端盖 End cap

5.6.7.3 水下切粒机 Underwater pelletizer

1. 水下切粒机外形 Underwater pelletizer appearance

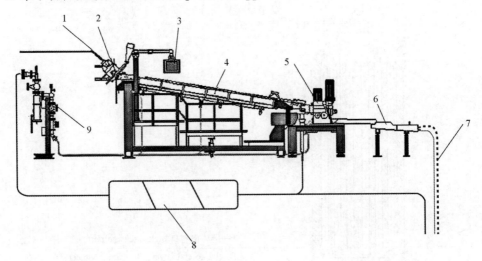

图 5.6.22 水下切粒机外形图 Outline drawing of underwater pelletizer

1—熔体进料口 Melt feed inlet；2—铸带头 Cast lead；3—红外监测探头 Infrared detector；4—冷却水槽 Cooling tank；
5—切割室 Cutting chamber；6—切片分离器 Slices separator；7—工艺水及切片 Process water and slice；
8—工艺水再生处理系统 Process water regenerative treatment system；9—水阀 Water valve

2. 水下切粒机结构 Underwater pelletizer structure

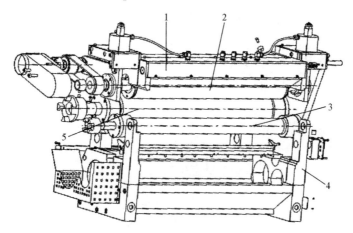

图 5.6.23 水下切粒机结构图 Structure drawing of underwater pelletizer
1—切割室盖 Cutting chamber cover；2—前引料轴 Front feeding shaft；
3—后引料轴 Rear feeding shaft；4—切割室 Cutting chamber；5—滚切刀 Slitting knife

5.6.7.4 脱水机 Dehydrator

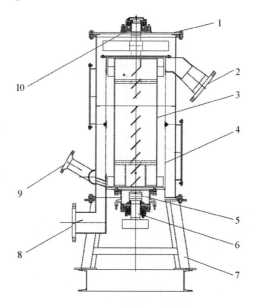

图 5.6.24 脱水机结构图 Structure drawing of dehydrator
1—上盖 Cover；2—出料口 Discharge opening；3—叶轮 Wheel；4—过滤网 Filter；
5—机械密封 Mechanical seal；6—轴 Shaft；7—支架 Support；8—出水口 Water outlet；
9—进料口 Feed port；10—轴承座 Bearing seat

5.7 羟胺肟化 Hydroxylamine oximation

5.7.1 反应设备 Reaction equipment

5.7.1.1 羟胺反应器 Hydroxylamine reactor

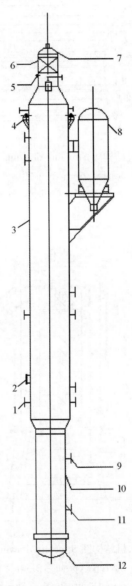

图 5.7.1 羟胺反应器结构图 Structure drawing of hydroxylamine reactor
1—无机液入口 Inorganic liquid inlet；2—人孔 Manhole；3—筒体 Cylinder；4—支座 Support；
5—工艺水入口 Process water inlet；6—除沫网 Demister mesh；7、12—封头 Head；
8—高位槽 High slot；9—循环水入口 Circulating water inlet；10—换热器 Heat exchanger；
11—循环水出口 Circulating water outlet

5.7.1.2 甲烷反应器 Methane reactor

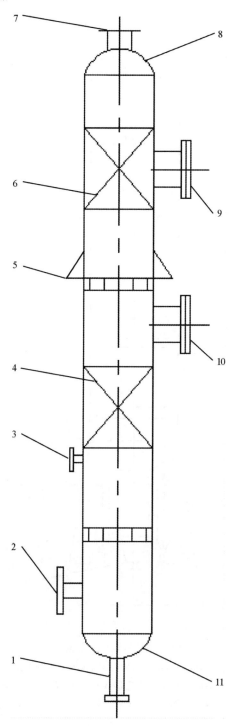

图 5.7.2 甲烷反应器结构图 Structure drawing of methane reactor
1—排液口 Discharge port；2—气体出口 Gas outlet；3—接管 Connecting pipe；4、6—催化剂 Catalyst；
5—支座 Support；7—气体进口 Gas inlet；8、11—封头 Head；9、10—人孔 Manhole

5.7.1.3 肟化反应器 Oximation reactor

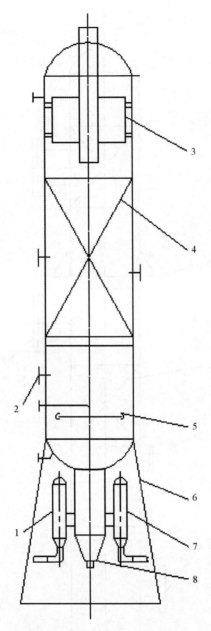

图 5.7.3 肟化反应器结构图 Structure drawing of oximation reactor
1—脉冲排出罐 Pulse discharge tank；2—人孔 Manhole；3—顶部分布器 Top distributor；
4—催化剂 Packing；5—底部分布器 Bottom distributor；6—裙座 Skirt；
7—脉冲吸入罐 Pulse inhalation tank；8—旋转阀 Rotary valve

5.7.1.4 氨氧化反应器 Ammoxidation reactor

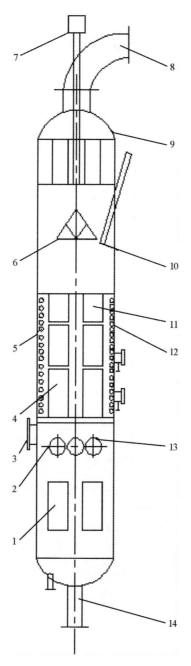

图 5.7.4 氨氧化反应器结构图 Structure drawing of ammoxidation reactor

1—预热器 Preheater；2—过热器 Superheater；3—人孔 Manhole；4—第二蒸发器 Second evaporator；
5—水冷室 Water cooling chamber；6—转臂 Jib；7—电机 Motor；8—气体入口 Gas inlet；9—封头 Head；
10—点火器 Igniter；11—第一蒸发器 First evaporator；12—水冷壁 Water cooling wall；
13—汽化器 Carburettor；14—气体出口 Gas outlet

5.7.1.5 氧化氮脱除器 Nitrogen oxide dehydrator

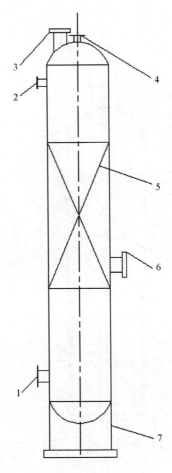

图 5.7.5 氧化氮脱除器结构图 Structure drawing of nitrogen oxide dehydrator
1—气体出口 Gas outlet；2—接管 Connecting pipe；3、6—人孔 Manhole；4—气体入口 Gas inlet；
5—填料 Packing；7—裙座 Skirt

5.7.2 塔设备 Column equipment

5.7.2.1 萃取塔 Extraction column

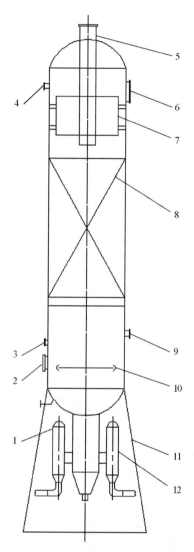

图 5.7.6 萃取塔结构图 Structure drawing of extraction column

1—脉冲排出罐 Pulse discharge tank；2、6—人孔 Manhole；3—手孔 Hand hole；
4—物料出口 Material outlet；5—出料口 Material inlet；7、10—分布器 Distributor；
8—填料 Packing；9—进料口 Feed inlet；11—裙座 Skirt；
12—脉冲吸入罐 Pulse suction tank

5.7.2.2 汽提塔 Stripping column

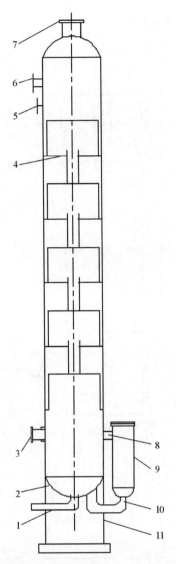

图 5.7.7 汽提塔结构图 Structure drawing of stripping column

1—出料口 Material outlet；2—封头 Head；3—回收液入口 Recycling liquid inlet；
4—塔盘 Tray；5—进料口 Feed inlet；6—人孔 Manhole；7—汽相出口 Vapor phase outlet；
8—汽相入口 Vapor phase inlet；9—再沸器 Reboiler；10—液相出口 Vapor phase outlet；11—裙座 Skirt

5.7.2.3 第一精馏塔 First distillation column

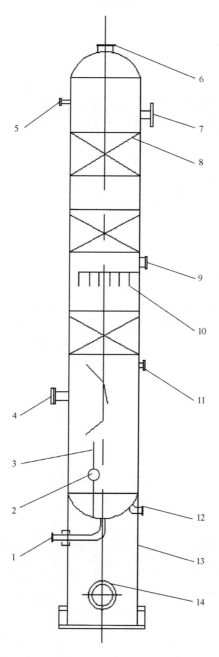

图 5.7.8 第一精馏塔结构图 Structure drawing of first distillation column

1—液相出口 Liquid phase outlet；2、7、11—人孔 Manhole；3—隔板 Partition；
4—气相入口 Gas phase inlet；5—回流口 Reflux port；6—气相出口 Gas phase outlet；
8—填料 Packing；9—进料口 Feed inlet；10—分布器 Distributor；12—出料口 Material outlet；
13—裙座 Skirt；14—检修口 Access hole

5.7.3 换热设备 Heat exchange equipment
5.7.3.1 气体加热器 Gas Heater

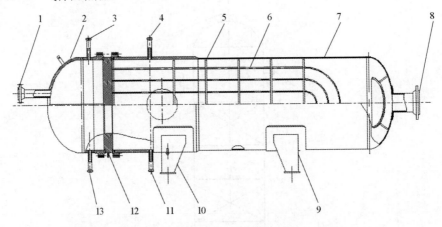

图 5.7.9 气体加热器结构图 Structure drawing of gas heater
1—蒸汽入口 Steam inlet；2—管箱 Tube box；3、4—放空口 Vent；5—折流板 Baffle；
6—换热管 Heat exchange tube；7—筒体 Shell；8—气体入口 Gas inlet；
9—活动支座 Movable support；10—固定支座 Fixed support；
11—排放口 Discharge port；12—管板 Tube sheet；13—冷凝液出口 Condensate outlet

5.7.3.2 汽提塔再沸器 Stripper reboiler

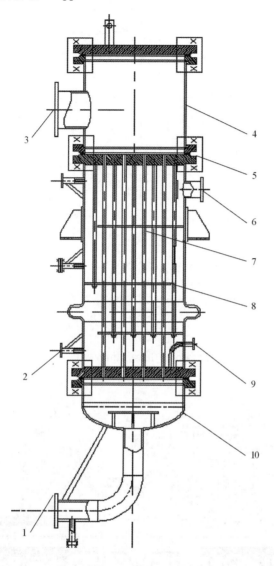

图 5.7.10 汽提塔再沸器结构图 Structure drawing of stripper reboiler

1—进料口 Feed port；2—排气口 Exhaust port；3—出料口 Discharge opening；4—管箱 Tube box；
5—管板 Tube sheet；6—蒸汽入口 Steam inlet；7—换热管 Heat exchange tube；8—折流板 Baffle；
9—蒸汽出口 Steam outlet；10—封头 Head

5.7.3.3 肟冷却器 Oxime cooler

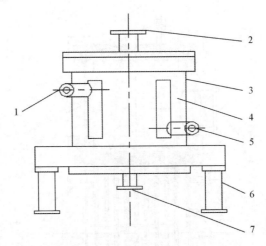

图 5.7.11 肟冷却器结构图 Structure drawing of oxime cooler
1—热端入口 Hot side inlet；2—冷端出口 Cold side outlet；3—壳体 Shell；4—螺旋板 Spiral plate；
5—冷端入口 Cold side inlet；6—支座 Support；7—热端出口 Hot side outlet

5.7.4 泵 Pump

5.7.4.1 羟胺进料泵 Hydroxylamine feed pump

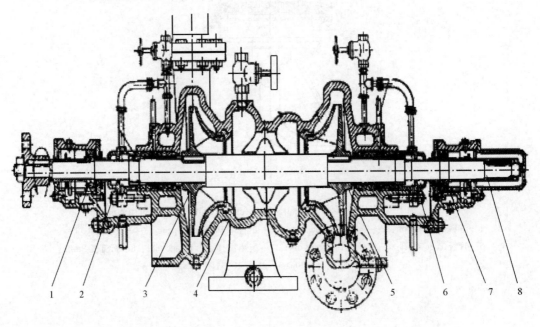

图 5.7.12 羟胺进料泵结构图 Structure drawing of hydroxylamine feed pump
1、7—轴承 Bearing；2、6—机械密封 Mechanical seal；3—次级叶轮 Secondary impeller；
4—泵体 Pump body；5—首级叶轮 First stage impeller；8—泵轴 Shaft

5.7.4.2 锅炉给水泵 Boiler feed pump

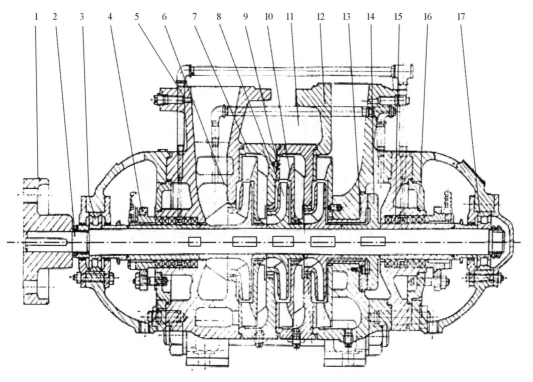

图 5.7.13 锅炉给水泵结构图 Structure drawing of boiler feed pump
1—联轴器 Coupling；2—轴 Shaft；3、17—轴承 Bearing；4—密封压盖 Outer flange；
5—泵入口 Pump inlet；6—密封环 Sealing ring；7—隔板 Clapboard；8—叶轮 Wheel；9—导叶 Guide vane；
10—导叶套 Guide vane sleeve；11—拉紧螺栓 Tighten bolt；12—泵出口 Pump outlet；
13—平衡座 Balanced sleeve；14—平衡盘 Balance disc；15—填料函 Stuffing box；
16—水冷室盖 Water-cooled chamber cover

5.7.4.3 催化剂计量泵 Catalyst metering pump

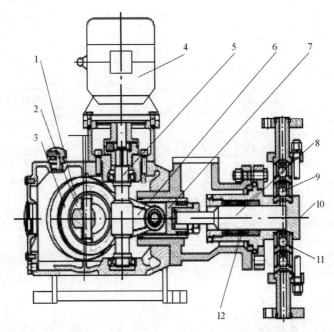

图 5.7.14 催化剂计量泵结构图 Structure drawing of catalyst metering pump
1—偏心轮 Eccentric；2—偏心轮销 Eccentric pin；3—蜗轮 Worm wheel；4—电机 Motor；5—蜗杆 Worm rod；
6—连杆 Connecting rod；7—十字头 Crosshead；8—柱塞 Plunger；9—出口阀 Outlet valve；
10—缸体 Cylinder；11—进口阀 Inlet valve；12—填料 Packing

5.7.5 容器与储罐 Container and tank

5.7.5.1 甲苯、肟储罐 Toluene & oxime tank

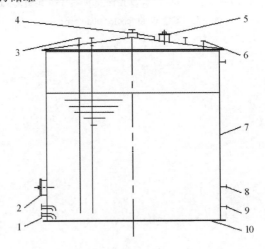

图 5.7.15 甲苯、肟储罐结构图 Structure drawing of toluene & oxime tank
1—排放口 Discharge port；2、5—人孔 Manhole；3、6—进料口 Feed inlet；
4—安全阀接口 Safety valve interface；7—筒体 Cylinder；8—备用口 Spare port；
9—出料口 Material outlet；10—底板 Bottom panel

5.7.5.2 羟胺反应器进料罐 Hydroxylamine reactor feed tank

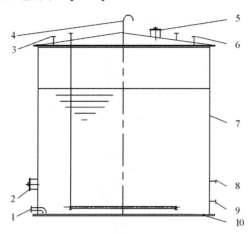

图 5.7.16 羟胺反应器进料罐结构图 Structure drawing of hydroxylamine reactor feed tank

1—排放口 Discharge port；2、5—人孔 Manhole；3—进料口 Feed inlet；4—呼吸口 Vent；
6—循环物料入口 Circulating material inlet；7—筒体 Cylinder；8—备用口 Spare port；
9—出料口 Material inlet；10—底板 Bottom panel

5.7.5.3 中和器 Neutralizer

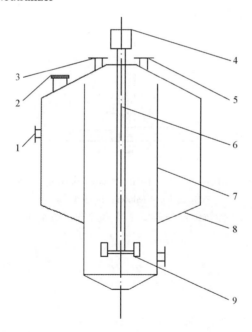

图 5.7.17 中和器结构图 Structure drawing of neutralizer

1—物料出口 Materials outlet；2—人孔 Manhole；3—呼吸口 Respiratory pipe；4—电机 Motor；
5—物料进口 Materials inlet；6—搅拌轴 Stirring shaft；7—隔板 Clapboard；
8—壳体 Shell；9—搅拌桨叶 Stirring blade

5.7.6 其它设备 Other equipments
5.7.6.1 催化剂搅拌器 Catalyst mixer

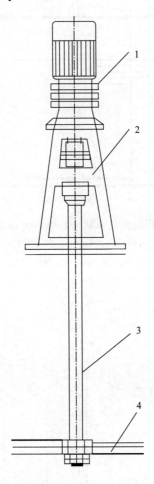

图 5.7.18 催化剂搅拌器结构图 Structure drawing of catalyst mixer
1—减速机 Speed reducer；2—支架 Support；
3—搅拌轴 Agitating shaft；4—搅拌桨叶 Stirring blade

5.7.6.2 传动装置 Pulse generator

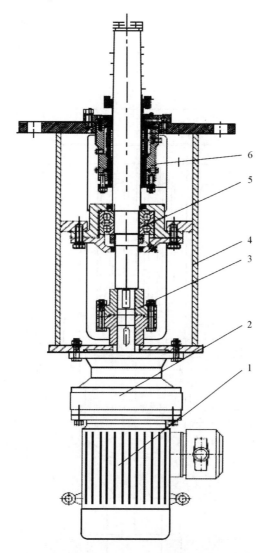

图 5.7.19 传动装置结构图 Structure drawing of pulse generator
1—电机 Motor；2—减速机 Speed reducer；3—联轴器 Coupling；4—支架 Support；
5—轴承 Bearing；6—机械密封 Mechanical seal

5.7.6.3 液氨过滤器 Liquid ammonia filter

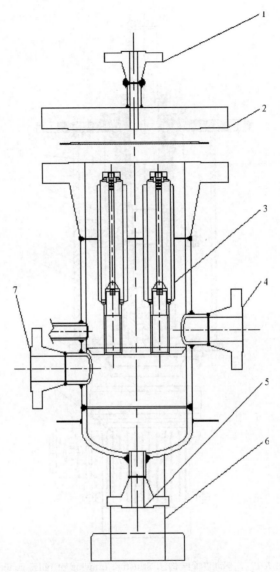

图 5.7.20 液氨过滤器结构图 Structure drawing of liquid ammonia filter
1—放空口 Vent；2—盖板 Cover plate；3—滤芯 Filter；4—液氨进口 Liquid ammonia inlet；
5—排放口 Discharge port；6—支座 Support；7—液氨出口 Liquid ammonia outlet

5.7.6.4 气氨过滤器 Ammonia gas filter

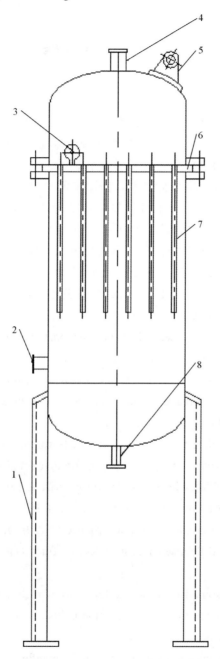

图 5.7.21 气氨过滤器结构图 Structure drawing of ammonia gas filter
1—支座 Support；2—气氨入口 Ammonia gas inlet；3—定位螺栓 Locating bolt；
4—气氨出口 Ammonia gas outlet；5—吊耳 Lug；6—管板 Tube sheet；
7—过滤元件 Filter element；8—排放口 Discharge port

6 芳烃 Arene

芳烃装置主要由二甲苯分馏装置、吸附分离装置、异构化装置、歧化及烷基转移装置、芳烃抽提装置等五套装置联合而成。二甲苯分馏装置是将重整装置脱戊烷塔底油、异构化汽提塔底油(脱庚烷塔)、歧化及烷基转移装置甲苯塔底油,经过精馏分离为C8A-OX-C8A+。吸附分离装置是将来自二甲苯分馏装置的C8A,经过模拟移动床进行吸附分离和精馏分离,得到PX产品。异构化装置是将吸附分离装置分离出的不含PX的C8A进行临氢异构反应和精馏分离,得到含PX的C8A。歧化及烷基转移装置是将TOL(C7A)和C9A在临氢环境中反应生成C8A,分馏后送二甲苯分馏装置进行分离处理。芳烃抽提装置是将二甲苯分馏装置重整油分馏塔顶组分进行萃取蒸馏,将芳烃和非芳烃分开。芳烃组分(主要是BZ和TOL)进入歧化及烷基转移装置进行回收,抽提副产品为NA。五套装置之间的主要物料关系如图6.0.1所示。

Aromatic plant is mainly composeds of xylene fractionation unit, adsorption separation unit, isomerization unit, disproportionation and transalkylation unit and aromatics extraction unit.

Xylene fractionation units rectify and separate depentanizer bottom oil of reformer, isomerization stripper bottom oil (de-heptane column) and bottom oil of disproportionation and transalkylation unit into C8A-OX-C8A+.

Adsorption separation unit gets PX product from adsorption separation and distillation separation of C8A getting from xylene fractionation unit through simulated moving bed.

Isomerization unit gets C8A with PX by hydroisomerization reaction and distillation separation of C8A without PX separating from adsorption separation unit.

Disproportionation and transalkylation unit produce C8A from the reaction of TOL(C7A) and C9A with hydrogen, and then separate it in Xylene fractionation unit after being fractionated.

The main function of aromatics extraction unit is to extractively distillate reformate of fractionation tower components from xylene fractionation unit so as to separate aromatic hydrocarbon product from non-aromatic hydrocarbon product.

The main material relationship among the five units mentioned above is illustrated in Figure 6.0.1.

6 芳烃 Arene

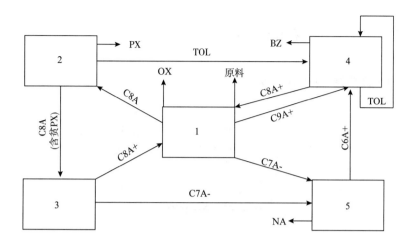

图 6.0.1 芳烃联合装置物料关系图
Diagram of aromatics material relationship

1—二甲苯分馏单元 Xylene fractionation unit；2—吸附分离单元 Adsorption separation unit；3—异构化单元 Isomerization unit；
4—岐化烷基转移单元 Disproportionation transalkylation unit；5—芳烃抽提单元 Microcoulometric yuan bill of lading

6.1 二甲苯分馏 Xylene fractionation

6.1.1 白土反应器 Clay reactor

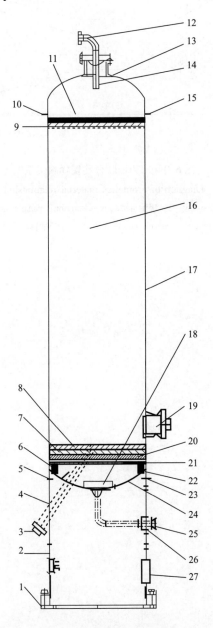

图 6.1.1 白土反应器结构图 Structure drawing of clay reactor
1—底板 Anchor；2—裙座 Skirt；3—白土卸料口 Clay discharge port；4—管线引出口 Pipeline outlet；
5、23—排气管 Exhaust pipe；6、9、11、22—瓷球 Porcelain ball；7、8、20、21—瓷砂 Porcelain sand；
10、15—保温支承圈 Insulation support ring；12—进料口 Feed inlet；13、19—人孔 Manhole；14—分配器 Distributor；
16—白土 Clay；17—筒体 Cylinder；18—出口挡板 Outlet damper；24—封头 Head；
25—出料口 Discharge outlet；26—补强圈 Reinforcing ring；27—检修口 Access hole

6.1.2 塔设备 Column equipment

6.1.2.1 二甲苯蒸馏塔 Xylene column

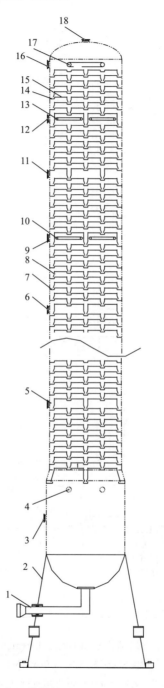

图 6.1.2 二甲苯蒸馏塔结构图 Structure drawing of xylene column

1—出料口 Discharge outlet；2—裙座 Skirt；3、5、6、11—人孔 Manhole；4—气相入口 Gas phase inlet；
7—受液盘 Liquid receiving pot；8—降液板 Downcomer plate；9、12—进料口 Feed inlet；
10、13—进料分布管 Distribution feed pipe；14—支承圈 Support ring；15—塔盘 Tray；16—回流口 Reflux inlet；
17—回流分布管 Reflux distribution pipe；18—气相出口 Gas phase outlet

6.1.2.2 二甲苯再蒸馏塔 Xylene rerun column

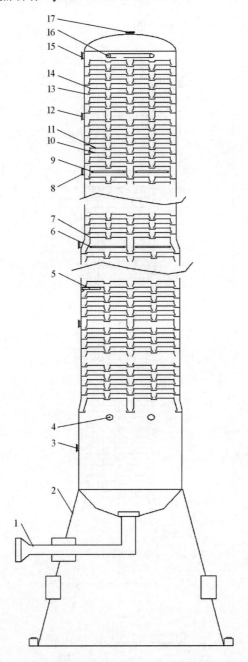

图 6.1.3 二甲苯再蒸馏塔结构图 Structure drawing of xylene rerun column

1—出料口 Discharge outlet；2—裙座 Skirt；3、12—人孔 Manhole；4—气相入口 Gas phase inlet；
5—支承梁 Supporting beam；6、9—进料分布管 Feed distribution pipe；7—降液板 Downcomer board；
8—进料口 Feed inlet；10、11—塔盘 Tray；13—受液盘 Liquid receiving pot；14—塔板支承件 Plates support；
15—回流入口 Reflux inlet；16—回流分布管 Reflux distribution pipe；17—气相出口 Gas phase outlet；

6.1.2.3 邻二甲苯塔 O-xylene column

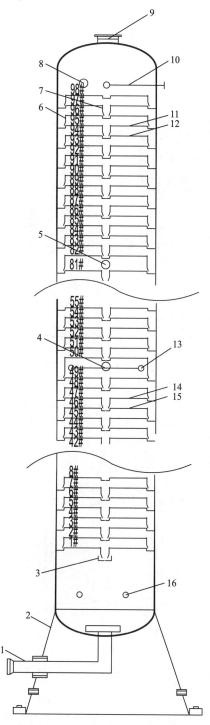

图 6.1.4 邻二甲苯塔结构图 Structure drawing of O-xylene column

1—出料口 Discharge outlet；2—裙座 Skirt；3—液封盘 Liquid seal disk；4、5、8—人孔 Manhole；
6—降液板支承件 Support of downcomer plate；7—降液板 Downcomer plate；9—气相出口 Gas phase outlet；
10—回流管 Return pipe；11、12、14、15—塔盘 Tray；13—进料口 Feed pipe；16—气相入口 Gas phase inlet

6.1.2.4 塔设备主要附件 Main accessories of column

塔设备的附件主要有塔盘、浮阀、进料管、塔盘支承件、卡子和降液管等。

Main accessories of column include tray, float valve, feeding pipe, tray support, clip parts and downcomer pipe, etc.

1. 塔盘 Tray

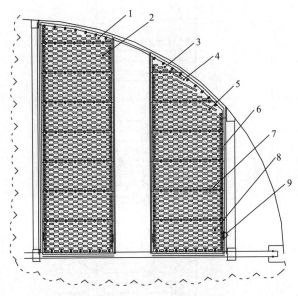

图 6.1.5　塔盘 Separated tray structure

1—弓形板 Arched plate；2—矩形板 Rectangular plate；3—角形板 Angled plate；
4—圆弧角形板 Arc-shaped angled plate；5—切角矩形板 Notching rectangular plate；
6—卡子 Clip；7—螺栓连接件 Bolt connector；8—通道板 Channel plate；9—把手 Hand grip

2. 回流分布管 Reflux distribution pipe

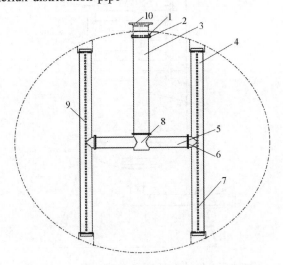

图 6.1.6　回流分布管 Reflux distribution pipe

1—螺栓 Bolt；2、6—法兰 Flange；3—引入管 Inlet pipe；4—分布管 Distribution pipe；
5—分流管 Manifold；7—U 形螺栓 U bolt；8—排液孔(开口向下) Liquid drainage hole (Opening downwards)；
9—分配孔(开口向下) Distribution hole (Opening downwards)；10—进料管 Feeding pipe

3. 进料分布管 Feed distribution pipe

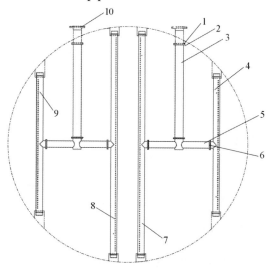

图 6.1.7　进料分布管 Feed pipe distribution

1—螺栓 Bolt；2、6—法兰 Flange；3—引入管 Inlet pipe；4、7—分布管 Distribution pipe；
5—分流管 Manifold；8—分配孔（开口向下）Distribution hole（Opening downwards）；
9—排液孔（开口向下）Liquid drainage hole（Opening downwards）；10—进料管 Feeding tube

4. 塔盘支承件 Tray support parts

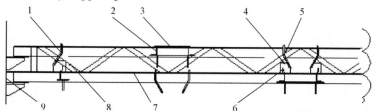

图 6.1.8　塔盘支承件 Tray support parts

1—两侧降液板 Downcomer side panels；2—偏心受液板支承梁 Eccentric liquid plate supporting beams；
3—偏心受液板 Eccentric liquid plate；4—固定板 Fixed plate；5—中间受液板 Middle liquid plate；
6—两侧入口堰 Inlet weir on both sides；7—塔盘 Tray；8—桁架 Girder；9—塔壁 Column wall

5. 卡子 Clip

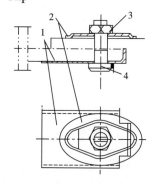

1—卡盘 Chuck；2—椭圆垫板 Elliptic gasket；3—螺母 Nut；4—圆头螺栓 Button head screw

图 6.1.9　卡子 Clip assemble structure

6. 降液管 Downcomer plate

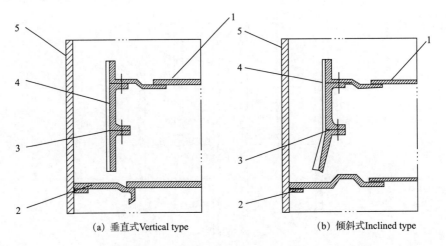

(a) 垂直式 Vertical type　　　　　(b) 倾斜式 Inclined type

图 6.1.10 降液管 Types of downcomer

1—塔盘 Tray；2—受液盘 Liquid receiving pot；3—紧固部件 Fastening parts；
4—降液管 Downcomer plate；5—塔壁 Column wall

7. 受液盘 Liquid receiving pot

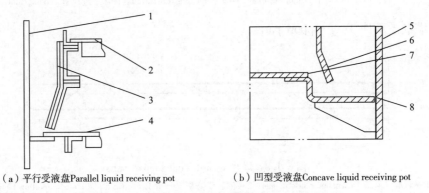

(a) 平行受液盘 Parallel liquid receiving pot　　　　　(b) 凹型受液盘 Concave liquid receiving pot

图 6.1.11 受液盘结构 Parallel liquid receiving structure

1、5—塔壁 Column wall；2、7—塔盘 Tray；
3、6—降液板 Downcomer plate；4、8—受液盘 Liquid receiving pot

8. 人孔 Manhole

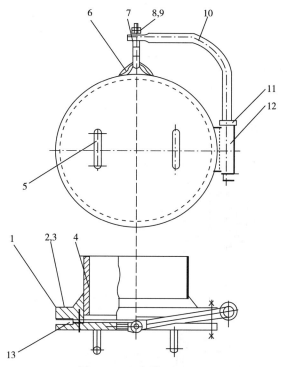

图 6.1.12　人孔 Manhole

1—法兰 Flange；2—紧固螺栓 Fastening bolt；3、9—螺母 Nut；
4—接管 Connecting pipe；5—把手 Handrail；6—吊耳 Lug；7—吊钩 Hook；8—垫圈 Pad；
10—悬臂 Cantilever；11—压盖 Gland；12—旋转轴 Rotation axis；13—人孔盖 Manhole cover

9. 裙座 Skirt

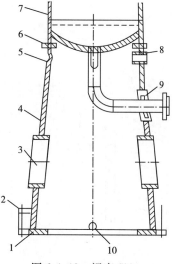

图 6.1.13　裙座 Skirt

1—基础环板 Foundation ring plate；2—螺栓座 Bolt seat；3—检修口 Access hole；4—裙座筒体 Skirt cylinder；
5—无保温时排气口 Air vent when no insulation；6—保温支承圈 Insulation support ring；7—塔体 Column body；
8—有保温时排气口 Air vent when insulation；9—引出管通道 Educing pipe channel；10—排液孔 Liquid drainage hole

10. 塔顶吊柱 Davit

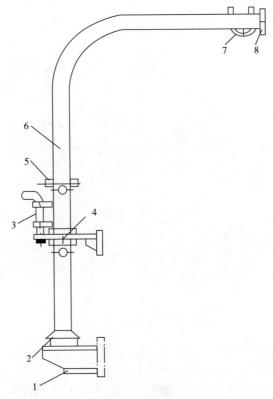

图 6.1.14　塔顶吊柱 Top davit structure

1—支架 Support；2—防雨罩 Rain cover；3—固定销 Fixed pin；4—导向板 Guiding plate；
5—手柄 Handle；6—吊柱管 Davit pipe；7—吊钩 Hook；8—挡板 Baffle

11. 保温结构 Insulation

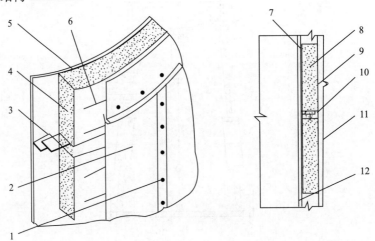

图 6.1.15　保温结构图 Structure drawing of column hot insulation

1、11—防潮层 Moisture proof layer；2、7—保护层 Protection layer；3、10—保温支承圈 Hot insulation support ring；
4、8—保温层 Hot insulation layer；5、12—塔外壁 Column outer wall；6、9—捆扎铁丝 Packing iron wire

6.1.3 加热炉 Heating furnace

6.1.3.1 二甲苯塔再沸炉 Xylene column reboiler furnace

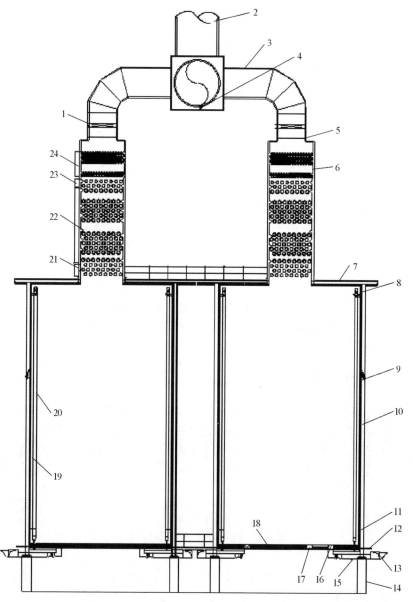

图 6.1.16 二甲苯塔再沸炉结构图 Structure drawing of xylene column heating furnace
1—烟道挡板 Flue baffle；2—独立烟囱烟道 Unseparated chimney flue；3—保温层 Insulation；4—空气预热器烟道 Removing air preheater flue；5—烟气采样口 Flue gas sampling port；6—吹灰蒸汽接口 Soot blowing steam interface；
7—炉管吊装口 Furnace tube hatch；8—炉管吊钩 Furnace tube hook；9—防爆门 Explosion-proof door；
10、11—辐射段看火门 Fire watch door of radiant section；12—灭火蒸汽管 Extinguishing steam pipe；
13—快开风门 Quick open damper；14—混凝土支柱 Concrete support column；15—风道 Duct；
16—燃烧器 Burner；17—辐射室人孔 Radiation chamber manhole；
18—厚耐火砖与绝热浇注料 Thick with heat insulation castable refractory；
19—厚陶瓷纤维毯 Thick ceramic fiber blanket；20—辐射炉管 Radiation furnace tube；
21—对流光管 Pipe；22、24—对流钉头管 Studded tube；23—预留管排 Reserved pipe line

6.1.3.2 加热炉主要附件 Main accessories of furnace

加热炉的主要附件有辐射段炉管、内支承件、外支承件、炉衬结构、燃烧器、吹灰器和烟道挡板等。

The main accessories of furnace are furnace tube, internal support, external support, lining structure, burners, soot blowers and flue baffle, etc.

1. 辐射段炉管 Furnace tube

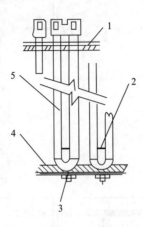

图 6.1.17 辐射段炉管 Furnace tube
1—炉肩 Furnace shoulder；2—炉管支架 Furnace tube holder；
3—底部管支架 Bottom tube holder；4—炉底 Bottom；5—立式(垂直)炉管 Vertical tube

2. 内支承件 Inner supports

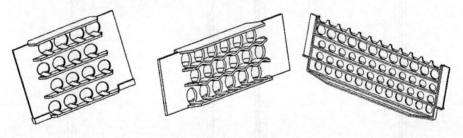

图 6.1.18 管板支承件 Tube sheet

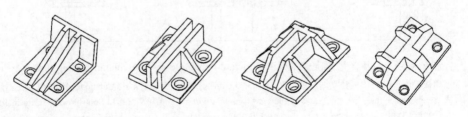

图 6.1.19 支架 Bracket

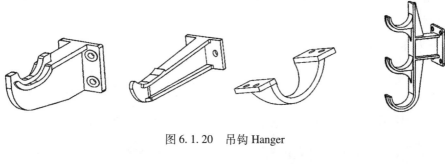

图 6.1.20 吊钩 Hanger

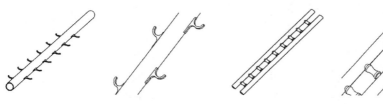

图 6.1.21 管吊架 Pipe hanging bracket

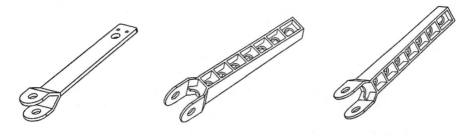

图 6.1.22 吊架 Hanging bracket

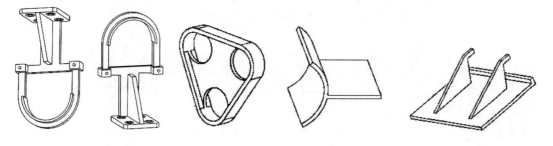

图 6.1.23 拉钩 Drag hook   图 6.1.24 管托 Tube bracket   图 6.1.25 托砖板 Shelf plate

3. 外支承件 General external supports

1) 弹簧 Spring

(1) 变力弹簧 Variable spring

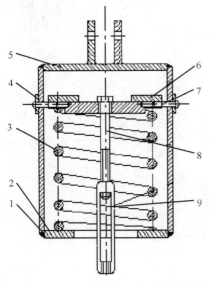
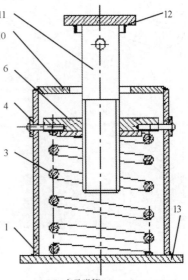

(a) 吊力弹簧 Lifting force spring　　　　(b) 支承弹簧 Supporting spring

图 6.1.26　变力弹簧 Variable spring

1—筒体 Cylinder；2—下盖板 Under cover plate；3—弹簧 Spring；4—加强板 Reinforcing plate；5—上盖板 Lug cover plate；6—中压板 Medium-pressure plate；7—固定销 Fixed pin；8—拉杆螺栓 Hanger bolt；9—花兰螺丝 Turnbuckle；10—顶板 Roof；11—承重柱 Load-bearing column；12—承重板 Loading board；13—底板 Under cover plate

(2) 恒力弹簧 Constant spring

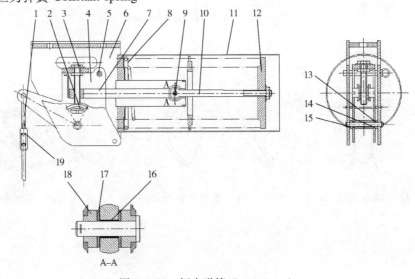

图 6.1.27　恒力弹簧 Constant spring

1—载荷螺栓 Loading bolt；2—位移指示牌 Displacement sign；3—调整螺栓 Adjusting bolt；4—回转框架 Rotary frame；5—固定销 Fixed pin；6—固定框架 Fixed frame；7—拉板 Pull plate；8—弹簧 Spring；9—联接轴 Connecting shaft；10—拉杆螺栓 Rod bolt；11—筒体 Cylinder；12—压板 Platen；13—主轴 Shaft；14—调整垫圈 Adjusting gasket；15—垫板 Plate；16—无润滑轴承 Non-lubricating bearings；17—尼龙垫片 Nylon gasket；18—滚轮 Roller；19—花篮螺栓 Turnbuckle

2）支吊架 Hangers

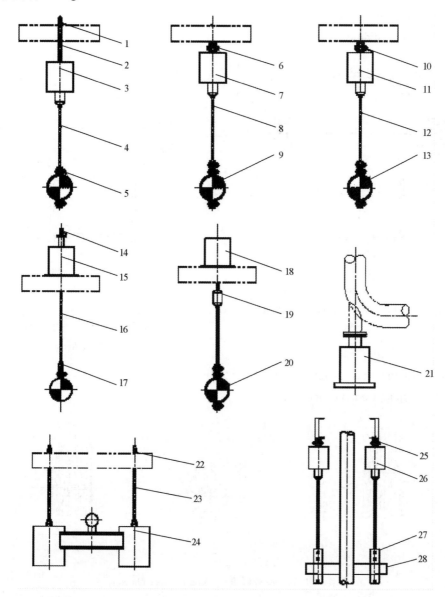

图 6.1.28　弹簧吊架 Spring hangers

1、14、22—螺母 Nut；2、16、23—双头螺纹吊杆 Double thread derrick；
3、27—A 型可变弹簧吊架 Variable spring type A；4、8、12—吊环型吊杆 Eyed rod；
5—双螺栓管夹 Double bold pipe clamp；6、25—U 形吊耳 Clevis；
7、26—B 型可变弹簧吊架 Variable spring type B；9、13、20—三螺栓管夹 Three bold pipe clamp；
10—吊板 Plate；11—D 型可变弹簧吊架 Variable spring type D；15—C 型可变弹簧吊架 Variable spring type C；
17—吊耳 Lug；18—E 型可变弹簧吊架 Variable spring type E；19—花篮螺母 Buckle；
21—F 型可变弹簧吊架 Variable spring type F；24—G 型可变弹簧吊架；；28—耳轴管

3) 支座 Bearing

(1) 固定支座 Fixed bracket

(2) 滑动支座 Sliding bearing

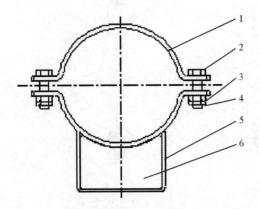 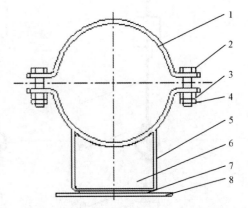

图 6.1.29　固定支座 Fixed bearing

1—管夹 Pipe clip；2—螺栓 Bolt；3—螺母 Nut；
4—扁螺母 Flat nut；5—支座 Support；6—筋板 Ribbed plate

图 6.1.30　滑动支座 Sliding bearing

1—管夹 Pipe clip；2—螺栓 Bolt；3—螺母 Nut；
4—扁螺母 Flat nut；5—支座 Support；
6—筋板 Ribbed plate；7—滑动板 Sliding board；
8—底板 Anchor

4. 炉衬 Lining construction and types for furnace

1) 陶瓷纤维毯 Ceramic fiber blanket

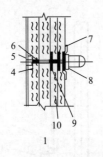 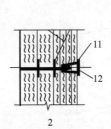 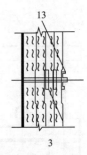

图 6.1.31　陶瓷纤维毯 Ceramic fiber blanket

1—压盖螺母固定结构 Stud anchor protected with ceramic cup nut；2—陶瓷杯转卡固定结构
Stud anchor protected with ceramic cup-lock filled with moldable ceramic fiber；3—转卡压盖固定结构
Stud anchor with turn clip；4—背衬 Back-up layer；5—专用螺母 Special nut；6—锚固钉 Twist stud anchor；
7—专用垫片 Special washer；8—陶瓷杯螺母 Ceramic cup nut；9—快速卡子 Speed clip；10—陶瓷纤维毯
Ceramic fiber blanket；11—陶瓷杯 Ceramic cup-lock；12—陶瓷杯盖 Ceramic cup-lock cover；13—旋转卡子 Turn clip

6 芳烃 Arene

2) 辐射室砌砖结构 Radiant chamber brick

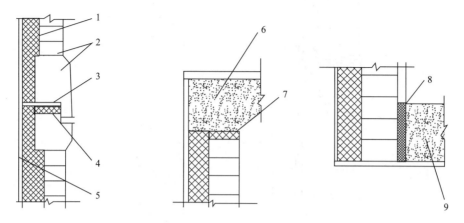

图 6.1.32 辐射室砌砖结构 Structure of radiant chamber brick
1—隔热层 Back-up layer；2—耐火砖 Firebrick；3—托板 Shelve for support；
4、7、8—陶瓷纤维毯 Ceramic fiber blanket；5—表面钢板 Surface steel plate；
6—炉顶衬里 Top lining；9—炉底衬里 Bottom lining

3) 拉砖 Tieback brick

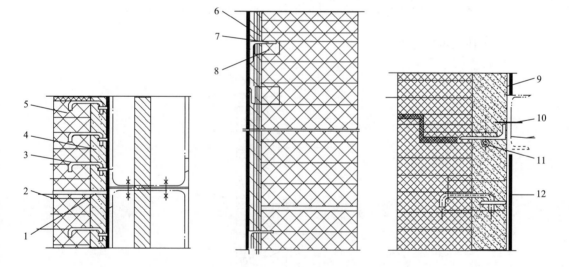

图 6.1.33 拉砖结构 I Structure of tieback brick I
1—托板 Shelve for support；2—陶瓷纤维毯 Ceramic fiber blanket；3—拉砖用锚固钉 Tieback brick anchor；
4—隔热层 Back-up layer；5—耐火砖 Firebrick；6—拉杆 Tie rod；7—垫圈 Washer；
8—拉杆座 Beam support for tie rod；9—挂(拉吊)砖架 Tieback brick hanger；
10—支承板 Lintel plate；11—螺栓和螺母 Bolt and nut；12—锚板 Tieback plate

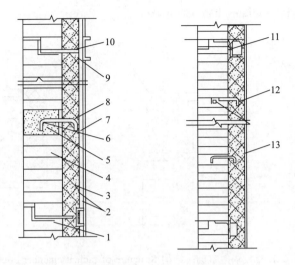

图 6.1.34 拉砖结构 Ⅱ Structure of tieback brick Ⅱ

1—陶瓷纤维毯 Ceramic fiber blanket；2—密封涂层和表面钢板 Seal coated or surface steel plate；
3—隔热层 Back-up layer；4—耐火砖 Firebrick；5—轻质耐热衬里 Light heat resistance lining；
6—拉杆 Tie rod；7—钢管拉杆座 Steel pipe tie rod seat；8—拉砖用锚固钉 Tieback brick anchor；
9—螺栓和螺母 Bolt and nut；10—托板 Shelve for support；11—角钢托板 Angle steel pallet；
12—角钢拉杆座 Angle steel tie rod seat；13—表面钢板 Surface steel plate

4) 挂砖 Hanging brick

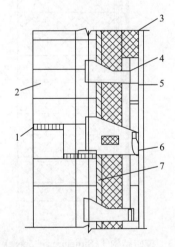

图 6.1.35 挂砖结构 Structure of hanging brick

1—陶瓷纤维毯 Ceramic fiber blanket；2—异型挂砖 Deformed tieback brick；
3—隔热层 Back-up layer；4—挂砖架 Brick hanging frame；5—表面钢板 Surface steel plate；
6—砖架梁 Brick girder；7—密封层 Sealing layer

6 芳烃 Arene

5) 吊砖 Hanging tieback brick

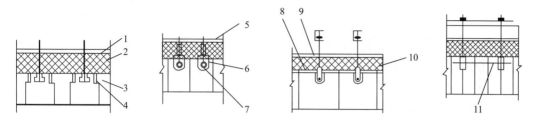

图 6.1.36 吊砖结构图 Structure drawing of hanging tieback brick

1—密封层 Sealing layer；2、10—隔热层 Back-up layer；3—耐火砖 Firebrick；4—吊砖架 Brick hoisting frame；
5—石棉板 Asbestos plate；6—吊杆 Suspender；7—螺栓 Bolt；8—石棉板 Asbestos plate；9—抹面层 Wipe surface；
10—竖吊杆 Suspender；11—拉杆 Tie rod

6) 浇注料炉衬 Castable lining

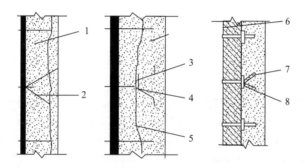

图 6.1.37 浇注炉衬结构图 Structure drawing of castable lining

1—浇注料 Castable；2—V 形锚固钉 V type clip castable anchor；3—捆绑铁丝 Bound metal wire；
4—Y 形锚固钉 Y type clip castable anchor；5—铁丝网 Wire mesh；6—隔热层 Back-up layer；
7—双层结构用 Y 型锚固钉 Y type clip anchor for double layer；8—自锁垫圈 Self-locking washer

5. 油气联合燃烧器 Combination gas and oil burner

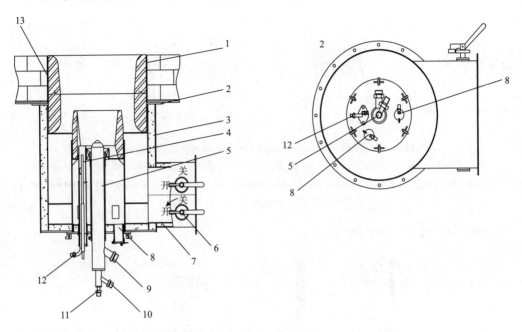

图 6.1.38　油气联合燃烧器结构图 Structure drawing of combination gas and oil burner

1—火盆砖 Brazier brick；2—螺栓 Bolt；3—筒体 Cylinder；4—长明灯 Long light；5—油气联合枪 Oil and gas joint rod；6—通风手柄 Ventilation handle；7—风道 Air channel；8—看火孔 Kiln eye；9—瓦斯入口 Gas inlet；10—雾化蒸汽入口 Atomizing steam inlet；11—燃料油入口 Fuel inlet；12—长明灯入口 Long light inlet；13—耐热纤维 Heat-resistant fiber

6. 吹灰器 Blower

1) 旋转式吹灰器 Rotary blower

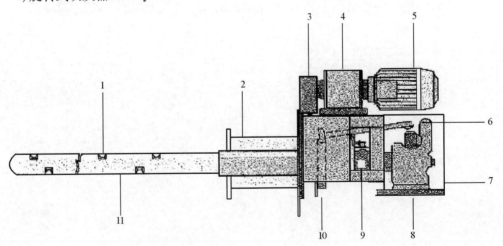

图 6.1.39　旋转式吹灰器结构图 Structure drawing of rotary blower

1—喷嘴 Nozzle；2—墙箱 Wall box；3—驱动链轮 Drive sprocket；4—齿轮箱 Reduction gearbox；5—电机 Motor；6—开阀机构 Valve operating lever；7—防护罩 Safety cover；8—动截止阀 Automatic isolating valve；9—限位开关 Limit switch；10—凸轮盘 Cam disc；11—喷嘴管（单列或双列）Rotating multi-jet element（single or twin row）

6 芳烃 Arene

2) 伸缩式吹灰器 Retractable blower

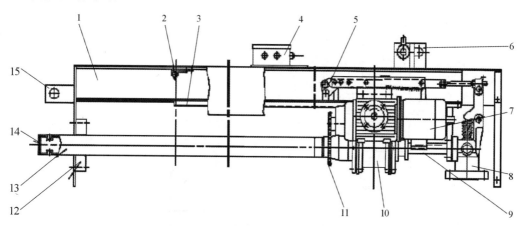

图 6.1.40 伸缩式吹灰器结构图 Structure drawing of Retractable blower

1—大梁 Girders；2—行程开关 Stroke switch；3—齿条 Rack；4—接线盒 Junction box；
5—开阀机构 Open valve structure；6—后部支架 Rear hanger；7—电机 Motor；
8—阀门 Valve；9—内管 Inner tube；10—行走箱 Walking Box；11—链轮 Sprocket；
12—前部托轮 Front supporting rollers group；13—外管 Outer tube；14—喷头 Nozzle；
15—前端板 Front plate

7. 烟道挡板 Flue baffle

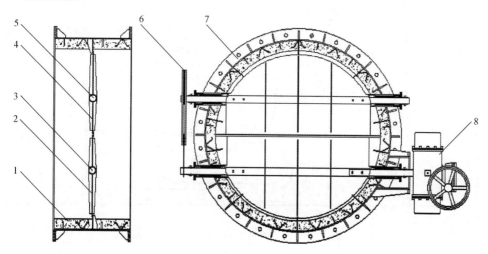

图 6.1.41 烟道挡板结构图 Structure drawing of flue baffle

1—筒体 Cylinder；2、4—挡板 Baffle；3、5—轴 Axes；6—风门调节机构 Baffle adjusting setup；
7—法兰 Flange；8—手动调节机构 Manual adjusting setup

6.1.3.3 余热回收系统 Waste heat recovery system

1. 余热回收系统 Waste heat recovery system

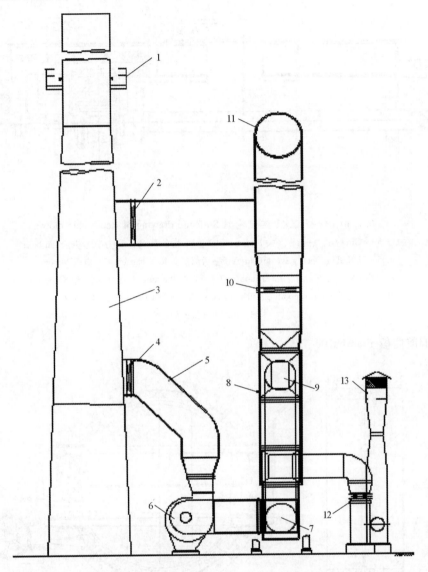

图 6.1.42 余热回收系统 Waste heat recovery system

1—操作平台 Operating platform；2—烟道旁路挡板 Flue bypass damper；3—烟囱 Chimney；
4—引风机出口挡板 Induced draft fan outlet baffle；5—绝热浇注材料 Insulation casting material；
6—引风机 Induced draft fan；7—引风机入口 Induced draft fan inlet；8—热管式空气预热器 Heat pipe air preheater；
9—至加热炉风道 To furnace duct；10—入口挡板 Entry barrier；11—加热炉烟道 Furnace flue；
12—鼓风机挡板 Blower baffle；13—空气入口 Air inlet

6 芳烃 Arene

2. 热管空气预热器 Oblique heat-pipe air preheater

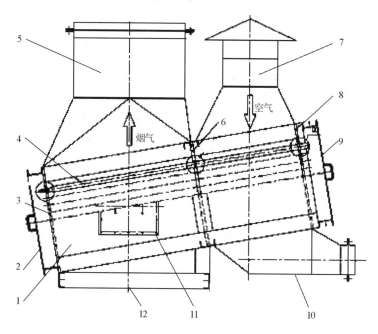

图 6.1.43 热管空气预热器 Oblique heat-pipe air preheater

1—箱体 Main body；2、9—安装门 Loading door；3—热端管板 Hot side tube plate；4—热管 Heat pipe；
5—烟气出口 Flue gas outlet；6—中间隔板 Middle tube plate；7—空气入口 Air inlet；
8—冷端管板 Cold side tube plate；10—空气出口 Air outlet；11—烟气入口 Flue gas inlet；12—支座 Support

6.1.4 换热设备 Heat exchange equipment

6.1.4.1 重整油塔蒸汽重沸器 Steam reboiler of reformate column

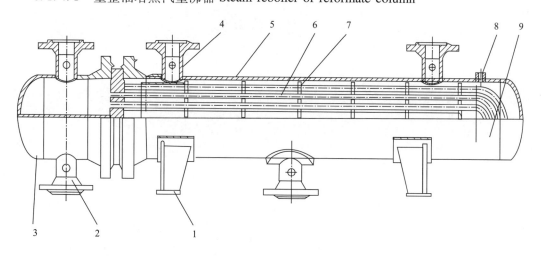

图 6.1.44 重整油塔蒸汽重沸器结构图 Structure drawing of steam reboiler of reformate column

1—支座 Support；2—管程接管 Tube pipe；3—管箱 Channel box；4—壳程接管 Shell pipe；5、9—筒体 Casing；
6—管束 Tube bundle；7—折流板 Impingement baffle；8—热电偶接口 Thermocouple interface

6.1.4.2 重整油塔顶产品后冷器 Reformate column overhead product aftercooler

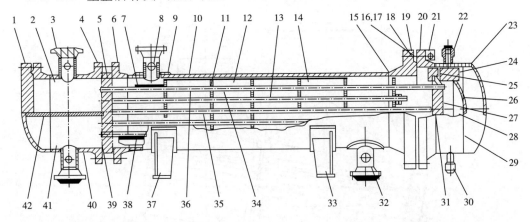

图 6.1.45 重整油塔顶产品后冷器结构图 Structure drawing of reformate column overhead product aftercooler

1—平盖 Flat cover；2、41—管箱 Channel box；3—接管法兰 Pipe flange；4—管箱法兰 Channel flange；
5—固定管板 Fixed tube sheet；6、19—筒体法兰 Shell hange；7—防冲挡板 Impingement baffle；
8—仪表接口 Instrument interface；9—补强圈 Reinforcing ring；10—筒体 Cylinder；11—折流板 Baffle plate；
12—旁路挡板 Bypass baffle；13—拉杆 Tie rod；14—定距管 Spacer tube；15—支持板 Supporting plate；
16—双头螺柱 Stud；17—螺母 Nut；18—外头盖垫片 Outside head cover gasket；20—外头盖法兰 Outer bonnet side flange；
21—吊耳 Lug；22—排气口 Exhaust port；23—外头盖封头 Outside head；24—浮头法兰 Floating head flange；
25—浮头垫片 Floating head gasket；26—浮头封头 Floating Head；27—浮动管板 Floating tubesheet；
28—浮头 Floating head；29—外头盖 Outer bonnet；30—排液口 Liquid outlet；31—钩圈 Hook ring；
32—接管 Connecting pipe；33—活动支座 Movable seat；34、36—换热管 Heat exchange tube；35—盲管 Dummy tube；
37—固定支座 Fixed support；38—滑道 Slideway；39—管箱垫片 Channel box gasket；
40—管箱筒体 Cylinder tube box；42—分程隔板 Pass partition plate

6.1.5 泵 Pump

6.1.5.1 重整油塔回流泵 Reformate tower reflux pump

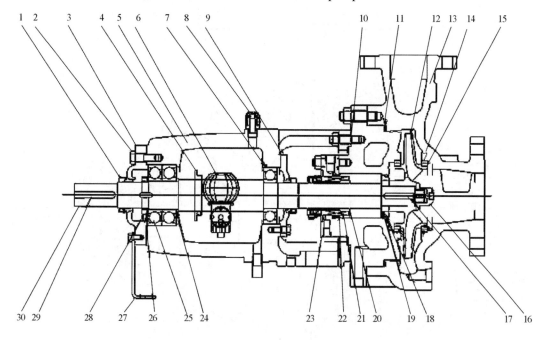

图 6.1.46 重整油塔回流泵结构图 Structure drawing of reformate tower reflux pump

1—防尘盘 Deflector；2—螺栓-螺母 Bolts and nuts；3、9—轴承压盖 Bearing cover；4—抛油盘 Oil flinger；
5—轴承箱 Bearing box；6—恒位油杯 Constant level oiler；7、25—球轴承 Ball bearing；8—放气塞 Air breather；
10—泵盖 Gland cover；11—密封垫 Grasket；12—叶轮 Wheel；13—泵体 Pump body；
14—泵体口环 Pump inlet ring；15—叶轮口环 Impeller wear ring；16—叶轮螺母 Impeller nut；17、29—键 Key；
18—喉部衬套 Throat bushing；19—定距轴套 Distance diece；20—机械密封轴套 Mechanical Sea Sleeve；
21—机械密封 Mechanical seal；22—O 形密封圈 O sealing ring；23—机封压盖 Seal end plate；
24—定距块 Distance piece；26—止动垫片 Gasket；
27—支架 Support；28—轴承螺母 Bearing nut；30—泵 Pump

6.1.5.2 二甲苯重沸炉泵 Xylene reboiler furnace pump

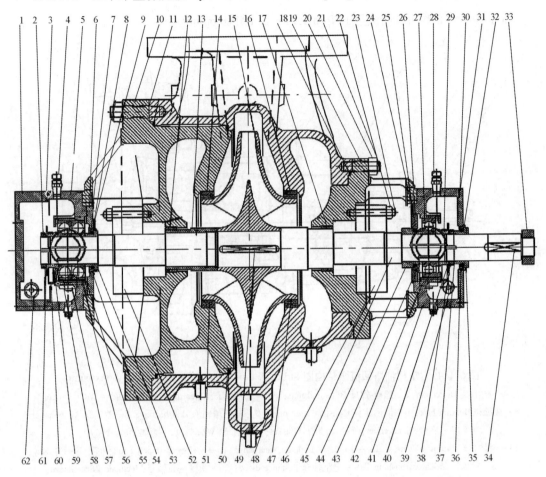

图 6.1.47 二甲苯重沸炉泵结构图 Structure drawing of xylene reboiler furnace pump

1、31—轴承盖 Bearing cover；2、37—轴承螺母 Bearing nut；3、7、26、30、36、54—O 形密封圈 O sealing ring；4、28—油过滤器 Oil filter；5、41—轴承箱 Bearing Box；6、29、59—轴承套 Intermediate bearing sleeve；8、9、11、18、24、25、50—垫片 Washer；10—泵压盖 Casing cover；12、17—轴套 Sleeve；13—叶轮螺母 Impeller nut；14、16—泵盖密封环 Casing wear ring；15—叶轮 Wheel；19、21、23—螺栓 Stud；20、22—六角螺母 Hexagon nut；27—滑动轴承 Anti-friction bearing；32、44、53—迷宫密封环 Labyrinth ring；33—轴螺母 Shaft nut；34、49—键 Key；35—迷宫密封内套 Labyrinth seal inner sleeve；38、61—锁紧垫圈 Lockwasher；39、62—冷却装置 Cooling device；40—甩油环 Oil ring；42—轴承座 Bearing seat；43、55、60—轴承压盖 Bearing gland；45—轴 Shaft；46、52—机械密封 Mechanical seal；47、51—叶轮口环 Impeller wear ring；48—蜗壳 Spiral case；56、58—止推轴承 Thrust bearing；57—油杯 Oiler

6.1.5.3 泵的主要零部件 Main components of the pump

泵的主要零部件有底座、轴承箱、轴、叶轮、密封环、隔板、衬套、轴套、平衡孔、平衡管、导叶和联轴器等。

The main components of the pump are pump base, bearing box, shaft, impeller, seal ring, partition, bushing, shaft sleeve, balance hole, balance tube, diffuser and coupling, etc.

1. 底座 Pump body base

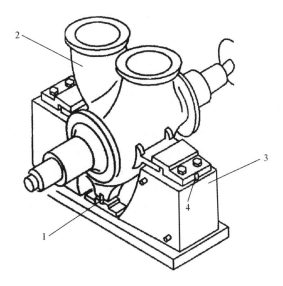

图 6.1.48　底座结构图 Structure drawing of pump body base
1—纵向导向键 Vertically oriented key；2—泵体 Pump body；3—底座 Base；
4—横向导向键 Horizontal orientation key

2. 轴承箱 Bearing box

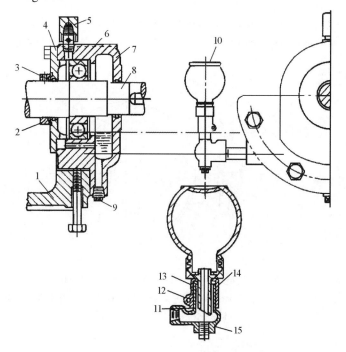

图 6.1.49　轴承箱结构图 Structure drawing of bearing box
1—箱体 Pump casing；2—挡水环 Retainer ring；3—排油槽 Oil scupper；4—轴承压盖 Bearing gland；5—通气帽 Vent cap；
6—滚动轴承 Rolling bearing；7—轴承箱 Bearing box；8—轴 Shaft；9—丝堵 Drain stopper；10—油杯 Oil cup；
11—液面 Level；12—锁紧螺母 Lock nut；13—通气孔 Vent；14—外套管 Outer bushing；15—内套管 Inner bushing

3. 泵轴 Pump shaft

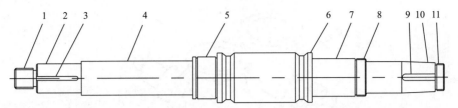

图 6.1.50　单级泵泵轴 Single-stage centrifugal pump shaft
1—叶轮螺母螺纹 Impeller nut thread；2—叶轮轴颈 Impeller shaft section；3、9—键槽 Key slot；
4—轴套轴颈 Sleeve shaft section；5—前轴承轴颈 Front bearing journal；6—轴肩 Shaft shoulder；
7—后轴承轴颈 Rear bearing journal；8—轴承螺母螺纹 Bearing nut thread；
10—半联轴器轴颈 Half coupling shaft；11—螺母螺纹 Nut thread

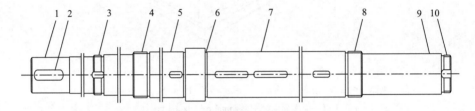

图 6.1.51　多级泵泵轴 Multi-stage centrifugal pump shaft
1—半联轴器轴颈 Half coupling shafts；2—键槽 Key slot；3、10—轴承螺母螺纹 Bearing nut thread；
4—轴套螺母螺纹 Sleeve nut thread；5—轴套轴颈 Sleeve shaft section；6—轴肩 Shaft shoulder；
7—叶轮轴段 Impeller shaft section；8—螺母螺纹 Nut thread；9—轴承轴颈 Bearing nut thread

4. 叶轮 Impeller

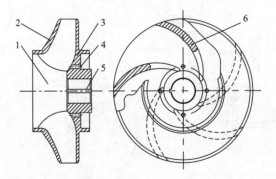

图 6.1.52　单吸闭式叶轮 Single-stage cantilever centrifugal pump impeller
1—轮毂 Wheel；2—前盖板 Front cover；3—后盖板 Rear cover；4—平衡孔 Balance hole；
5—键槽 Key slot；6—叶片 Vane

6 芳烃 Arene

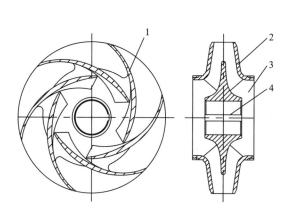

图 6.1.53 双吸闭式叶轮 Double suction impeller
1—叶片 Blade；2—盖板 Cover；
3—轮毂 Wheel；4—键槽 Key slot

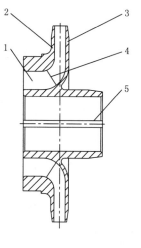

图 6.1.54 多级离心泵叶轮
Multi-stage centrifugal pump impeller
1—轮毂 Wheel；2—前盖板 Front cover；
3—后盖板 Rear cover；4—叶片 Vane；
5—键槽 Key slot

5. 密封环 Seal ring

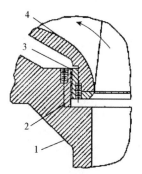

(a) 普通圆柱型 General cylindrical type

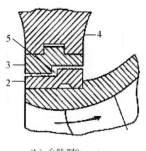

(b) 台阶型 Step type

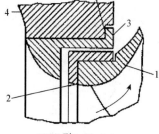

(c) 双L型 Double L type

图 6.1.55 密封环结构类型 Types of seal ring
1—壳体 Pump casing；2—壳体密封环 Casing sealing ring；
3—叶轮密封环 Impeller labyrinth seal ring；
4—叶轮 Wheel；5—过流室 Overflow room

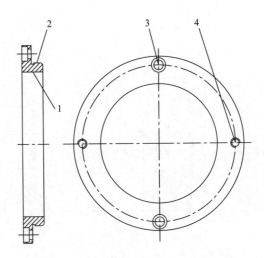

图 6.1.56 壳体密封环 Casing seal ring

1—承磨面 Grinding surface；2—配合面 Mating surface；
3—固定螺栓孔 Fixed screw hole；4—顶丝螺孔 Jack screw hole

6. 衬套 Bushing

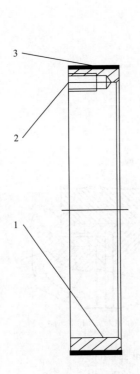

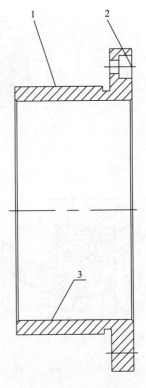

图 6.1.57 叶轮密封环 Impeller seal ring

1—配合面 Mating surface；
2—固定螺栓孔 Fixed screw hole；
3—承磨面 Grinding surface

图 6.1.58 隔板衬套 Diaphragm bushing

1—配合面 Mating surface；
2—固定螺栓孔 Fixed screw hole；
3—承磨面 Grinding surface

7. 轴套 Shaft sleeve

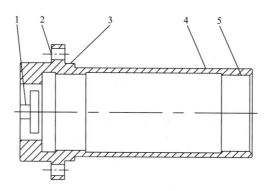

图 6.1.59　单级离心泵轴套 Sleeve of single-stage centrifugal pump

1—键槽 Key slot；2—固定螺栓孔 Fixed screw hole；3—密封座定位止口 Position spigot of seal block；
4—密封面 Seal surface；5—配合面 Mating surface

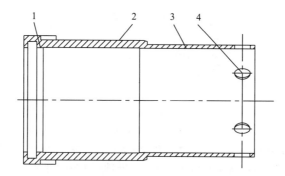

图 6.1.60　多级离心泵轴套 Sleeve of multi-stage centrifugal pump

1、3—配合面 Mating surface；2—密封面 Seal surface；4—防转孔 Anti-rotation role

8. 平衡装置 Balancing device

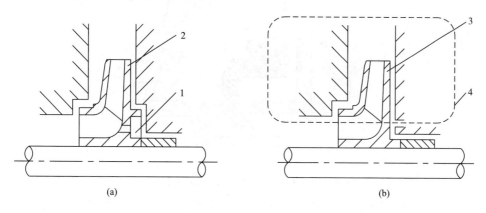

图 6.1.61　平衡孔和平衡管 Balance hole and balance tube

1—平衡孔 Balance hole；2、3—叶轮 Wheel；4—平衡管 Balance pipe

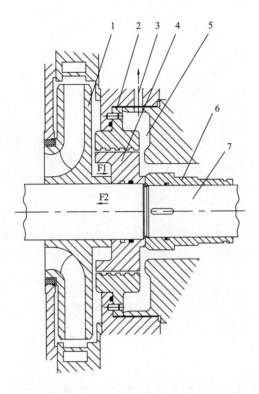

图 6.1.62 平衡鼓 Balance drum

1—末级叶轮 Final stage impeller；2—平衡套 Balance sleeve；3—平衡口 Balance port；
4—平衡鼓 Balance drum；5—平衡室 Balance room；6—轴套 Sleeve；7—轴 Shaft；
F1—平衡力 Balance force；F2—轴向力 Axial force

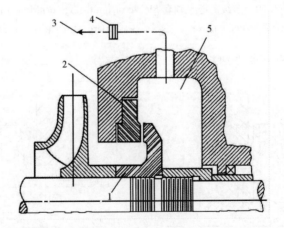

图 6.1.63 平衡盘 Balance disk

1—平衡盘 Balance disk；2—平衡座 Balance block；3—平衡管 Balance pipe；
4—限流孔板 Restriction orifice；5—平衡室 Balance room

9. 导叶 Diffuser

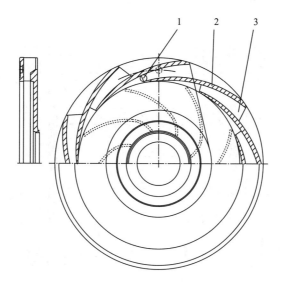

图 6.1.64 导叶 Diffuser

1—防转销 Hole for anti-rotating pin；2—叶片 Vane；3—流道 Flow channel

10. 隔板 Partition

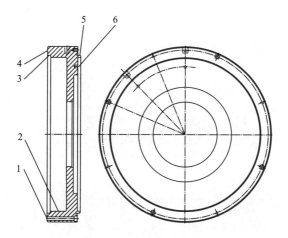

图 6.1.65 隔板 Partition

1—穿杆螺栓孔 Bolt hole for pole through；2—导叶配合面 Diffuser mating surface；
3、5—定位止口 Position spigot；4—密封面 Seal surface；
6—防转销孔 Hole for anti-rotating pin

11. 弹性膜片联轴器 Elastic membrane coupling

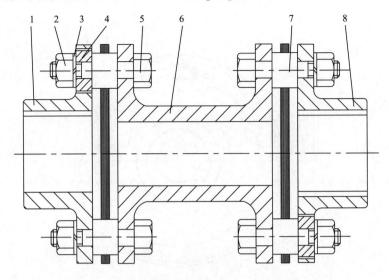

图 6.1.66 弹性膜片联轴器 Elastic membrane coupling

1、8—半联轴节 Half coupling；2—螺母 Nut；3—弹簧垫圈 Spring washer；
4—补偿垫块 Compensation pad；5—精密螺栓 Precision bolt；
6—中间节 Middle section；7—膜片 Membrane

12. 密封 Seal

1) 单端面机械密封 Single mechanical seal

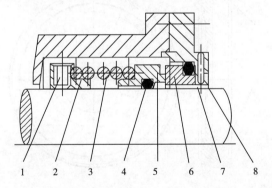

6.1.67 单端面机械密封结构图 Structure drawing of single mechanical seal

1—紧固螺钉 Tightening screw；2—弹簧座 Spring seat；3—弹簧 Spring；4、7—密封圈 seal ring；
5—动环 Rotating ring；6—静环 Static ring；8—防转销 Anti-rotating pin

2）双端面机械密封 Double mechanical seal

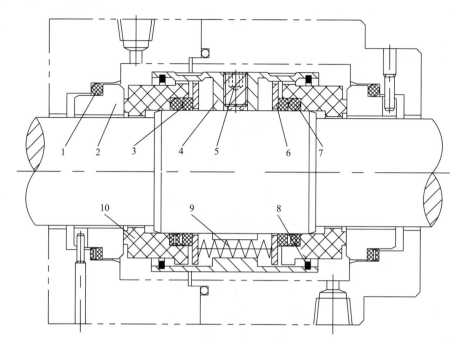

图 6.1.68 双端面机械密封结构图 Structure drawing of double mechanical seal
1—静环密封圈 Static seal ring；2—静环 Static ring；3—挡圈 Retaining ring；4—弹簧座 Spring seat；
5—紧定螺钉 Tightening screw；6—浮动环 Thrust ring；7—动环密封圈 Rotating seal ring；
8—卡环 Clamp ring；9—弹簧 Spring；10—动环 Rotating ring

3）双端面密封辅助系统 Double seal auxiliary system

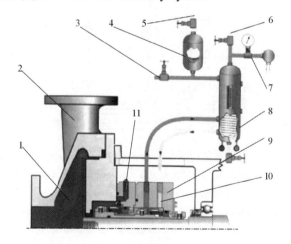

图 6.1.69 双端面密封辅助系统图 System diagram of double seal auxiliary system
1—介质 Medium；2—泵体 Pump body；3—阻塞液补充口 Blocking fluid supply port；
4—气胆式蓄压器 Bladder-type accumulator；5—气胆式充气连接口 Inflatable bladder-type interface；
6—泄压口 Relief port；7—压力表 Pressure gauge；8—密封油罐 Seal tank；9—机械密封 Mechanical seals；
10—阻塞液入口 Blocking fluid inlet；11—密封自冲洗液 Sealing self-flushing fluid

6.2 吸附分离 Adsorption separation

6.2.1 吸附塔 Adsorption column

6.2.1.1 吸附塔结构 Structure of adsorption column

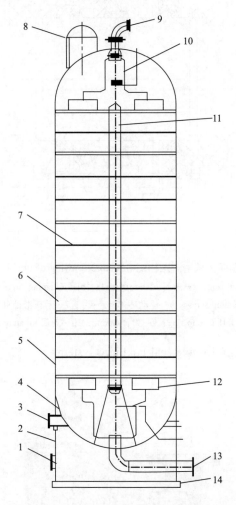

图 6.2.1 吸附塔结构图 Structure drawing of adsorption column
1—检修口 Access hole；2—裙座 Skirt；3—吸附剂出口 Adsorbent outlet；4—封头 Head；5—筒体 Cylinder；
6—吸附剂 Adsorbent；7—塔盘 Tray；8—顶部人孔 Top manhole；9—入料口 Feed inlet；
10—入料分布管 Feed distribution pipe；11—中心管 Central pipe；12—液体收集器 Liquid collector；
13—出料口 Feed outlet；14—底板 Anchor

6.2.1.2 吸附塔分解结构 Adsorption column breakdown structure

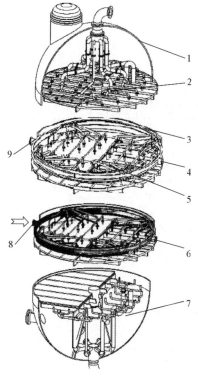

1—顶封头 Cover head；2—顶封头分布管 Cover head distribution pipe；3、6—环形室 Annular chamber；4—抽出/抽余出料口 Discharging \ raffinating outlet；5—中间格栅及分布管 Middle grille and its distributing tube；7—底封头及收集管 Bottom head and its distribution pipe；8、9—进料口 Feed inlet

图 6.2.2 吸附塔分解结构图
Exploded view of adsorption column

6.2.1.3 吸附塔主要部件 Main component of adsorption column

1. 顶封头 Cover head

1）顶封头分布管 Cover head distribution pipe

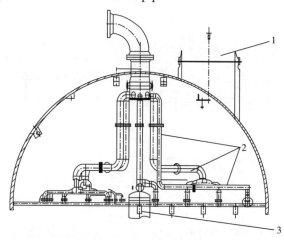

图 6.2.3 顶封头分布管 Cover head distribution pipe
1—封头人孔 Manhole head；2—顶封头分布管 Cover head distribution pipe；
3—中心管 Central tube

2) 顶封头俯视图 Top view of cover head

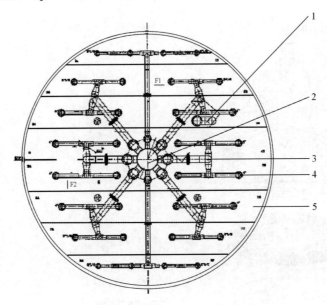

图 6.2.4 顶封头俯视图 Top view of cover head
1—平衡管 Balance pipe；2—分配器 Distributor；
3—进料管 Feeding pipe；4—进料口 Feed inlet；5—格栅 Grille；

3) 顶封头剖面 Cover head profile

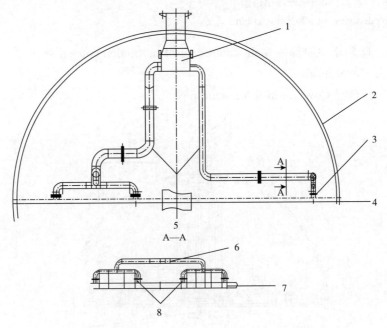

图 6.2.5 顶封头剖面图 Sectional view of cover head
1—分配器 Distributor；2—塔壁 Column wall；3—进料口 Feed inlet；4、7—顶格栅 Top grille；
5、6—进/出料管 Feeding inlet and outlet tube；8—格栅板接口 Grill plate interface

6 芳烃 Arene

4）顶封头配管 Cover head pipe

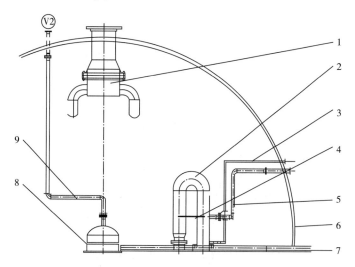

图 6.2.6　顶封头配管图 Piping diagram of cover head

1—分配器 Distributor；2—平衡管 Balance pipe；3—采样口 Sampling port；4—加强筋 Stiffener；
5—封头冲洗管 Head flush pipe；6—塔壁 Column wall；7—顶格栅 Top grille；8—中心管 Central tube；
9—中心管冲洗管 Flushing out center tube；

2. 底封头 Bottom head

1）底封头仰视图 Bottom view of bottom head

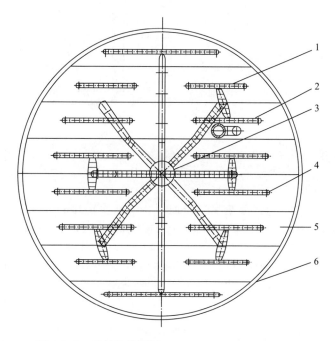

图 6.2.7　底封头仰视图 Bottom view of bottom head

1—物料出口管 Material outlet pipe；2—平衡管 Balance pipe；3—物料收集管 Material collection tube；
4—喷嘴 Nozzle；5—底格栅 Bottom grille；6—塔壁 Column wall

2) 底封头剖面图 Sectional view of bottom head

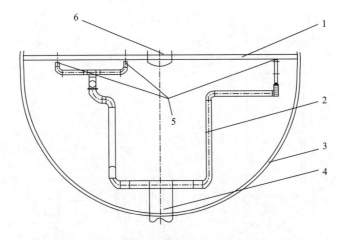

图 6.2.8　底封头剖面图 Sectional view of bottom head
1—底格栅 Bottom grille；2—物料出口管 Material outlet pipe；3—塔壁 Column wall；
4—物料收集管 Material collection tube；5—喷嘴 Nozzle；6—中心管 Central tube

3) 底封头管线图 Bottom head pipes

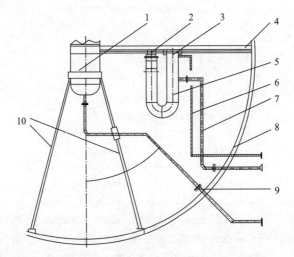

图 6.2.9　底封头管线图 Structure drawing of bottom head pipes
1—中心管 Central tube；2—平衡管入口 Balance tube inlet；3—平衡管出口 Balance tube outlet；
4—底格栅 Bottom grille；5—平衡管 Balance pipe；6—采样管 Sampling tube；
7—封头冲洗管 Head flush inlet pipe；8—塔壁 Column wall；9—中心管冲洗管 Flushing center tube；
10—中心管支座 Center tube support base

3. 环形室 Annular chamber

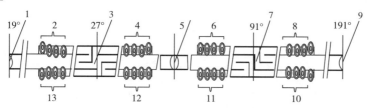

图 6.2.10　环形室结构图 Structure drawing of annular chamber

1、9—抽出液出料口 Extract discharging outlet；2、4、6、8—抽余液出料口 Residual liquid discharging outlet；
3、7—环形室 Annular chamber；5—进料口 Feed inlet；10、11、12、13—解析剂进口 Analytical agent inlet

4. 床层格栅 Bed grille

1) 中间格栅板 Middle grating plate

图 6.2.11　中间格栅板外形图 Outline drawing of middle grating plate

1—格栅板 Grating plate；2、5—出料口 Outlet；3—进料口 Inlet；
4—中心管 Center pipe

2) 奇数层格栅板 Odd layer grating plate

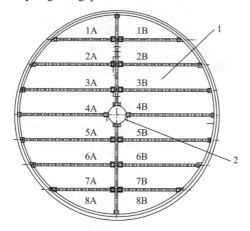

图 6.2.12　奇数层格栅板 Odd layer grating plate

1—格栅板 Grating plate；2—中心管 Central pipe

3）偶数层格栅 Even layer grating

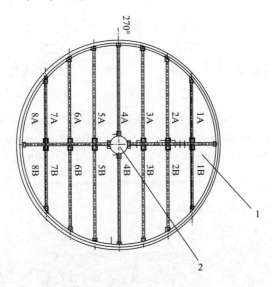

图 6.2.13　偶数层格栅板 Even layer grating

1—格栅板 Grating plate；2—中心管 Central pipe

4）顶格栅板 Top grating plate

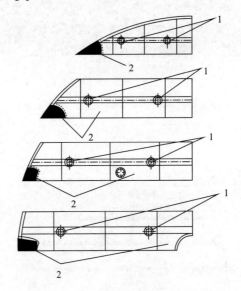

图 6.2.14　顶格栅板 Top grating plate

1—进料管 Feeding pipe；2—格栅板 Grating plate

5)底格栅板 Bottom grating plate

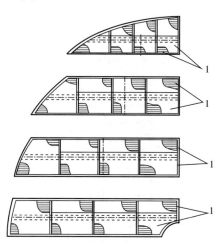

图 6.2.15　底格栅板 Bottom grating plate
1—格栅板 Grating plate

6)中间格栅板 Middle grating plate

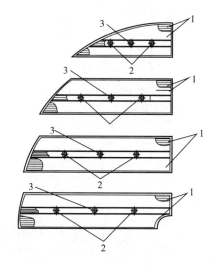

图 6.2.16　中间格栅板 Middle grating plate
1—格栅板 Grating plate；2—出料口 Outlet；3—进料口 Inlet

6.2.2 分馏塔 Fractionator

6.2.2.1 抽出液塔 Aspiration column

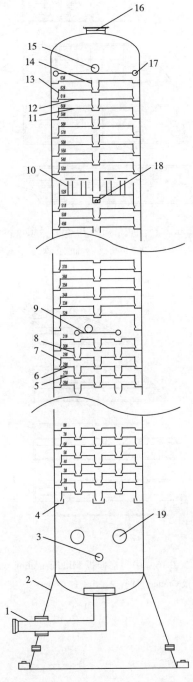

图 6.2.17 抽出液塔结构图 Structure drawing of aspiration column

1—液相出口 Liquid outlet；2—裙座 Skirt；3、15—人孔 Manhole；4—受液盘 Liquid receiving pot；
5、6、11、12—塔盘 Tray；7、13—降液板 Downcomer plate；8、14—降液板支承件 Downcomer plate weldments；
9—进料口 Feed inlet；10—集油箱 Oil collection tank；16—气相出口 Gas phase outlet；17—回流入口 Reflux inlet；
18—中间抽出口 Intermediate pumping outlet；19—汽相入口 Gas phase inlet

6.2.2.2 抽余液塔 Raffinate column

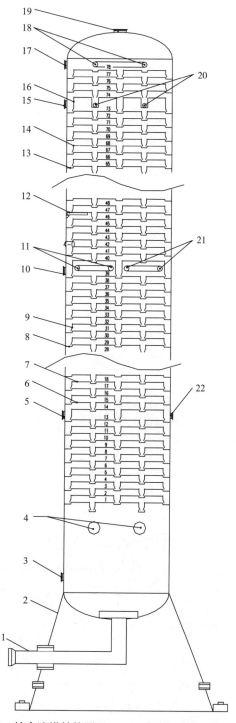

图 6.2.18 抽余液塔结构图 Structure drawing of raffinate column

1—液相出口 Liquid outlet；2—裙座 Skirt；3、5、10、15、17—人孔 Manhole；4—汽相入口 Gas phase inlet；6、7—塔盘 Tray；8—受液盘 Liquid receiving pot；9、14、16—降液板 Downcomer plate；11、21—进料分布管 Feed distribution pipe；12—桁架梁 Lattice girder；13—降液板支承件 Support of downcomer plate；18—回流分布管 Reflux distribution pipe；19—气相出口 Gas phase outlet；20—侧线出口 Side outlet；22—安全阀接口 Safety valve interface；

6.2.2.3 PX 成品塔 PX finished product column

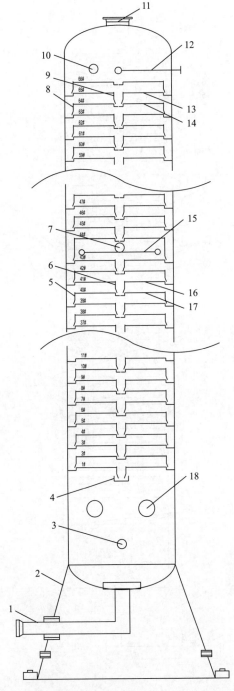

图 6.2.19 PX 成品塔结构图 Structure drawing of PX finished product column
1—液相出口 Liquid outlet；2—裙座 Skirt；3、7、10—人孔 Manhole；4—液封盘 Liquid seal disk；
5、6、8、9—降液板 Downcomer board；11—气相出口 Gas phase outlet；12—塔顶回流口 Overhead reflux opening；
13、14、16、17—塔盘 Tray；15—进料口 Feed inlet；18—气相入口 Gas phase inlet

6.2.3 回流罐 Reflux tank

6.2.3.1 PX 成品塔回流罐 Reflux drum of PX finished product column

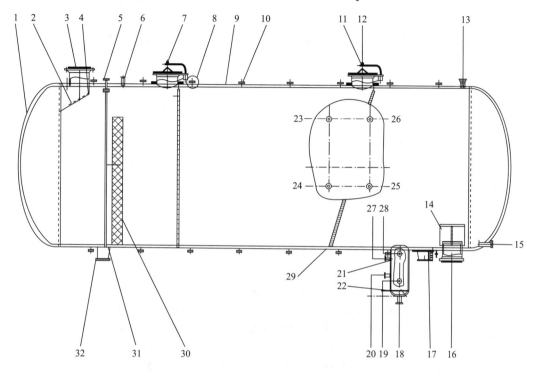

图 6.2.20 PX 成品塔回流罐结构图 Structure drawing of PX finished product column

1、22—封头 Head；2—防冲挡板 Impingement baffle；3、5—进料口 Feed inlet；4—导流板 Deflector；
6、8—放空口 Vent；7、11—人孔 Manhole；9—筒体 Cylinder；10、15—备用口 Alternate opening；
12—人孔盖转臂 Manhole cover jib；13—气体出口 Gas outlet；14—防涡器 Vortex breaker；16—油出口 Oil outlet；
17—固定支座 Fixed support；18—水包出口 Water drum outlet；19、28—玻璃板液位计 Glass liquid level meter；
20、25、26、27—液位计接口 Liquid level gauge interface；21—水包筒体 Water drum chamber；
23、24—变送器接口 Transmitter interface；29—梯子 Ladder；30—丝网除沫器 Wire mesh demister；
31—立板 Vertical plate；32—活动支座 Shifting bearing

6.2.3.2 脱庚烷塔回流罐 Deheptane reflux tank

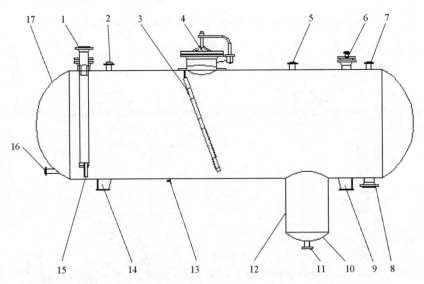

图 6.2.21 脱庚烷塔回流罐结构图 Structure drawing of deheptane reflux tank
1—进料口 Feed inlet；2—压力平衡口 Pressure balance port；3—梯子 Ladder；4—人孔 Manhole；
5—放空口 Vent；6—压力计接口 Manometer interface；7—气体出口 Gas outlet；
8—出料口 Discharge outlet；9、14—支座 Support；10、17—封头 Head；11—排水口 Drain；
12—筒体 Cylinder；13—保温支承圈 Insulation support ring；
15—加强角钢 Stiffening angle；16—备用口 Alternate opening

6.2.3.3 二甲苯塔回流罐 Xylene reflux tank

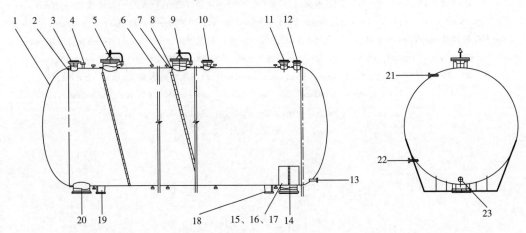

图 6.2.22 二甲苯塔回流罐结构图 Structure drawing of xylene reflux tank
1—封头 Head；2—筒体 Cylinder；3—旁路接口 Bypass interface；4—放空口 Vent；5、9—人孔 Manhole；
6—保温支承圈 Reinforcing ring wing plate；7、16—补强圈 Reinforcing ring web plate；8—梯子 Ladder；
10—安全阀接口 Safety valve interface；11—平衡线接口 Balance line interface；12—气体出口 Gas outlet；
13、23—备用口 Alternate opening；14—油出口 Oil outlet；15—筋板 Reinforcing plate；
17—防涡器 Vortex breaker；18—固定支座 Fixed support；19—活动支座 Shifting bearing；
20—进料口 Feed inlet；21、22—液位计接口 Liquid level gauge interface

6 芳烃 Arene

6.2.4 换热设备 Heat exchange equipment

6.2.4.1 抽出液塔再沸器 Liquid extraction column reboiler

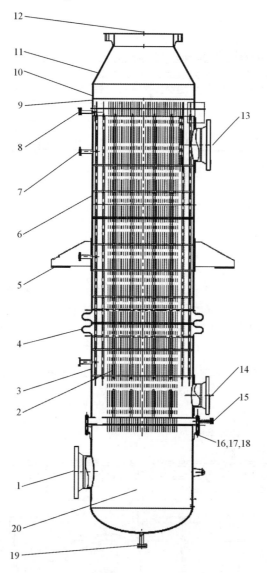

图 6.2.23 抽出液塔再沸器结构图 Structure drawing of liquid extraction column reboiler

1—管程入口 Tube inlet；2—换热管 Heat exchange tube；3、6—筒体 Cylinder；4—膨胀节 Expansion joint；5—支座 Support；7—放空口 Vent；8—排气口 Exhaust port；9—保温支承圈 Insulation support ring；10—筒节 Tube section；11—折边锥形封头 Toriconical head；12—管程出口 Tube outlet；13—壳程入口 Shell inlet；14—壳程出口 Shell outlet；15—排液口 Liquid outlet；16—螺栓 Bolt；17—螺母 Nut；18—垫片 Gasket；19—排污口 Outfall；20—管箱 Channel box

6.2.4.2 PX 成品塔再沸器 PX product column reboiler

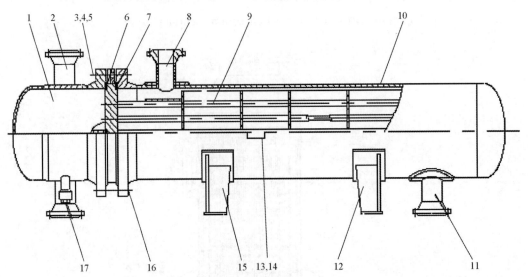

图 6.2.24 PX 成品塔再沸器结构图 Structure drawing of PX finished product column reboiler
1—管箱 Channel box；2、17—管程接口 Tube interface；3、4、5—螺栓组件 Bolt assembly；
6—管箱垫片 Channel box gasket；7—筒体垫片 Channel box side gasket；8、11—壳程接口 Shell interface；
9—管束 Tube bundle；10—筒体 Casing；12、15—支座 Support；13—铭牌架 Name plate frame；14—铭牌 Nameplate；
16—带肩双头螺栓 Shoulder stud

6.2.4.3 吸附分离进料再沸器 Adsorption separation feed reboiler

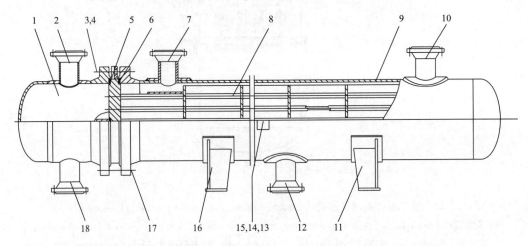

图 6.2.25 吸附分离进料再沸器结构图 Structure drawing of adsorption separation feed reboiler
1—管箱 Channel box；2—管程入口 Tube inlet；3—螺母 Nut；4—双头螺栓 Double end bolts；
5—管箱垫片 Channel box gasket；6—筒体垫片 Channel box side gasket；7、10—壳程出口 Shell outlet；
8—管束 Tube bundle；9—筒体 Cylinder；11—活动支座 Movable saddle；12—壳程入口 Shell inlet；
13—标牌架 Sign rack；14—铭牌 Nameplate；15—铭牌钉 Nameplate nail；16—固定支座 Fixed support；
17—带肩双头螺栓 Shoulder stud；18—管程出口 Tube outlet

6.2.4.4 对二甲苯产品后冷器 Paraxylene product aftercooler

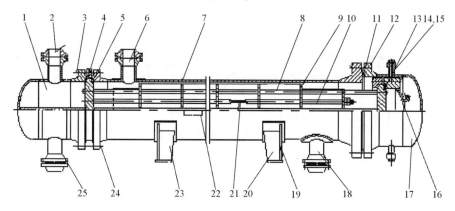

图 6.2.26 对二甲苯产品后冷器结构图 Structure drawing of paraxylene product aftercooler

1—管箱 Channel box；2—冷却水出口 Cooling water outlet；3—管箱法兰 Channel box flange；4—管板 Tube sheet；
5—筒体垫片 Channel box side gasket；6、18—壳程接口 Shell interface；7—筒体 Cylinder；8—管束 Tube bundle；
9—折流板 Baffle；10、21—拉杆 Tie rod；11—外头盖垫片 Outside head cover gasket；
12—外头盖法兰 Outside head flange；13—钩圈 Hook ring；14—浮头盖垫片 Floating head cover gasket；
15、24—双头螺栓 Stud；16—浮头 Floating head；17—外头盖 Outer head cover；
19—接地板 Ground plate；20、23—支座 Support；22—铭牌 Sign；
25—冷却水入口 Cooling water inlet

6.2.4.5 解吸剂再蒸馏塔重沸器 Desorbent rerun column reboiler

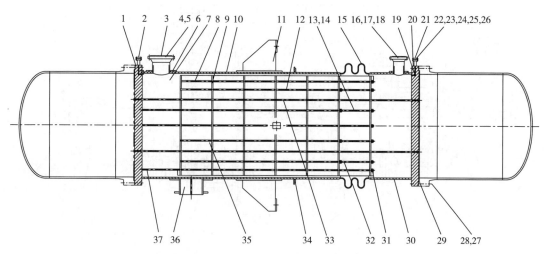

图 6.2.27 解吸剂再蒸馏塔重沸器结构图 Structure drawing of desorbent rerun column reboiler

1、19—管板 Tube sheet；2、21—放空口 Vent；3—蒸汽入口 Steam inlet；4、16、20、22—法兰 Flange；
5、17—接管 Connecting pipe；6、18—补强圈 Reinforcing ring；7—防冲挡板 Impingement baffle；
8、12、32、35—定距管 Spacer tube；9—折流板 Baffle plate；10、30—筒体 Cylinder；11—支座 Support；
13、37—拉杆 Tie rod；14、23、25、28、31—螺母 Nut；15—膨胀节 Expansion joint；16—蒸汽出口 Steam outlet；
24、27—双头螺栓 Stud；26、29—波齿垫 Corrugated pad；33—换热管 Heat exchange tube；
34—保温支承圈 Insulation support ring；36—吊耳 Lug

6.2.4.6 吸附分离白土精制冷却器 Adsorption separation clay refining cooler

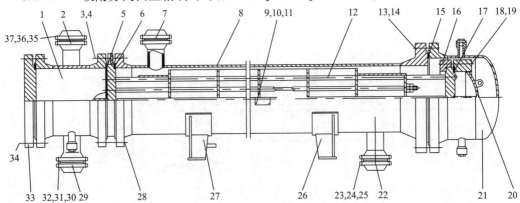

图 6.2.28 吸附分离白土精制冷却器结构图 Structure drawing of adsorption separation clay refining cooler

1—管箱 Channel box；2—管程出口 Tube outlet；3—带肩双头螺栓 Shoulder stud；4、14、19、23、30、35—螺母 Nut；
5、6、15、17、25、32、37—波齿垫 Corrugated pad；7—壳程入口 Shell inlet；8—筒体 Cylinder；9—标牌 Sign；
10—标牌用钉 Nail of sign；11—标牌架 Sign rack；12—管束 Tube bundle；13、18、24、28、31、34、36—双头螺栓 Stud；
16—钩圈 Hook ring；20—浮头 Floating head；21—外头盖 Outer head cover；22—壳程出口 Shell outlet；
26、27—支座 Support；29—管程入口 Tube inlet；33—平盖 Flat cover

6.2.5 泵 Pump

6.2.5.1 抽余液塔底泵 Raffinate column bottom pump

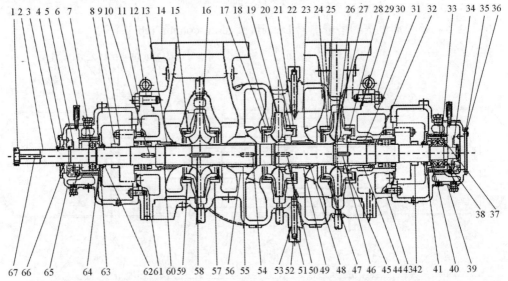

图 6.2.29 抽余液塔底泵结构图 Structure drawing of raffinate column bottom pump

1、22—锁紧螺母 Locknut；2—轴 Shaft；3、62、66—迷宫密封环 Labyrinth seal ring；4、36—轴承压盖 Bearing cover；
5、34—抛油环 Flinger；6—通气孔 Vent；7—径向轴承 Radial bearing；8、42—机械密封 Mechanical seal；
9、11、30—螺母 Nut；10、46、50、54、58、67—键 Key；12—吊耳 Lug；13—狭口衬圈 Throat bushing；
14、43、61—泵体 Pump body；15、18、20、24、28—叶轮口环 Impeller wear ring；16、19、27—叶轮 Wheel；
17、21、25、45、57—泵体口环 Pump inlet ring；23、47、56、59—级间隔板 Interstage partition；
26、35、64—密封垫圈 Sealing gasket；29—金属缠绕垫 Spiral wound gasket；31、44、48、49、55—衬套 Bushing；
32—卡簧 Snap spring；33—推力轴承 Thrust bearing；37—轴承垫圈 Bearing washer；38、66—轴承螺母 Bearing nut；
39—推力轴承箱 Thrust bearing box；40—轴承隔离圈 Bearing spacer；41、64—挡油板 Oil baffle plate；
51—定位螺母 Adjusting nut；52—螺帽 Blind nut；53—垫圈 Gasket；60—蜗壳 Spiral case；
63—导流板 Guide plate；65—径向轴承箱 Radial bearing shell

6 芳烃 Arene

6.2.5.2 吸附分离注水泵 Adsorption separation injection pump

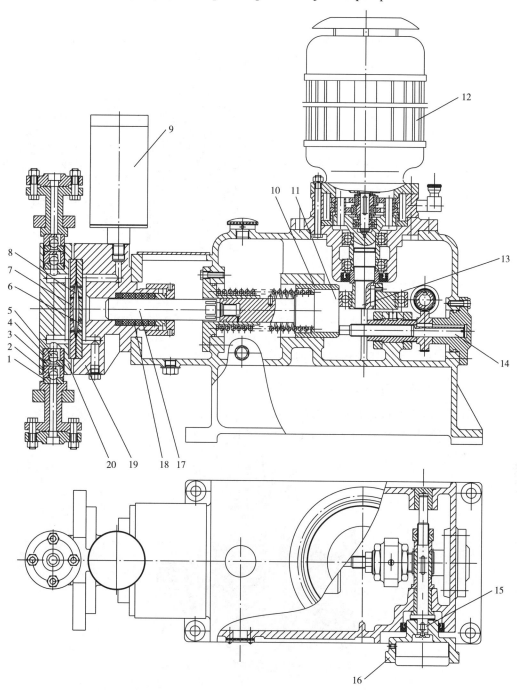

图 6.2.30 吸附分离注水泵结构图 Structure drawing of adsorption separation injection pump

1—下阀座 Lower valve seat；2—下导向套 Lower guide sleeve；3—上阀座 Upper valve seat；
4—上导向套 Upper guide sleeve；5—阀球 Valve ball；6—前限制板 Front limit plate；7—后限制板 Rear limit plate；
8—隔膜 Membrance；9—安全补油阀 Security filling valve；10—导向套 Guide sleeve；11—十字头 Crosshead；
12—电机 Motor；13—轴 Output shaft；14—调节杆 Regulating bar；15—调节轴 Regulating shaft；
16—调节转盘 Knob；17—柱塞 Plunger；18—密封环 Seal ring；19—前缸头 Front cylinder head；
20—后缸头 Rear cylinder head

·625·

6.2.5.3 解吸剂地下罐泵 Desorbent underground tank pump

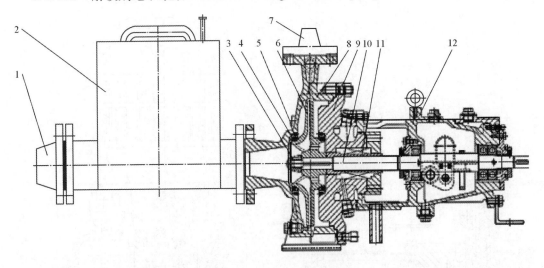

图 6.2.31 解吸剂地下罐泵结构图 Structure drawing of desorbent underground tank pump

1—吸入口 Intake port；2—双射流自吸罐 Double jet self-priming tank；3—叶轮螺母 Impeller nut；
4—叶轮密封环 Impeller labyrinth seal ring；5—泵体密封 Pump seal；6—叶轮 Wheel；7—排出口 Outlet；
8—泵体 Pump body；9—泵盖 Pump cover；10—泵轴 Shaft；
11—机械密封 Mechanical seal；12—轴承箱 Bearing box

6.2.6 特种阀门 Special valves

6.2.6.1 程控阀 Program control valve

1. 程控阀外形 Outline of program control valve

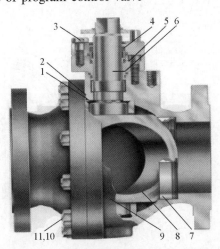

图 6.2.32 程控阀外形图 Outline drawing of program control valve

1—次阀体密封 Secondary body seal；2—主阀体密封 Main body seal；3—密封螺栓 Sealing bolt；
4—次阀杆密封 Secondary stem seal；5—主阀杆密封 Main stem seal；6—阀杆 Valve stem；
7—球阀座 Ball seat；8—内球体 Inner ball body；9—阀体 Valve body；
10—螺母 Nut；11—螺栓 Screw

· 626 ·

2. 程控阀结构 Program control valve structure

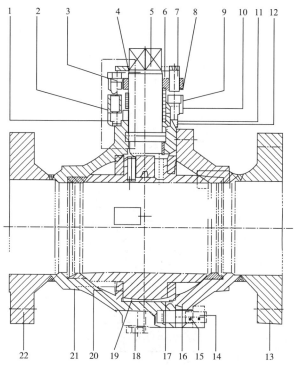

图 6.2.33 程控阀结构图 Structure drawing of program control valve

1—六角螺栓 Hexagon head screw；2、7—销子 Pin；3、14—六角螺母 Hexagon head screw；4—卡环 Snap ring；5—阀杆 Valve stem；6—限位盘 Stop disk；8—调整盖 Gland adjustment；9、15—螺栓 Bolt；10、11—密封腔 Gland；12—螺母 Nuts；13、22—阀体 Valve body；16—密封环 Sealing ring；17—O形密封圈 O sealing ring；18—锁紧螺母 Locknut；19—内球体 Inner ball；20—球体 Ball；21—球阀座 Ball seat

3. 程控阀机构 Program control valve structure

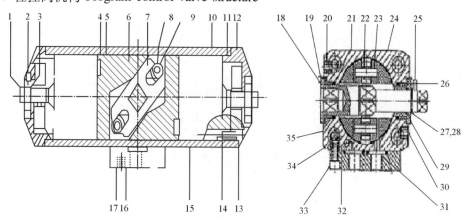

图 6.2.34 程控阀机构图 Structure drawing of program control valve

1—调整螺栓 Adjusting screw；2—六角螺母 Hexagon nut；3、5、11、29、31—O形密封圈 O sealing ring；4—导板 Guide strip；6—活塞 Piston；7—双摇杆机构 Double rocker；8—垫板 Strip；9—导向护圈 Guard ring for guide；10—阀体 Valve body；12—阀盖 Valve cover；13—内六角螺栓 Socket head cap screw；14、20、30、34—螺纹嵌件 Threaded insert；15、18—加油脂口 Cylinder grease；16、25—螺纹销 Threaded pin；17—连接板 Connecting plate；19—定心环 Centering ring；21—轴承箱 Bearing box；22—圆柱销 Straight pin；23—轴承 Bearing；24—铭牌 Nameplate；26—隔套 Spacer bush；27、28—定位环 Locating ring；32—插销 Plug；33—内六角螺栓 Socket head cap screw；35—轴套 Sleeve

4. 程控阀主要零部件 Main accessories of program control valve

1) 阀杆密封 Stem sealing

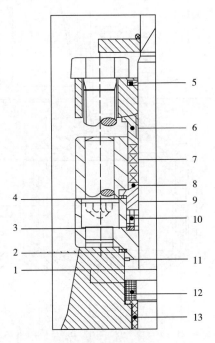

图 6.2.35　阀杆密封 Stem sealing

1、8—止推环 Thrust ring；2、4、10、12—密封圈 Sealing ring；3—垫片 Gasket；
5—滑动片 Wiper；6—阀杆密封腔 Gland；7—密封环 Sealing ring；
9—轴承密封圈 Bearing seal；11—O 形密封圈 O sealing ring；
13—法兰衬套 Flange bushing

2) 阀球座 Valve ball seat

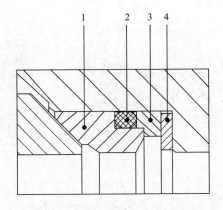

图 6.2.36　阀球座 Valve ball seat

1—密封圈 Sealing ring；2—O 形密封圈 O sealing ring；
3—止推环 Thrust ring；4—轴承密封圈 Bearing seal ring

3) 泄漏检测孔 Leakage test bore

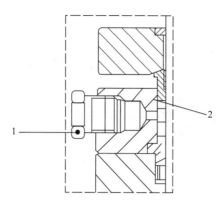

图 6.2.37 泄漏检测孔 Leakage test bore
1—紧固螺栓 Fastening screw;
2—中间密封圈 Intermediate ring

6.2.6.2 旋转阀 Rotary valve
1. 旋转阀外形 Rotary valve outline

图 6.2.38 旋转阀外形图 Outling drawing of rotary valve
1—棘轮手臂 Ratchet arm; 2—压力控制油管口 Pressure control nozzle; 3—拱顶 Vault;
4—定子板 Stator plate; 5—床层管线接口 Bed pipeline interface;
6—工艺管线接口 Process pipeline interface

2. 旋转阀结构 Rotary valve structure

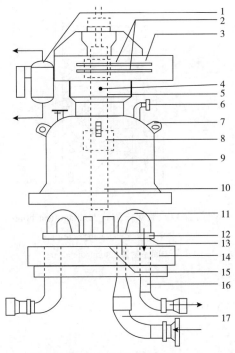

图 6.2.39　旋转阀结构图 Structure drawing of rotary valve

1—密封油罐 Sealing tank；2—棘轮手臂 Ratchet arm；3—棘轮 Ratchet；4—上部轴 Upper shaft；
5—机械密封 Mechanical seal；6—压力控制油接口 Pressure control oil interface；7—吊耳 Lug；
8—联轴器 Coupling；9—下部轴 Lower shaft；10—转子板轴套 Rotor plate sleeve；
11—转子板跨接管线 Rotor plate cross-over line；12—转子板 Rotor plate；
13—密封垫片 Seal gasket；14—定子板 Stator plate；15—定子板底座 Stator plate base；
16—床层管线 Bed pipeline；17—工艺管线 Process pipeline

3. 旋转阀主要零部件 Main accessories of rotary valve

1）定子板 Stator plate

（1）定子板外形 Stator plate outline

图 6.2.40　定子板外形图 Outline drawing of stator plate

1—螺栓孔 Bolt hole；2、5—床层管线接口 Bed pipeline interface；
3—槽道 Channel；4—轴套 Sleeve

（2）定子板结构 Structure of stator plate

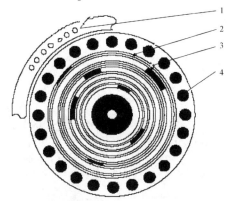

图 6.2.41　定子板结构图 Structure drawing of stator plate

1—螺栓孔 Bolt hole；2—槽道 Channel；3—工艺管线接口 Process piping interface；
4—床层管线接口 Bed pipeline interface

2）棘轮 Ratchet

（1）棘轮外形 Ratchet outline

图 6.2.42　棘轮外形图 Outline drawing of ratchet

1—轮轴 Spines shaft；2—棘轮 Ratchet；3—棘轮臂 Ratchet arm

（2）棘轮结构 Ratchet structure

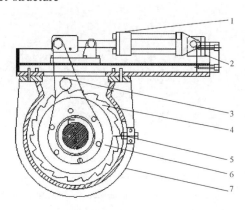

图 6.2.43　棘轮结构图 Structure drawing of ratchet

1—液压油缸 Hydraulic cylinder；2—支架 Support；3—棘轮臂 Ratchet arm；4—棘轮 Ratchet wheel；
5—限位开关 Limit switch；6—棘轮背 Ratchet wheel back；7—棘轮壳 Ratchet casing

3）转子板 Rotor plate

（1）转子板外形 Rotor plate outline

图 6.2.44　转子板外形图 Outline drawing of rotor plate
1—跨接管线 Crossover pipe；2—膨胀节 Expansion joint；
3—转子板 Rotor plate；4—定子板 Stator plate

（2）转子板底表面 Bottom surface of the rotor plate

图 6.2.45　转子板底表面图 Bottom surface view of the rotor plate
1—床层管线接口 Bed pipeline interface；2—备用接口 Alternate interface；
3—工艺管线接口 Process pipeline interface；4—沉头螺栓 Sunk Screw；
5—聚四氟乙烯垫片 PTFE gasket；6—轴套 Sleeve；7—槽道 Channel

6.3 歧化-烷基转移 Disproportionation-transalkylation

6.3.1 歧化反应器 Disproportionation reactor

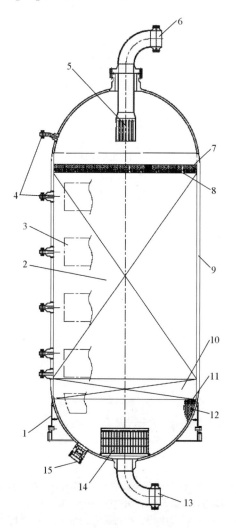

图 6.3.1 歧化反应器结构图 Structure drawing of disproportionation reactor
1—裙座 Skirt；2、10—催化剂床层 Catalyst bed；3—热电偶支承件 Thermo well support；
4—热电偶接口 Thermocouple interface；5—入口扩散器 Inlet divider；6—反应器入口 Reactor inlet；
7、12—大瓷球 Big porcelain ball；8、11—小瓷球 Small porcelain ball；9—筒体 Reactor shell；
13—反应器出口 Reactor outlet；14—出口收集器 Outlet collecting screen；15—出料口 Discharge port

6.3.2 塔设备 Column device

6.3.2.1 苯塔 Benzene column

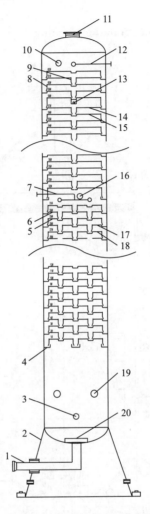

图 6.3.2 苯塔结构图 Structure drawing of benzene column

1—液相出口 Liquid phase outlet；2—裙座 Skirt；3、10、16—人孔 Manhole；
4—受液盘 Liquid receiving pot；5—塔板固定件 Plate fixed pieces；6、14、15、17、18—塔板 Plate；
7—进料口 Feed inlet；8—降液板 Downcomer board；9—降液板固定件 Fixed pieces of downcomer board；
11—气相出口 Gas phase outlet；12—回流入口 Reflux inlet；13—抽出口 Pumping outlet；
19—汽相入口 Vapor phase inlet；20—防涡器 Vortex breaker

6.3.2.2 歧化稳定塔 Disproportionation stabilizer column

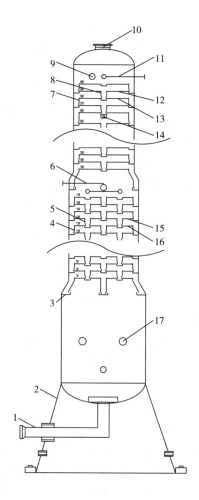

图 6.3.3 歧化稳定塔结构图 Structure drawing of disproportionation stabilizer column

1—液相出口 Liquid phase outlet；2—裙座 Skirt；3—受液盘 Liquid receiving pot；
4、7、8—降液板 Downcomer board；5—降液板固定件 Fixed pieces of downcomer board；
6—进料口 Feed inlet；9—人孔 Manhole；10—气相出口 Gas phase outlet；11—回流入口 Reflux inlet；
12、13、15、16—塔板 Plate；14—抽出口 Pumping outlet；17—气相入口 Gas phase inlet

6.3.3 歧化循环氢压缩机 Disproportionation recycle hydrogen compressor

歧化循环氢压缩机为垂直剖分型三级离心式压缩机，主要零部件有叶轮、轴、平衡装置、推力盘、联轴器和密封等。

Disproportionation recycle hydrogen compressor is vertical split-type three-stage centrifugal compressor. The main components are impeller, shaft, balancing device, thrust plate, coupling and seal, etc.

6.3.3.1 歧化循环氢压缩机 Disproportionation recycle hydrogen compressor

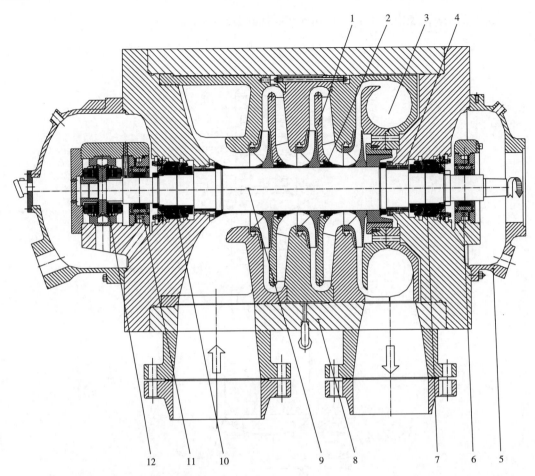

图 6.3.4 歧化循环氢压缩机结构图 Structure drawing of disproportionation recycle hydrogen compressor

1—叶轮 Wheel；2、4—迷宫密封 Labyrinth seal；3—蜗壳 Volute；5—轴承箱 Bearing box；6、11—径向轴承 Rodial bearing；7、10—干气密封 Dry gas seal；8—机体 Compressor body；9—主轴 Shaft；12—推力轴承 Thrust bearing

6.3.3.2 压缩机主要零部件 Main components of compressor

1. 叶轮 Impeller

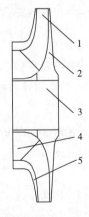

图 6.3.5 叶轮结构图 Structure drawing of impeller

1—叶轮出口 Impeller outlet；2—后盖板 Rear cover；3—轴 Shaft；4—叶轮入口 Impeller inlet；5—前盖板 Front cover

2. 主轴 Main shaft

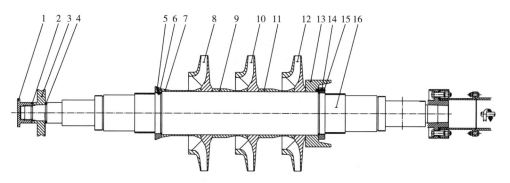

图 6.3.6　主轴结构图 Structure drawing of main shaft

1—半联轴器 Half；2—止动圈 Retaining ring；3—齿轮 Gear；4—调整环 Adjustment ring；
5、15—螺栓 Screw；6、14—螺母 Nut；7、9、11—隔套 Spacer；
8、10、12—叶轮 Wheel；13—平衡鼓 Balance disk；16—主轴 Shaft

3. 平衡鼓 Balance disk

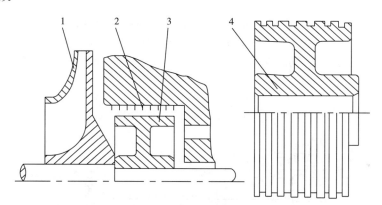

图 6.3.7　平衡鼓结构图 Structure drawing of balance disk

1—末级叶轮 Final stage impeller；2—平衡鼓气封 Balance disc gas seal；
3，4—平衡鼓 Balance disc

4. 推力盘 Thrust plate

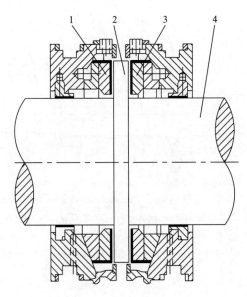

图 6.3.8 推力盘结构图 Structure drawing of thrust disk
1—副推力轴瓦 Secondary thrust bearing；2—推力盘 Thrust disc；
3—主推力轴瓦 Main thrust bearing；4—轴 Shaft

5. 联轴器 Coupling

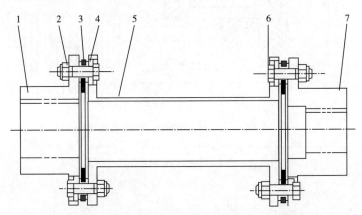

图 6.3.9 联轴器结构图 Structure drawing of coupling
1、7—半联轴器 Half-coupling；2、6—螺栓、Bolt；3—膜片 Membrane；
4—螺母 Nut；5—中间短节 Middle swage nipple

6. 机壳 Housing case

图 6.3.10 机壳解体图 Exploded view of housing case

1—轴承箱 Bearing box；2—轴承 Bearing；
3—压盖 Cover；4—转子 Rotor；5—壳体 Casing

7. 干气密封 Dry gas seal

1) 干气密封外形 Dry gas seal outline
2) 干气密封端面 End face of dry gas seal

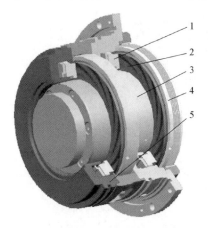 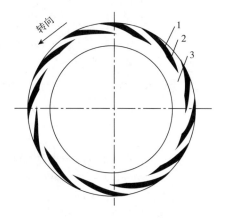

图 6.3.11 干气密封外形图
Outline drawing of dry gas seal

1—静环 Static ring；
2—动环 Rotating ring；
3—轴套 Sleeve；
4—弹性组件 Elastic components；
5—压盖 Cover

图 6.3.12 干气密封端面结构图
Structure drawing of end face of dry gas seal

1—对数螺旋槽 Logarithmic spiral groove；
2—密封堰 Sealing weir；
3—密封坝 Sealing dam

3）双端面干气密封结构 Double-end dry gas seal structure

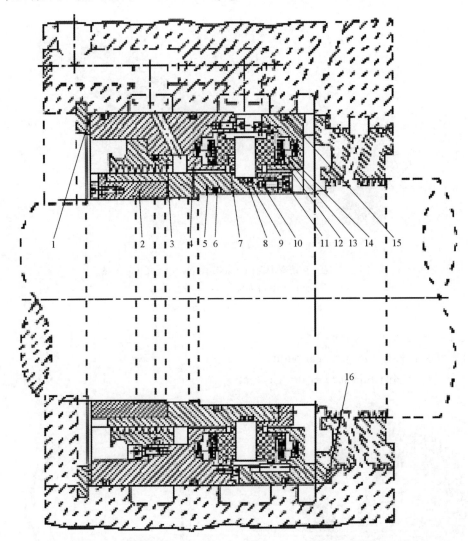

图 6.3.13　双端面干气密封结构图 Structure drawing of double-end dry gas seal
1、16—调整垫片 Adjusting pad；2—螺母 Nut；3—迷宫密封环 Comb seal ring；
4、14—弹簧座 Spring seat；5—轴套 Sleeve；6、8、15—O 形密封圈 O sealing ring；
7—静环 Static ring；9—动环定心弹簧 Rotating ring centering spring；10—动环 Rotating ring；
11—动环压紧套 Rotating ring clamp sleeve；12—推环 Thrust ring；13—弹簧 Spring

6.3.4 歧化进料罐 Disproportionation feed tank

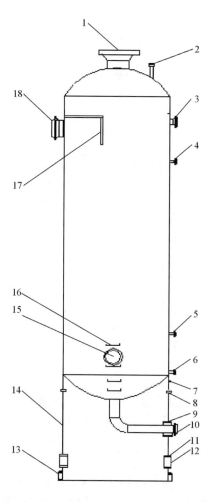

图 6.3.14 歧化进料罐结构图 Structure drawing of disproportionation feed tank
1—气体出口 Gas outlet；2—放空口 Vent；3—安全阀接口 Safety valve interface；
4、5—磁浮子液位计接口 Interface of magnetic floater liquid level meter／
Interface of double flange differential pressure transmitter；
6—备用口 Utility interface；7—保温支承圈 Insulation support ring；
8—排气口 Exhaust port；9、11—引出口 Extraction port；10—液体出口 Liquid outlet；
12—检修口 Access hole；13—地脚螺栓 Anchor bolt；14—裙座 Skirt；15—人孔 Manhole；
16—梯子 Ladder；17—防冲挡板 Impingement baffle；18—进料口 Feed inlet

6.3.5 歧化加热炉 Disproportionation furnace

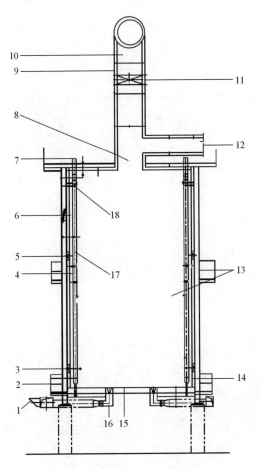

图 6.3.15 歧化反应加热炉结构图
Structure drawing of disproportionation reaction furnace

1—快开风门 Quick open damper；2—护栏 Guardrail；3—看火窗 Fire watch window；
4—炉膛热电偶套管、管壁热电偶套管 Furnace thermowell, thermowell wall；
5—看火门 Fire watch door；6—防爆门 Pressure relief door；
7—工艺介质出/入口 Process media inlet / outlet；
8—对流室 Convection chamber；9—测压管 Piezometer；
10—烟道口 Chimney；11—烟道挡板 Baffle；
12—烟道分支口 Joint chimney branch；13—辐射室 Radiation chamber；
14—蒸汽接口 Steam pipe interface；15—辐射室人孔 Manhole of radiation chamber；
16—燃料气入口，长明灯，看火孔 Fuel gas inlet, Long light fuel
gas inlet, Furnace bottom fire watch hole；
17—炉管 Furnace tube；18—炉管吊架 Furnace tube hook

6 芳烃 Arene

6.3.6 换热设备 Heat exchanger equipment

6.3.6.1 稳定塔顶气冷却器 Stabilizer column overhead gas cooler

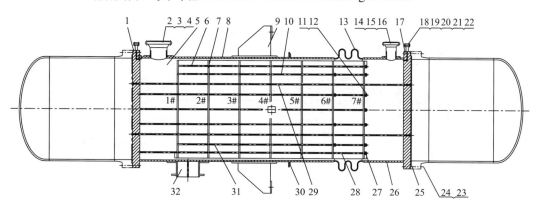

图 6.3.16 稳定塔顶气冷却器结构图 Structure drawing of stabilizer column overhead gas cooler

1、17—管板 Tube sheet；2、14、18、19—法兰 Flange；3、15—接管 Connecting pipe；4、16—补强圈 Reinforcing ring；5—防冲挡板 Impingement baffle；6、10、31—定距管 Spacer tube；7—折流板 Baffle plate；8、26—筒体 Cylinder；9—支座 Support；11、28—拉杆 Tie rod；12、21、24、27—螺母 Nut；13—膨胀节 Expansion joint；20、23—双头螺栓 Stud；22、25—波齿垫 Corrugated pad；29—换热管 Heat exchange tube；30—保温支承圈 Insulation support ring；32—吊耳 Lug

6.3.6.2 苯产品冷却器 Benzene product cooler

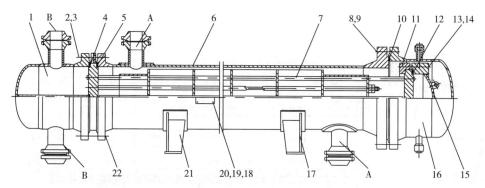

图 6.3.17 苯产品冷却器结构图 Structure drawing of benzene product cooler

1—管箱 Channel box；2、9、14—螺母 Nut；3—带肩双头螺栓 Shoulder stud；4—管箱侧垫片 Channel box gasket；5—筒体侧垫片 Channel box side gasket；6—筒体 Casing；7—管束 Tube bundle；8、13、22—双头螺栓 Stud；10—外头盖垫片 Outside head cover gasket；11—钩圈 Hook ring；12—浮头盖垫片 Floating head cover gasket；15—浮头 Floating head；16—外头盖 Outer head cover；17、21—支座 Support；18—铭牌架 Name plate frame；19—铭牌 Nameplate；20—铭牌钉 Nameplate nail；A—壳程接口 Shell interface；B—管程接口 Tube interface

· 643 ·

6.3.6.3 白土处理器进料/出料换热器 Clay processor feed/effluent exchanger

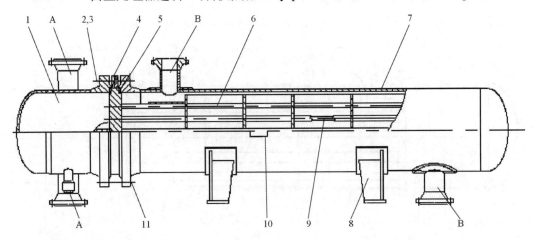

图 6.3.18 白土处理器进料/出料换热器结构图 Structure drawing of clay processor feed/effluent exchanger
1—管箱 Channel box；2—螺母 Nut；3—双头螺栓 Stud；4—管板 Tube sheet；
5—筒体侧垫片 Channel box gasket；6—管束 Tube bundle；7—筒体 Cylinder；
8—支座 Support；9—拉杆 Tie rod；10—铭牌 Sign；11—带肩双头螺栓 Shoulder stud；
A—管程接口 Tube interface；B—壳程接口 Shell interface

6.3.6.4 苯塔重沸/甲苯塔顶冷凝器 Benzene column reboiler / toluene overhead condenser

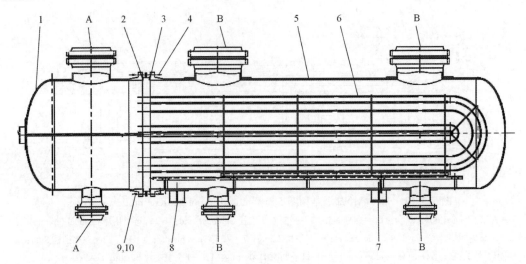

图 6.3.19 苯塔重沸/甲苯塔顶冷凝器结构图
Structure drawing of benzene column reboiler / toluene overhead condenser
1—管箱 Channel box；2—管板 Tube sheet；3—筒体侧垫片 Channel box gasket；
4、9—带肩双头螺栓 Shoulder stud；5—筒体 Casing；6—管束 Tube bundle；
7—活动支座 Sliding support；8—固定支座 Fixed support；
10—螺母 Nut；A—管程接口 Tube interface；B—壳程接口 Shell interface

6.3.6.5 歧化进料板式换热器 Disproportionation feed / product plate heat exchanger

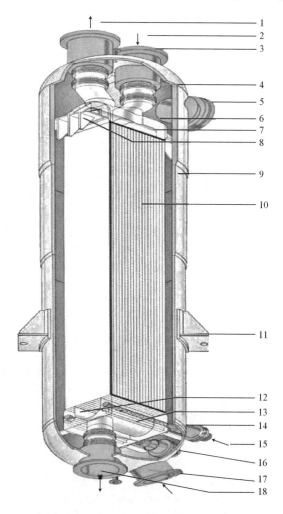

图 6.3.20 歧化进料板式换热器结构图
Structure drawing of disproportionation feed / product plate heat exchanger
1—出料口 Feed outlet；2—产物入口 Product inlet；3—放空口 Vent；
4—热端膨胀节 Hot end expansion joint；5—人孔 Manhole；6—进料管箱 Feeding tube box；
7—产物入口管箱 Product inlet tube box；8—板束支承 Plate beam support；9—筒体 Cylinder；
10—焊接板束 Welding plate beam；11—支座 Support；12—产物出口管箱 Product outlet tube box；
13—文丘里管 Venturi tube；14—喷雾杆 Spray bar；15—液体进料口 Liquid feed inlet；
16—冷端膨胀节 Cold end expansion joint；17—循环氢气入口 Recycle hydrogen inlet；
18—产物出口 Product outlet

6.3.7 泵 Pump

6.3.7.1 歧化反应进料泵 Disproportionation feed pump

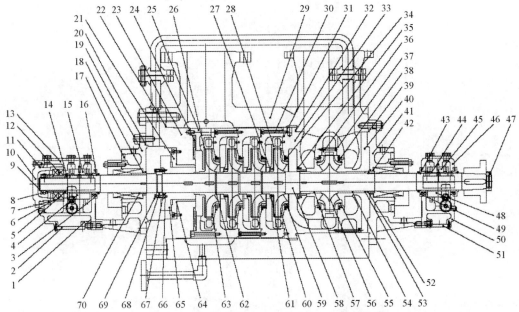

图 6.3.21 歧化反应进料泵结构图 Structure drawing of disproportionation feed pump

1—后轴承箱 Bearing housing；2、17—冷却室盖 Cooling chamber cover；3、49—防尘盖 Dust cover；
4、48—下轴瓦 Lower bearing sheel；5—轴承外套 Bearing cover；6—轴承内套 Bearing inner sleeve；
7—油环套 Oil ring set；8、44—油环 Oil ring；9—螺母 Nut；10—轴承端盖 Bearing cover；
11、16、43—O 形密封圈 O sealing ring；12—球轴承 Ball bearing；13—后轴承盖 Rear bearing cover；
14—调整垫 Adjustable pad；15、45—上轴瓦 Upper bearing bush；18、42—衬套 Bushing；
19—后轴承托架 Rear bearing bracket；20、31、37、39、53、64、66—柔性石墨垫 Flexible graphite gasket；
21—平衡套 Balancing sleeve；22、32—缠绕垫 Wound gasket；23—泵盖 Pump cover；24—碟簧 Disc spring；
25—吐出环 Discharge ring；26—末级导叶 Final stage diffuser；27—导叶套 Diffusor sleeve；28—导叶 Diffuser；
29—外筒体 Outer cylinder；30—中段 Middle piece；33—左吸水室 Left suction chamber；
34—中段密封环 Middle seal ring；35—吸入室衬套 Suction chamber bush；36—首级导叶 First stage diffuser；
38—吸入函体密封环 Suction box body seal ring；40—右吸水室 Right suction chamber；
41—前轴承座 Front bearing bracket；46—前轴承盖 Front bearing cover；47—锁紧螺母 Lock nut；
50—前轴承箱 Front bearing body；51—前轴承箱冷却室 Cooling cover of front bearing body；
52—定距螺套 Distance screw；54—右定距螺套 Right distance screw；55—首级叶轮密封环 First stage impeller seal ring；
56—首级叶轮 First stage impeller；57—左定距螺套 Left distance screw；58—轴 Shaft；
59—次级叶轮密封环 Second stage impeller seal ring；60—次级叶轮 Second stage impeller；61—轮毂密封环 Wheel seal ring；
62、68—分半卡环 Impeller half snap ring；63—末级叶轮 Final stage impeller；65—平衡盘 Balance disk；
67—压环 Compression ring；69—卡环套 Snap ring sets；70—喉部轴套 Throat sleeve

6.3.7.2 苯产品泵 Benzene products pump

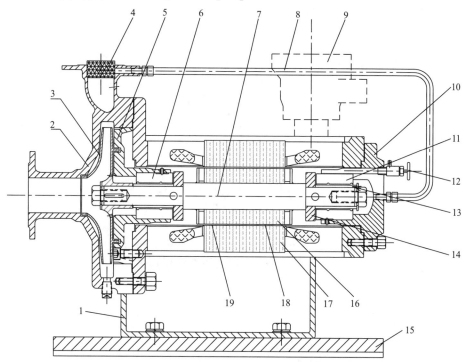

图 6.3.22 苯产品泵结构图 Structure drawing of benzene products pump

1—机架 Frame；2—泵体 Pump body；3—叶轮 Wheel；4—过滤器 Filter；5—端盖 End cover；6、11—轴承 Bearing；
7—轴 Shaft；8—循环管 Circulating pipe；9—接线盒 Connecting box；10—压盖 Gland；12—排气阀 Exhaust valve；
13—轴套 Sleeve；14—推力盘 Thrust disc；15—底座 Base；16—转子 Rotor；17—定子 Stator；
18—转子屏蔽套 Rotor shield sleeve；19—定子屏蔽套 Stator shield pipe

6.3.7.3 压缩机润滑油泵 Compressor oil pump

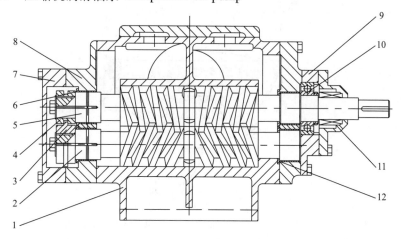

图 6.3.23 压缩机润滑油泵 Compressor oil pump

1—泵体 Pump body；2—从动螺杆 Driven screw；3—从动轮 Driven gear；4—主动轮 Driving gear；
5—主动螺杆 Driving screw；6、12—衬套 Bush；7—后盖 Rear cover；8—后端盖 Rear end cover；
9—轴承座 Bearing seat；10—轴承 Bearing；11—机械密封 Mechanical seal

6.4 异构化 Isomerization

6.4.1 异构化反应器 Isomerization reactor

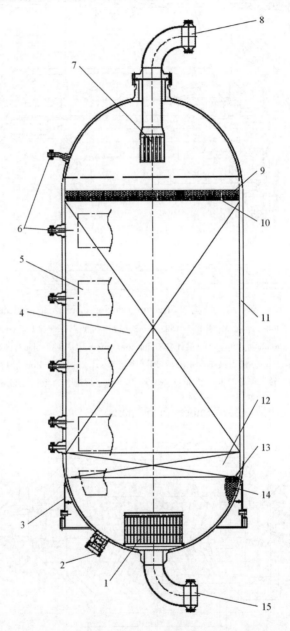

图 6.4.1 异构化反应器结构图 Structure drawing of isomerization reactor

1—出口收集器 Outlet collecting screen；2—卸料口 Discharge meatus；3—裙座 Brackets (or skirt)；
4、12—催化剂床层 Catalyst bed；5—热电偶支承件 Thermocouple supports；6—热电偶接口 Thermocouple interface；
7—入口扩散器 Inlet divider；8—反应器入口 Reactor inlet；9、14—大瓷球 Big porcelain ball；
10、13—小瓷球 Small porcelain ball；11—筒体 Reactor shell；15—油气出口 Oil and gas outlet

6.4.2 塔设备 Column equipment

6.4.2.1 异构化脱庚烷塔 De-heptane isomerization column

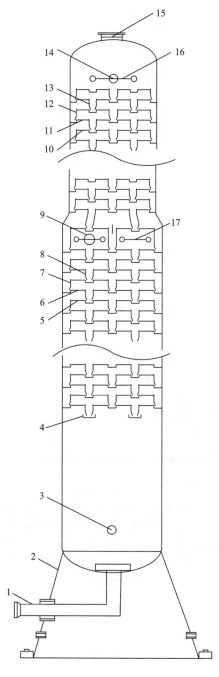

图 6.4.2 异构化脱庚烷塔结构图 Structure drawing of deheptane isomerization column

1—出料口 Discharge outlet；2—裙座 Skirt；3、9、14—人孔 Manhole；4—受液盘 Liquid receiving pot；
5、10—塔板 Plate；6、11—降液板固定件 Downcomer plate fixed pieces；7、12—降液板 Downcomer plate；
8、13—塔板固定件 Plate fixed pieces；15—气相出口 Gas phase outlet；16—回流入口 Overhead reflux inlet；
17—进料口 Feed inlet

6.4.2.2 异构化汽提塔 Isomerization stripping column

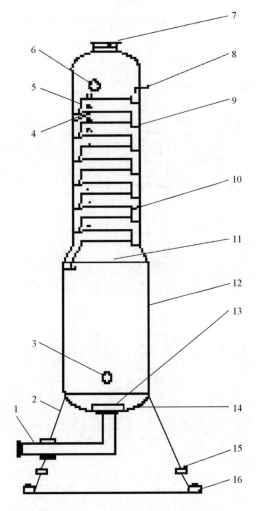

图 6.4.3 异构化汽提塔结构图 Structure drawing of isomerization stripping column

1—出料口 Bottom discharge outlet; 2—裙座 Skirt; 3、6—人孔 Manhole; 4—塔板 Plate;
5—降液板 Downcomer plate; 7—气相出口 Overhead gas outlet; 8—进料口 Feed inlet;
9—受液盘 Liquid receiving pot; 10—塔板支承圈 Plate bearing ring; 11—锥形筒体 Tapered cylinder;
12—筒体 Cylinder; 13—出口挡板 Outlet damper; 14 封头 Head; 15—检修口 Access hole;
16—地脚螺栓 Anchor bolt

6.4.3 异构化循环氢压缩机 Isomerization recycle hydrogen compressor

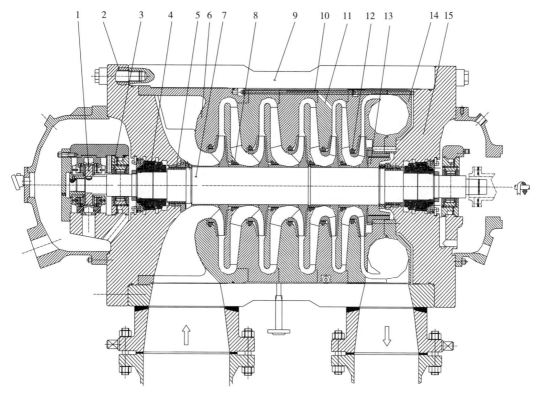

图 6.4.4 异构化循环氢压缩机结构图 Structure drawing of isomerization recycle hydrogen compressor

1—推力轴承 Thrust bearing；2—左端盖 Left end cap；3—可倾瓦轴承 Tilting pad bearing；
4—干气密封 Dry gas seal；5—轴端密封 Shaft end seal；6—入口隔板 Entrance partition；7—主轴 Shaft；
8—壳体级间密封 Sealing between casing；9—机壳 Housing；10—螺杆 Screw；11—隔板 Partition；
12—叶轮级间密封 Impeller interstage seal；13—平衡鼓 Balance disk；14—出口隔板 Outlet partition；
15—右端盖 Right end cap

6.4.4 异构化分液罐 Isomerization sub tank

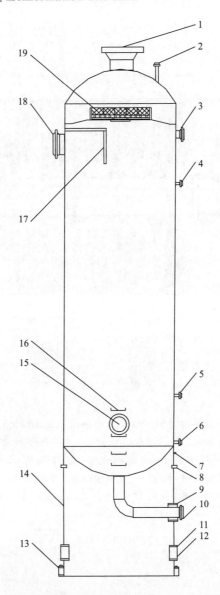

图 6.4.5 异构化分液罐结构图 Structure drawing of isomerization sub tank
1—气体出口 Gas outlet；2—放空口 Vent；3—安全阀接口 Safety valve interface；
4、5—磁浮子液位计口/双法兰差压变送器口 Port of magnetic floater liquid level meter /
Port of double flange differential pressure transmitter；6—公用接口 Utility interface；
7—保温支承圈 Insulation support ring；8—排气口 Exhaustion port；
9、11—引出管接口 Extraction pipe interface；10—液体出口 Liquid outlet；12—检修口 Access hole；
13—地脚螺栓 Anchor bolt；14—裙座 Skirt；15—人孔 Manhole；16—梯子 Ladder；
17—防冲挡板 Impingement baffle；18—进料口 Feed inlet；
19—丝网除沫器 Bottom-mounted wire mesh demister

6 芳烃 Arene

6.4.5 塔顶换热器 Overhead heat exchanger

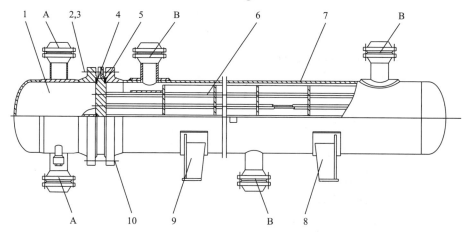

图 6.4.6 塔顶换热器结构图 Structure drawing of overhead heat exchanger

1—管箱 Channel box；2—螺母 Nut；3—双头螺栓 Double end bolts；4—管箱侧垫片 Channel box gasket；
5—筒体侧垫片 Channel box side gasket；6—管束 Tube bundle；7—筒体 Cylinder；8—活动支座 Sliding support；
9—固定支座 Fixed support；10—带肩双头螺栓 Shoulder stud；A—管程接口 Tube interface；
B—壳程接口 Shell interface

6.4.6 异构化进料泵 Isomerization feed pump

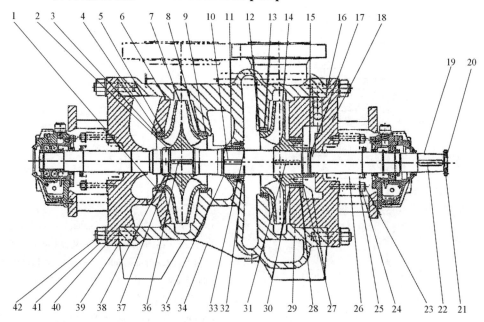

图 6.4.7 异构化进料泵结构图 Structure drawing of isomerization feed pump

1、3、8—首级叶轮口环 First stage impeller ring；2、17、33—O 形密封圈 O sealing ring；4、9—首级泵体口环 First stage pump ring；
5、32—内沉六角螺栓 Socket head cap screw；6、16、29、39—金属缠绕垫 Spiral wound gasket；7—首级叶轮 First stage impeller；
10、22、31、36—键 Key；11、35—级间轴套 Interstage bushing；12、27—次级叶轮口环 Second stage impeller wear ring；
13、28—次级泵体口环 Second stage pump ring；14—次级叶轮 Second stage impeller；15—叶轮螺母 Impeller nut；
18—排放泵盖 Discharge pump cover；19—泵轴 Pump shaft；20、24、41—螺母 Nut；21、25—锁紧垫圈 Lock washer；
23、42—螺栓 Bolt；26—机械密封 Mechanical seal；30—泵盖 Pump cover；34—泵体 Pump body；
37—分半环 Half ring；38—分半环固定套 Fixed sleeve of half ring；40—吸入泵盖 Suction pump cover

6.5 芳烃抽提 Aromatics extraction

6.5.1 塔设备 Column equipment

6.5.1.1 抽提蒸馏塔 Extractive distillation column

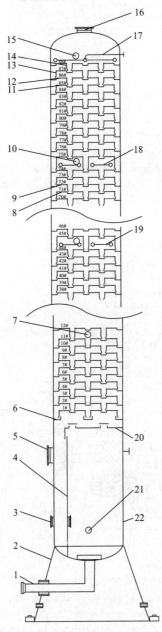

1—液相出口 Liquid phase outlet；2—裙座 Skirt；3—出料口 Discharge port；4—隔板 Partition；5—气相入口 Gas phase inlet；6—受液盘 Liquid receiving pot；7、10、15、21—人孔 Manhole；8、11—塔板 Plate；9、12—塔板支承圈 Plates bearing ring；13—降液板 Downcomer plate；14—降液板固定件 Fixed pieces of downcomer plate；16—气相出口 Gas phase inlet；17、18—贫溶剂口 Lean solvent opening；19—进料口 Feed inlet；20—集液箱 Fixed piece of oil sump tank；22—筒体 Cylinder

图 6.5.1 抽提蒸馏塔结构图
Structure drawing of extractive distillation column

6.5.1.2 溶剂回收塔 Solvent-recovery column

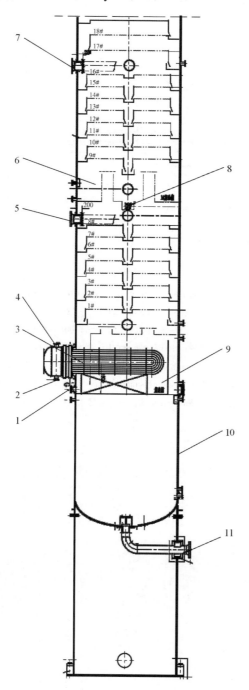

图 6.5.2 回收塔结构图 Structure drawing of solvent-recovery column

1—排液口 Liquid outlet；2—蒸汽出口 Steam outlet；3—U 形管再沸器 U-tube reboiler；
4—蒸汽入口 Steam inlet；5—中段进料口 Middle feed opening；6、9—集液箱 Collecting tank；
7—进料口 Feed inlet；8—中段出料口 Middle outlet；10—筒体 Cylinder；
11—塔底出料口 Bottom discharge outlet

6.5.2 非芳烃蒸馏塔再沸器 Non-aromatic hydrocarbon distillation column reboiler

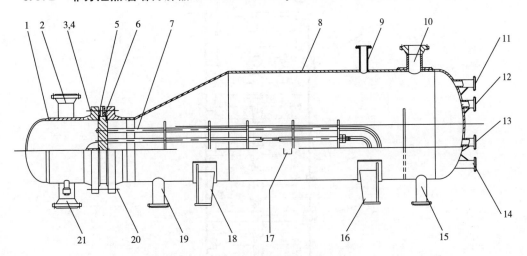

图 6.5.3 非芳烃蒸馏塔再沸器结构图
Structure drawing of non-aromatic hydrocarbon distillation column reboiler

1—管箱 Channel box；2、21—管程接口 Tube interface；3—螺母 Nut；4—双头螺栓 Isometric stud；
5—管箱侧垫片 Channel box gasket；6—筒体侧垫片 Channel box side gasket；7—管束 Tube bundle；
8—筒体 Casing；9—排气口 Exhaust port；10—物料出口 Reboiler material outlet；
11、14—液位计接口 Liquid level gauge interface；12、13—液位控制口 Level gauge control interface；
15—排放口 Discharge port；16—活动支座 Sliding support；17—铭牌 Nameplate；18—固定支座 Fixed support；
19—物料入口 Material inlet；20—带肩双头螺栓 Shoulder stud

6.5.3 泵 Pump

6.5.3.1 贫溶剂泵 Lean solvent pump

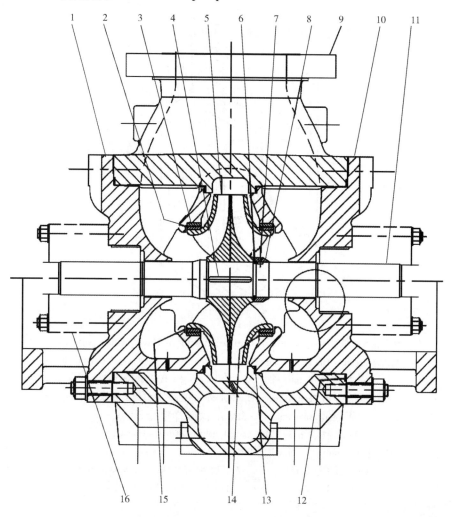

图 6.5.4 贫溶剂泵结构图 Structure drawing of lean solvent pump

1、10—泵盖 Pump cover；2—叶轮密封环 Impeller labyrinth seal ring；3—泵体密封环 Pump inlet ring；
4—键 Key；5—叶轮 Wheel；6—分半环 Half ring；7、14、15—紧固螺栓 Tightening screw；
8—分半环固定套 Half ring fixed sleeve；9—泵体 Pump body；
11—轴 Shaft；12、13—垫片 Gasket；16—机械密封 Mechanical seal

6.5.3.2 爪式真空泵 Claw vacuum pump

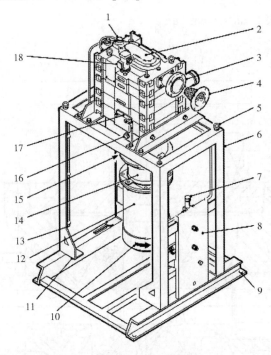

图 6.5.5 爪式真空泵外形图 Outline drawing of claw vacuum pumps

1—冷却液注入器 Coolant filling cap；2—轴承盖 Bearing cover；3—泵入口 Pump inlet；
4—入口过滤器 Inlet Filter；5—吊耳 Lug；6—固定架 Bracket；
7—温度调节控制器 Temperature adjustment control；8—维修面板 Access panel；
9—底座 Base seat；10—旋转方向箭头 Rotating arrow；
11—固定架下部横构件 Lateral member of lower bracket；
12—电机 Motor；13—冷却液溢出管接口 Coolant overflow pipe interface；
14—联轴器保护罩 Coupling cap；15—油位目视镜 Oil level visual mirror；
16—注油口 Oil filling port；17—清洗管 Cleaning pipe；
18—开关 Hot toggle hook switch

7 合成氨与尿素 Synthetic ammonia and Urea

7.1 合成氨 Synthetic ammonia

合成氨生产设备主要有转化炉、甲烷化反应器、合成塔反应器、换热设备、压缩机组和泵等。

Synthetic ammonia production equipment includes mainly reformer, methanation reactor, synthesis column reactors, heat-exchange equipment, compressors and pumps

7.1.1 反应设备 Reaction equipment

7.1.1.1 一段转化炉 One passage changed stove

在合成氨厂，一段转化炉在生产过程中的主要作用是将原料气(天然气)里的烃(主要是甲烷)与水蒸气反应，转化成氢气、一氧化碳和二氧化碳等气体，转化反应是在转化炉中进行的。一段转化炉分为辐射段和对流段两部分。辐射段是将原料烃在炉管内进行转化反应，对流段则是回收烟道气剩余热量预热工艺介质。

In synthetic ammonia plant, one passage change stove major role in the process is made the raw material gas (natural gas) in the hydrocarbons (mainly methane) and steam react converted into hydrogen, carbon monoxide and carbon dioxide gases, the conversion reaction is carried out in the conversion furnace. The change stove divided into two parts, radiant section and the convection section. Radiant section is the raw hydrocarbon conversion reactions within the furnace tube, the convection section is recycled flue gas residual heat and preheat the process medium.

1. 一段转化炉 One-stage onverter

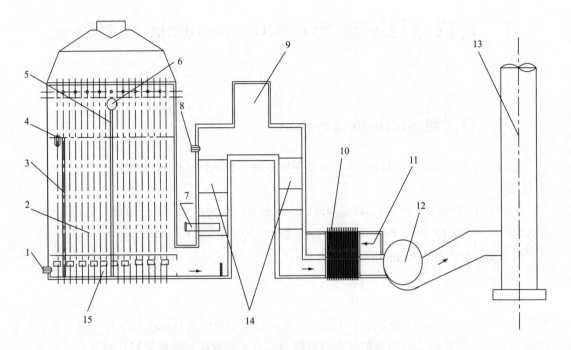

图 7.1.1　一段转化炉结构图 Structure drawing of one passage changed stove

1—烟道火嘴 Flue fire mouth；2—辐射段 Radiant section；
3—炉管 Furnace tube；4—炉顶火嘴 Top burner fire mouth；
5—上升管 Ascension pipe；6—输气总管 Gas manifold；
7—对流段风道 Convection section air flue；
8—对流段火嘴 Convection section fire mouth；
9、15—烟道 Flue；10—热管空气预热器 Heat pipe；
11—引风机风道 Blower duct；
12—引风机 Blowers；13—烟囱 Chimney；
14—对流段 Convection section

7 合成氨与尿素 Synthetic ammonia and Urea

2. 一段转化炉主要部件 One passage changed stove unit

1）转化管 Conversion tube
2）上升管 Riser pipe

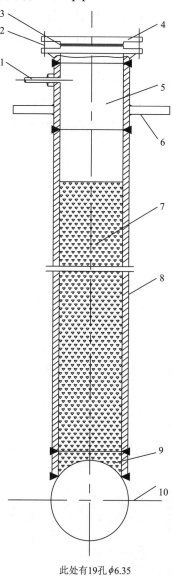

此处有19孔 ϕ6.35

图 7.1.2 转化管
Structure drawing of conversion tube

1—猪尾管 Pigtail；2—双头螺栓 Stud；3—垫片 Gasket；4—转化管盲板 Conversion o tube blind plate；5—转化管延伸段 Conversion tube extension；6—吊耳 Lifting lug；7—触媒 Accelerant；8—转化管 Conversion tube；9—接头转化管 Conversion pipe fittings；10—下集气管 Under manifold pipe

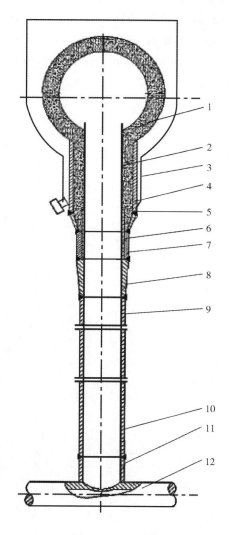

图 7.1.3 上升管
Structure drawing of riser pipe

1—氧化铝空心球+耐热砼 Alumina hollow ball + Heat resistant concrete；2—内衬管 Lined pipe；3—水夹套 Water jacket；4—承压管 Pressure pipe；5、7、8、9—外伸段 Extension segment；6—陶瓷纤维 Ceramic fiber；10—受热段 Heating section；11—短管 Tube socket；12—下集气管 Under manifold pipe

7.1.1.2 加氢反应器 Hydrogenation reactor

加氢反应器是将原料烃中的有机硫及其他组分，在 300~400℃钴钼催化剂的作用下，与加入的氢气进行转化反应。主要反应为有机硫转化为硫化氢、不饱和烯烃加氢饱和、含氮有机化合物脱氮(生成氨气和烃类)及含氧有机物脱氧(生成水和烯烃)等。

A hydrogenation reactor is used to carry out hydragenation react of organic sulfur in the raw material hydrocarbon and other components with hydrogen. under the action of a cobalt-molybdenum catalyst of300~400℃, The main reactions are the conversion of organic sulfur to hydrogen sulfide, hydrogenation of the unsaturated olefin saturation, nitrogen-nitrogen (ammonia and hydrocarbon generation) and the oxygen-containing organic compounds deoxy (to form water and an olefin).

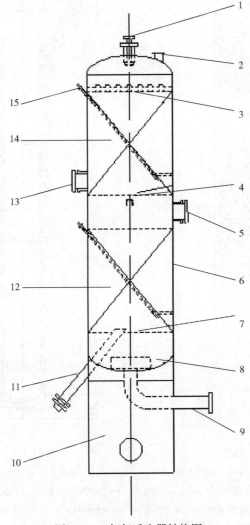

图 7.1.4 加氢反应器结构图
Structure drawing of hydrogenation reactor
1—气体进口 Gas inlet；2、13—人孔 Manhole；3—格栅板 Grate plate；4—碎屑捕集器 Debris trap；
5、11—出料口 discharge hole；6—筒体 Barrel；7—浮动丝网 A floating screen；8—填充瓷球 Filling ball；
9—气体出口 Gas outlet；10—裙座 Skirt；12、14—催化剂 Catalyst；15—温度计插管 Thermometer intubation

7 合成氨与尿素 Synthetic ammonia and Urea

7.1.1.3 脱硫反应器 Desulfurization reactor

氧化锌脱硫是目前工业上采用的脱硫效果最好的一种方法，其脱除硫化氢的能力很强，经氧化锌脱硫净化后的气体中的硫化氢含量可降到 0.1ppm 以下。

Zinc oxide desulfurization is currently the best process in the industry, due to its strong ability to remove hydrogen sulfide, hydrogen sulfide content in the gas can be reduced to 0.1ppm or less

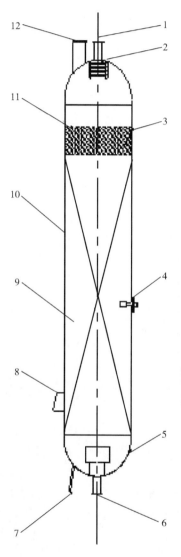

图 7.1.5 脱硫反应器结构图 Structure drawing of desulfurization reactor

1—气体入口 Gas inlet；2—进口挡板 Import baffle；
3—氧化铝瓷球 Alumina Ceramic Balls；4—取样口 sample connection；
5—封头 Head；6—气体出口 Gas outlet；
7—卸料口 discharge opening；8—支座 Support；
9—催化剂 Padding；10—筒体 Barrel；
11—隔栅板 Grating；12—人孔 Manhole

7.1.1.4 甲烷化反应器 Methanation reactor

甲烷化反应的目的是将工艺气中碳的氧化物转变为甲烷，除去碳的氧化物，即除去合成氨催化剂的毒物。

The purpose of the methanation reaction is to convert carbon oxides in the process gas to methane, thus carbon oxides removed, i.e., to remove the synthetic ammonia catalyst poisons

7.1.1.5 变换反应器 Shift reactor

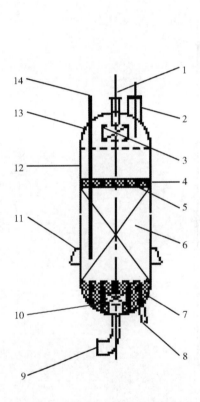

图 7.1.6 甲烷化反应器结构图
Structure drawing of methanation reactor
1—气体入口 Gas inlet；2—人孔 Manhole；3—气体捕集器 Gas trap；4、7—氧化铝瓷球 Alumina ceramic balls；5—筛网 Mesh；6—催化剂 Catalyst；8—出料口 Discharge hole；9—气体出口 Gas outlet；10—气体捕集器 Gas trap；11—支座 Support；12—筒体 Barrel；13—封头 Head；14—温度计插管 Thermometer intubation

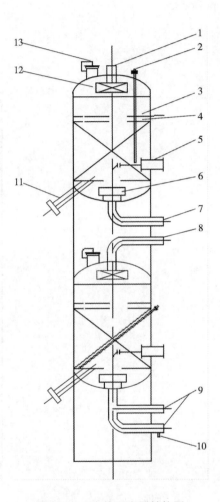

图 7.1.7 变换反应器结构图
Structure drawing of shift reactor
1—气体入口 Gas inlet；2—温度计接口 Thermometer mouth；3—格板 Grid；4—丝网 Wire mesh；5—蒸汽入口 Steam inlet；6—过滤器 Filter；7—中变口 Export；8—低变进口 Import；9—气体出口 Gas outlet；10—排水口 Drainage outlet；11—出料口 Discherge outlet；12—分布器 Distributor；13—人孔 Manhole

7 合成氨与尿素 Synthetic ammonia and Urea

7.1.1.6 合成塔反应器 Synthesis column reactor

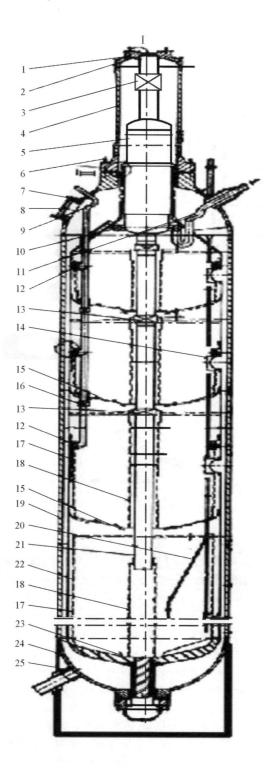

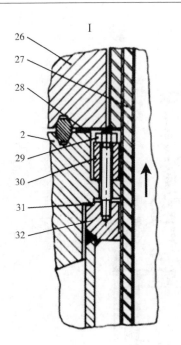
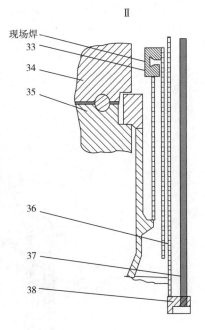

图 7.1.8 合成塔反应器结构图 Structure drawing of synthesis column reactor

1—接管 Connecting pipe；2、24—封头 Head；3、13、16—膨胀节 Expansion joint；4、19—筒体 Barrel；
5—换热器 Heat exchanger；6、29—螺栓、螺母 Bolt, nut；7—人孔盖 Manhole cover；8—人孔 Manhole；
9、11—挠性管 flexible tube；10—环管 tube loop；12—冷激环管 Chill pipe；14—检查口 Inspection hole；
15—床层间人孔 Manhole between beds；17—外集气筒 External gas housing；18—内集气筒 Internal gas housing；
20—热电偶套管 Thermowell；21—中心管 Center tube；22—触媒筐 Catalyst baskets；23—耐热混凝土 Resistant concrete；
25—裙座 Skirt；26—出口管线法兰 Outlet pipeline flange；27—保温套 Insulation sleeve；28—石棉垫 Asbestos pad；
30—活套 Looper；31—钢包垫 Ladle pad；32—法兰环 Flange ring；33—换热器支承环 Exchanger supporting ring；
34—主法兰 Main flange；35—封头凸缘 Head flange；36—换热器筒体 Heat exchanger barrel；
37—换热管 Heat exchanger tube；38—管板 Tube sheet；39—冷激管 Cold shock tube；
40—丝网 Screen；41—压紧环 Compression ring

7 合成氨与尿素 Synthetic ammonia and Urea

7.1.2 塔设备 Column equipment

7.1.2.1 汽提塔 Stripper
7.1.2.2 CO_2 吸收塔 CO_2 absorber tower

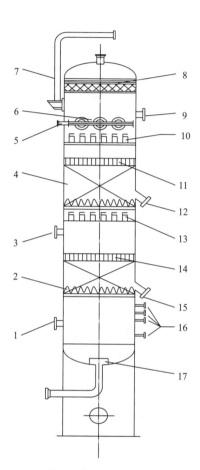

图 7.1.9 汽提塔结构图
Structure drawing of stripper

1、3、9—人孔 Manhole；2—填料支承 Packing support；4—填料 Packing；5—进料口 Feed inlet；6—分布器 Distributor；7—塔顶吊柱 Top davit 8—丝网除沫器 Screen demister；10、13—塔盘 Tray；11、14—填料压栅 Packing pressure gate；12、15—出料口 Discharge hole；16—玻璃管液面计 Glass level gauge；17—防涡器 Vortex breaker

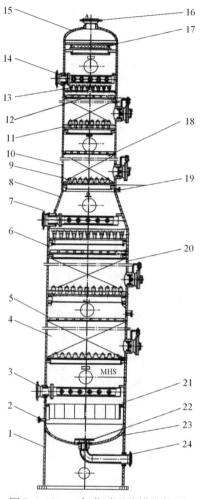

图 7.1.10 二氧化碳吸收塔结构图
Structure drawing of CO_2 absorber tower

1—裙座 Skirt；2—蒸汽出口 Steam outlet；3—低变气进口 Lower variable gas inlet；4、18、20—填料 Packing layer；5—格栅板 Grate plate；6—下筒体 Lower barrel；7—半贫液管 Semi-lean solution tube；8—锥体 Cone section；9、11—填料支承板 Packing support plate；10—上筒体 Upper the cylinder；12—塔盘 Trays；13—液体分布器 Liquid distributor；14—贫液管 Poor liquid tube；15—上封头 Upper the head；16—脱碳气出口 Decarbonization gas outlet；17—丝网除沫器 Screen demister；19—气体取样口 Gas sampling mouth；21—消漩器 Anti-vortex device；22—防涡器 Vortex breaker；23—下封头 Lower head；24—富液出口 Rich liquid outlet pipe

· 667 ·

7.1.2.3 CO₂ 再生塔 CO₂ regeneration tower
7.1.2.4 洗涤塔 Scrubber tower

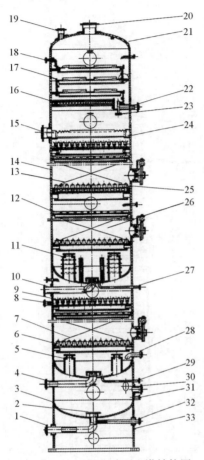

图 7.1.11 二氧化碳再生塔结构图
Structure drawing of carbon dioxide regeneration tower
1、32—贫液出口 Poor liquid outlet pipe; 2—防涡器 Vortex breaker; 3—下封头 Lower head; 4—凝液总管 Condensate manifold; 5、11—聚液池 Take liquid pool; 6、25—填料支承板 Packing support plate; 7、13、26—填料 Packing; 8—回流液进口 Reflux inlet tube; 9—液体分布器 Liquid distributor; 10—半贫液出口 Semi-lean liquid outlet; 12—格栅板 Grate plate; 14—筒体 Cylinder; 15—富液进口 Rich liquid inlet; 16—丝网除沫器 Mesh layer demister; 17—塔板 Plates; 18—冷凝液进口 Condensate inlet; 19—安全阀接口 Safety valve interface; 20—二氧化碳气出口 Carbon dioxide gas outlet; 21—上封头 Upper head; 22—回流液出口 Reflux exports; 23—回流液收集槽 Reflux collection tank; 24—富液布液器 Rich liquid distributor; 27—除沫器 Demister; 28—塔底回流液进口 Bottom reflux inlet; 29—排液口 Drain mouth; 30—再沸器回流口 Reboiler return mouth; 31—低压蒸汽进口 Low pressure steam inlet; 33—裙座 Skirt

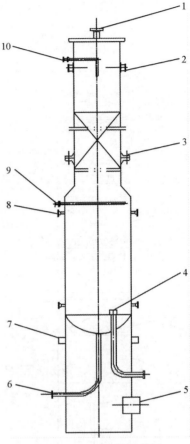

图 7.1.12 洗涤塔结构图
Structure drawing of scrubber tower
1—氨气出口 Ammonia exports; 2—吊耳 Lug; 3—法兰 Flange; 4—防涡器 Vortex breaker; 5—检修口 Access hole; 6—液体排出口 Liquid outlet; 7—排气口 Exhaust port; 8—液位计接口 Liquid level gauge interface; 9—氨气进口 Ammonia gas inlet; 10—水进口 Water inlet

7 合成氨与尿素 Synthetic ammonia and Urea

7.1.3 换热设备 Heat exchanger equipment

7.1.3.1 废热锅炉 Waste heat boiler

废热锅炉是合成氨生产中的重要设备之一，它利用二段转化工艺气的热量产生高压蒸汽。

Waste heat boiler is one of the important equipment in synthetic ammonia production, which uses two-stage transformation process gas heat to generate high-pressure steam.

1. 刺刀管式废热锅炉 A bayonet tube waste heat boiler
2. 固定管板式废热锅炉 Fixed tube sheet waste heat boiler

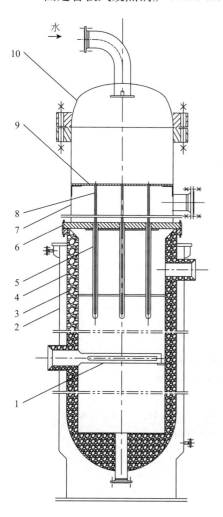

图 7.1.13 刺刀管式废热锅炉结构图 Structure drawing of a bayonet tube waste heat boiler
1—气体分布器 Gas distributor; 2—水夹套 Water jacket; 3—隔热层 Insulation layer; 4—筒体 Barrel; 5—外管 Outer tube; 6—大管板 Big tube plate; 7—管箱 Tube box; 8—内管 Within tube; 9—小管板 Small tube plate; 10—封头 Head

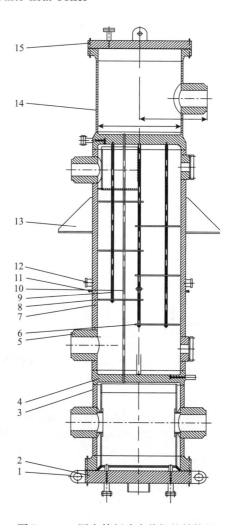

图 7.1.14 固定管板式废热锅炉结构图
Structure drawing of fixed tube sheet waste heat boiler
1—吊耳 Lug; 2—下平盖 Lower flat cover; 3—下管箱 Down tube box; 4—下管板 Lower tube plate; 5、12—接管 Connecting pipe; 6、8—折流板 Baffles; 7—筒体 Cylinder; 9—定距管 Spacer tube; 10—换热管 Heat exchange tube; 11—保温支承圈 Insulation support ring; 13—支座 Support; 14—上管箱 Upper tube box; 15—上平盖 Upper flat cver

3. 椭圆形管板废热锅炉 Oval tube sheet heat boiler

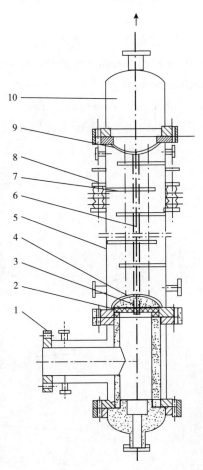

图 7.1.15 椭圆形管板废热锅炉结构图
Structure drawing of oval tube sheet waste heat boiler
1—三通 Tee；2—保护板 Protection board；3—保护套管 Protective sleeve tube；
4—下管板 Lower tube plate；5—筒体 Barrel；6—换热管 Heat transfer tube；7—折流板 Baffles；
8—膨胀节 Expansion joints；9—上管板 Upper tube plate；10—汽包 Steam drum

4. 卧式废热锅炉 Horizontal waste heat boiler

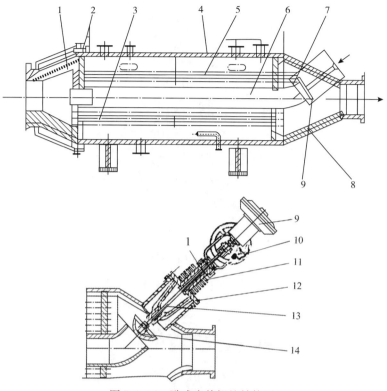

图 7.1.16 卧式废热锅炉结构图
Structure drawing of horizontal waste heat boiler

1—耐热混凝土 Resistant concrete；2—管板 Tube sheet；3—换热管 Heat exchange tube；
4—筒体 Barrel；5—拉杆 Tie rod；6—中心管 Center tube；7—弯管 Elbow tube；
8—锥形封头 Conical head；9—气动薄膜执行机构 Pneumatic diaphragm actuator；
10—手轮机构 Hand wheel；11—散热片 Fins；
12—阀盖 Valve cover；13—阀杆 Valve stem；
14—阀头 Valve head

5. "三菱"型废热锅炉 "Mitsubishi" type waste heat boiler

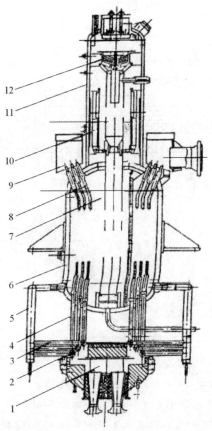

图 7.1.17 "三菱"型废热锅炉结构图 Structure drawing of "Mitsubishi" type waste heat boiler
1—气体入口分配器 Gas inlet distributor；2—环形集流管 Ring header tube；3—连接管 Communication pipe；4—双套管 Double tube；5—下降管 Downcomer；6—锅炉筒体 Boiler barrel；7—中央降水管 Center downcomers；8—螺旋蒸发管 Spiral evaporator tubes；9—集气箱 Set gas tank；10—汽水分离器 Vapour-water separator；11—汽包 Drum；12—丝网捕集器 Mesh trap

6. 薄管板废热锅炉 Thin tube plate waste heat boiler

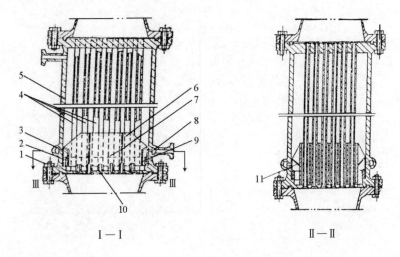

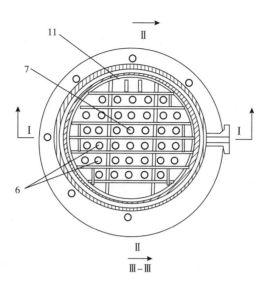

图 7.1.18　薄管板废热锅炉结构图
Structure drawing of thin tube plate waste heat boiler

1—气体入口分配器 Gas inlet distributor；2—环形集流管 Ring header tube；
3—连接管 Communication pipe；4—双套管 Double tube；
5—锅炉筒体 Barrel；6—下降管 Downcomer；
7—中央降水管 Center downcomers；8—螺旋蒸发管 Spiral evaporator tubes；
9—集气箱 Set gas tank；10—汽水分离器 Vapour-water separator；
11—汽包 Drum

7. 盘管式废热锅炉 Tube waste heat boiler

1) 四头单层盘管式废热锅炉 Four single-coil waste heat boiler
2) 四头双层盘管式废热锅炉 Four double-coil waste heat boiler

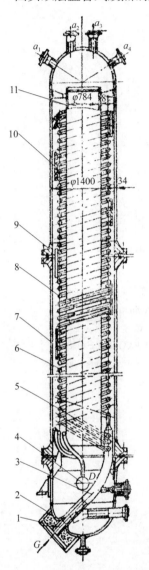 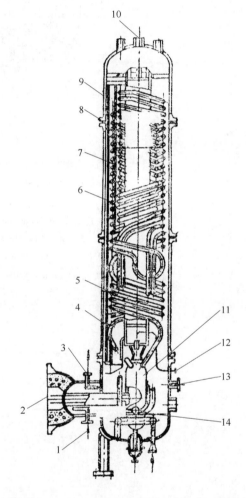

图 7.1.19 四头单层盘管式废热锅炉结构图
Structure drawing of four single-coil waste heat boiler
1—套管 Casing；2—加强凸缘 Strengthen flange；3—进气总管 Intake manifold；4—出气总管 Outlet manifold；5、6、8、10—传热管 Heat transfer tubes；7—筒体 Barrel；9—筒体法兰 Shell hange；11—中心管 Central tube

图 7.1.20 四头双层盘管式废热锅炉结构图
Structure drawing of four double-coil waste heat boiler
1—水入口 Water inlet；2—裂解气入口 Pyrolysis gas inlet；3—喷嘴 Nozzle；4、5、6、7—传热管 Heat transfer tubes；8—筒体法兰 Shell hange；9—筒体 Barrel；10—蒸汽出口 Steam outlet；11—分离罐 Separation tank；12—软化水入口 softened water inlet；13—温度计接口 Thermometer interface；14—裂解气出口 Cracking gas outlet

7 合成氨与尿素 Synthetic ammonia and Urea

8. 倒置插入式废热锅炉 Inverted plug-in waste heat boiler

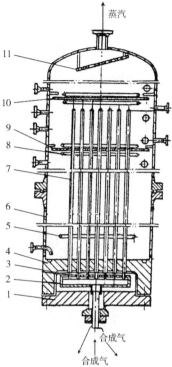

图 7.1.21 倒置插入式废热锅炉结构图
Structure drawing of inverted plug-in waste heat boiler

1—端盖 Cover；2—分气盒 Minute gas box；3—内套管 Inner sleeve；4—管板 Tube sheet；5—蒸汽管 Steam pipe；6—筒体 Barrel；7—外套管 Jacket tube；8—液面排污器 Liquid discharge device；9—清水器 Water device；10—汽水粗分器 Soda coarse splitter；11—汽水分离器 Vapour-water separator

9. U 形管废热锅炉 U-tube waste heat boiler
Horizontal U-shaped pipe waste heat boiler

1）组合式 U 形管废热锅炉 Combined U-tube waste heat boiler

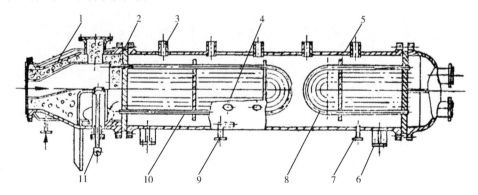

图 7.1.22 组合式 U 形管废热锅炉结构图 Structure drawing of Combined U-tube waste heat boiler

1—水汽夹套 Vapor jacket；2—管板 Tube sheet；3—蒸汽上升管 Steam rising pipe；4—水下降管接口 Water drop tube interface；5—支承板 Support plate；6—支座 Support；7—排污口 Outfall；8、10—U 形管 U-shaped tube；9—蒸汽进口 Steam imports；11—旁路阀 Bypass valve

2)气管式 U 形管废热锅炉 Weasand U-tube waste heat boiler

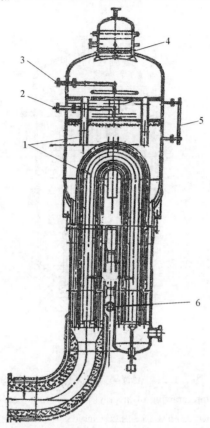

图 7.1.23 气管式 U 形管废热锅炉结构图 Structure drawing of weasand U-tube waste heat boiler
1—降水管 Downcomer；2—排污管 Discharge pipe；3—给水管 Water supply pipe；4—汽水分离器 Steam-water separator；5—液位计 Liquid level meter；6—平衡管 Balance pipe

7 合成氨与尿素 Synthetic ammonia and Urea

3) 水管式 U 形管废热锅炉 Pipe U-tube waste heat boiler

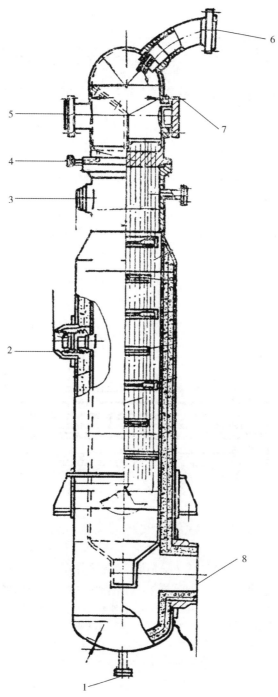

图 7.1.24 水管式 U 形管废热锅炉结构图 Structure drawing of pipe U-tube waste heat boiler
1—壳程排污口 Shell drain outlet；2—水蒸气入口 Steam inlet；3—工艺气出口 Process gas outlet；
4—管箱排污口 Channel drain outlet；5—水蒸气出口 Steam outlet；6—水入口 Water inlet；
7—人孔 Manhole；8—工艺气入口 Process gas inlet

7.1.3.2 二氧化碳再生塔再沸器 Carbon dioxide regeneration tower reboiler

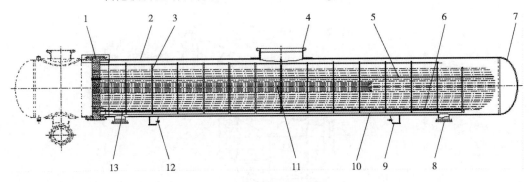

图 7.1.25 二氧化碳再生塔再沸器结构图
Structure drawing of Carbon dioxide regeneration tower reboiler

1—管板 Tube sheet；2—筒体 Cylinder；3—支承板 Supporting plate；4—法兰 Flange；
5—U 形换热管束 U heat exchange bunble；6—拉杆 Tie rod；7—封头 Head；8、13—接管 Connecting pipe；
9—支座 Support；10—滑道 Chute；11—铭牌 Nameplate；12—接地板 Ground plate

7.1.3.3 锅炉给水换热器 Boiler feed water heat exchanger

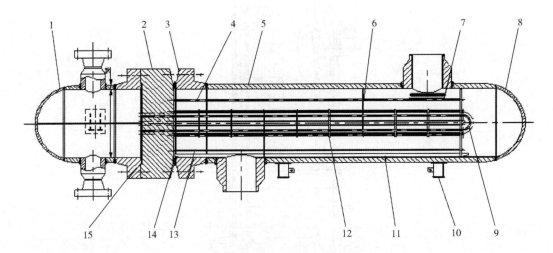

图 7.1.26 锅炉给水换热器结构图
Structure drawing of boiler feed water heat exchanger

1—管箱 Tube box；2—管板 Tube sheet；3—筒体法兰 Shell hange；4、13—拉杆 Tie rod；
5—筒体 Cylinder；6—折流板 Baffles；7—防冲挡板 Impingement baffle；8—封头 Head；
9—换热管束 Tube bundle；10—支座 Support；11—滑道 Chute；12—支承板 Support plate；
14—垫片 Gasket；15—管箱法兰 Head flange

7 合成氨与尿素 Synthetic ammonia and Urea

7.1.3.4 甲烷化原料气加热器 Methanation feed gas heater

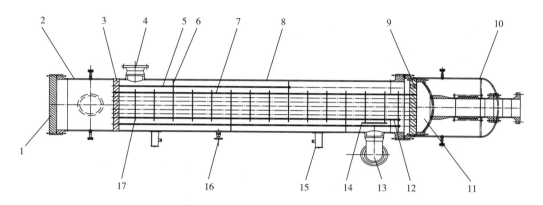

图 7.1.27　甲烷化原料气加热器结构图　Structure drawing of methanation feed gas heater
1—平盖 Flat cover；2—左管箱 Left tube box；3—管板 Tube sheet；4、13、16—接管 Connecting pipe；
5—定距管 Spacer tube；6、17—折流板 Baffles；7—换热管 Heat exchange tube；8—筒体 Barrel；
9—浮动管板 Floating tube sheet；10—右管箱 Right tube box；
11—浮动管箱 Float tube box；12—拉杆 Tie rod；
14—防冲挡板 Impingement baffle；15—支座 Support

7.1.3.5 表面冷凝器 Surface condenser

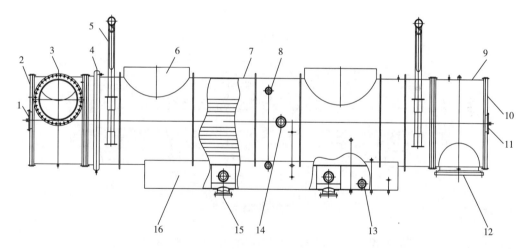

图 7.1.28　表面冷凝器结构图　Structure drawing of surface condenser
1—前端管箱 Front end channel；2—前端管箱盖 Front end channel cover；
3—循环水入口 Circulating water inlet；4—膨胀节 Expansion joint；
5—吊耳 Lifting lug；6—蒸汽入口 Steam inlet；7—筒体 Barrel；
8—液位计接口 Liquid level gauge interface；9—后端管箱 Back end channel；
10—后端管箱盖 Back end channel cover；11—检修口 Access hole；
12—循环水出口 Circulating water outlet；
13—泵回流管接口 Pump return pipe interface；
14—抽气器接口 Aspirator interface；
15—泵接口 Pump interface；16—热水槽 Hot water tank

7.1.3.6 空气压缩机段间冷却器 Inter cooler air compressor section

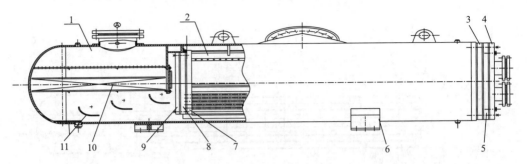

图 7.1.29 空气压缩机段间冷却器结构图 Structure drawing of inter cooler air compressor section

1—筒体 Barrel；2—管束 Tube bundles；3—前管板 Front tube plate；
4—前端盖 The front cover；5—前水室 Before the water chamber；
6—支座 Support；7—后管板 Back tube plate；8—后水室 The water chamber；
9—后端盖 End cap；10—除沫器 Demister；11—分布板 Distribution board

7.1.4 泵 Pump

7.1.4.1 贫液泵 Poor pump

1. 贫液泵 1 Poor pump 1

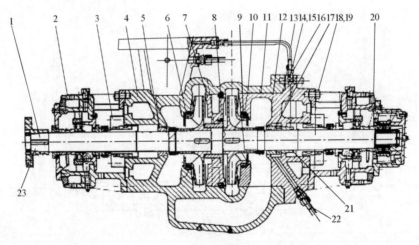

图 7.1.30 贫液泵 1 结构图 Structure drawing of poor pump 1

1—泵轴 Pump shaft；2—轴承箱 Bearing box；3—机械密封 Mechanical seals；4—泵体 Pump body；5、17—衬套 Bush sleeve；6、9—密封环 Pump seal ring；7—隔板 Bulkhead；8、10、12—螺钉 Screws；11—泵盖 Pump cover；13—垫片 Gasket；14、18—螺栓 Stud；15、19—螺母 Nuts；16—平衡管 Balance pipe；20—止推轴承 Thrust bearing；21—轴套 Sleeve；22—接管 Connecting pipe；23—半联轴器 Half coupling

2. 贫液泵 2 poor pump 2

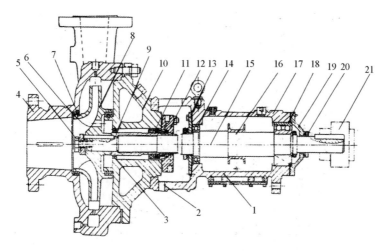

图 7.1.31　贫液泵 2 结构图 Structure drawing of poor pump 2

1—轴承箱 Bearing box；2—放油口 Oil drain；3—减压套 Decompression sets；4—泵体 Pump；5—叶轮锁紧螺母 Impeller screws；6—叶轮密封环 Impeller sealing ring；7—壳体密封环 Casing sealing ring；8—叶轮 Wheel；9—轴套 Sleeve；10—泵盖 Pump cover；11—机械密封 Mechanical seals；12—压盖 Gland；13—后轴承盖 After bearing cover；14—径向轴承 Rodial bearing；15—泵轴 Shaft；16—甩油环 Flinger；17—托架 Bracket；18—止推轴承 Thrust bearing；19—前轴承盖 Front bearing cap；20—挡油环 Deflector plate；21—联轴器 Couplings

3. 贫液泵主要零部件 Main components depleted pump

1) 密封结构 Sealing

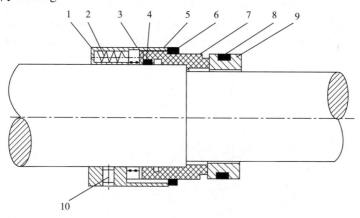

图 7.1.32　密封结构图 Structure drawing of sealing

1—弹簧座 Spring seat；2—弹簧 Spring；3—推环 Push ring；4、8—O 形密封环 O sealing ring；5—背环 Back ring；6—卡簧 Spring card；7—动环 Rotating ring；9—静环 Stationary ring；10—固定螺钉

2) 调速器 Govemor

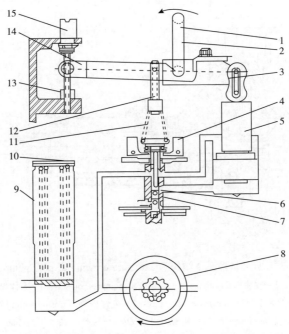

图 7.1.33　调速器结构图 Structure drawing of governor
1—端轴 End shaft；2—滑块 Slider；3—端轴杠杆 Shaft lever；
4—飞锤 Flyweight；5—油动机活塞 Oil motive piston；6—控制面板 Control Panel；
7—滑阀套 Slide valve sleeve；8—油泵 Pump；9—堵压器 Pressure regulator；
10—调速柱塞 Speed plunger；11—调速弹簧 Governor spring；12—调速螺杆 Speed screw；
13—高速限位档块 Speed limit stopper；14—反馈杠杆 Feedback lever；15—调速螺栓 Speed screw

7.1.4.2 锅炉给水泵 Boiler feed water pump

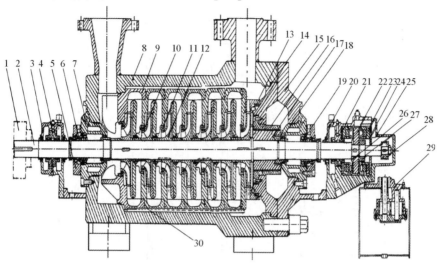

图 7.1.34 锅炉给水泵结构图 Structure drawing of boiler feed water pump

1—联轴器 Coupling；2—泵轴 Shaft；3、20—轴承箱 Bearing box；4—径向轴承 Rodial bearing；5—压盖 Gland；6—机械密封 Mechanical seals；7—箱体 Box；8—泵体 Pump；9—隔板 Partition；10—叶轮密封环 Impeller labyrinth seal ring；11—隔板密封 Partition mouth ring；12—叶轮 Wheel；13、14—垫片 Gasket；15、25—端盖 Cover；16—平衡盘密封 Balance disc seal；17—平衡盘 Balance disc；18—锁紧螺母 lock nut；19—轴套 Sleeve；21—调整垫片 Adjusting pads；22—止推盘 Thrust plate；23—止推轴承 Thrust bearing；24—外防护板 Outer shield；26、27—止推盘螺母 Ended；28—传动蜗轮 Drive worm；29—主油泵 Main pump；30—定位器 Locator

7.1.4.3 半贫液泵 Semi-depleted pump

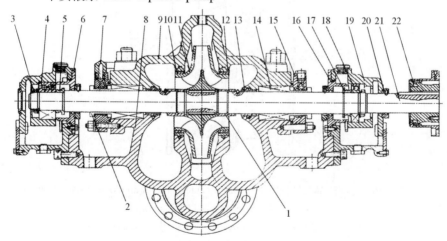

图 7.1.35 半贫液泵结构图 Structure drawing of semi-depleted pump

1—轴套螺母 Sleeve nut；2—机械密封轴套 Mechanical seal sleeve；3—轴承锁紧螺母 Bearing locknuts；4—止推轴承 Thrust bearing；5—甩油环 Flinger；6、16—轴承箱端盖 Bearing box cover；7—接管 Connecting pipe；8—下壳体 Lower housing；9—上壳体 Upper housing；10—叶轮密封环 Impeller labyrinth seal ring；11—叶轮 Wheel；12—壳体密封环 Casing sealing ring；13—减压套 Decompression sets；14—机械密封 Mechanical seals；15—密封压盖 Mechanical seal gland；17—油环套筒 Oil ring sleeve；18—球轴承 Ball bearings；19—轴承箱 Bearing box；20—偏导器盘 Deflector plate；21—泵轴 Shaft；22—联轴器 Coupling

7.1.4.4 冷氨产品泵 Cold ammonia products pump

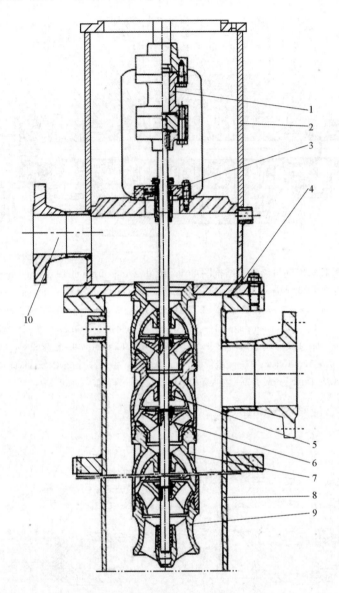

图 7.1.36 冷氨产品泵结构图 Structure drawing of cold ammonia products pump
1—联轴器 Couplings；2—支架 Support；
3—机械密封 Mechanical seals；4—O 形密封圈 O sealing ring；
5—石墨轴承衬套 Graphite bearing bush；
6—串接叶轮 Tandem impeller；
7—串接壳体 Tandem housing；8—泵体 Shell；
9—首级泵体 Heads housing；
10—泵入口 Pump inlet

7 合成氨与尿素 Synthetic ammonia and Urea

7.1.4.5 冷凝液泵 Condensate pump

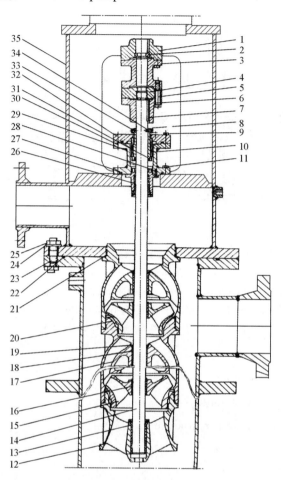

图 7.1.37　冷凝液泵结构图 Structure drawing of condensate pump

1—电机联轴器 Motor coupling；2—卡圈 Collar；

3—联轴器短节 A short section of the coupling；

4—调节螺母 Adjusting nut；5、9、11—螺栓 Bolt；

6—泵联轴器 Pump coupling；7—键 Key；

8—驱动环 Drive ring；10—机械密封 Mechanical seals；

12—首级壳体 Heads housing；

13、18、26—聚四氟乙烯轴套 Teflon bearing bush；

14—泵轴 Shaft；15—首级叶轮 First stage impeller；

16—锁紧锥套 Locking sleeve；

17—串接壳体 Tandem housing；19—挡圈 Ring；

20—串接叶轮 Tandem impeller；

21、22、27、31—O 形密封圈 O sealing ring；

23—双头螺栓 Studs；24—垫片 Gasket；

25—螺母 Nut；28、33、34—螺钉 Screws；

29—轴套 Sleeve；30—箱体 Box；

32—密封压盖 Gland；35—支架 Support

7.1.4.6 甲烷化冷凝液回收泵 Methanation condensate recovery pumps

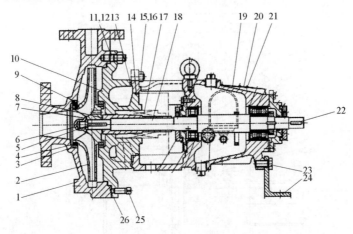

图 7.1.38 甲烷化冷凝液回收泵结构图 Structure drawing of methanation condensate recovery pumps
1—泵体 Pump；2—叶轮 Wheel；3—壳体密封环 Casing sealing ring；4—叶轮密封环 Impeller labyrinth seal ring；5、22—键 Key；6—叶轮螺母 Impeller nut；7、8—O 形密封圈 O sealing ring；9—螺钉 Screws；10—泵盖 Pump cover；11、15、23、25—螺栓 Bolt；12、16—螺母 Nuts；13—泵体 Pump body；14—轴承箱 bearing housing；17—轴封 Seal；18—泵轴 Pump shaft；19—标牌 Signs；20—转向牌 Turn card；21—销钉 Pin；24—支架 Support；26—垫片 Gasket

7.1.5 压缩机 Compressor

7.1.5.1 空压机 Air Compressor

1. 空压机结构 Air compressor structure

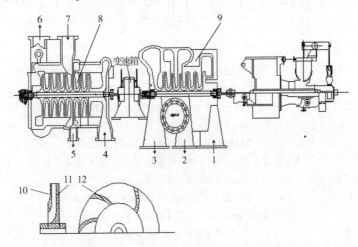

图 7.1.39 空压机结构图 Structure drawing of air compressor structure
1——段入口 First level entrance；2——段出口 First level export；3—二段出口 Sec export；4—三段入口 Three level entrance；5—三段出口 Three level outlet；6—四段出口 Four level export；7—四段入口 Four level entrance；8—高压缸 High-pressure cylinder；9—低压缸 Low-pressure cylinder；10—盖板 Cover；11—叶轮 Wheel；12—叶片 Vanes

· 686 ·

7 合成氨与尿素 Synthetic ammonia and Urea

2. 空压机低压缸 Air compressor low-pressure cylinder

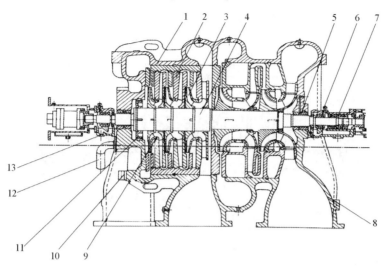

图 7.1.40 空压机低压缸结构图 Structure drawing of air compressor low-pressure cylinder
1—上壳体 Upper housing；2—隔板 Partition；3—叶轮 Wheel；4—主轴 Shaft；5—轴端密封 Shaft seal；6—径向轴承 Rodial bearing；7—止推轴承 Thrust bearing；8—下壳体 Lower housing；9—叶轮密封 Impeller seal；10—滑销 Sliding pin；11—平衡鼓 Balance disc；12—平衡管 Balance pipe；13—排油管 Discharge pipe

3. 空压机高压缸 High-pressure cylinder air compressor

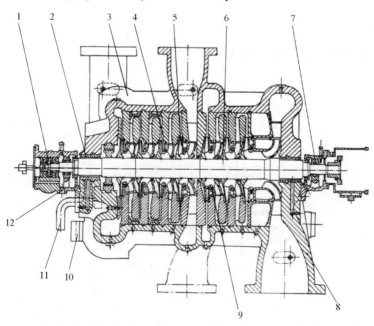

图 7.1.41 空压机高压缸结构图 Structure drawing of high-pressure cylinder air compressor
1—止推轴承 Thrust bearing；2—轴端密封 Shaft seal；3—缸盖 Cylinder cover；4—隔板 Spindle；5—叶轮 Wheel；6—主轴 Shaft；7—径向轴承 Rodial bearing；8—壳体 Housing；9—叶轮密封 Impeller seal；10—滑销 Sliding pin；11—平衡管 Balance pipe；12—排油管 Discharge pipe

4. 空压机汽轮机 Compressor turbine

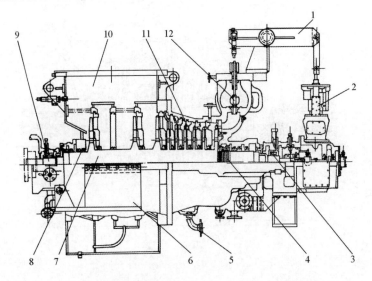

图 7.1.42 空压机汽轮机结构图 Structure drawing of compressor turbine
1—提板阀臂 Lift plate valve arm；2—调速器 Speed regulator；
3—径向轴承 Rodial bearing；4、8—汽封 Gland sealing；
5—排油管 Drain line；6—支承板 Support plate；
7—主轴 Shaft；9—后轴承 Spindle；
10—排汽缸 Exhaust casing；11—隔板 Partition；
12—提板阀 Lift plate valve

5. 空压机增速器 Air compressor speed increaser

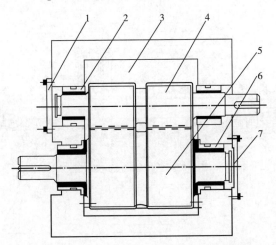

图 7.1.43 空压机增速器结构图 Structure drawing of air compressor speed increaser
1—小齿轮端盖 Pinion cover；2—小齿轮轴承 Pinion bearing；
3—齿轮箱 Box；4—小齿轮 Small gear；5—大齿轮 Large gear；
6—大齿轮轴承 Large gear bearing；7—大齿轮端盖 Gear cover

6. 空压机伺服机构

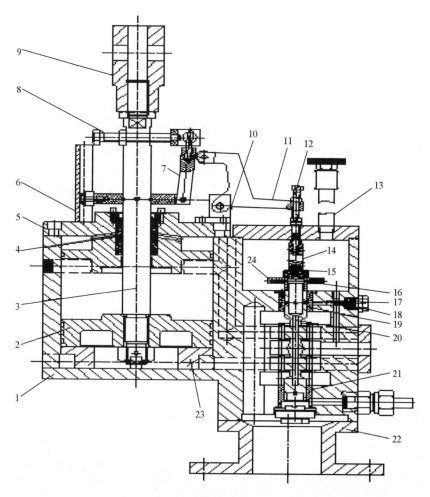

图 7.1.44 空压机伺服机构结构图 Structure drawing of air compressor servo

1—油缸壳体 Cylinder housing；2—活塞 Piston；

3—活塞杆 Rod；4—密封 Seal；5—油缸上盖 Cylinder cover；

6—导轨 Rail；7—反馈板 Feedback board；8、12—调节螺栓 Adjusting bolt；

9—叉接头 Fork joints；10—支座 Support；11—弯角杠杆 Angled lever；

13—放空管 Vent pipe；

14—弹簧 Spring；15—推力轴承 Thrust bearing；

16—转盘 Turntable；17—螺塞 Plug；18—调节阀 Regulating valve；

19—上套筒 Upper sleeve；20—滑阀 Slip valves；

21—下套筒 Lower sleeve；22—回油底盘 Back to the oil pan；

23—定距套 Fixed pitch sleeve；24—喷油口 Injection port

7. 空压机主汽阀 Air compressor main steam valve

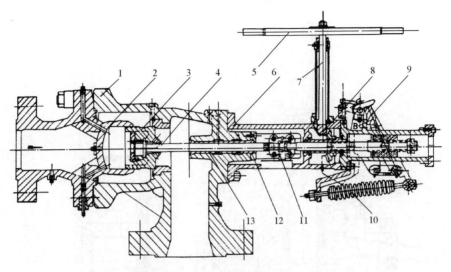

图 7.1.45 空压机主汽阀结构图 Structure drawing of air compressor main steam valve
1—阀体 Valve body；2—阀座 Valve seat；3—阀头 Valve head；
4—阀杆 Valve stem；5—手轮 Hand wheel；6—支架 Support；
7—传动杆 Drive rod；8—伞齿轮 Bevel gear；9—弹簧杠杆 Spring lever；
10—弹簧 Spring；11—联轴器 Couplings；12—压盖 Gland；13—填料函 Stuffing box

7.1.5.2 原料气压缩机 Feed gas compressor

1. 原料气压缩机结构 Feed gas compressor structure

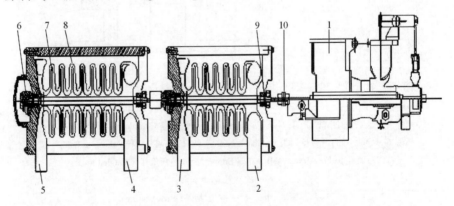

图 7.1.46 原料气压缩机结构图 Structure drawing of feed gas compressor structure
1—蒸汽出口 Steam outlet；2—低压缸入口 Cylinder inlet；
3—低压缸出口 Low pressure cylinder outlet；4—高压缸入口 High-pressure cylinder inlet；
5—高压缸出口 High-pressure cylinder outlet；
6—止推轴承 Thrust bearing；
7—高压缸叶轮 High pressure cylinder impeller；
8—高压缸隔板 High pressure cylinder partition；
9—径向轴承 Rodial bearing；10—联轴器 Coupling

2. 原料气压缩机低压缸 Feed gas compressor Low-pressure cylinder

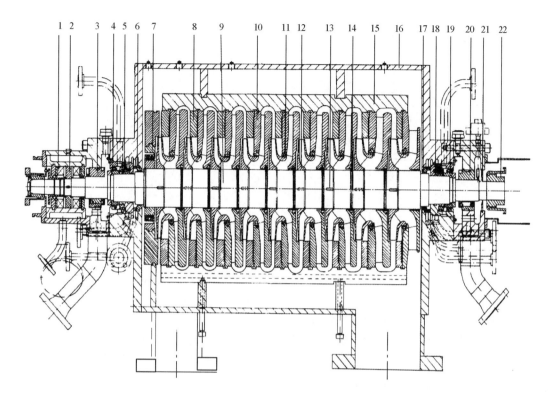

图 7.1.47　原料气压缩机低压缸结构图
Structure drawing of feed gas compressor low-pressure cylinder
1—推力轴承 Thrust bearing；2—推力盘 Thrust plate；
3、20—径向轴承 Radial bearing；4、19—动环 Rotating ring；
5、18—静环 stationary ring；6、17—轴端汽封 Shaft Seal；
7—平衡鼓 Balance drum；8—扩压器 Diffuser；
9—主轴 Shaft；10—隔板 Partitions；
11、14—密封环 Seal ring；12—叶轮 Wheel；
13—键 Key；15—内缸体 Internal cylinder；
16—外缸体 Outer cylinder；21—挡油环 Oil baffle plate；
22—半联轴器 Coupling

3. 原料气压缩机高压缸 Feed gas compressor high pressure cylinders

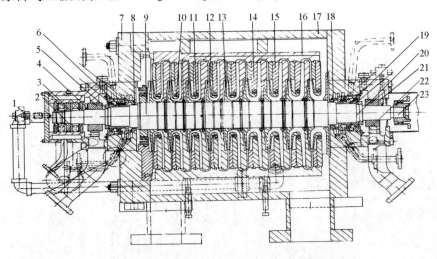

图 7.1.48　原料气压缩机高压缸结构图
Structure drawing of feed gas compressor high pressure cylinders

1、23—半联轴器 coupling half；2、22—测振探头 Thrust plate vibration measurement probe；3—止推轴承 Thrust bearing；4、21—径向轴承 Rodial bearing；5—动环 Rotating ring；6、19—静环 stationary ring；7—端盖 Cover；8、18—轴端密封 Shaft seal；9、20—主轴 Shaft；10、11—密封环 Seal ring；12—扩压器 Diffuser；13—叶轮 Wheel；14、15—隔板 Partition；16—内缸体 Internal cylinder；17—外缸体 Outer cylinder

4. 原料气压缩机汽轮机 Feed gas compressor turbine

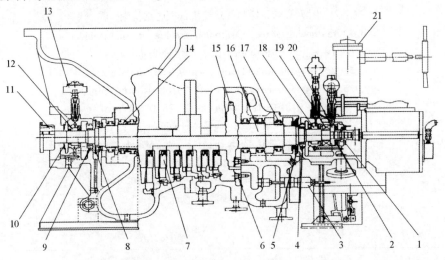

图 7.1.49　原料气压缩机汽轮机结构图 Structure drawing of feed gas compressor turbine
1—轴位移测试盘 Shaft displacement test disc；2—副推力轴承 Vice-thrust pad；3—手阀 Hand valve；4、9、10—挡油环 Oil block；5、8—挡汽环 Stop steam ring；6—喷嘴环 Nozzle ring；7—隔板 Partition；11—半联轴器 Half coupling；12、18—径向轴承 Rodial bearing；13—热电偶 Thermocouples；14、17—汽封 Steam seal；15—叶轮 Wheel；16—主轴 Shaft；19—主推力轴承 Main thrust bearing；20—止推盘 Thrust plate；21—主汽阀 Main steam valve

7 合成氨与尿素 Synthetic ammonia and Urea

5. 原料气压缩机机械密封 Feed gas compressor mechanical seal

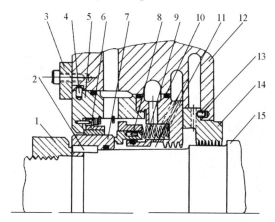

图 7.1.50 原料气压缩机机械密封结构图 Structure drawing of feed gas compressor mechanical seal
1—螺母 Nut；2—传动销 Drive pin；3—挡片 Block pieces；4、13—防转销 Anti-rotation pin；
5—衬套座 Bushing seat；6—衬套 Bushing；7—动环 Rotating ring；8—静环 Static ring；
9—弹簧 Spring；10—O 形密封圈 O sealing ring；11—密封函 Sealed box；12—缸体 Cylinder；
14—迷宫密封 Labyrinth seals；15—轴 Shaft

6. 原料气压缩机污油分离器 Feed gas compressor sump oil separator

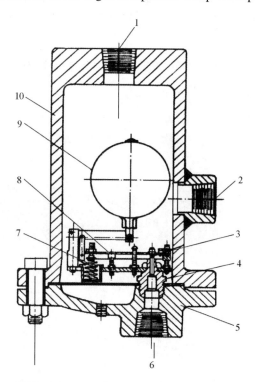

图 7.1.51 原料气压缩机污油分离器结构图 Structure drawing of feed gas compressor sump oil separator
1—排气口 Exhaust port；2—进气口 The air intake；3、4—阀座 Valve seat；5—端盖 Cover；
6—排油口 Oil discharge port；7—弹簧 Spring；8—杠杆 Lever；9—浮球 Float；10—壳体 Case

7. 原料气压缩机透平调速器 Feed gas compressor turbine governor

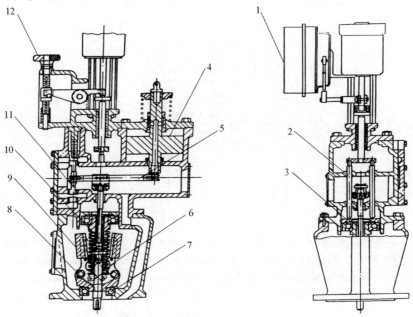

图 7.1.52　原料气压缩机透平调速器结构图 Structure drawing of feed gas compressor turbine governor
1—气动马达 Air motor；2—刀刃 Blade；3、7—滚动轴承 Rolling；4—油动机活塞 Oil motive piston；5—杠杆 Lever；6—弹簧 Spring；8—传动轴 Drive shaft；9—飞锤 Flyweight；10—调速器杆 Governor rod；11—错油门滑阀 Wrong throttle slide valve；12—手动同步器 Manual synchronizer

8. 原料气压缩机透平自动主汽阀 Automatic feed gas compressor turbine main steam valve

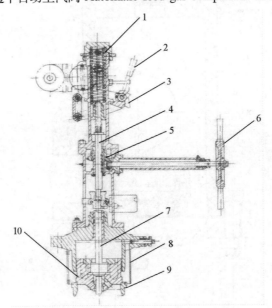

图 7.1.53　原料气压缩机透平自动主汽阀结构图
Structure drawing of automatic feed gas compressor turbine main steam valve
1—弹簧 Spring；2—手动停车手柄 Manual parking handle；3—挂钩 Hook；4—丝杠 Screw；5—伞齿轮 Bevel gear；6—手轮 Hand wheel；7—阀杆 Stem；8—滤网 Strainer；9—阀座 Valve seat；10—阀头 Valve head

7 合成氨与尿素 Synthetic ammonia and Urea

7.1.5.3 合成气压缩机 Synthesis gas compressor

1. 合成气压缩机结构 Synthesis gas compressor structure

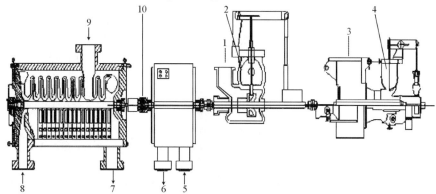

图 7.1.54 合成气压缩机结构 Structure drawing of synthesis gas compressor structure

1—中压蒸汽出口 Pressure steam outlet；2—高压透平 The high-pressure turbine；3—背压蒸汽出口 Back pressure steam outlet；4—中压透平 Intermediate pressure turbine；5—低压缸入口 Low-pressure cylinder inlet；6—低压缸出口 Low pressure cylinder outlet；7—高压缸出口 High pressure cylinder outlet；8—高压缸入口 High pressure cylinder inlet；9—循环段入口 Loop segment entry；10—联轴器 Coupling

2. 合成气压缩机中压冷凝式汽轮机 Synthesis gas compressor pressure condensing steam turbine

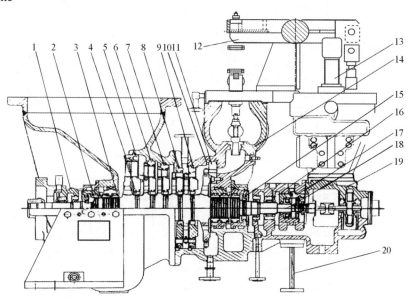

图 7.1.55 合成气压缩机中压冷凝式汽轮机结构图
Structure drawing of synthesis gas compressor pressure condensing steam turbine

1—后轴承箱 Bearing housing；2—后支架 Back bracket；3—低压端汽封 Low side steam seal；4—隔板汽封 Diaphragm Gland；5—第五级隔板 Fifth Class divisions；6—第四级隔板 Fourth class divisions；7—第三级隔板 Third stage separator；8—第二级隔板 Second class divisions；9—第一级隔板 First stage separator；10—楔块 Wedges；11—喷嘴 Nozzle；12—调节阀 Regulating valve；13—调速器 Governors；14—汽缸 Cylinder；15—高压端汽封 High pressure steam seal end；16—轴承座 Bearing support；17—推力轴承 Thrust bearing；18—转子 Rotor；19—前轴承箱 Front bearing housing；20—前支架 Front bracket

3. 合成气压缩机高压背压式汽轮机 High-pressure synthesis gas compressor back pressure turbine

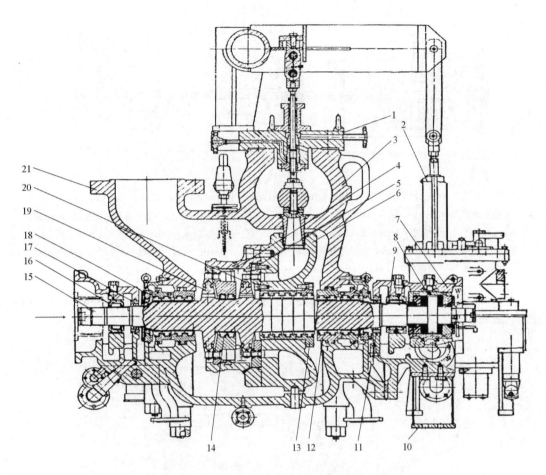

图 7.1.56 合成气压缩机高压背压式汽轮机结构图
Structure drawing of high-pressure synthesis gas compressor back pressure turbine

1—调节阀 Regulating valve；2—油动机 Oil motive；3—外汽缸 Outer cylinder；
4—隔板 Partition；5—喷嘴 Nozzle；6—喷嘴室 Nozzle chamber；
7—前轴承箱 Front bearing housing；8—推力轴承 Thrust bearing；
9—前轴承 Front bearing；10—支座 Support；11—前油封 Front oil seal；
12—高压端汽封 High pressure side steam seal；
13—内缸汽封 Inner cylinder steam seal；14—隔板汽封 Diaphragm gland；
15—转子 Rotor；16—后轴承 After bearing；17—后油封 Rear oil seal；
18—低压端汽封 Low side steam seal；19—第二级叶轮 Second impeller；
20—第一级叶轮 First stage impeller；21—排汽口 Exhaust hood

7 合成氨与尿素 Synthetic ammonia and Urea

4. 合成气压缩机密封油泵 Synthesis gas compressor seal oil pump

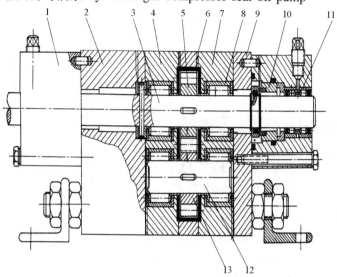

图 7.1.57 合成气压缩机密封油泵结构图 Structure drawing of synthesis gas compessor seal oil pump
1—机械密封箱体 Mechanical seal box；2、4、6、7、9—泵体 Pump；3—主动轴 Drive shaft；
5—主动齿轮 L-gear；8—主轴承 Main bearing；10—机械密封 Mechanical seals；
11—辅助轴承 Auxiliary bearing；12—从动轴 Driven shaft；13—从动齿轮 Driven gear

5. 合成气压缩机蓄压器 Syngas compressor pressure accumulator

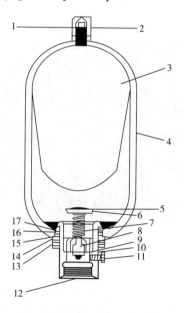

图 7.1.58 合成气压缩机蓄压器结构图
Structure drawing of syngas compressor pressure accumulator
1—气阀 Valve；2—阀帽 Bonnet；3—气囊 Airbag；4—壳体 Housing；5—膜片 Membrane；
6—阀头 Valve head；7—弹簧 Spring；8—活塞 Piston；9、13—锁紧螺母 Locknut；10—柱塞 Plunger；
11—管塞 Pipe plug；12—出口 Outlet；14—隔套 Spacer；15—O 形密封圈 O sealing ring；
16—垫圈 Washer；17—防挤压环 Anti-extrusion ring

7.1.5.4 氨冷冻机 Ammonia refrigeration

1. 氨冷冻机结构 Ammonia refrigeration machine structure

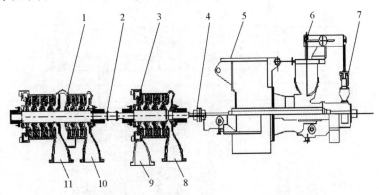

图7.1.59 氨冷冻机结构图 Structure drawing of ammonia refrigeration machine structure

1—高压缸 High-pressure cylinder；2、4—联轴器 Coupling；3—低压缸 Low-pressure cylinder；5—透平 Turbine；6—提板阀 Mention plate valve；7—伺服机构 Servo；8—低压缸入口 Low pressure cylinder inlet；9—低压缸出口 Low pressure cylinder outlet；10—高压缸入口 High pressure cylinder inlet；11—高压缸出口 High pressure cylinder outlet

2. 氨冷冻机低压缸 Low-pressure cylinder ammonia refrigeration machine

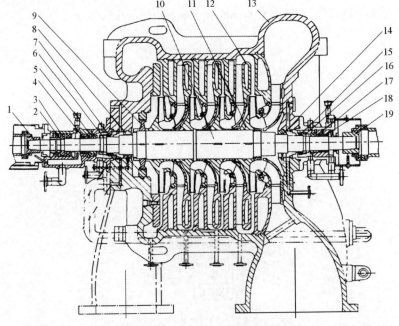

图7.1.60 氨冷冻机低压缸结构图
Structure drawing of low-pressure cylinder ammonia refrigeration machine

1、19—半联轴器 Coupling halves；2—推力轴承 Thrust bearing；3—推力盘 Thrust plate；4、18—径向轴承 Rodial bearing；5、17—减压衬套 Decompression bush；6、16—动环 Rotating ring；7、15—静环 Stationary ring；8、14—气封 Air sealing；9—平衡鼓 Balance disc；10—主轴 Shaft；11—叶轮 Wheel；12—隔板 Partition；13—壳体 Housing

3. 氨冷冻机高压缸 Ammonia Refrigeration high pressure cylinder

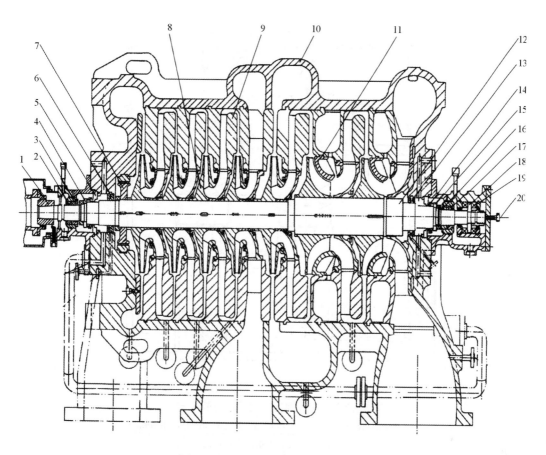

图 7.1.61　氨冷冻机高压缸结构图
Structure drawing of ammonia refrigeration high pressure cylinder

1—联轴器 Coupling；2、17—径向轴承 Rodial bearing；
3、16—减压衬套 Vacuum liner；4、15—动环 Rotating ring；
5、14—静环 Stationary ring；6、13—气封 Gas seal；
7—平衡鼓 Balance disc；8—叶轮 Wheel；9—隔板 Partition；
10—机壳 Housing；11—盖环 Cover ring；12—主轴 Shaft；
18—推力盘 Thrust plate；19—推力轴承 Thrust bearing；
20—轴位移探头 Shaft displacement probe

4. 氨冷冻机增速箱 Ammonia Refrigeration growth Box

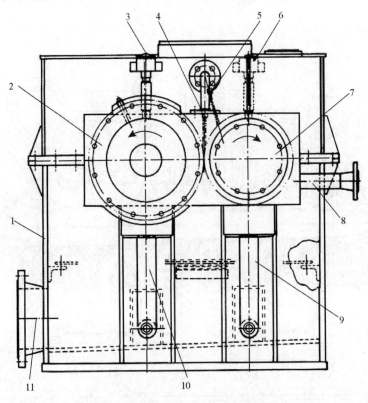

图 7.1.62　氨冷冻机增速箱结构图
Structure drawing of ammonia refrigeration growth box
1—箱体 Box；2—主动齿轮 Driving gear；
3，6—探针 Probe；4、5—喷油嘴 Injectors；
7—从动齿轮 Driven gear；8—进油管 Oil inlet；
9、10—轴承回油管 Bearing oil return pipe；
11—总回油管 Total oil return pipe

5. 氨冷冻机干气密封 Ammonia refrigeration machine dry gas seals

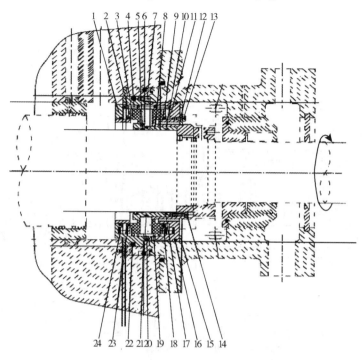

图 7.1.63 氨冷冻机干气密封结构图
Structure drawing of ammonia freeze drier group gas sealed

1、11—弹簧座 Spring seat；2、9—推环 Push ring；
3—静环 Stationary ring；4、7—防转销 Anti-rotation pin；
5—传动销 Drive pin；6—动环 Rotating ring；8—轴套 Sleeve；
10—压紧套 Compression sleeve；
12—定位板 Positioning plate；
13、14—连接螺钉 Connection screws；
15—垫片 Gasket；
16、17、18、21、22、23—O 形密封圈 O sealing ring；
19—卡圈 Clamp；20—螺钉 Screws；24—弹簧 Spring

6. 氨冷冻机驱动汽轮机 Ammonia refrigeration machine driven turbine

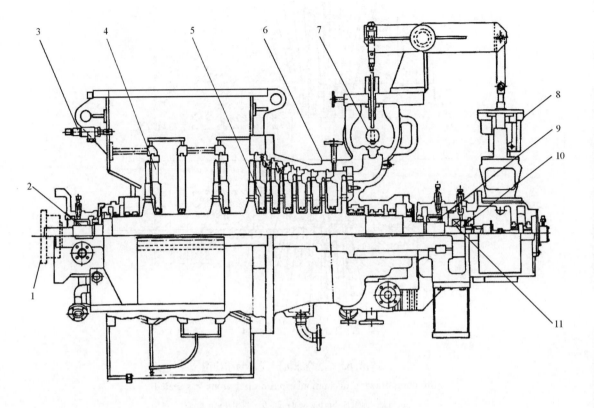

图 7.1.64 氨冷冻机驱动汽轮机结构图
Structure drawing of ammonia freeze drier drive steam turbine
1—联轴器 Couplings；2、9—径向轴承 Rodial bearing；
3—警报阀 Alarm valve；4—隔板 Partition；
5—叶轮 Wheel；6—壳体 Shell；
7—调节阀 Regulating valve；
8—油动机 Oil motivation；
10—推力盘 Thrust plate；
11—推力轴承 Thrust bearings

7 合成氨与尿素 Synthetic ammonia and Urea

7. 氨冷冻机止推盘 Ammonia refrigeration thrust plate

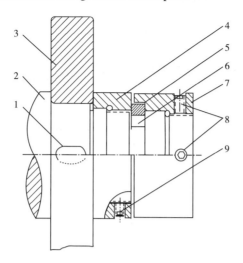

图 7.1.65 氨冷冻机止推盘结构图 Structure drawing of ammonia freeze drier stop shove plate
1—键 Key；2—轴 Shaft；3—推力盘 Thrust plate；4、7—螺母 Nut；5—中分楔紧环 Carve wedge ring；
6—楔形环槽 Wedge ring groove；8、9—锁紧螺钉 Locking screw

8. 氨冷冻机汽轮机主汽阀 Ammonia refrigeration machine turbine main steam valve

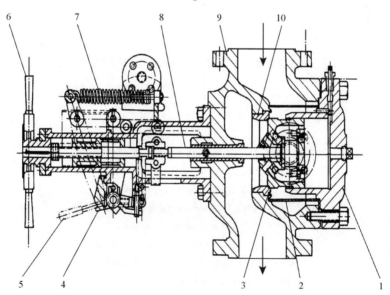

图 7.1.66 氨冷冻机汽轮机主汽阀结构图
Structure drawing of ammonia freeze drier steam turbine lead steam valve
1—端盖 Cover；2—滤网 Filter；3—阀头 Valve head；4—挂钩 Hook；
5—停车手柄 Parking handle；6—手轮 Hand wheel；7—弹簧 Spring；8—阀杆 Stem；
9—阀体 Body；10—阀座 Valve seat

7.1.6 液氨罐 Ammonia tank

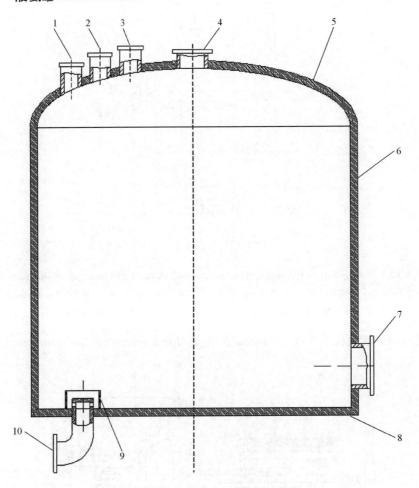

图 7.1.67 液氨罐结构图
Structure drawing of liquit ammonia storage tank

1—冷冻氨入口 Chilled ammonia inlet；
2—真空阀接口 Vacuum valve interface；
3—气体氨出口 Ammonia gas out；
4—上人孔 Upper manhole；5—罐顶 Tank top；
6—罐壁 Tank wall；7—下人孔 Lower manhole；
8—罐底 Bottom of the tank；9—挡板 Baffle；
10—冷冻氨出口 Export of frozen ammonia

7 合成氨与尿素 Synthetic ammonia and Urea

7.2 尿素 Urea

7.2.1 反应设备 Reaction equipment

7.2.1.1 脱氢反应器 Dehydrogenation reactor

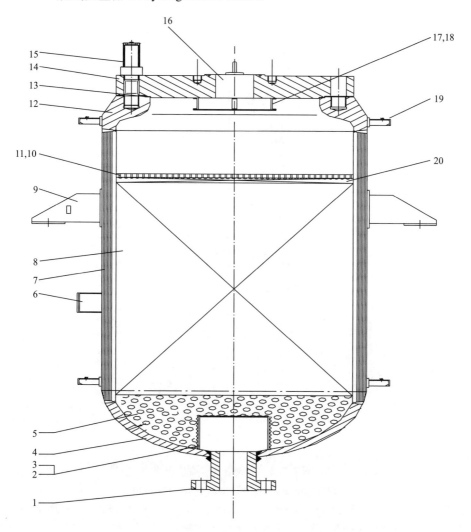

图 7.2.1 脱氢反应器结构图 Structure drawing of dehydrogenation reactor

1—气体出口 Gas outlet；2—多孔挡板 Porous baffle；3、10—丝网 Screen；4—封头 Head；
5、20—瓷球 Aluminum ball；6—铭牌 Nameplate；7—筒体 Cylinder；8—催化剂 Catalyst；
9—支座 Support；11—催化剂压板 Catalyst clamp；12—筒体连接体 Cylinder connecting；
13—齿形垫 Tooth profile pad；14—平盖 Flat；15—螺栓 Bolt；16—气体入口 Gas inlet；
17—支承筋 Strut；18—挡板 Baffle；19—保温支承圈 Insulation bearing ring

7.2.1.2 合成塔反应器 Synthetic tower reactor

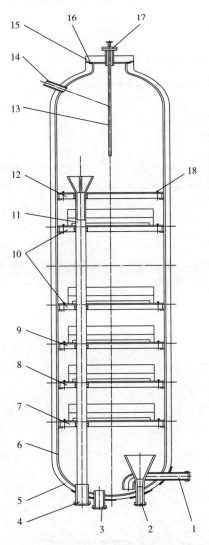

图 7.2.2 合成塔反应器结构图
Structure drawing of synthetic tower reactor

1—气体进口 Gas inlet；2—甲铵液出口 Methyl ammonium liquid outlet；3—液体进口 Liquid outlet；4—尿素溶液出口 Urea solution outlet；5—封头 Head；6—衬里 Lining；7、8、9、10、12—塔盘 Column tray；11—溢流管 Overflow pipe；13—钴源套管 Cobalt source casing pipe；14—气体出口 Gas outlet；15—平盖 Flat cover；16—齿形垫 Tooth profile pad；17—钴源套管法兰 Cobalt source casing pipe flange；18—钩头螺栓 Hooked bolt

7 合成氨与尿素 Synthetic ammonia and Urea

7.2.2 塔设备 Column equipment

7.2.2.1 精馏塔 Distillation column

1. 精馏塔结构 Structure of distillation rectifying tower

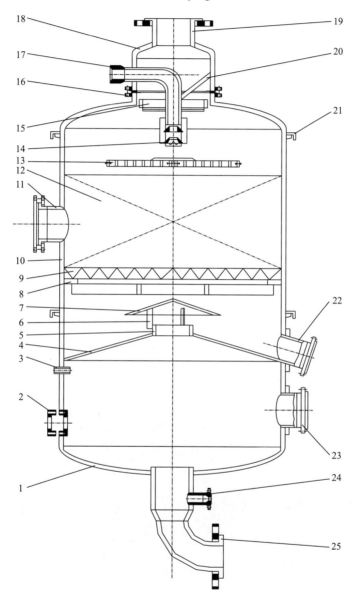

图 7.2.3 精馏塔结构图 Structure drawing of distillation tower

1、18—封头 Head；2—液位计接口 Liquid level gauge interface；3、24—液位变送器接口 Level transmitter interface；4—锥顶 Cone top；5—升气管 Riser；6—支承板 Support plate；7—升气帽 Liter gas cap；8—格栅板 Grid；9—波纹板 Corrugated board；10—筒体 Barrel；11—手孔 Hand hole；12—填料 Filler；13—分布器 Distributor；14—喷头 Nozzle；15—挡板 Backstop；16—封头法兰 End socket flange；17—尿液入口 Urea liquid inlet；19—气体出口 Gas outlet；20—筋板 Rib plate；21—保温支承圈 Insulation bearing ring；22、25—尿液出口 Urea liquid outlet；23—尿液返回口 Urea fluid return port

2. 精馏塔主要零部件 The major parts of the rectifyingtower

1）栅板 Grid tray

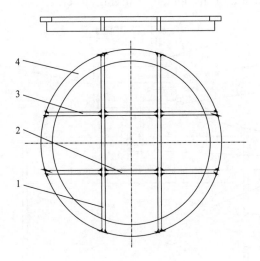

图 7.2.4　栅板 Grid tray

1、2、3—筋条 Ribs；4—钢板 Steel plate

2）分布器 Sparger

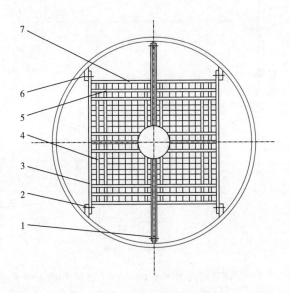

图 7.2.5　分布器 Sparger

1、3、4、5、7—栅条 Grid；

2—支耳 Support lug；

6—螺栓 Bolt

7 合成氨与尿素 Synthetic ammonia and Urea

7.2.2.2 造粒塔 Prilling tower

1. 造粒塔结构 Structure of prilling tower

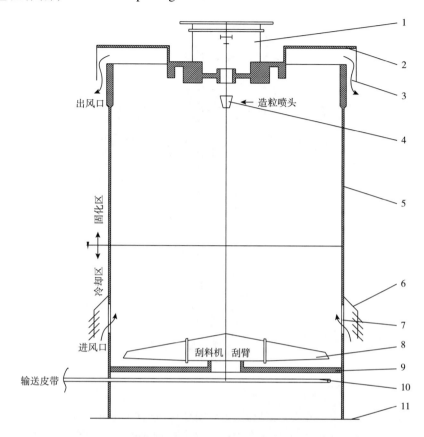

图 7.2.6 造粒塔结构图 Structure drawing of prilling tower

1—造粒间 Prilling case；2—塔头 Tower head；
3—出风口 Air outlet；4—造粒喷头 Prilling sprayer；
5—筒体 Barrel；6—风口护板 Wind gap defensive plate；
7—进风口 Air inlet；8—刮料机 Scraper；
9—刮料平台 Scraper platform；
10—输送皮带 Conveying belt；
11—塔底板 Tower bottom plate

2. 造粒塔主要零部件 The major parts of the prilling tower

1）造粒喷头 Prilling sprayer

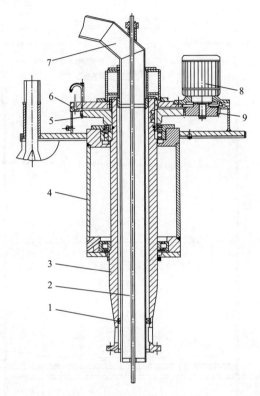

图 7.2.7　造粒喷头结构图 Structure drawing of prilling sprayer

1—定位环 Positioning ring；2—液位指示管 Level indicator tube；3—传动轴 Drive shaft；
4—支架 Support；5—从动齿轮 Large gear；6—齿轮箱盖 Gear case cover；7—下料管 Lower feeding tube；
8—电机 Motor；9—主动齿轮 Driving gear

2）刮料机刮臂 Scraper scrape arm

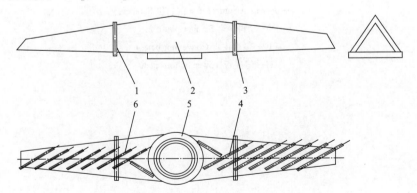

图 7.2.8　刮料机刮臂结构图 Structure drawing of scraper scrape arm

1、3—连接板 Connecting plate；2—刮臂 Scraping arm；
4、6—刮板 Scraping plate；5—圆锥罩 Cone cover

3）液力联轴器 Hydraulic coupling

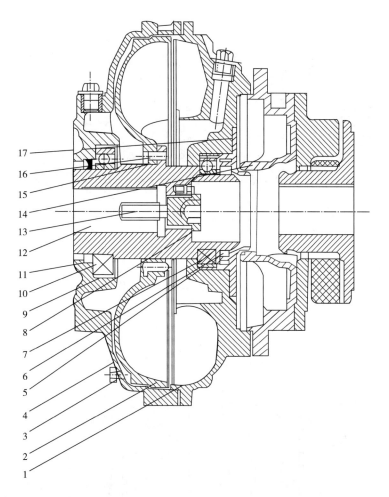

图 7.2.9 液力联轴器结构图 Structure drawing of hydraulic coupling

1—泵叶轮 Pump impeller；2—从动轮 Driven wheel；
3—可熔丝堵 Fusible plug；4、14—壳体 Shell；
5、11—密封环 Sealing ring；6、9—波型弹簧 Wave spring；
7、10—滚动轴承 Rolling bearing；8—锁紧销 Locking pin；
12—从动轴 Driven shaft；13—固定螺栓 Fixing bolt；
15—垫圈 Washer；16—铆钉 Rivet；
17—轴承压盖 Bearing gland

4）减速器 Reducer

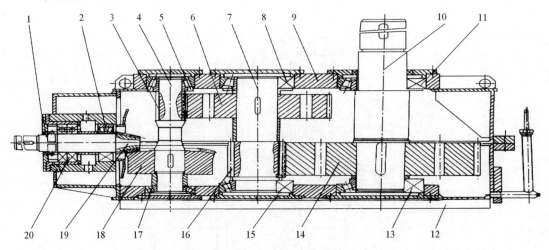

图 7.2.10　减速器结构图 Structure drawing of reducer

1—输入轴 Input shaft；2、5、8、11、13、15、17、20—轴承 Gearing；3、6、14、16—齿轮 Gear；4、7—中间轴 Middle shaft；9—上箱体 Upper case；10—输出轴 Output shaft；12—下箱体 Lower case；18、19—锥齿轮 Bevel gear

5）轴承箱 Bearing box

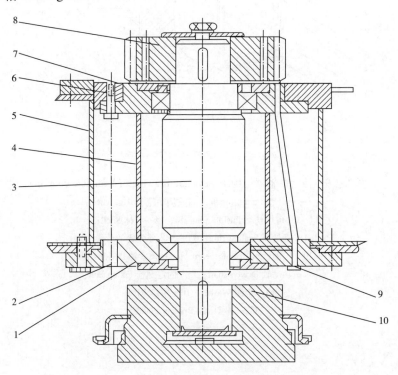

图 7.2.11　轴承箱结构图 Structure drawing of bearing box

1—下轴承座 Lower bearing seat；2—通孔 Through hole；3—主动轴 Main drive shaft；4—箱体 Box；5—机架 Rack；6—固定螺栓 Fixing bolt；7—上轴承座 Upper bearing seat；8—主动齿轮 Main drive pinion；9—润滑脂注入口 Lubricating grease conduit；10—半联轴器 Coupling half

7 合成氨与尿素 Synthetic ammonia and Urea

7.2.2.3 解析塔 Analytic tower

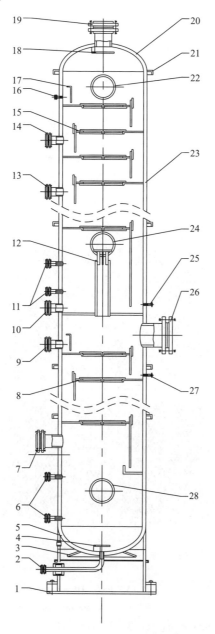

图 7.2.12 解析塔结构图 Structure drawing of analytic tower
1—裙座 Skirt；2、10—液体出口 Liquid outlet；3—锥形环 Conical ring；
4、17、18—挡板 Baffle；5—下封头 Lower head；6、11—液位计接口 Liquid level gauge interface；
7—蒸汽入口 Steam inlet；8、15—塔盘 Trays；9、14、16、25—液体进口 Liquid inlet；
12—升气管 Riser pipe；13—气体进口 Gas inlet；
19—气相出口 Vapor outlet；20—上封头 Upper head；
21—保温支承圈 Insulation bearing ring；
22、24、26、28—人孔 Manhole；23—筒体 Barrel；
27—温度计接口 Thermometer interface

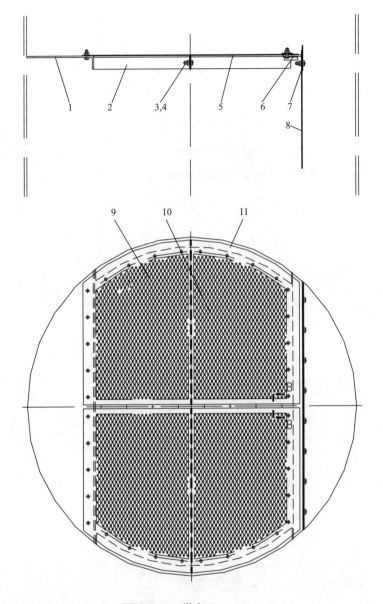

图 7.2.13 塔盘 Tower tray

1—受液盘 Liquid receiving disc；2—支承梁 The support beam；3—螺母 Nut；4—螺栓 Bolt；
5—支承圈 Support ring；6—支承板 Support plate；7—堰板 Weir plate；
8—降液板 Down-flow plate；9、11—塔板 Tray；10—卡子 Clip

7 合成氨与尿素 Synthetic ammonia and Urea

7.2.2.4 汽提塔 Stripping tower
1. 汽提塔结构 Structure of stripping Column

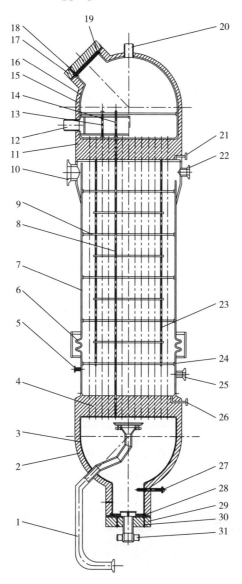

图 7.2.14　汽提塔结构图 Structure drawing of stripping tower

1—二氧化碳入口 Carbon dioxide inlet；2—下封头 Lower head；3、16—衬里 Lining；
4—下管板 Lower tube plate；5—惰性气体出口 Inert gas outlet；6—膨胀节 Expansion joint；
7—筒体 Barrel；8—汽提管 Stripping tube；9—折流板 Baffle；10—蒸汽进口 Steam inlet；
11—上管板 Upper tube plate；12—合成反应液进口 Synthesis reaction liquid inlet；13—定距杆 Distance bar；
14—管束 Tube bundle；15—上封头 Upper head；17—上端盖 Upper cover；18、30—检漏口 Leak hunting hole；
19、28—齿形垫 Tooth pad；20—气体出口 Gas outlet；21—放空口 Vent；22—安全阀接口 Safety valve interface；
23—拉杆 Tie rod；24—螺母 Nut；25—蒸汽出口 Steam outlet；26—排放口 Outfall；
27—液位计接口 Liquid level gauge interface；29—下端盖 Lower cover；31—汽提液出口 Stripping liquid outlet export

· 715 ·

2. 汽提塔主要零部件 The major parts of the stripping tower

1) 分布管 Distribution pipe

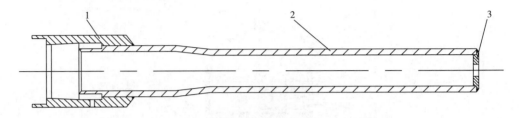

图 7.2.15　分布管 Structure drawing of distribution pipe

1—分布头 Distribution head；2—分布管 Distribution pipe；3—端盖 End cap

2) 定距杆 Distance bar

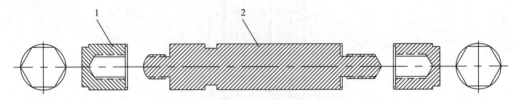

图 7.2.16　定距杆 Structure drawing of distance bar

1—定距帽 Distance cap；2—定距杆 Distance bar

3) 膨胀节 Expansion joint

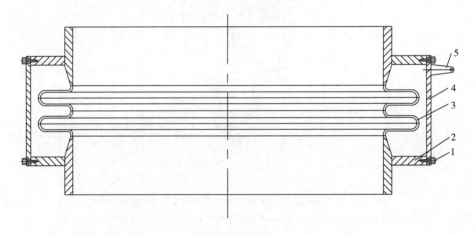

图 7.2.17　膨胀节 Structure drawing of expansion joint

1—螺栓 Bolt；2—护圈 Retainer；
3—波形膨胀节 Waveform expansion joint；
4—护板 Guard；5—吊耳 Lug

7.2.3 换热设备 Heat-exchange equipment
7.2.3.1 高压甲铵冷凝器 High pressure ammonia condenser
1. 高压甲铵冷凝器结构 Structure of high pressure ammonium condenser

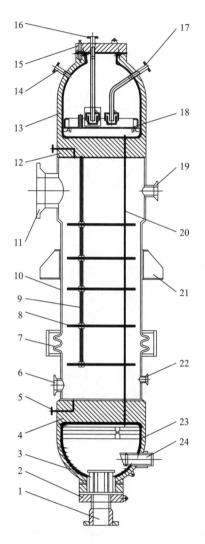

图7.2.18 高压甲铵冷凝器结构图 Structure drawing of high pressure ammonium condenser
1—液相出口 Liquid phase outlet；2—下盖板 Lower cover plate；3—下封头 Lower head；
4—下管板 Lower tube plate；5—检漏口 Leak hole；6—冷凝液进口 Condensate inlet；
7—膨胀节 Expansion joint；8—折流板 Baffle；9—拉杆 Tie rod；
10—筒体 Barrel；11—蒸汽出口 Steam outlet；12—上管板 Upper tube plate；
13—上封头 Upper head；14—气体出口 Gas outlet；15—上盖板 Upper cover plate；
16、17—液体进口 Liquid inlet；18—分布板 Distribution plate；
19—安全阀接口 Safety valve interface；20—换热管 Heat exchange tube；
21—支座 Support；22—蒸汽入口 Steam port；23—衬里 Lining；
24—气相出口 Vapor outlet

2. 高压甲铵冷凝器主要零部件 Major part of the high-pressure methyl ammonium condenser

1）气液分布器 Gas-liquid distributor

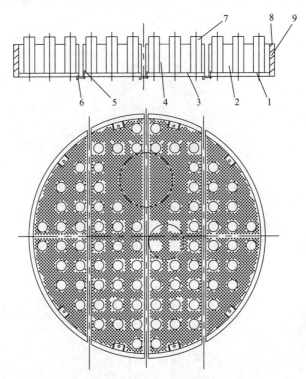

图 7.2.19　气液分布器 Structure drawing of gas-liquid distributor

1、3—分布板 Distribution plate；2、4—弧形板 Curved plate；
5—螺栓 Bolt；6—支承角钢 Bearing Angle；7—分布管 Distribution pipe；
8—勾头螺栓孔板 Eave tile bolt hole plate；9—筋板 Rib plate

2）膨胀节 Expansion joint

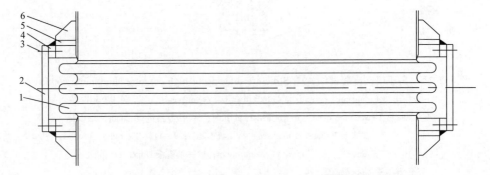

图 7.2.20　膨胀节 Structure drawing of expansion joint

1—波形膨胀节 Waveform expansion joint；2—护板 Guard；
3—螺栓 Bolt；4—圈板 Cycle board；
5—环板 Ring plate；6—筋板 Rib plate

7.2.3.2 高压甲铵洗涤器 High-pressure methyl ammonium scrubber

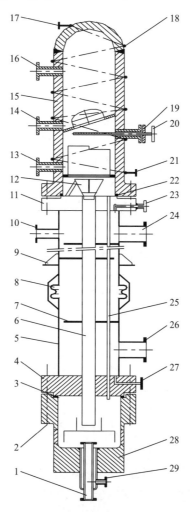

图 7.2.21 高压甲铵洗涤器结构图 Structure drawing of high pressure methyl ammonium scrubber

1—气体入口 Gas inlet；2—主螺栓 Main bolt；3、22—齿形垫 Tooth pad；4—下管板 Lower tube plate；
5—筒体 Shell；6—中心管 Center pipe；7—折流板 Baffle；8—膨胀节 Expansion joint；9—支座 Support；
10—安全阀接口 Safety valve interface；11—上管板 Upper tube plate；12—锥形口 Bell；
13—甲铵液出口 Methyl ammonium liquid outlet；14、20—气体出口 Gas outlet；
15—上管箱 Upper tube box；16—气体入口 Gas inlet；17—伴热线入口 Heating pipe interface；
18—伴热线 Heat pipe 19—出口管连接口 Outlet pipe connecting port；21—伴热线出口 Heater outlet；
23—放空口 Vent；24—冷凝液入口 Condensate inlet；25—换热管 Heat exchange tube；
26—冷凝液出口 Condensate outlet；27—排放口 Outfall；28—下管箱 Down tube box；
29—甲铵液入口 Methyl ammonium liquid inlet

7.2.4 泵 Pump

7.2.4.1 高压氨泵 High pressure ammonia pump

高压氨泵的作用是将来自合成氨车间送来的原料液氨(压力 2.0~2.4MPa，温度 40℃)加压至 16MPa，再经氨加热器加热到 55~75℃ 后送入高压喷射器。

The role of the high-pressure ammonia pump is to press raw ammonia from the ammonia unit is from 2.0 – 2.4Mpa, 40℃ to 16MPa, heat it by heater to 55 ~ 75℃ and then send it into the high-pressure injection devices.

1. 高压氨泵外形 Outline of high pressure ammonia pump

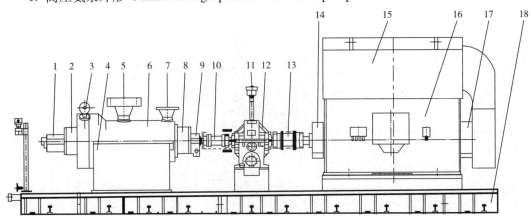

图 7.2.22　高压氨泵外形图 Outline of high pressure ammonia pump

1、9、14、17—轴承箱 Bearing box；2、8—机械密封 Mechanical seal；3—平衡管 Balance pipe；4—泵支座 Pump support base；5—泵出口 Pump outlet；6—泵体 Pump body；7—泵入口 Pump inlet；10、13—联轴器 Coupling；11—呼吸口 Breathing cap；12—增速器 Growth box；15—通风罩 Hood；16—电机 Motor；18—底座 Base board

2. 高压氨泵结构 Structure of high pressure ammonia pump

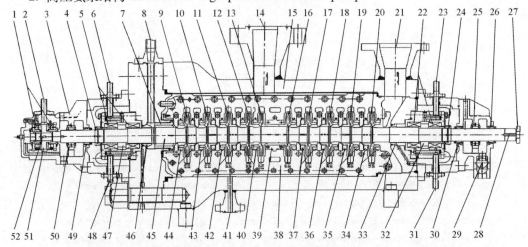

图 7.2.23　高压氨泵结构图 Structure drawing of high pressure ammonia pump

1—推力轴承 Thrust bearing；2—润滑油注入口 Lubricating oil return pipe；3—轴承箱 Bearing box；4、24—轴套卡环 Cutting sleeve；5、22—静环 Stationary ring；6、23—冲洗水出口 Washing water outlet；7、33—减压套 Decompression sleeve；8—平衡管 Balance pipe；9、10、11、12、13、16、17、18、19、20—叶轮 Wheel；14—泵出口 Pump outlet；15—壳体 Pump shell；21—泵入口 Pump inlet；25、50—径向轴承 Rodial bearing；26、30、49—油封 Oil seal；27—半联轴器 Coupling；28、52—键 Key；29—轴承箱回油视镜 Bearing box oil return sight glass；31、48—冲洗水入口 Washing water inlet；32、47—动环 Rotating ring；34、35、36、37、40、42、43、44—密封环 Seal ring；38—内缸体 Internal cylinder；39—中间隔板 Middle clapboard；41—排凝口 Condensate drain mouth；45—转子 Rotor；46—平衡鼓 Balance drum；51—推力盘 Thrust plate

7.2.4.2 高压甲铵泵 High pressure methyl ammonium pump
1. 高压甲铵泵外形 Outline of high pressure methyl ammonium pump

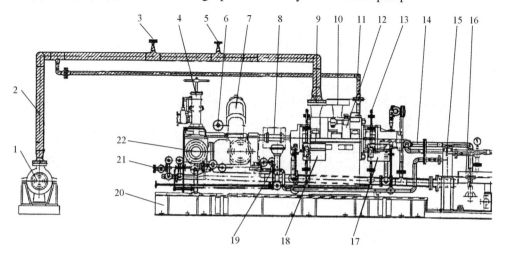

图 7.2.29　高压甲铵泵外形图 Outline of high pressure methyl ammonium pump

1—增压泵 Booster；2—增压泵出口管 Booster export pipeline；3—排气口 Exhaust port；
4—速关阀 Quick closing valve；5—仪表接口 Instrument interface；6—排气管 Exhaust pipe；
7—汽轮机 Turbine；8—联轴器 Couplings；9—泵入口 Pump inlet；
10—泵出口 Pump outlet；11—平衡管 Balance pipe；12—泵体 Pump body；
13—冲洗水管 Wash tube；14—润滑油管 Lubricant pipe；15—油过滤器 Oil filter；
16—油冷却器 Oil cooler；17、18—流量计 Flowmeter；19—调节阀 Regulating valve；
20—底座 Base support；21—控制油管 Oil controlling pipelines；22—蒸汽入口 Steam inlet

2. 高压甲铵泵结构 High pressure methyl ammonium pump

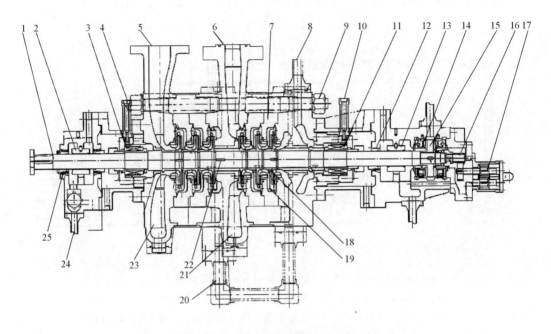

图 7.2.25 高压甲铵泵结构图
Structure drawing of high pressure methyl-ammonium pump

1—轴 Shaft；2、13—径向轴承 Rodial bearing；3、11—机械密封 Mechanical seals；
4、10—机械密封冲洗水管 Mechanical seal flush water tube；5—泵入口 Pump inlet；
6—泵出口 Pump outlet；7—级间隔板 Stage interval plate；8—平衡管 Balance pipe；
9—螺栓 Bolt；12、25—油封 Oil seal；14—推力轴承 Thrust bearing；15—推力盘 Thrust plate；
16—齿轮 Gear；17—轴头泵齿轮 Spindle head pump gear；18—叶轮 Wheel；
19—导流套 Deflector sets；20—连接管 Communication pipe；
21—出口蜗壳 Outlet volute；22—键 Key；23—入口蜗壳 Inlet volute；
24—回油管线 Return oil pipeline

7.2.5 压缩机 Compressor

7.2.5.1 二氧化碳压缩机 Carbon-dioxide gas compressor

二氧化碳压缩机是尿素装置的主要设备之一，其作用是将增压至 0.25MPa 的二氧化碳气体加压到 14.4MPa，送入汽提塔。

Carbon-dioxide gas compressor is one of the major equipment of urea plant, the role of it is to pressure the carbon dioxide from 0.25MPa to 14.4MPa, then send it to the stripping coloumn.

7 合成氨与尿素 Synthetic ammonia and Urea

1. 二氧化碳压缩机外形 Outline of carbon-dioxide gas compressor unit

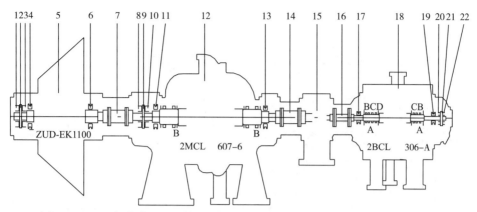

图 7.2.26 二氧化碳压缩机外形图 Outline of carbon-dioxide gas compressor unit

1、10、20—副推力轴承 Deputy thrust bearing；2、9、21—推力盘 Thrust plate；3、8、22—主推力轴承 Main thrust bearing；4、6、11、13、17、19—径向轴承 Rodial bearing；5—汽轮机 Turbine；7、14、16—联轴器 Couplings；12—低压缸 Low-pressure cylinder；15—增速器 Speed increaser；18—高压缸 High-pressure cylinder

2. 二氧化碳压缩机主要零部件 The main components of carbon dioxide compressor

1）低压缸 Low pressure cylinder

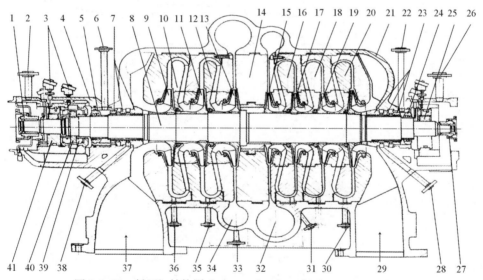

图 7.2.27 低压缸结构图 Structure drawing of low pressure cylinder

1、27—联轴器 Coupling；2、26—排油气管 Oil discharge exhaust pipe；3—轴承测温探头 Bearing temperature probe；4—轴承箱 Bearing box；5、24—油封 Oil seal；6、22—排气管 exhaust pipe；7—迷宫密封 Labyrinth seal；8—主轴 Shaft；9、11、13、15、17、19—叶轮 Wheel；10、12、16、18、20—级间隔板 Clapboard；14—中间隔板 Middle spacing board；21—壳体 Shell；23—汽封 Steam seal；25、41—径向轴承 Radial bearing；28、38—平衡管 Balance pipe；29—二段入口 Two-section inlet；30—二段出口 Two-section outlet；31—一段出口 One-section outlet；32、34—涡壳 Volute；33、35、36—排凝管 Condensate drain pipe；37—一段入口 One-section inlet；39—推力轴承 Thrust bearing；40—推力盘 Thrust plate

2）变速箱 Gearbox

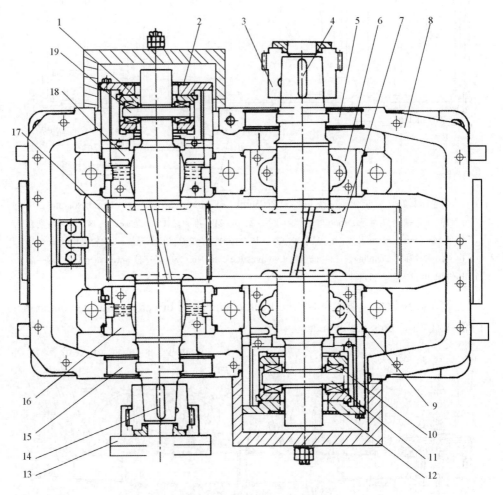

图 7.2.29　变速箱结构图 Structure drawing of gearbox

1、10—推力轴承 Thrust bearing；2、5、12、15—油封 Oil seal；

3、13—联轴器 Coupling；4、14—键 Key；

6、9、16、18—径向轴承 Radial bearing；

7—低速齿轮 Low speed shaft gear；

8—箱体 Housing；11、19—推力盘 Thrust plate；

17—高速齿轮 High speed gear

7 合成氨与尿素 Synthetic ammonia and Urea

3) 高压缸 High pressure cylinder

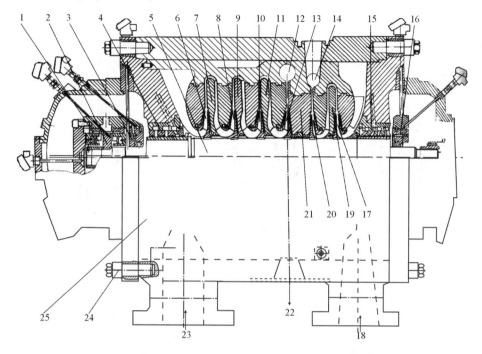

图 7.2.29 高压缸结构图 Structure drawing of high pressure cylinder

1—推力轴承 Thrust bearing；2—推力盘 Thrust plate；3、16—径向轴承 Radial bearing；
4、15—迷宫密封 Labyrinth seal；5—轴 Shaft；
6、8、10、13、17、20—叶轮 Wheel；7、9、11、19、21—级间隔板 Clapboard；
12、14、22—出口 Outlet；18、23—入口 Inlet；24—螺栓 Bolt；25—缸体 Cylinder

4) 低压缸转子 Low pressure cylinder rotor

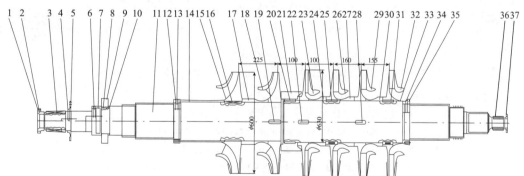

图 7.2.30 低压缸转子结构图 Structure drawing of low pressure cylinder rotor

1、37—半联轴器 Coupling half；2—螺钉 Bolt；3、8、16、19、20、22、25、28、31、36—键 Key；
4、6、13、34—紧固螺钉 Fastening screw；5—轴向位移指示盘 Axial displacement indicator disc；
7—止推盘紧固螺母 Thrust collar fastening nut；9—止推盘 Thrust plate；
10—调距环 Pitch ring；11—主轴 Shaft；12、35—压紧螺母 Compression nut；
14—固定套 Fixed sets；15、18、23、26、29、32—叶轮 Wheel；
17、24、27、30、33—叶轮套 Wheel set；21—平衡鼓 Balance druml

5）低压缸径向轴承 Low pressure cylinder radial bearing

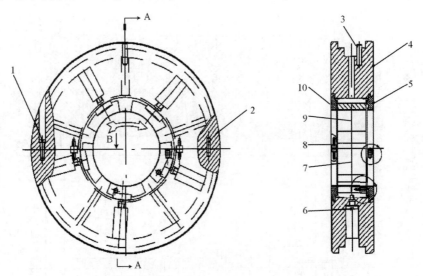

图 7.2.31　低压缸径向轴承结构图
Structure drawing of low pressure cylinder radial bearing

1、8—螺钉 Screw；2—定位销 Locating pin；
3—圆柱销 Cylindrical pin；4—轴承外壁 Bearing outer wall；
5、10—控油环 Oil control ring；6—垫圈 Washer；7、9—扇形面 Sector

6）高压缸转子 High pressure cylinder rotor

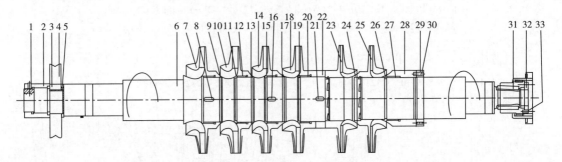

图 7.2.32　高压缸转子结构图
Structure drawing of high pressure cylinder rotor

1、29、30、33—紧固螺钉 Fastening screw；2—止推盘螺母 Thrust collar nut；
3、8、12、16、20、21、27、31—平键 Flat key；4—止推盘 Thrust collar；
5—调距环 Adjusting ring；6—主轴 Shaft；
7、10、14、18、23、25—叶轮 Wheel；
9、13、17、24、26—叶轮套 Impeller set；
11、15、19—级间定位环 Interstage positioning ring；
22—级间汽封 Staged gland sealing；
28—平衡鼓 Balancing drum；
32—连接盘 Connecting plate

7 合成氨与尿素 Synthetic ammonia and Urea

7)汽轮机速关阀 Steam turbine high pressure accident stop valve

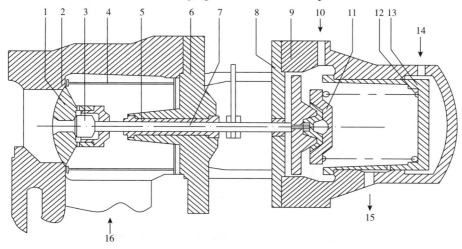

图 7.2.33 汽轮机速关阀结构图 Structure drawing of steam turbine high pressure accident stop valve

1—阀蝶 Butterfly valve；2—阀座 Valve seat；3—卸载阀 Unloading valve；4—蒸汽滤网 Steam strainer；
5—套筒 Sleeve；6—阀盖 Valve cover；7—阀杆 Stem；8—支座 Support；
9—油缸 Cylinder；10—速关油入口 Bump oil inlet；11—弹簧座 Spring seat；
12—弹簧 Spring；13—活塞 Piston；14—启动油入口 Start inlet；
15—油排出口 Oil discharge；16—蒸汽进口 Steam inlet

8)汽轮机高压提板式群阀 Steam turbine high pressure raise plate-type group valve

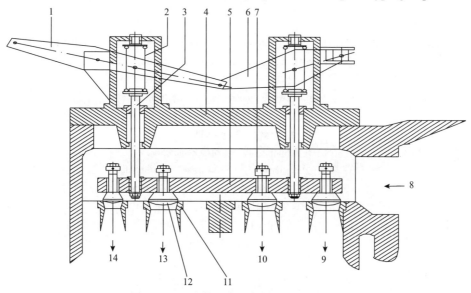

图 7.2.34 汽轮机高压提板式群阀结构图
Structure drawing of steam turbine high pressure raise plate-type group valve

1—前杠杆 Front lever；2—弹簧 Spring；3—提升杆 Lift lever；4—阀盖 Valve cover；
5—提升板 Riser；6—后杠杆 Back lever；7—罩型螺母 Cover nut；8—蒸汽进口 Steam inlet；
9、10、13、14—蒸汽出口 Steam outlet；11—阀座 Valve seat；12—阀头 Valve head

9）汽轮机油动机 Steam turbine oil-drive motor

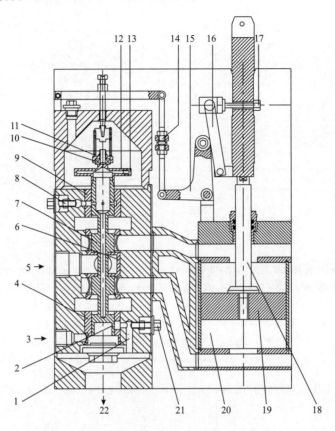

图 7.2.35　汽轮机油动机结构图 Structure drawing of steam turbine oil-drive motor

1—泄油孔 Vent hole；2—放油口 Drain hole；
3—进油口 Oil inlet；4、7、9—套筒 Sleeve；
5—动力油口 Power oil mouth；6—滑阀体 Slide valve；
8、21—调节阀 Regulating valve；
10—推力球轴承 Thrust ball bearing；
11—弹簧 Spring；12—转动盘 Rotating disk；
13—喷油孔 Injection hole；14—调节螺母 Adjusting nut；
15—弯角杠杆 Angled lever；16—反馈板 Feedback board；
17—调节螺栓 Adjusting bolt；18—活塞杆 Rod；
19—活塞 Piston；20—油缸 Cylinder；
22—排油口 Oil outlet

7 合成氨与尿素 Synthetic ammonia and Urea

10）汽轮机低压速关阀 Steam turbine low pressure accident stop valve

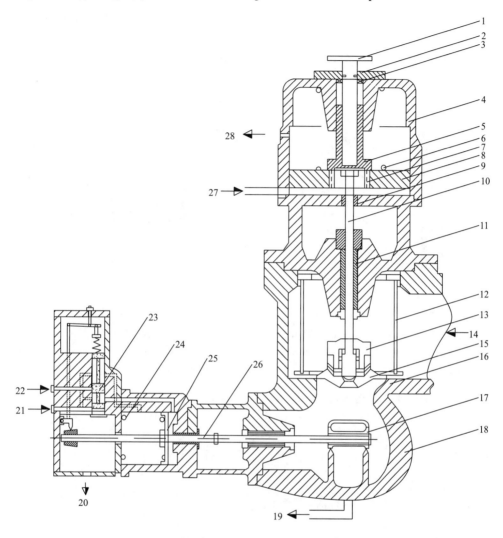

图 7.2.36　汽轮机低压速关阀结构图
Structure drawing of steam turbine low pressure accident stop valve

1—手轮 Hand wheel；2—阀盖 Valve cover；3—轴承 Bearing；
4—油缸 Cylinder；5、13—螺纹衬套 Threaded bushing；
6、24—弹簧 Springs；7—泄油孔 Vent hole；8、25—活塞 Piston；
9、11—衬套 Bush；10、26—阀杆 Valve rod；12—蒸汽滤网 Steam strainer；
14—蒸汽入口 Steam inlet；15、17—阀头 Valve head；16—阀体 Valve housing；
18—阀锥体 Valve cone；19—蒸汽出口 Steam outlet；20、28—油出口 Oil outlet；
21—油入口 Oil inlet；22—动力油入口 Power oil inlet；23—滑阀 Slide valve；
27—速关油入口 Bump oil mouth

7.2.5.2　二氧化碳增压机 Carbon dioxide supercharger

二氧化碳增压机的作用是将来自合成氨车间脱碳装置的二氧化碳气体增压至 0.25MPa 后送入二氧化碳压缩机继续加压。

The role of carbon dioxide supercharger is pressurizing the carbon dioxide from synthesis ammonia decarburization device to 0.25MPa, and then send it to carbon-dioxide gas compressor for continued inflating.

1. 二氧化碳增压机外形 Outline of carbon dioxide supercharger group

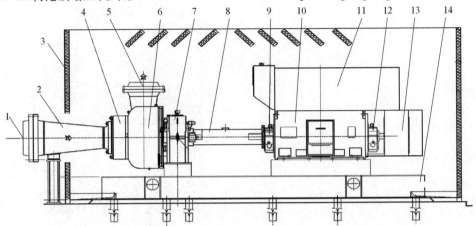

图 7.2.37 二氧化碳增压机外形图 Outline of carbon dioxide supercharger group
1—增压机入口 Supercharger inlet; 2—过滤器 Filter; 3—消音室 Noise room; 4—导叶 Guide vanes; 5—增压机出口 Turbocharger outlet; 6—壳体 Shell; 7—增速器 Speed increaser; 8—联轴器 Coupling; 9、12—轴承箱 Bearing box; 10—电机 Motor; 11—通风罩 Hood; 13—风扇罩 Fan cover; 14—底座 Base support

2. 二氧化碳增压机结构 Carbon dioxide supercharger

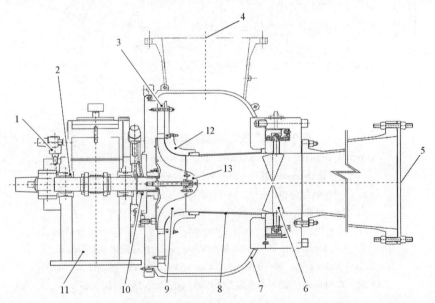

图 7.2.38 二氧化碳增压机结构图 Structure drawing of carbon dioxide supercharger
1—轴向位移探头 Axial displacement probe; 2—推力轴承 Thrust bearing; 3—扩压器 Diffuser; 4—增压机出口 Turbocharger outlet; 5—增压机入口 Supercharger inlet; 6—入口导叶 Inlet guide vanes; 7—壳体 Shell; 8—输送器 Conveyor; 9—叶轮 Wheel; 10—迷宫密封 Labyrinth seals; 11—支座 Support; 12—入口隔板 Inlet baffle; 13—叶轮锁紧螺母 Impeller lock nut

7.2.6 其它设备 Other equipments

7.2.6.1 高压喷射器 High pressure ejector

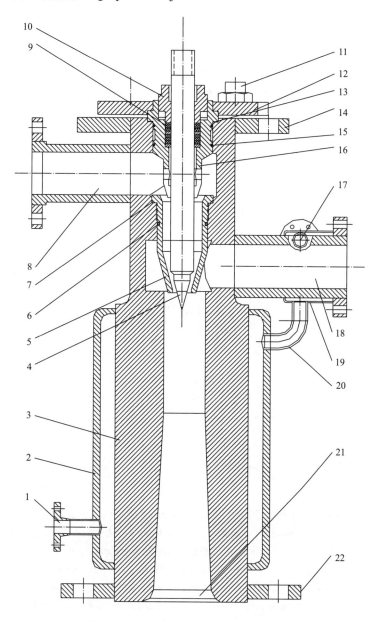

图 7.2.39 高压喷射器结构图 Structure drawing of high pressure ejector

1—夹套蒸汽出口 Jacket steam outlet；2—夹套 Jacket；3—机体 Compressor body；
4—调节杆 Adjustment lever；5—喷嘴 Nozzle；6、7、15—密封环 Seal ring；8—液氨入口 Ammonia entrance；
9—填料 Filler；10—填料压盖 Stuffing gland；11—填料压盖螺栓 Packing gland bolts；12—填料函压盖 Stuffing box gland；
13—垫片 Gasket；14，22—法兰 Flange；16—填料函 Stuffing box；17—夹套蒸汽进口 Jacketed steam inlet；
18—甲铵进口 Potassium inlet；19—接管 Connecting pipe；20—弯管 Elbow pipe；
21—甲铵出口 Potassium outlet

7.2.6.2 高压角阀 High pressure angle valve

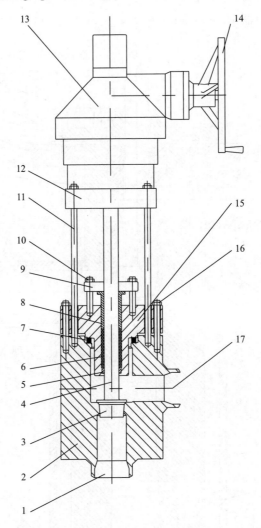

图 7.2.40　高压角阀结构图 Structure drawing of high pressure angle valve

1—阀入口 Valve inlet；2—阀体 Body；3—阀芯 Spool；
4—阀杆 Stem；5—填料底环 Packing the bottom ring；
6—填料 Packing；7—垫片 Gasket；8—填料压环 Packing pressure ring；
9—填料压盖 Stuffing gland；10、16—螺栓 Bolt；11—导向杆 Guide bar；
12—导向盘 Guide plate；13—伞齿轮 Bevel gear；14—手轮 Hand wheel；
15—阀盖 Valve cover；17—阀出口 Valve outlet

8 煤化工 Cool chemical industry

8.1 粉煤化工 Powdered coal chemical industry

8.1.1 反应设备 Reaction equipment

8.1.1.1 预变换炉 Pre-shift Converter

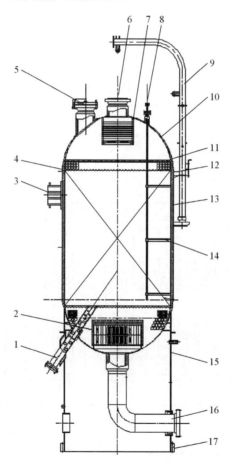

图 8.1.1 预变换炉结构图 Structure drawing of pre-shift Converter
1—卸料口 Discharging hole；2—气体捕集器 Outlet gas trap；3—吊耳 Lifting lug；4—金属丝网 Wire mesh；
5—人孔 Manhole；6—粗煤气入口 Raw gas inlet；7—气体分布器 Gas distributor；
8—热电偶接口 Thermocouple interface；9—塔顶吊柱 Top davit；10—封头 Head；11—格栅板 Grating；
12—保温支承圈 Insulating ring；13—筒体 Cylinder；14—支架 Support；15—裙座 Skirt；
16—变换气出口 Shift gas outlet；17—接地板 Ground plate

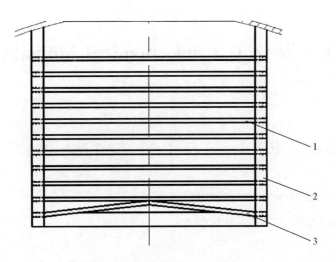

图 8.1.2 气体分布器 Gas distributor

1—环板 Annular plate；
2—筋板 Reinforcing plate；
3—锥形挡板 Conical baffle

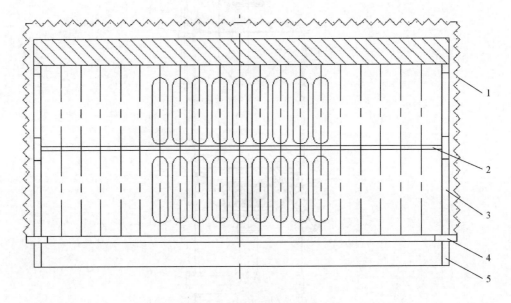

图 8.1.3 气体捕集器 Gas trap

1—金属丝网 Wire mesh；
2—加强环 Reinforcing ring；
3—筒体 Cylinder；
4—环板 Annular plate；
5—支承 Support

8 煤化工 Cool chemical industry

8.1.1.2 甲烷化炉 Methanation furnace

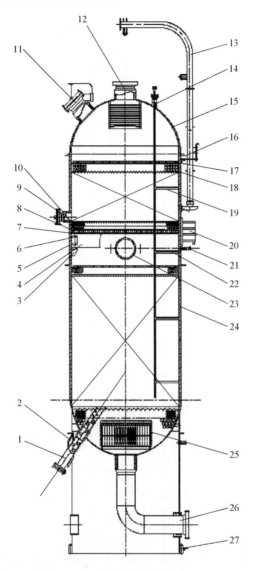

图 8.1.4 甲烷化炉结构图 Structure drawing of methanation furnace
1—催化剂出口 Discharged catalyst mouth；2—筋板 Reinforcing plate；
3、19、22—支架 Support；4—支承梁 Supporting beam；
5—连接板 Junction plate；6—螺栓 Bolt；7—支承圈 Support ring；
8—支承栅板 Support grid；9、18—金属丝网 Wire mesh；
10—脱硫剂出口 Desulfurization agent outlet；11、23—人孔 Manhole；
12—合成气入口 Syngas entrance；13—吊柱 Davit；14—热电偶接口 Thermocouple mouth；
15—封头 Head；16—保温支承圈 Insulating ring；17—压栅 Hold-down grid；
20—吊耳 Lifting lug；21—取样口 Sample connection；
24—筒体 Shell；25—气体捕集器 Outlet gas trap；26—合成气出口 Synthesizer；
27—接地板 Ground plate

8.1.2 塔设备 Tower equipment

8.1.2.1 二氧化碳吸收塔 Carbon dioxide absorbing tower

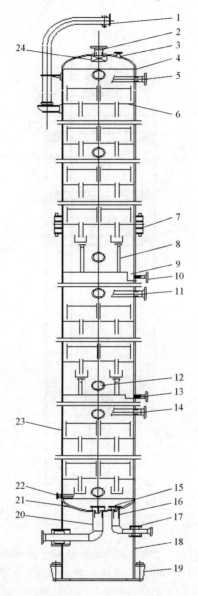

图 8.1.5 二氧化碳吸收塔结构图 Structure drawing of carbon dioxide absorbing tower

1—吊柱 Davit；2—净化气出口 Purified gas outlet；3—放空口 Vent；
4—保温支持圈 Insulation support ring；5—贫甲醇入口 Poor methanol inlet；
6—塔盘 Tray；7—吊耳 Lifting lug；8—升气管 Standpipe；9—集油箱 Oil sump tank；
10、13、16、20—富甲醇出口 Rich methanol outlet；
11、14—富甲醇入口 Rich methanol inlet；12—人孔 Manhole；
15—防涡器 Vortex breaker；17—补强板 Reinforcement plate；18—裙座 Skirt；
19—接地板 Ground plate；21—封头 Head；22—脱硫气入口 Desulfurization agent inlet；
23—筒体 Shell；24—除沫器 Demister

8.1.2.2 冷凝液汽提塔 Condensate stripping tower

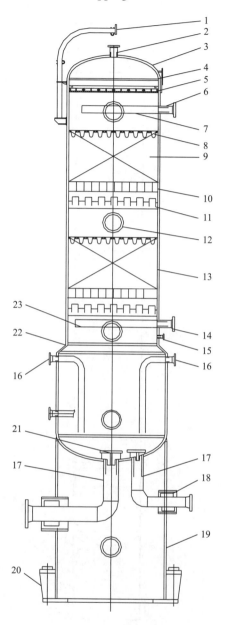

图 8.1.6 冷凝液汽提塔结构图 Structure drawing of condensate stripping tower

1—塔顶吊柱 Top davit；2—酸性气出口 Acidic gas exports；3—封头 Head；4—吊耳 Lifting lug；5—丝网除沫器 Mesh Demister；6—冷凝液入口 Condensate inlet；7—进料分布管 Feed distributed pipe；8—液体分布器 Liquid distributor；9—填料 Packing；10—填料支承 Packing support；11—集油箱 Oil sump tank；12—人孔 Manhole；13—筒体 Shell；14—低压蒸汽入口 Low pressure steam inlet；15—闪蒸蒸汽入口 Flash steam inlet；16—冷凝液出口 Condensate outlet；17—循环液入口 Circulating fluid inlet；18—补强板 Reinforcement plate；19—裙座 Skirt；20—接地板 Ground plate；21—防涡器 Vortex breaker；22—保温支承圈 Insulation support ring；23—气体分布器 Gas distributor

8.1.2.3 硫化氢吸收塔 Hydrogen sulfide absorption tower

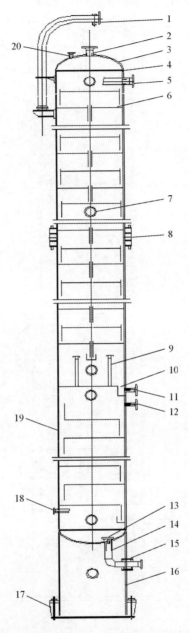

图 8.1.7 硫化氢吸收塔结构图 Structure drawing of hydrogen sulfide absorption tower
1—塔顶吊柱 Top davit；2—脱硫气出口 Desulfurization agent inlet；3—封头 Head；4—保温支承圈 Insulation support ring；5—主洗甲醇入口 Main washing methanol inlet；6—塔盘 Tray；7—人孔 Manhole；8—吊耳 Lifting lug；9—升气管 Standpipe；10—集油箱 Oil sump tank；11—主洗甲醇出口 Main washing methanol outlet；12—预洗甲醇入口 Prewash methanol inlet；13—防涡器 Vortex breaker；14—预洗甲醇出口 Prewash methanol outlet；15—补强板 Reinforcement plate；16—裙座 Skirt；17—接地板 Ground plate；18—变换气入口 Shift gas inlet connection pipe；19—筒体 Shell；20—放空口 Vent

8.1.3 换热设备 Heat-exchange equipment

8.1.3.1 低变出口换热器 Low variable outlet heat exchanger

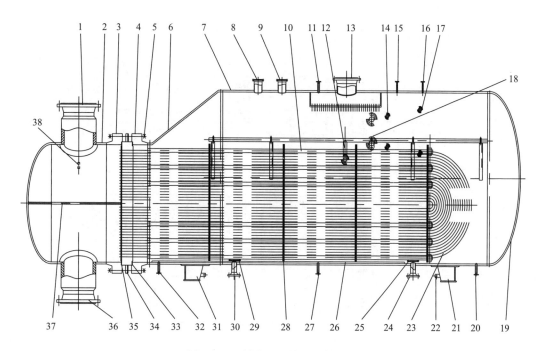

图 8.1.8 低变出口换热器结构图
Structure drawing of low variable outlet heat exchanger

1—变换气入口 Shift gas inlet；2—管箱 Channel；
3—管箱法兰 Channel flange；4—筒体法兰 Shell hange；
5—双头螺栓、螺母 Double-head stud bolts/nuts；
6—偏心锥筒 Eccentric conical shell；7—筒体 Shell；
8、9—安全阀接口 Safety valve interface；
10—定距管 Spacer tube；11—放气口 Gas vent；
12、27、32—排放口 Outlet；13—低压蒸汽出口 Low pressure steam outlet；
14、18—自控液位计接口 Automatic level gauge mouth；
15—就地压力计接口 Pressure gauge mouth；
16—自控压力计接口 Automatic pressure gauge mouth；
17—就地液计接口 Liquid level gauge mouth；19—封头 Head；
20—排液口 Liquid outlet；21—滑动支座 Sliding saddle；
22—接地板 Ground plate；23—U形换热管 U-shaped heat exchange tube；
24、30—低压水入口 Low pressure water inlet；
25、29—防冲挡板 Impingement baffle；
26—滑道 Slide；28—支承板 Support plate；
31—固定支座 Fixed support；33—筒体侧垫片 Channel side gasket；
34—定位环 Locking lug；35—管箱侧垫片 Channel gasket；
36—变换气出口 Shift gas outlet；37—分程隔板 Pass partition plate；
38—吊耳 Lifting lug

8.1.3.2 贫/富甲醇换热器 Poor/rich methanol heat exchanger

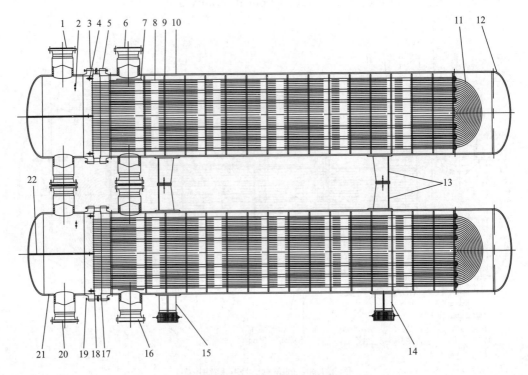

图 8.1.9 贫/富甲醇换热器结构图
Structure drawing of poor/rich methanol heat exchanger

1—富甲醇入口 Rich methanol outlet；
2—管箱吊耳 Channel lifting lug；
3—管箱法兰 Channel flange；
4—管板 Tube sheet；
5—筒体法兰 Shell hange；6—贫甲醇入口 Poor methanol inlet；
7—防冲挡板 Impingement baffle；
8—定距管 Spacer tube；
9—折流板 Baffle plate；10—筒体 Shell；
11—U 形换热管 U-shaped heat exchange tube；
12—封头 Head；13、14—活动支座 Mobile saddle；
15—固定支座 Fixed support；16—贫甲醇出口 Poor methanol outlet；
17—筒体侧垫片 shell side gasket；
18—管箱侧垫片 Channel gasket；
19—双头螺栓、螺母 Double-head stud bolts/nuts；
20—富甲醇出口 Rich of methanol export；21—管箱 Channel；
22—分程隔板 Pass partition plate

8.1.3.3 中压废热锅炉 Medium pressure waste heat boiler

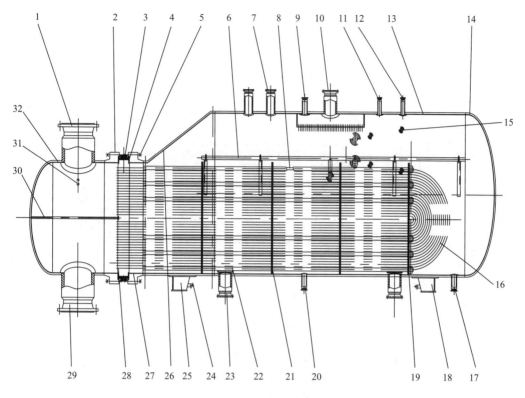

图 8.1.10 中压废热锅炉结构图
Structure drawing of medium pressure waste heat boiler

1—变换气入口 Shift gas inlet；2—管箱法兰 Channel flange；
3—密封垫 Sealing gasket；4—筒体侧垫片 Shell pass side gasket；
5—双头螺栓、螺母 Double-head stud bolts/nuts；
6—定距管 Spacer tube；7—安全阀接口 Safety valve interface；
8—拉杆 Tie rod；9—放空口 Vent；
10—中压蒸汽出口 Medium pressure steam outlet；
11—就地压力计接口 Pressure gauge mouth；
12—自控压力计接口 Automatic pressure gauge mouth；
13—筒体 Shell；14—封头 Head；
15—就地液面计口 Liquid level gauge mouth；
16—U 形换热管 U-shaped heat exchange tube；
17—排液口 Liquid outlet；18—活动支座 Mobile saddle；
19、21—支承板 Support plate；20—排污口 Drain outlet；
22—防冲挡板 Impingement baffle；
23—冷凝液入口 Condensate inlet；24—接地板 Ground plate；
25—固定支座 Fixed support；26—偏心锥壳 Eccentric conical shell；
27—筒体法兰 Shell hange；28—管箱侧垫片 Channel gasket；
29—变换气出口 Shift gas outlet；30—分程隔板 Pass partition plate；
31—吊耳 Lifting lug；32—管箱 Channel

8.1.3.4 低压废热锅炉 Low pressure waste heat boiler

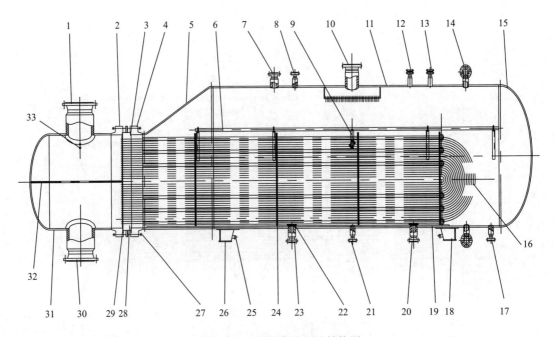

图 8.1.11　低压废热锅炉结构图
Structure drawing of low pressure waste heat boiler

1—变换气入口 Shift gas inlet；
2—管箱法兰 Channel flange；
3—筒体侧垫片 Shell pass side gasket；
4—筒体法兰 Shell hange；
5—偏心锥筒 Eccentric conical shell；
6—定距管 Spacer tube；7—安全阀接口 Safety valve interface；
8—放空口 Vent；9、21—排污口 Drain outlet；
10—低压蒸汽出口 Low pressure steam outlet；
11—筒体 Shell；12—就地压力计接口 Pressure gauge mouth；
13—自控压力计接口 Automatic pressure gauge mouth；
14—液位计接口 Liquid level gauge interface；15—封头 Head；
16—U 形换热管 U-shaped heat exchange tube；
17—排液口 Liquid outlet；18—活动支座 Mobile saddle；
19—滑道 Slide；20、23—锅炉水入口 Boiler water inlet；
22—防冲挡板 Impingement baffle；24—支承板 Support plate；
25—接地板 Ground plate；26—固定支座 Fixed support；
27—双头螺栓、螺母 Double-head stud bolts/nuts；
28—管板 Tube sheet；29—管箱侧垫片 Channel side gasket；
30—变换气出口 Shift gas outlet；
31—管箱 Channel cylinder；
32—分程隔板 Pass partition plate；
33—吊耳 Lifting lug

8.1.4 泵 Pump

8.1.4.1 废热锅炉给水泵 Waste heat boiler feed pump

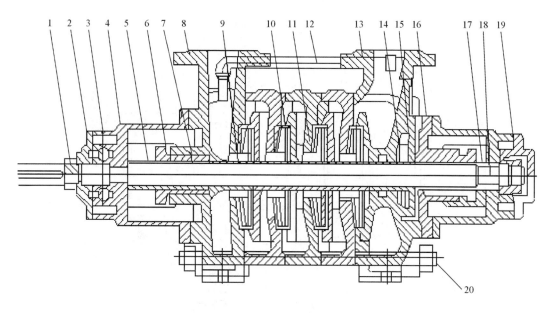

图 8.1.12 废热锅炉给水泵结构图
Structure drawing of waste heat boiler feed pump

1—轴套螺母 Shaft sleeve nut;
2—轴承盖 Bearing cap;
3—轴承 Bearing;
4—轴承箱 Bearing box;
5、17—轴套 Sleeve;
6—密封压盖 Packing gland;
7—机械密封 Mechanical seal;
8—进水口 Suction water inlet;
9—密封环 Seal ring;
10—叶轮 Wheel;
11—隔板 Partition;
12—平衡管 Balance pipe;
13—出水口 Discharge water outlet;
14—平衡座 Balance seat;
15—平衡盘 Balance plate;
16—泵盖 Pump cover;
18—轴 Shaft;
19—锁紧螺母 Locknut;
20—壳体连接螺栓 Shell connecting bolts

8.1.4.2 硫化氢吸收塔进料泵 Hydrogen sulfide absorption tower feed pump

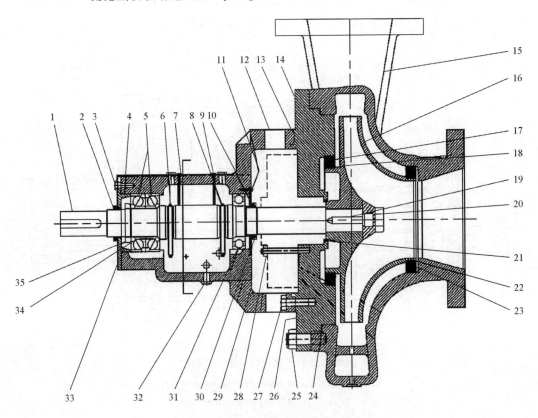

图 8.1.13 硫化氢吸收塔进料泵结构图
Structure drawing of hydrogen sulfide absorption tower feed pump

1—轴 Shaft；2、30—油封 Oil seal；3—端盖螺栓 End cover bolt；
4—O 形密封圈 O sealing ring；5—止推轴承 Thrust bearing；
6、8—甩油环 Flinger；7、9—抛油环挡圈 Ring oiler stop pin；
10—径向轴承端盖 Radial bearing end cover；
11—螺栓 Bolt；12、29—轴承箱 Bearing box；
13、24—垫片 gasket；14—泵盖 Seal chamber cover；
15—泵出口 Pump outlet chamber；
16、18—叶轮 Wheel；17—耐磨环 Wear ring；
19—泵轴 Impeller shaft；20—叶轮锁紧螺母 Impeller lock nut；
21—叶轮衬套 Impeller bush；22—叶轮密封环 Impeller labyrinth seal ring；
23—泵体耐磨环 Pump body wear ring；25、27、28—泵体螺栓 Pump body bolt；
26—泵体 Pump body；31—径向轴承 Radial bearing；32—排油孔 Oil outlet；
33—止推轴承盖 Thrust bearing cover；
34—止推轴承锁紧垫片 Thrust bearing locking gasket；
35—止推轴承锁紧螺母 Thrust bearing locking nut

8.1.4.3 液氨泵 Ammonia pump

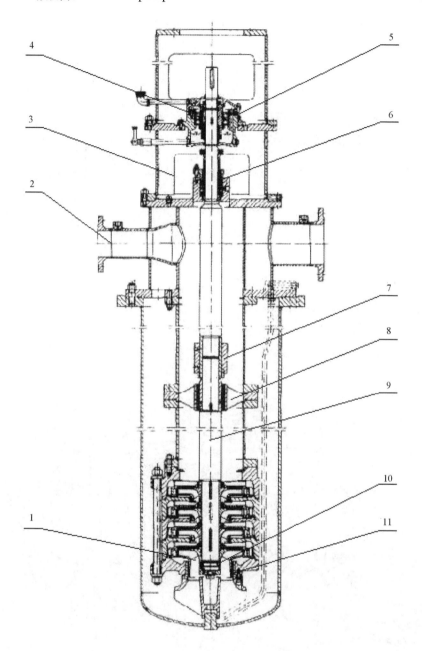

图 8.1.14 液氨泵结构图 Structure drawing of ammonia pump
1—叶轮 Wheel；2—泵出口 Outlet；3—填料箱体 Packing box；
4—向心推力球轴承 Radial thrust ball bearing；
5—轴承箱 Bearing box；6—轴封 Shaft seal；
7—联轴器 Pump coupling；8—中间支承 Intermediate support；
9—泵轴 Shaft；10—定位环 Locating ring；
11—轴承 Bearing

8.1.5 压缩机 Compressor

8.1.5.1 氨气螺杆压缩机 Ammonia screw compressor

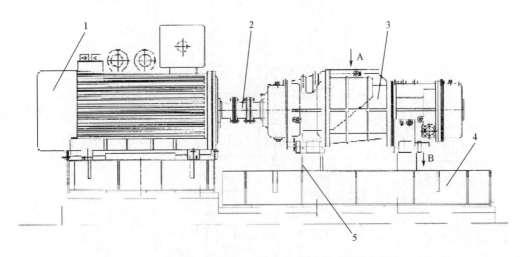

图 8.1.15　氨气螺杆压缩机外形图 Outline drawing of ammonia compressor
1—电动机 Motor；2—联轴器 Coupling；3—螺杆压缩机 Screw compressor；
4—底座 Base；5—支承 Support；A—吸气 Suction；B—排气 Exhaust

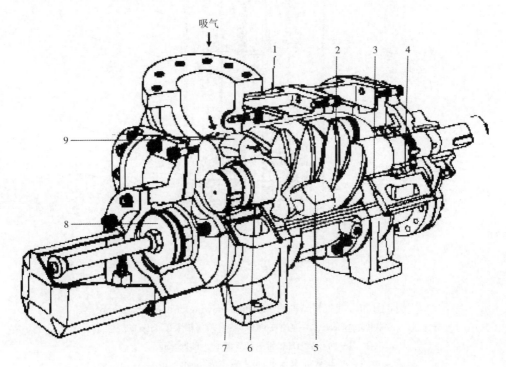

图 8.1.16　氨气螺杆压缩机结构图 Structure drawing of ammonia compressor
1—机体 Compressor body；2—阳螺杆 Male rotor；3—滑动轴承 Sliding bearing；4—滚动轴承 Roller bearing；
5—调节滑阀 Regulating slide valve；6—密封 Seal；7—平衡活塞 Balance piston；
8—调节滑阀控制活塞 Regulating slide valve control piston；9—阴螺杆 Female rotor

8.1.5.2 硫化氢循环压缩机 Hydrogen sulphide circulation compressor

1. 压缩机布置图 Layout of compressor

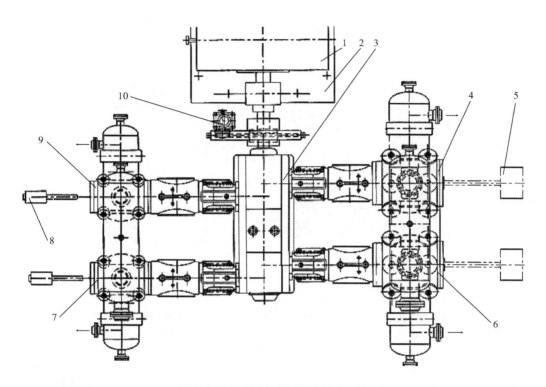

图 8.1.17 硫化氢循环压缩机布置图
Layout drawing of hydrogen sulphide circulation compressor

1—电动机 Motor;
2—保护罩 Shielding component;
3—机身 Frame component;
4、6—一级气缸 Level 1 cylinder component;
5—一级活塞 Level 1 piston component;
7、9—二级气缸 Level 2 cylinder component;
8—二级活塞 Level 2 piston component;
10—盘车装置 Turning component

2. 压缩机外形 Outline of compressor

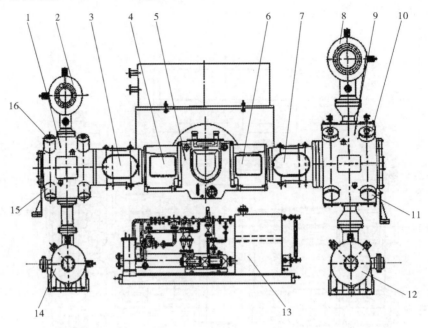

图 8.1.18 硫化氢循环压缩机外形图 Outline drawing of hydrogen sulphide circulation compressor
1—二级气缸 Level 2 cylinder component；2—二级进气缓冲器 Level 2 intake buffer；3—二级接筒 Level 2 coupling；4—二级中体 Level 2 midbody；5—机身 Frame component；6——级中体 Level 1 midbody；7——级接筒 Level 1 coupling；8——级进气缓冲器 Level 1 intake buffer；9——级气缸 Level 1 cylinder component；10——级吸气阀 Level 1 air suction valve；11——级排气阀 Level 1 exhaust valve；12——级排气缓冲器 Level 1 exhaust buffer；13—油站 Oil station；14—二级排气缓冲器 Second exhaust buffer；15—二级排气阀 Level 2 exhaust valve；16—二级吸气阀 Level 2 air suction valve

3. 压缩机结构 Structure of compressor

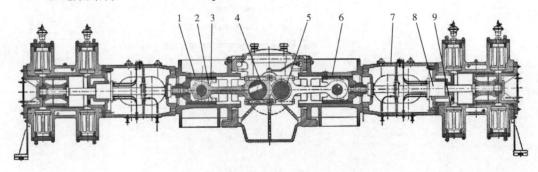

图 8.1.19 硫化氢循环压缩机结构图
Structure drawing of hydrogen sulphide circulation compressor
1、5—十字头 Cross component；2、4—连杆 Connecting rod component；
3—曲轴 Crankshaft component；6—接筒 Coupling component；
7—连接体 Connecting body；8—填料函 Packing；
9——级气缸 Level 1 cylinder component

8.1.5.3 氨制冷离心压缩机 Ammonia refrigeration centrifugal compressor

1. 压缩机高压缸 Compressor high pressure cylinder

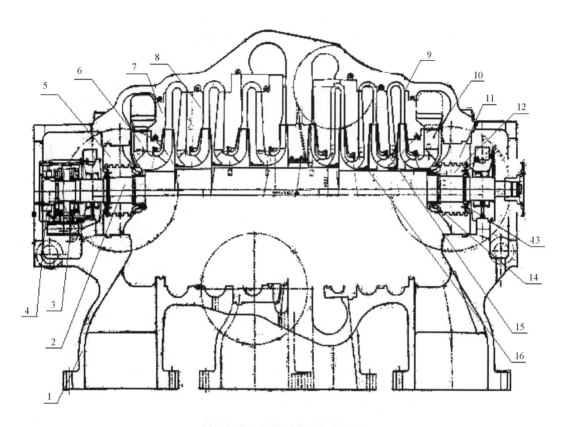

图 8.1.20 压缩机高压缸结构图
Structure drawing of compressor high pressure cylinder

1—缸体 Cylinder；2—转子 Rotor；
3—径向轴承 Radial bearing；
4—止推轴承 Thrust bearing；
5、11—密封体 Seal body；
6—入口导流器 Inlet fluid director；
7—入口蜗壳 Inlet volute；
8—级间隔板 Interstage partition；
9—出口蜗壳 Outlet volute；
10—出口导流器 Outlet fluid director；12—干气密封 Dry gas seal；
13—气封 Gas seal；
14—主轴气封 Main shaft gas seal；
15—叶轮密封环 Impeller labyrinth seal ring；
16—级间迷宫密封 Interstage labyrinth seal

2. 压缩机低压缸 Compressor low pressure cylinder

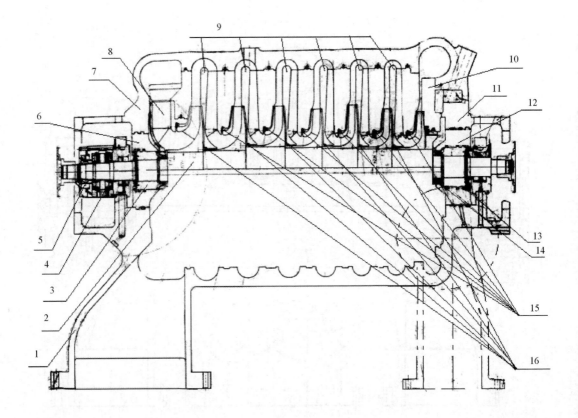

图 8.1.21 压缩机低压缸结构图
Structure drawing of compressor low pressure cylinder

1—缸体 Cylinder;
2—转子 Rotor;
3、12—干气密封 Dry gas seal;
4—径向轴承 Radial bearing;
5—推力轴承 Thrust bearing;
6、11—密封体 Seal body;
7—入口蜗壳 Inlet volute;
8—入口导流器 Inlet fluid director;
9—级间隔板 Interstage partition;
10—出口蜗壳 Outlet volute;
13—气封 Gas seal;
14—主轴密封 Main shaft seal;
15—叶轮密封环 Impeller labyrinth seal ring;
16—级间迷宫密封 Interstage labyrinth seal

8 煤化工 Cool chemical industry

3. 盘车装置 Turning equipment

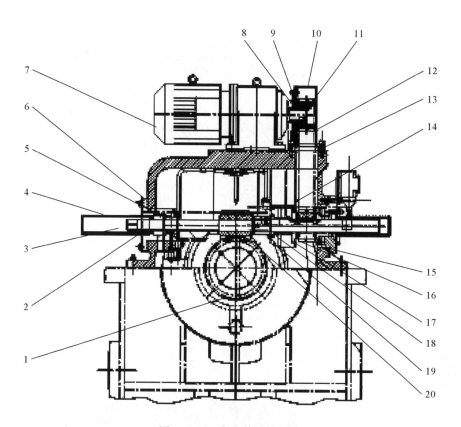

图 8.1.22 盘车装置结构图
Structure drawing of turning equipment

1—蜗轮 Worm wheel；
2、8—油封 Oil seal；
3—轴 Shaft；4—轴帽 Shaft cap；
5、16、19—轴承 Bearing；6、15—轴承座 Bearing seat；
7—电机 Motor；9—轴承盖 Bearing cover；
10—链轮罩 Sprocket cover；
11—主动链轮 Drive sprocket；
12—齿形链 Tooth shape chain；
13—调整垫 Adjustable pad；
14—喷嘴 Nozzle；
17—从动链轮 Passive sprocket；
18—止推轴承 Thrust bearing；
20—蜗杆 Worm rod

4. 干气密封 Dry gas seal

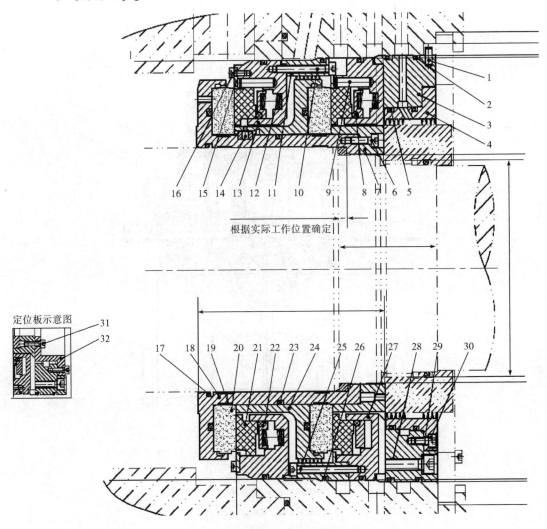

图 8.1.23 干气密封结构图 Structure drawing of dry gas seal

1、9—圆柱销 Cylindrical pin;

2、5、8、17、23、26—O 形密封圈 O sealing ring;

3—梳齿座 Comb seat; 4—隔离梳齿环 Isolation comb ring;

6、10、16、25、28、30、31—螺栓 Screw; 7—压紧套 Compress sleeve;

11—中间梳齿环 Middle comb ring; 12—弹簧 Spring;

13—介质侧弹簧座 Medium side spring seat;

14—传动键 Driving key; 15—限位环 Stop collar;

18—轴套 Sleeve; 19—公差环 Tolerance ring; 20—动环 Rotating ring;

21—静环 Static ring; 22—推环 Drive ring; 24—动环座 Rotating ring seat;

27—大气侧弹簧座 Atmospheric side spring seat;

29—垫片 Gasket; 32—定位板 Locating plate

8 煤化工 Cool chemical industry

8.1.6 其它设备 Other equipments

8.1.6.1 减温器 Desuperheater

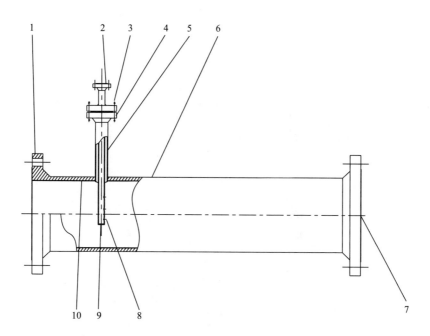

图 8.1.24 减温器结构图 Structure drawing of desuperheater

1—主管入口 Main pipe inlet；
2—减温水入口 Desuperheating water inlet；
3—双头螺栓、螺母 Double-head stud bolts/nuts；
4—减温套管法兰 Desuperheating water tube socket；
5—减温水套管 Desuperheating water tube socket；
6—减温器集箱 Desuperheater header；
7—主管出口 Main pipe flange；
8—喷嘴 Nozzle；
9—笛形喷嘴 Flute nozzle；
10—内衬 Lining

8.1.6.2 精密过滤器 Ultrafilter

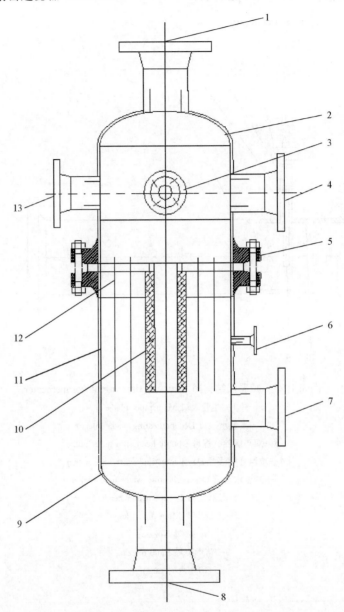

图 8.1.25 精密过滤器结构图 Structure drawing of ultrafilter

1—反洗进水口 Backwashing water inlet；2—封头 Head；
3、6—压力计接口 Differential pressure gauge mouth；
4—冷凝液出口 Condensate outlet；
5—螺栓 Bolt；7—冷凝液进口 Condensate inlet；
8—排污口 Drain outlet；
9—保温层 Insulation support ring；
10—滤芯 Filter element；
11—筒体 Shell；12—滤板 Filter plate；
13—反洗进气口 Backwashing inlet

8.2 水煤浆化工 Coal water slurry chemical industry

8.2.1 反应设备 Reaction equipment

8.2.1.1 气化炉 Gasifier

1. 气化炉结构 Structure of gasifier

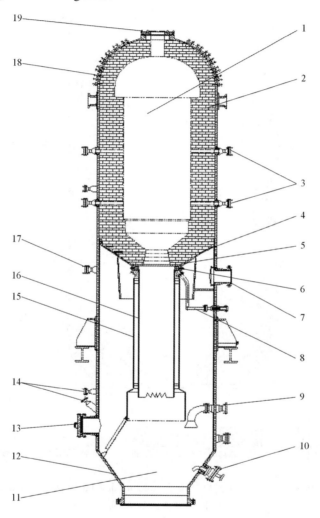

图 8.2.1 气化炉结构图 Structure drawing of gasifier

1—气化室 Vaporizer；2—气化炉内衬砖 Gasifier lining brick；3、18—热电偶接口 High temperature thermocouple installation flange；4—托砖板 Brick supporter plate；5—镶环 Insert ring；6—急冷环 Quick-cooling ring；7—工艺气出口 Process gas outlet pipe；8—急冷水管 Chilled water pipe；9—气化炉黑水出口 Gasifier dirty water outlet pipe；10—锁斗循环水接管 Lock hopper circulating water nozzle flange；11—急冷室 Chilled water chamber；12—气化炉壳 Gasifier shell；13—人孔 Manhole；14—液位计接口 Liquid level gauge interface；15—抽引管 Draft tube；16—水环管 Water ring pipe；17—置换氮气管法兰 Nitrogen replacement tube flange；19—烧嘴连接法兰 Burner connecting flange

2. 气化炉衬里 Gasifier lining

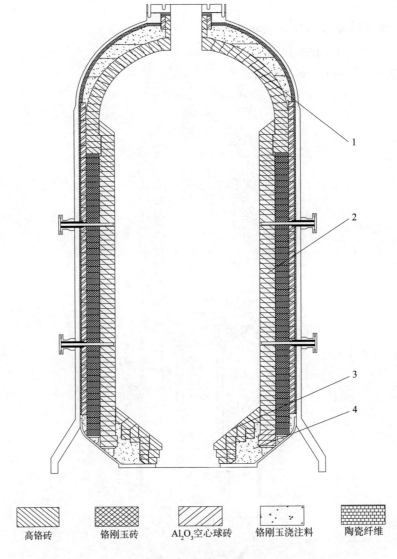

图 8.2.2 气化炉衬里结构图 Structure drawing of gasifier lining
1—拱顶砖 Crown brick；
2—筒体砖 Cylinder brick；
3—锥底砖 Cone base brick；
4—浇注料 Castable

图例：高铬砖；铬刚玉砖；Al_2O_3 空心球砖；铬刚玉浇注料；陶瓷纤维

8 煤化工 Cool chemical industry

3. 烧嘴 Burner

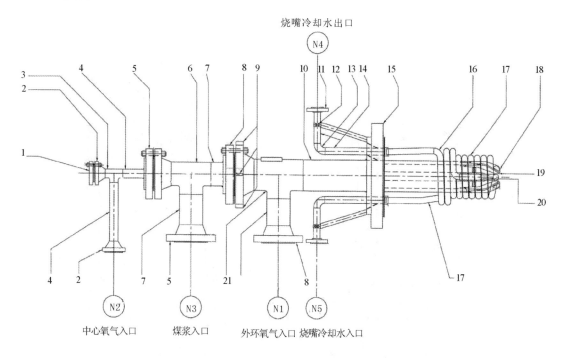

图 8.2.3 烧嘴结构图 Structure drawing of burner
1—中心氧管法兰盖 Center oxygen tube flange cover；
2—中心氧管法兰 Center oxygen tube flange；
3—中心氧管三通 Center oxygen tube tee；
4—中心氧直管 Center oxygen straight pipe；
5—煤浆管法兰 Coal slurry pipe flange；
6—煤浆管三通 Coal slurry pipe tee；
7—煤浆直管 Coal slurry straight pipe；
8—外环氧管法兰 Exo-epoxide pipe flange；
9—吊耳 Lifting lug；
10—外环氧直管 Exo-epoxide straight pipe；
11—烧嘴冷却水管 Burner cooling water pipe flange；
12—冷却水管支架 Cooling water pipe support；
13—冷却水管弯头 Cooling water elbow；
14—冷却水直管 Cooling water straight pipe；
15—烧嘴法兰 Burner flange；
16—冷却水外盘管 Cooling water outer-coil；
17—冷却水内盘管 Cooling water inner-coil；
18—外环氧喷嘴 Exo-epoxide nozzle；
19—煤浆喷嘴 Coal slurry nozzle；
20—中心氧喷嘴 Center oxygen nozzle；
21—外环氧管三通 Exo-epoxide pipe tee

8.2.1.2 第一中温变换炉 First mean temperature transformation furnace
8.2.1.3 低温变换炉 Low temperature transformation furnace

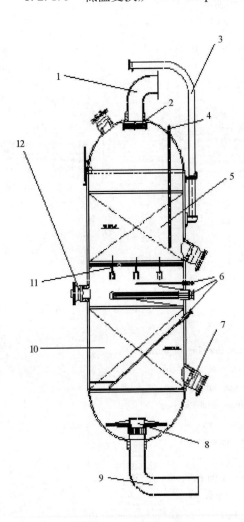

图 8.2.4 第一中温变换炉结构图
Structure drawing of first mean temperature transformation furnace

1—反应物入口 Product inlet；2—除沫器 Demister；3—塔顶吊柱 Top davit；4，6—热电偶 Thermal couple；5，10—催化剂床层 Catalyst bed；7—催化剂卸料口 Catalyst unloading port；8—出口收集器 Outlet collector；9—反应物出口 Product outlet；11—催化剂支承盘 Catalyst supporting plate；12—人孔 Manhole

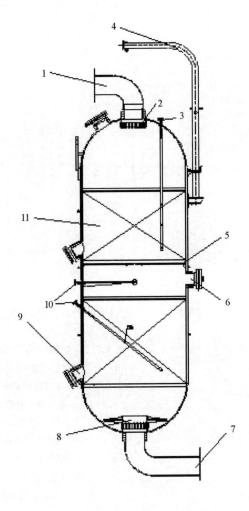

图 8.2.5 低温变换炉结构图
Structure drawing of low temperature transformation furnace

1—反应物入口 Product inlet；2—除沫器 Demister；3，10—热电偶 Thermal couple；4—塔顶吊柱 Top davit；5—催化剂支承盘 Catalyst supporting plate；6—人孔 Manhole；7—反应物出口 Product outlet；8—出口收集器 Outlet collector；9—催化剂卸料口 Catalyst unloading port；11—催化剂床层 Catalyst bed

8 煤化工 Cool chemical industry

8.2.1.4 甲烷化炉 Methanation furnace

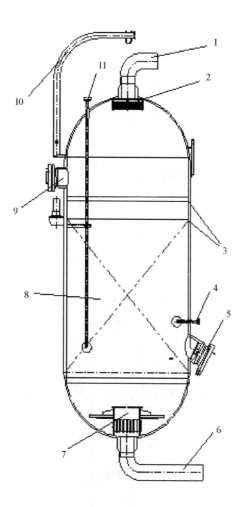

图 8.2.6 甲烷化炉结构图
Structure drawing of methanation furnace

1—反应物入口 Product inlet；
2—除沫器 Demister；
3—钢丝网 Wire mesh；
4,11—热电偶 Thermal couple；
5—催化剂卸料口 Catalyst unloading port；
6—出料口 Product outlet；
7—出口收集器 Outlet collector；
8—甲烷化床层 Methanation bed；
9—人孔 Manhole；
10—塔顶吊柱 Top davit

8.2.2 塔设备 Column equipment

8.2.2.1 合成气洗涤塔 Syngas scrubber

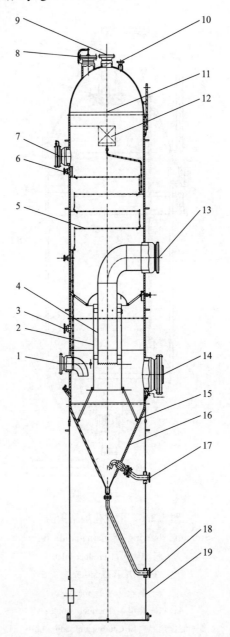

图 8.2.7 合成气洗涤塔结构图 Structure drawing of syngas scrubber

1—灰水出口 Grey water outlet；2—下降管 Downcomer；3—冷凝液出口 Condensate outlet；4—上升管 Rise tube；5—塔盘 Tray；6—冷凝液进口 Condensate inlet；7，8，14—人孔 Manhole；9—合成气入口 Syngas outlet；10—安全阀接口 Safety valve interface；11—旋风分离器 Cyclone separator；12—除沫器 Demister；13—合成气出口 Syngas outlet；15—支承板 Support plate；16—锥底 Cone bottom；17、18—黑水出口 Dirty water outlet；19—裙座 Skirt

8.2.2.2 脱硫塔 Purification desulfurization column
8.2.2.3 脱碳塔 Purification decarburization column

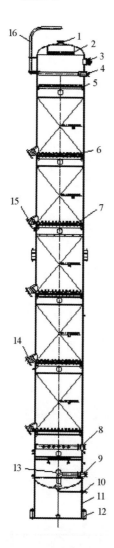

图 8.2.8 脱硫塔结构图
Structure drawing of purification
desulfurization column

1—脱硫气出口 Desulfuriztion gas export；2—顶部除沫器 Top demister；3、13—人孔 Manhole；4—溶液进口 Solution import；5—溶液分布器 Solution distributor；6、7—填料支承 Packing support；8—变换气进口 Transformation gas imports；9—溶液出口 Solution export；10—底部排液口 Bottom outlet；11—裙座 Skirt；12—地脚螺栓 Anchor bolt；14、15—卸料口 Discharge hole；16—塔顶吊柱 Top davit

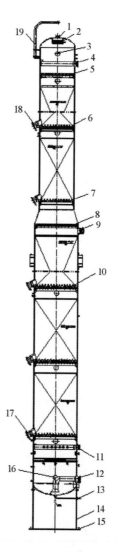

图 8.2.9 脱碳塔结构图
Structure drawing of purification
decarburization column

1—脱碳气出口 Desulfuriztion gas export；2—顶部除沫器 Top demister；3、9、16—人孔 Manhole；4—溶液进口 Solution import；5、8—溶液分布器 Solution distributor；6、7、10—填料支承 Packing support；11—脱碳气进口 Decarburization gas imports；12—溶液出口 Solution export；13—底部排液口 Bottom outlet；14—裙座 Skirt；15—地脚螺栓 Anchor bolt；17、18—卸料口 Discharge hole；19—塔顶吊柱 Top davit

8.2.2.4 精馏塔 Purification rectifying column

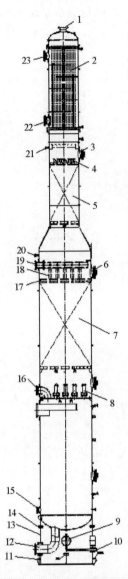

图 8.2.10 精馏塔结构图 Structure drawing of purification rectifying column

1—气氨出口 Gas ammonia outlet；2—塔顶冷却器 Upper cooler；
3、6、9—人孔 Man hole；4、18—液体分布器 Liquid distributor；
5、7—规整填料 Structured packing；8—升气帽 Rise gas cap；
10—底部排液口 Bottom outlet；11—地脚螺栓 Anchor bolt；
12—液位平衡口 Level balance mouth；13—裙座 Skirt；
14—排气口 Exhaust port；15—稀氨水出口 Dilute aqua ammonia export；
16—氨水出口 Ammonia water export；17—填料压板 Packing clamp；
19—浓氨水进口 Strong aqua import；20—测温口 Temperature mouth；21—取样口 Sampling mouth；
22—循环水进口 Circulating water inlet；23—循环水出口 Circulating water outlet

8 煤化工 Cool chemical industry

8.2.3 换热设备 Heat-exchange equipment

8.2.3.1 甲烷化加热器 Purification methanation heater

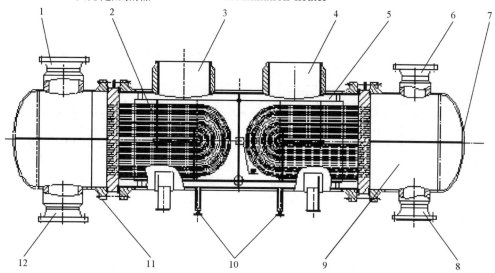

图 8.2.11 甲烷化加热器结构图 Structure drawing of purification methanation heater

1—合成氨脱碳气出口 Synthetic ammonia decarburizing gas outlet；
2—换热管束 Tube bundle；3—变换气进口 Transformation gas inlet；
4—变换气出口 Transformation gas outlet；5—筒体 Barrel；
6—氢气净化气出口 Hydrogen purification air outlet；7—封头 Head；
8—氢气净化气进口 Hydrogen purification air inlet；9—管箱 Channel；
10—排放口 Discharge outlet；11—管箱法兰 Channel flange；
12—合成氨脱碳气进口 Synthetic ammonia decarburizing gas inlet

8.2.3.2 中温换热器 Purification medium temperature heat exchanger

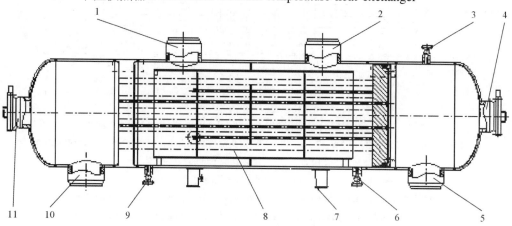

图 8.2.12 中温换热器结构图
Structure drawing of purification medium temperature heat exchanger

1—水煤气出口 Water gas outlet；2—水煤气进口 Water gas inlet；3—放空口 Vent；
4、11—人孔 Manhole；5—变换气出口 Transformation gas outlet；6、9—排凝口 Condensate discharge outlet；
7—支座 Support；8—管束 Tube bundle；10—变换气入口 Transformation gas inlet

· 763 ·

8.2.3.3 增压机后冷却器 Cooler after booster

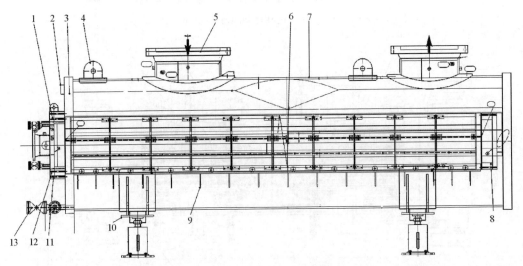

图 8.2.13 增压机后冷却器结构图 Structure drawing of cooler after booster

1、3、8—封头 Head；2—螺栓 Bolt；4—吊耳 Lifting lug；5—入口 Inlet；
6—管束 Tube bundle；7—筒体 Cylinder；9—折流板 Baffle plate；10—弹簧支承 Spring support；
11、12—封头螺栓、螺母 Head bolt；13—排凝阀 Condensate drain outlet

8.2.3.4 灰水加热器 Grey water heater

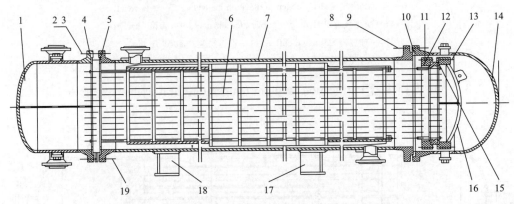

图 8.2.14 灰水加热器结构图 Structure drawing of grey water heater

1—管箱 Channel；2、8—螺母 Nut；3、9、13、19—双头螺栓 Stud bolt；4—管箱垫片 Channel gasket；
5—筒体法兰 Shell flange；6—管束 Tube bundle；7—筒体 Shell；
10—外头盖垫片 Outer cover gasket；11—钩圈法兰 Hook ring flange；
12—浮头管板 Floating tuke plate 14—外头盖 Outer cover；15—浮头垫片 Floating head gasket；
16—浮头盖 Floating head cover；17—活动支座 Sliding support seat；
18—固定支座 Fixed support

8.2.3.5 贫富液换热器 NHD solution plate heat exchanger

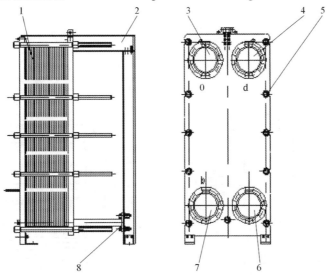

图 8.2.15 贫富液换热器结构图
Structure drawing of NHD solution plate heat exchanger

1—板片 Plate；2—支承梁 Support beam；3—富液入口 Rich fluid inlet；4—贫液出口 Lean fluid outlet；5—板片紧固螺栓 Plate fastening bolt；6—富液出口 Rich fluid outlet；7—贫液入口 Lean fluid inlet；8—支承紧固螺栓 Support fastening bolt

8.2.4 容器与罐 Storage tank and vessel

8.2.4.1 激冷水过滤器 Chilled water filter

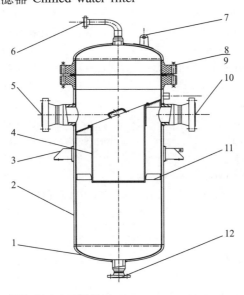

图 8.2.16 激冷水过滤器结构图 Structure drawing of chilled water filter

1—封头 Head；2—筒体 Cylinder；3—支座 Support；4—过滤筒 Filter cartridge；5—激冷水出口 Chilled water outlet；6—冲洗水进口 Wash water inlet；7—吊耳 Lifting lug；8—封头法兰 head flange；9—螺栓 Bolt；10—激冷水进口 Chilled water inlet；11—挡板 Baffle；12—排放口 Discharge outlet

8.2.4.2 锁斗 Lock hopper
8.2.4.3 水煤气分离罐 Water gas separator

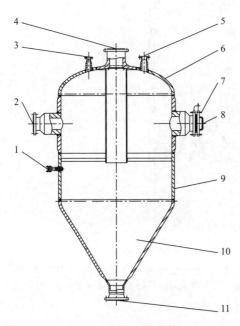

图 8.2.17 锁斗结构图
Structure drawing of lock hopper
1—温度计接口 Thermometer mouth；
2—冲洗水入口 Wash water inlet；
3—泄压口 Pressure relief opening；
4—渣水入口 Slag water inlet；
5—冲洗水出口 Circulating water outlet；
6—封头 Head；
7—螺栓 Bolt；
8—人孔 Manhole；
9—筒体 Cylinder；
10—锥形室 Cone charmer；
11—渣水出口 Slag water outlet

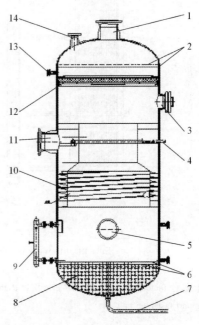

图 8.2.18 水煤气分离罐结构图
Structure drawing of water gas separator
1—水煤气出口 Water gas outlet；
2，6—不锈钢丝网 Stainless wire screen；
3、5—人孔 Manhole；
4—热电偶 Thermal couple；
7—排液口 Discharge liquid interface；
8—防旋盘 Prevents rotation plate；
9—液位计 Liquid level meter；
10—气体螺旋分布器 Gas screw distribution device；
11—水煤气进口 Water gas inlet；
12—除沫器 Demister；
13—压力测量口 Pressure measurement interface；
14—安全阀接口 Safety valve interface

8 煤化工 Cool chemical industry

8.2.5 特种阀门 Special valves

8.2.5.1 氧气切断阀 Oxygen Shut-off valve

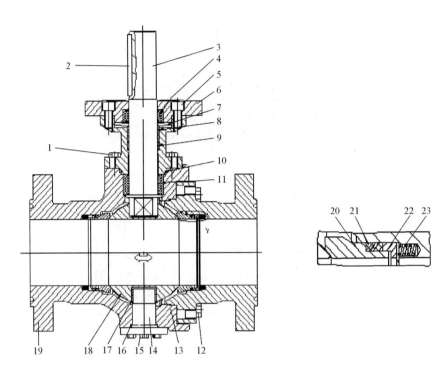

图 8.2.19 氧气切断阀结构图 Structure drawing of oxygen Shut-off valve

1、5、12、15—螺栓 Bolt；2—键 Key；

3—阀杆 Valve rod；4、11、17—轴套 Sleeve；

6—轴承盖 Bearing cap；

7—板簧 Plate spring；

8—压盖垫圈 Gland gasket；

9—阀杆填料 Valve stem packing；

10、13、16—密封圈 Seal ring；

14—下轴 Lower shaft；

18—阀球 Valve ball；

19—阀体 Valve body；

20—阀座 Valve seat；

21—阀座密封件 Valve seat sealing element；

22—挡圈 Closing ring；

23—旋转弹簧 Rotating spring

8.2.5.2 锁斗阀 Lock hopper valve

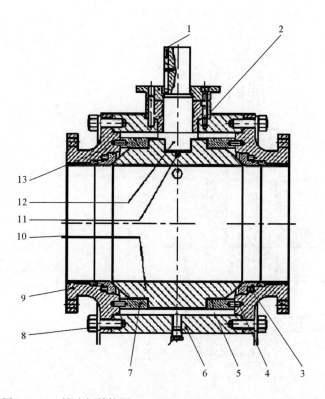

图 8.2.20 锁斗阀结构图 Structure drawing of lock hopper valve

1—键 Key；
2—压盖螺钉 Gland screw；
3—弹簧座 Spring seat；
4—耳板销 Ear plate pin；
5—轴承隔板 Bearing clapboard；
6—阀体 Valve body；
7—耳板 Ear plate；
8—螺栓 Bolt；
9—法兰端面 End face；
10—阀球 Valve ball；
11—防转弹簧 Anti-rotating spring；
12—阀杆 Valve rod；
13—阀座 Valve seat

8.2.5.3 煤浆放料阀 Coal slurry feeding valve

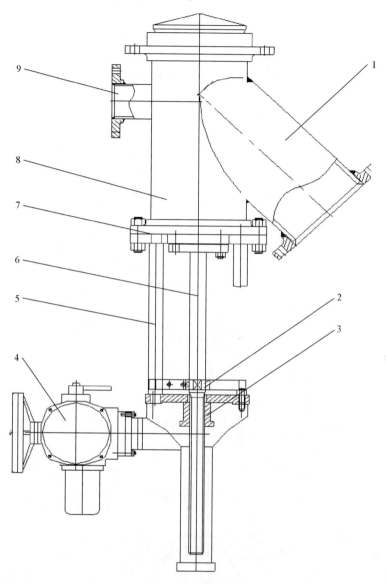

图 8.2.21 煤浆放料阀结构图
Structure drawing of coal slurry feeding valve
1—下料口 Feed opening;
2—填料 Padding;
3—填料压盖 Stuffing gland;
4—电动头 Electric head;
5—支架 Support;
6—阀杆 Valve rod; 7—法兰 Flange;
8—阀体 Valve body;
9—冲洗水口 Wash water opening

8.2.5.4 黑水角阀 Dirty water angle valve

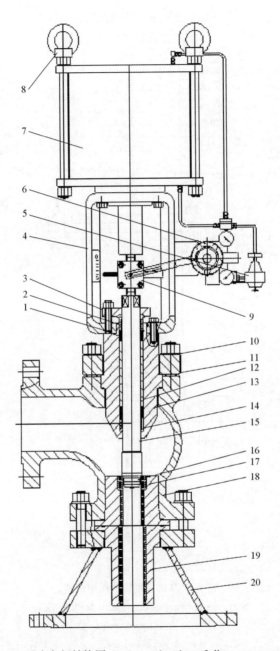

图 8.2.22 黑水角阀结构图 Structure drawing of dirty water angle valve

1—填料 Padding；2—上导向管 Upper guide tube；3—填料压盖 Stuffing gland；4—气缸支架 Air cylinder support；5—过滤减压器 Filter decompressor；6—定位器 Positioner；7—气缸 Air cylinder；8—吊环 Lifting ring；9—对夹组件 Butt Clamp component；10—螺栓 Bolt；11—阀盖 Valve cover；12—填料隔离套管 Padding isolation bushing；13—填料函 Stuffing box；14—下导向管 Bottom guide tube；15—阀杆 Valve rod；16—阀头 Valve head；17—阀座 Valve seat；18—阀体 Valve body；19—过渡段 Transition segment；20—扩散段 Diffuser

8.2.5.5 轨道球阀 Track ball valve

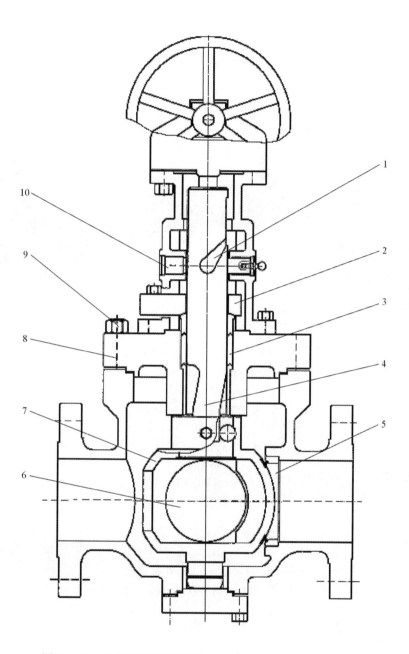

图 8.2.23 轨道球阀结构图 Structure drawing of track ball valve
1—导向槽 Guideway；
2—填料压盖 Stuffing gland；
3—填料 Padding；4—阀杆 Valve rod；
5—阀座 Valve seat；6—阀球 Valve ball；
7—阀体 Valve body；8—端盖 End cap；
9—螺栓 Bolt；10—导向销 Guide pin

8.2.6 泵 Pump

8.2.6.1 低压煤浆泵 Low pressure coal slurry pump

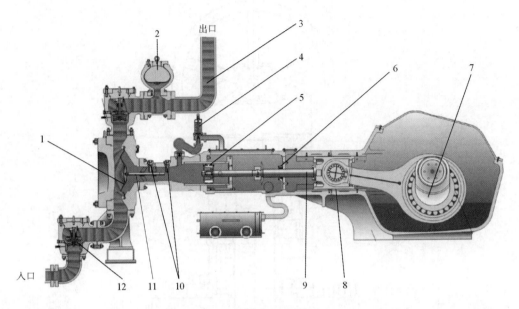

图 8.2.24　低压煤浆泵结构图
Structure drawing of low pressure coal slurry pump

1—隔膜 Membrane；
2—出口缓冲罐 Outlet buffer tank；
3—煤浆 Coal slurry；
4—安全泄压阀 Safety relief valve；
5—活塞 Piston；
6—油封 Oil seal；
7—曲轴 Crank shaft；
8—滑块十字头 Slider cross；
9—活塞杆 Piston rod；
10—探头 Probe；
11—探杆 Feeler lever；
12—单向阀 Check valve

8.2.6.2 高压煤浆泵 High pressure coal slurry pump

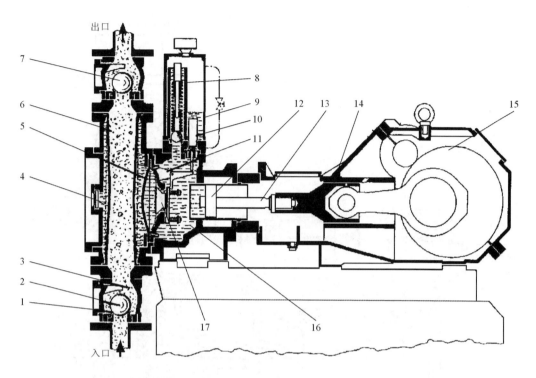

图 8.2.25 高压煤浆泵结构图 Structure drawing of high pressure coal slurry pump

1—阀况监测器 Valvecondition monitor；
2—入口单向阀 Inlet check valve；
3—煤浆 Coal slurry；
4—二次液压液 Secondary hydraulic fluid；
5—隔膜 Membrane；
6—软管 Soft tube；
7—出口单向阀 Outlet check valve；
8—安全阀 Safety valve；
9—液压油箱 Hydraulic fluid chamber；
10—补/排油阀 Replenishing and discharging oil valve；
11—曲柄顶杆 Crank plunger；
12—活塞 Piston；
13—活塞杆 Piston rod；
14—滑块 Sliding block；
15—曲轴 Crankshaft；
16—液压油 Hydraulic oil；
17—隔膜室控制盘 Membrane chamber control panel

8.2.6.3 沉降槽给料泵 Settling tank feeding pump

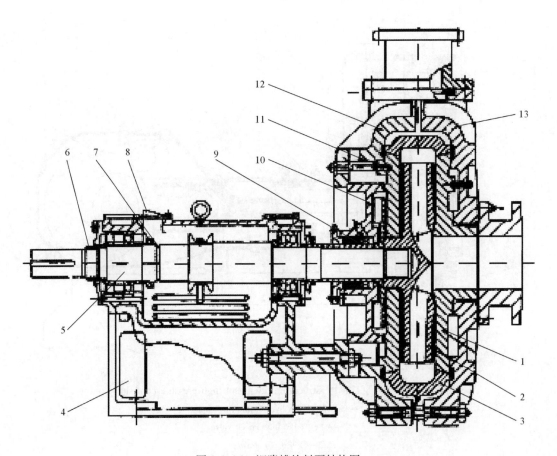

图 8.2.26　沉降槽给料泵结构图
Structure drawing of settling tank feeding pump
1—前护板 Front backplate；
2—叶轮 Wheel；
3—护套 Jacket；
4—托架体 Bracket body；
5—泵轴 Shaft；
6—锁紧螺母 Locknut；
7—定位螺母 Set screw nut；
8—轴承箱 Bearing box；
9—密封函 Sealing box；
10—盖板 Cover plate；
11—紧固螺栓 Fastening bolt；
12—后泵壳 Rear pump case；
13—前泵壳 Front pump case

8.2.6.4 脱碳贫液泵透平 Decarburization barren solution pump turbine

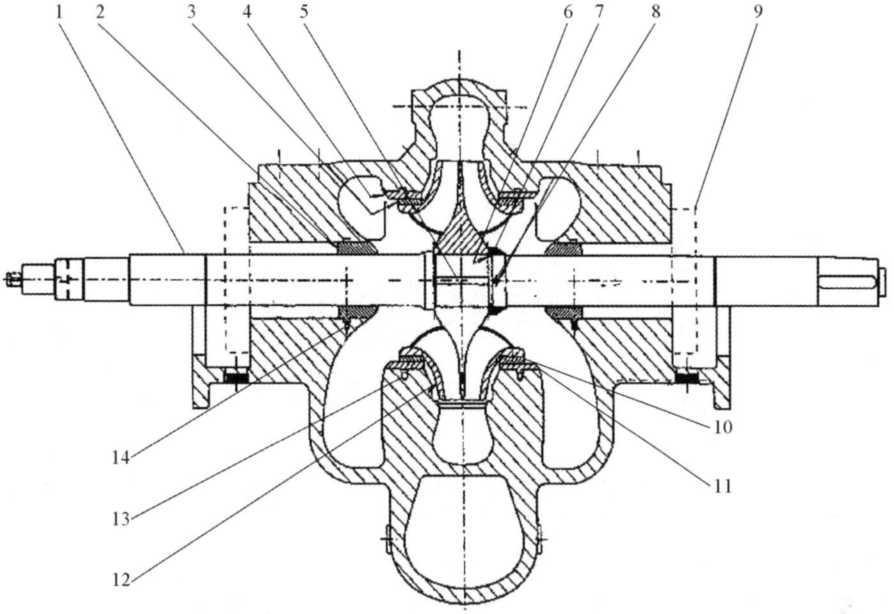

图 8.2.27 脱碳贫液泵透平结构图
Structure drawing of decarburization barren solution pump turbine
1—泵轴 pump shaft；2—喉部套管 Laryngeal casing；3—叶轮环 Impeller ring；4—缸体销 Cylinder pin；
5—叶轮键 Impeller key；6—止推环 Thrust ring；7—箍圈 Cranse；8—定位螺钉 Positioning screw；
9—机械密封 Mechanical seal；10—垫片 Gasket；11—弹簧垫圈 Spring washer；
12—压盖垫圈 Gland washer；13—键槽 Keyway；14—轴套 Sleeve

8.2.7 磨煤机 Coal pulverizer

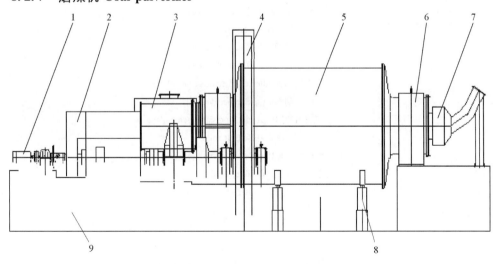

图 8.2.28 磨煤机结构图 Structure drawing of coal pulverizer
1—辅助传动装置 Auxiliary drive device；2—驱动装置 Drive device；3—出料口 Feed outlet；
4—传动齿轮 Driving gear；5—筒体 Cylinder；6—主轴承 Main bearing；
7—进料器 Feeder；8—手动液压顶 Manual hydraulic roof；9—底座 Base

8.2.8 真空过滤机 Vacuum filter

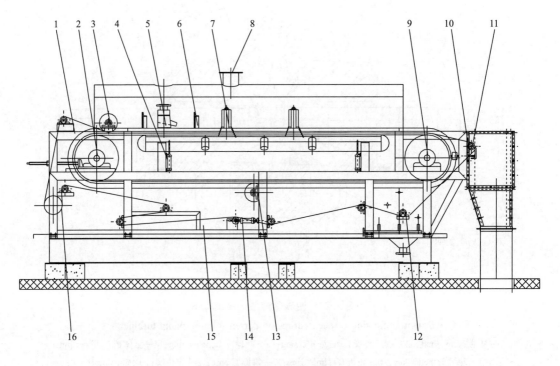

图 8.2.29　真空过滤机结构图 Structure drawing of vacuum filter

1—机架 Rack；2—从动辊 Driven roller；
3—压布辊 Press-roller；
4—真空箱支承装置 Vacuum chamber support device；
5—加料装置 Feeding device；
6—隔板 Clapboard；
7—滤饼淋洗装置 Filter cake leaching device；
8—手孔 Hand hole；9—主动辊 Drive roll；
10—滤布改向辊 Filter cloth redirection roller；
11—刮料装置 Scraper device；
12—滤布压带辊 Filter cloth pressure belt roller；
13—皮带托辊 Belt roller；
14—滤布纠偏辊 Filter cloth correction rollers；
15—滤布 Filter cloth；
16—滤布张紧辊 Filter cloth stretch roll

8.2.9 煤浆搅拌器 Coal slurry blender

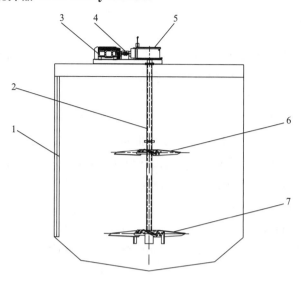

图 8.2.30 煤浆搅拌器结构图 Structure drawing of coal slurry blender
1—槽 Duct；2—传动轴 Transmission shaft；3—电机 Motor；4—联轴器 Coupling；
5—减速箱 Reducer casing；6—上桨叶 Upper blade；7—下桨叶 Bottom blade

8.2.10 捞渣机 Slag conveyor

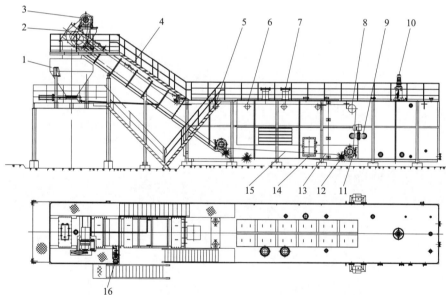

图 8.2.31 捞渣机外形图 Structure drawing of slag conveyor
1—渣斗 Slag hopper；2—拖动机构 Drag mechanism；3—驱动机构 Drive mechanism；
4—斜梯 Inclined ladder；5—断链检测导轮轴 Broken link test guide wheel shaft；
6—前导轮轴 Front guide wheel shaft；7—进料口 Feed inlet；8—后导轮轴 Rear guide wheel shaft；
9—蝶阀 Butterfly valve；10—灰浆池搅拌器 Ash slurry pond stirrer；11—内导轮 Guide roller；
12—轴封水管 Shaft seal water pipe；13—刮板 Scraper；14—人孔 Manhole；15—链条 Chain；
16—捞渣机液压站 Slag conveyor hydraulic pressure station

8.3 MTO 化工 MTO chemical industry

MTO 是甲醇转化制烯烃的简称，MTO 装置就是利用煤、天然气或煤层气为原料制成的甲醇在催化剂的作用下进一步转化为低碳烯烃，包括反应再生、急冷水洗、烟气能量利用和回收、反应取热、再生取热等部分。后续为烯烃的精制分离，包括碱洗、干燥、压缩、制冷、脱碳、炔烃前加氢等。

MTO is the abbreviation of Methanol To Olefins, MTO device converts the methanol, which is made by the use of coal, natural gas and coalbed methane as raw material and isfurthev transformed into low carbon olefin under the action of catalyst. MTO comprises reactor-regenerator, quenched water wash, the use and recycling of the flue gas energy, reaction heat removal, regenerative heat removal. The following process is the refined of the olefins, which comprises caustic wash, dry, compression, refrigeration, decarbonization, hydrogenation before alkyne.

8.3.1 反应与再生设备 Reactor-regenerator system

8.3.1.1 反应设备 Reaction equipment

MTO 反应器有密相床层反应器和快速流化床型反应器两种型式。

There are two types MTO reactor, the dense phase bed reactor and the fast fluidized bed reactor.

8 煤化工 Cool chemical industry

1. 密相床层反应器 Dense phase bed reactor

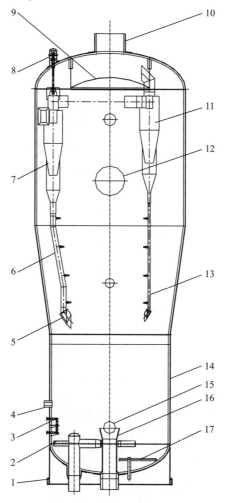

图 8.3.1 密相床层反应器结构图 Structure drawing of dense phase bed reactor

1—裙座 skirt；2—甲醇分布器 Methanol distributor；
3—内取热器 Inside heat；
4—再生催化剂入口 Regen rated CAT inlet；
5—旋分料腿翼阀 Trickle valves；
6——级旋分料腿 First stage cyclone dipleges；
7——级旋风分离器 First stage cyclones；
8—旋分吊座 Cyclone support；
9—内集气室 Internal collection chamber；
10—反应油气出口 Reactor oil gas outlet；
11—二级旋风分离器 Second stage cyclones；
12—装卸孔 Handling holeHandling hole；
13—二级旋分料腿 Second stage cyclone dipleges；
14—反应器筒体 Reactor shell；15—人孔 Manhole；
16—待生催化剂出口 Spent CAT outlet；
17—催化剂装卸口 CAT filing / withdrawal

2. 快速流化床型反应器 Fast fluidized bed reactor

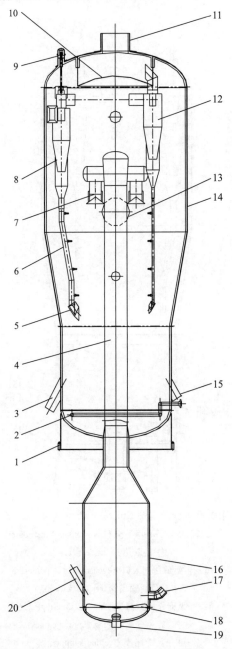

图 8.3.2　快速流化床型反应器结构图 Structure drawing of fast fluidized bed reactor
1—裙座 skirt；2—流化蒸汽管 fluidization steam pipe；3—外循环管出口 Circulation CAT outlet；
4—内提升管 Internal riser；5—旋分料腿翼阀 Trickle valves；6——级旋分料腿 cyclone dipleges；
7—快速分离器 Fast separator；8——级旋风分离器 First stage cyclones；9—旋分吊座 Cyclone support；
10—内集气室 Interal colletion chamber；11—反应油气出口 Reactor oil gas outlet；
12—二级旋风分离器 Second stage cyclones；13—装卸口 Handling hole；14—反应器筒体 Reactor shell；
15—外取热器入口 Cooler inlet；16—快速反应器筒体 Fast reactor shell；17—再生催化剂入口 Regen erated CAT inlet；
18—甲醇分布器 Methanol distributor；19—甲醇进料口 Methanol inlet；20—外循环管入口 Circulation CAT inlet

8.3.1.2 再生设备 Regeneration equipment

MTO 的再生器有鼓泡床型和烧焦罐型两种型式：

MTO regenerator have two types, the bubble column regenerator and the coke burning tank regenerator.

鼓泡床型再生器由主风分布管、催化剂流化床和多级旋风分离器等部分组成，如图 8.3.3 所示。

The bubble column regenerator comprises the main air distribution, CAT fluidized bed and multi stage cyclone, as shown in Fig. 8.3.3.

烧焦罐型再生器由主风分布管、烧焦罐、大孔分布板、稀相段和多级旋风分离器等部分组成，如图 8.3.4 所示。

The coke burning tank regenerator comprises the main air distribution, the coke burning tank, wide bore distribution plate, dilute-phase zone and multi stage cyclone, as shown in Fig. 8.3.4.

1. 鼓泡床型再生器 Bubble column regenerator

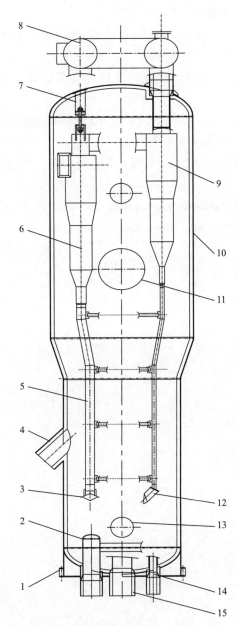

图 8.3.3　鼓泡床型再生器结构图 Structure drawing of bubble column regenerator
1—裙座 Skirt；2—主风分布管 Main air distributor；3—防倒锥 Inverted cone；
4—外取热器催化剂入口 Exterior cooler CAT inlet；
5——级旋分料腿 First stage cyclones dipleges；
6——级旋风分离器 First stage cyclones；7—旋分吊挂 Cyclones support；
8—烟气集合管 Flue gas collection pipe；9—二级旋风分离器 Second stage cyclones；
10—再生器筒体 Regenerator shell；11—装卸口 Handling hole；12—旋分料腿翼阀 Trickle valves；
13—人孔 Manhole；14—待生剂入口 Spent CAT inlet；15—再生催化剂出口 Regen CAT outlet

2. 烧焦罐型再生器 Coke burning tank regenerator

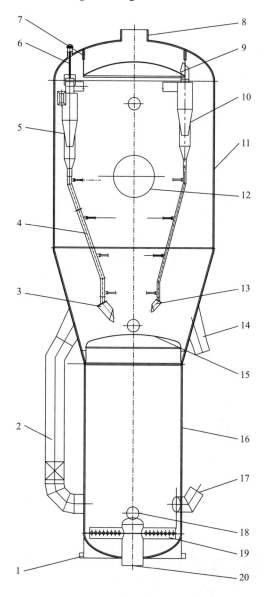

图 8.3.4 烧焦罐型再生器结构图 Structure drawing of coke burning tank regenerator
1—裙座 Skirt；2—催化剂循环管 Circ ulation CAT tube；3——级旋分料腿翼阀 Cyclones dipleges trickle valves；
4——级旋分料腿 First stage cyclones dipleges；5——级旋风分离器 First stage cyclones；
6—旋分吊挂 Cyclones support；7—内集气室 Interal colletion chamber；
8—烟气出口 Flue gas outlet；9—二级旋风分离器出口管 Second stage cyclones outlet tube；
10—二级旋风分离器 Second stage cyclones；11—再生器筒体 Regenerator shell；
12—装卸口 Handling hole；13—二级旋分料腿翼阀 dipleges trickle valves；
14—再生催化剂出口 Regenerator CAT outlet；15—烟气分布板 Flue gas distributor；
16—烧焦罐筒体 Coke burning tank shell；17—待生催化剂入口 Spent CAT inlet；
18—人孔 Manhole；19—主风分布管 Main air distribution；20—主风入口 Main air inlet

8.3.1.3 反应与再生设备附件 Accessories of reactor-regenerator equipment

反应与再生设备附件主要包括一、二级旋风分离器、内取热器、进料分布器、主风分布管、汽提器和旋分料腿翼阀等。

Accessories of the reactor-regenerator equipment include first and second stage cyclones separator, inside heater, feed distributor for CAT, main air distributor, stripper, trickle valves and so on.

1. 旋风分离器 Cyclone separator

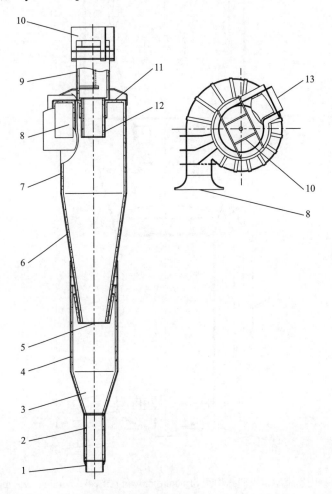

图 8.3.5　旋风分离器结构图 Structure drawing of Cyclone separator

1—旋分料腿 Cyclone dipleg；
2—旋分料腿耐磨段 Wear-resistant tube of cyclone dipleg；
3—灰斗锥体 Dust hopper cone section；
4—灰斗 Dust hopper；5—排尘口 Drain outlet；
6—旋分锥体 Cyclone cone section；7—旋分筒体 Cyclone cylinder；
8—旋分入口 Cyclone inlet；9—旋分出口管 Cyclone outlet tube；
10—旋分吊挂 Cyclones support；11—旋分顶板 Cyclone top plate；
12—旋分升气管 Cyclone exit pipe；13—旋分出口 Cyclone outlet

2. 气体分布器 Air distributor

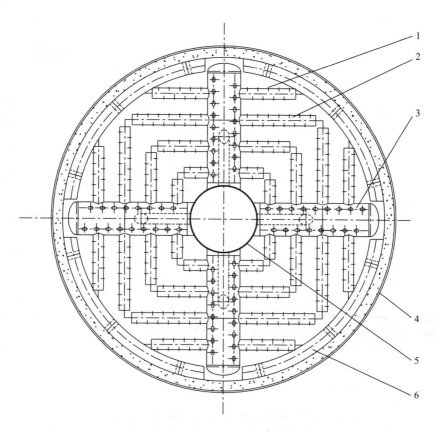

图 8.3.6　气体分布器结构图 Structure drawing of Inlet distributor for air

1—喷嘴 Distributor jet；
2—分支管 Branch arm；
3—支管 Branch pipe；
4—两器筒体 Reator or Regenerator shell；
5—进料主管 Main inlet pipe for gas；
6—边缘挡板 Edge baffle

3. 内取热器 Internal heater

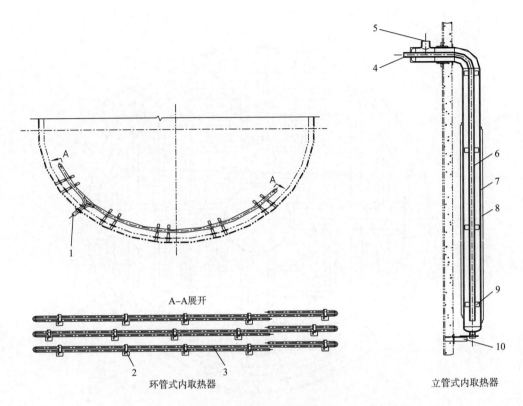

图 8.3.7 内取热器结构图 Structure drawing of internal heater

1—内取热器入口管 Inlet pipe of internal heater;
2—支架 Support;
3—环式内取热管 ring pipe of internal heater;
4—介质入口 Gas and CAT inlet;
5—介质出口 Gas and CAT outlet;
6—内管 Inside tube;
7—外管 Outside tube;
8—翅片 Fin;
9—导向片 Guide rib;
10—导向架 Guide frame

4. 汽提器 Stripper

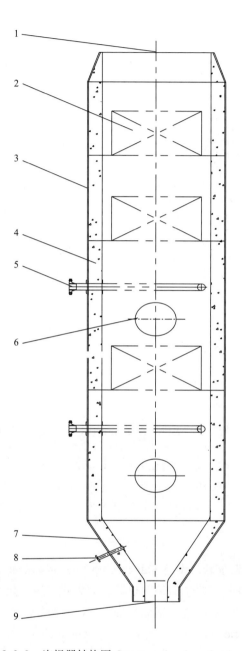

图 8.3.8 汽提器结构图 Structure drawing of stripper

1—催化剂入口 CAT inlet；
2—汽提格栅 Stripper grids；
3—筒体 Shell；4—衬里 Lining；
5—汽提环管 Ring pipe of stripper；
6—人孔 Manhole；7—锥段 Cone；
8—松动风口 Loose tuyere；
9—催化剂出口 CAT outlet

5. 旋分料腿翼阀 Trickle valves

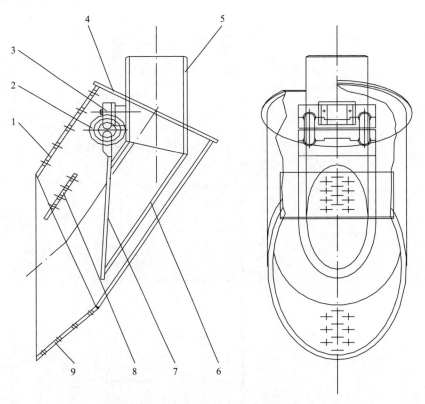

图 8.3.9　旋分料腿翼阀结构图 Structure drawing of Trickle valves
1—外套筒 Outer sleeve；2—吊环 Rings；
3—固定板 Fixed plate；4—端板 End plate；
5—旋分料腿 Cyclone dipleg；6—料腿弯管 Dipleg elbow；
7—阀板 Valve plate；
8—限位挡板 Limit baffle plate；
9—斜歪套筒 Oblique slanting sleeve

8.3.1.4　反应再生系统其他设备 Other equipment of reactor-regenerator system

反应再生系统其他设备主要包括外取热器、立管式三级旋风分离器、卧管式三级旋风分离器、孔板降压器、烟气水封罐、辅助燃烧室、CO 焚烧炉和单动滑阀等。

Other equipment of the reactor-regenerator system include external cooler、vertical tube third stage cyclones separator、horizontal tube third stage cyclones separator、orifice step-down transformer、flue gas water-sealed tank、sub-chamber、CO incinerator、single slide valve and so on.

8 煤化工 Cool chemical industry

1. 外取热器 Exteral cooler

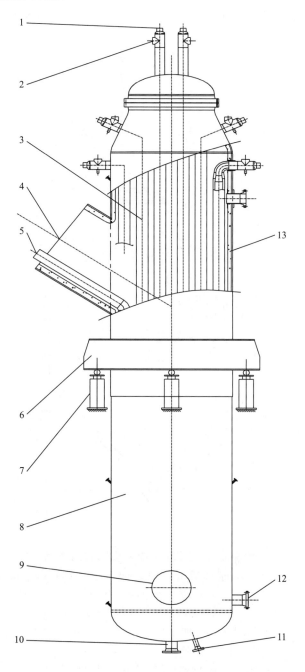

图 8.3.10 外取热器结构图 Structure drawing of exteral cooler
1—冷介质入口管 Inlet pipe of quenching medium；2—热介质出口管 Outlet pipe of thermal medium；
3—换热管束 Tube bundle；4—催化剂入口管 CAT inlet pipe；
5—催化剂出口管 CAT outlet pipe；6—支座 Support；7—弹簧支架 Spring support；
8—筒体 Shell；9—人孔 Manhole；10—流化风入口 Inlet of fluidization wind；
11—催化剂卸料口 CAT discharge outlet；12—烟气返回口 Flue gas reflux pipe；
13—隔热耐磨衬里 Heat-insulation and wear-resistant lining

2. 三级旋风分离器 Third stage cyclones separator

1) 立管式三级旋风分离器 Vertical tube third stage cyclones separator

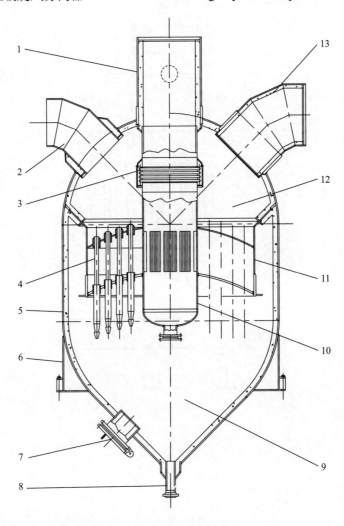

图 8.3.11　立管式三级旋风分离器结构图
Structure drawing of Vertical tube third stage cyclones separator

1—烟气入口管 Flue gas inlet；

2、13—净化烟气出口管 Clean flue gas outlet；

3—膨胀节 Expansion joint；4—分离单管 Separating tube；

5—筒体 Shell；6—裙座 Skirt；

7—人孔 Manhole；8—排尘口 Drain outlet；

9—积尘室 Dust chamber；

10—中心管 Center tube；

11—吊筒 Hanging shell；

12—集气室 Collection chamber

2) 卧管式三级旋风分离器 Horizontal tube third stage cyclones separator

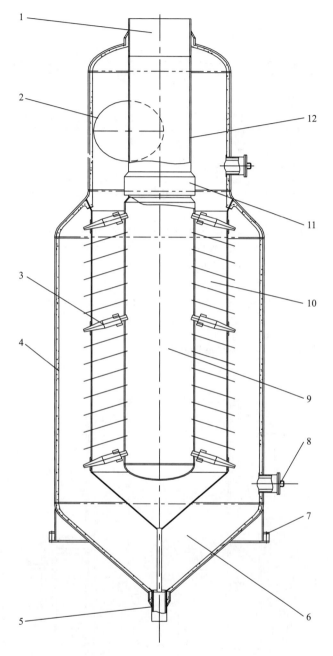

图 8.3.12　卧管式三级旋风分离器结构图
Structure drawing of horizontal tube third stage cyclones separator
1、12—净化烟气出口管 Clean flue gas outlet；2—烟气入口管 Flue gas inlet；
3—分离单管 Cyclone；4—筒体 Shell；5—排尘口 Drain outlet；6—积尘室 Dust chamber；
7—裙座 Skirt；8—人孔 Manhole；9—集气室 Collection chamber；
10—气体分配室 Gas distributing chamber；11—轴向位移膨胀节 Axial expansion bollows

3. 孔板降压器 Orifice step-down transformer

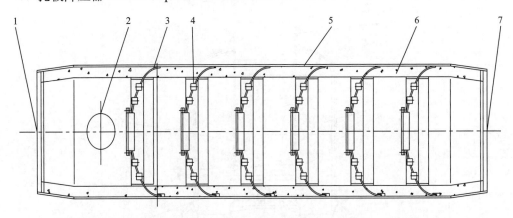

图 8.3.13 孔板降压器结构图 Structure drawing of orifice step-down transformer
1—烟气入口管 Flue gas inlet；2—人孔 Manhole；
3—降压孔板 Decompression orifice plate；
4—喷嘴 Nozzle；5—筒体 Shell；
6—隔热耐磨衬里 Heat-insulation and wear-resistant lining；
7—烟气出口管 Flue gas outlet

4. 烟气水封罐 Flue gas water-sealed tank

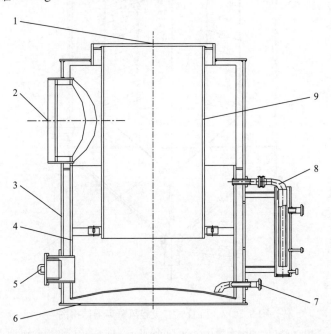

图 8.3.14 烟气水封罐结构图 Structure drawing of flue gas water-sealed tank
1—烟气入口管 Flue gas inlet；2—烟气出口管 Flue gas outlet；3—筒体 Shell；
4—内套筒 Inside sleeve；5—人孔 Manhole；6—底板 Bottom plate；7—进水口 Water inlet；
8—液位控制组件 Liquid level control component；9—中心管 Center tube

8 煤化工 Cool chemical industry

5. 辅助燃烧室 Sub-chamber

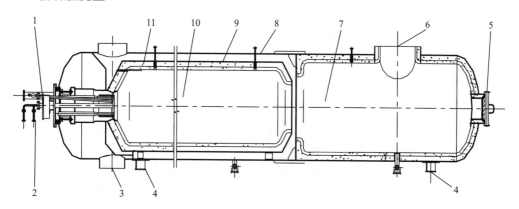

图 8.3.15 辅助燃烧室结构图 Structure drawing of Sub-chamber
1—一次风入口 Inlet of the primary air; 2—燃料入口 Inlet of the fuel;
3—二次风入口 Inlet of the secondary air;
4—固定/滑动支座 Fixed/sliding saddle;
5—人孔 Manhole; 6—热空气出口 Outlet of the hot air;
7—混合室 Mixing chamber; 8—外壳体 Outer shell;
9—内套筒 Inside shell; 10—燃烧室 Combustion chamber; 11—耐火砖 Fire brick

6. CO 焚烧炉 CO incinerator

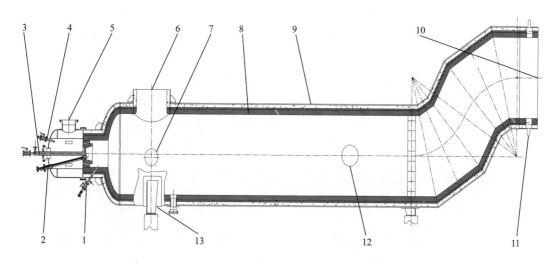

图 8.3.16 CO 焚烧炉结构图 Structure drawing of CO incinerator
1—火焰检测器接口 Flame detector interface;
2—点火装置 Ignition device;
3—燃料油入口 Fuel inlet; 4—看火孔 Kiln eye;
5—空气入口 Air inlet; 6—烟气入口 Flue gas inlet;
7—焚烧空气入口 Inlet of air incinerator; 8—耐火材料 Fireproofing lining;
9—壳体 Shell; 10—烟气出口 Flue gas outlet;
11—膨胀节 Expansion joint; 12—人孔 Manhole;
13—支座 Support

7. 单动滑阀 Single slide valve

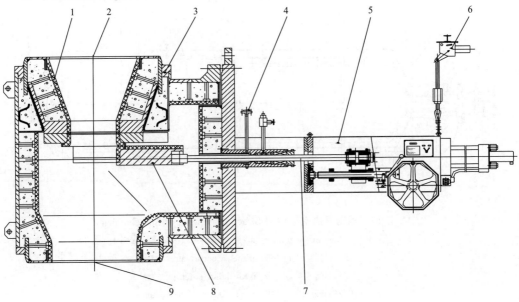

图 8.3.17 单动滑阀结构图 Structure drawing of single slide valve

1—高耐磨隔热双层衬里 Re-abrasion resistant and heat-insulation lining；2—介质入口 Gas and CAT inlet；
3—阀体 Valve body；4—蒸汽吹扫口 Steam purge port；5—控制箱 Control box；6—弹簧吊架 Sping hanger；
7—阀杆 Valve bar；8—阀板 Valve plate；9—介质出口 Gas and CAT outlet

8. 反应再生系统各种衬里结构 Structure of all kinds of concrete of the reactor-regenerator system

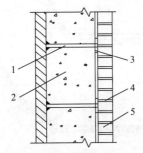

（a）龟甲网隔热耐磨双层衬里
Hexsteel insulation abrasion resistaut two lager lining

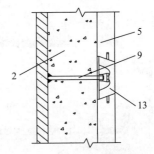

（b）无龟甲网隔热耐磨双层衬里
No hexstde insulation wear-resistaut two lager lining

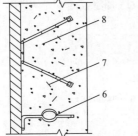

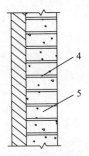

（c）隔热耐磨单层衬里
Thermal insulation and wear resistant single layer lining

（d）龟甲网高耐磨单层衬里
Hexsted insalation wear-resistant single layer lining

8 煤化工 Cool chemical industry

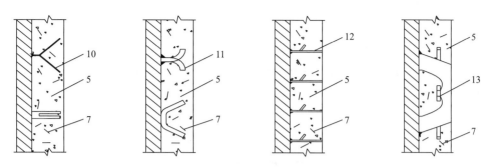

（e）无龟甲网高耐磨单层衬里 No hexsted insulation wear resistant single layer lining

图 8.3.18　反应再生系统各种衬里结构图
Structure drawing of Structure of all kinds of lining of the reactor-regenerator system

1—柱形锚固钉 Column type anchor nail；2—隔热混凝土 Heat-insulation concrete；3—端板 End-plate；
4—龟甲网 Hexsteel；5—高耐磨混凝土 High abrasion concrete；6—Ω 形锚固钉 Ω type anchor nail；
7—钢纤维 Steel fiber；8—隔热耐磨混凝土 Heat-insulation and abrasion concrete；
9—柱型螺栓 Column type bolt；10—Y 形锚固钉 Y type anchor nail；11—V 形锚固钉 V type anchor nail；
12—S 形锚固钉 S type anchor nail；13—侧拉型圆环 Lateral pulling type ring

9. 油气管线高温蝶阀 High temperature butterfly valve of oil and gas pipeline

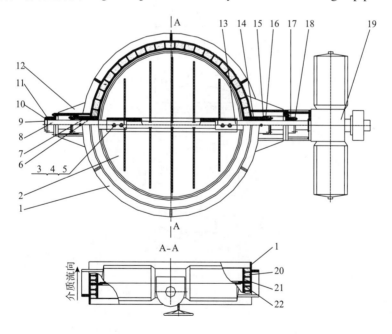

图 8.3.19　油气管线高温蝶阀结构图
Structure drawing of High temperature butterfly valve of oil and gas pipeline

1—阀体 Valve body；2—阀板 Valve plate；3—定位螺栓 Position bolt；4—螺母 Nut；5—方垫 Square washer；
6—轴承 Bearing；7—分流环 Splitter lip；8—下阀轴 Bottom shaft；9—轴盖 Cover；10—螺栓 Bolt；
11、17—轴套 Sleeve；12—筋板 Wed plate；13—上阀轴 Top shaft；14—挡环 Check ring；
15—柔性石墨环 Flexible graphite ring；16—衬套 Bush；18—卡环 Clasp；19—气动执行机构 Pneumatic actuator；
20—隔热衬里 Heat-insulated cining；21—耐磨衬里 Wear-resistant cining；22—锚固件 Anchor

8.3.2 塔设备 Column equipment

MTO 装置塔设备主要包括急冷-水洗塔和污水汽提塔。

MTO column equipment mainly contain the urgent cold-water washing column and the sewage stripper.

1. 急冷-水洗塔 Urgent cold-water washing column

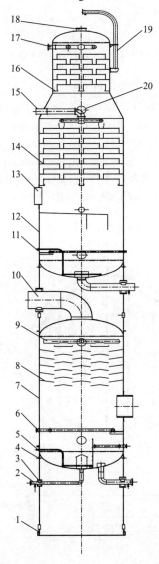

图 8.3.20　急冷-水洗塔结构图 Structure drawing of urgent cold-water washing column
1—裙座 Skirt；2—塔底催化剂抽出口 Bottom CAT outlet；3—引出管 Outlet pipe；4—裙座排气管 Sirt vent；5—塔底蒸汽入口 Steam inlet；6—进水口 Water inlet；7—急冷塔筒体 Urgent cold column shell；8—人字挡板 Herringbone baffle；9—急冷水返口 Urgent cold water outlet；10—急冷油气出口 Urgent cold gas outlet；11—上部塔底蒸汽吹扫口 Top Steam inlet；12—水洗塔筒体 Water washing column shell；13—急冷油气入口 Urgent cold gas inlet；14—塔盘 Tray；15—水洗塔下返口 Outlet at the bottom of the water washing column；16—受液盘 Liquid receiving plate；17—水洗塔上返口 Outlet at the top of the water washing column；18—油气出口 Gas outlet；19—塔顶吊柱 Top davit；20—人孔 Manhole

2. 污水汽提塔 Sewage stripper

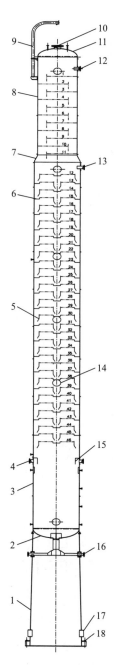

图 8.3.21 污水汽提塔结构图 Structure drawing of Sewage stripper

1—裙座 Skirt；2—底封头 Bottom head；3—下部筒体 Bottom shell；4、15—汽提蒸汽入口 Stripped vapor inlet；5—降液板 Downcomer tray on both sides；6—受液盘 Downcomer tray in the middle；7—锥段 Cone；8—上部筒体 Upper cylinder；9—塔顶吊柱 Top davit；10—汽提气出口 Stripping gas outlet；11—顶封头 Top head；12—塔顶回流口 Top column reflux hole；13—污水进料口 Sewage feed inlet；14—人孔 Manhole；16—重沸器接口 Reboiler hole；17—检修口 Access hole；18—地脚螺栓座 Anchor bolt seat

9 其它化工设备 Other chemical equipment

9.1 空分 Air separation

空分是利用深度冷冻的物理方法使空气液化，从空气中分离出氧、氮等气体。

Air separation is a facility using physical method of deep-freezing to liqudify the air and to separate the oxygen, nitrogen and other gases from the air.

9.1.1 塔设备 Column

9.1.1.1 双高精馏塔 Double-height rectifying column

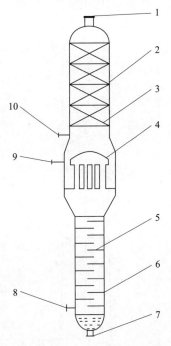

图 9.1.1 双高精馏塔结构图 Structure drawing of double-height rectifying column
1—氮气出口 Nitrogen outlet；2。6—筒体 Shell；
3—规整填料 Structured packing；4—冷凝蒸发器 Evaporator-condenser；
5—塔板 Tray；7—排液口 Bottom column fluid-discharge tube；
8—空气入口 Air inlet；
9—接管 Connecting pipe；
10—氧气出口 Oxygen outlet

9.1.1.2 单高精馏塔 Single-height rectifying column

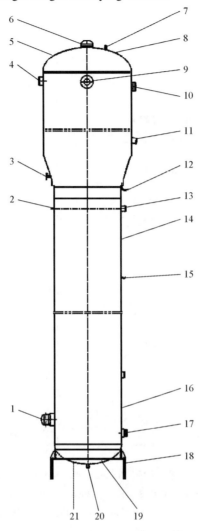

图 9.1.2 单高精馏塔结构图 Structure drawing of single-height rectifying column
1—空气进口管 Air inlet；
2、15、19—分析检测管 Analysis and detection tube；
3—液空分析及下塔液位接管 Liquid air analysis and bottom column liquid level connection；
4—氮气出口 Nitrogen outlet；5—封头 Head；6—液空蒸汽出口 Liquid air steam outlet；
7—不凝汽出口管 Non-condensing steam outlet；
8—液空压力及液面计 Liquid air pressure and liquid level gauge upper connection；
9—吊耳 Lifting lug；10—液空进口 Liquid and air inlet；11—液氮出口管 Liquid nitrogen outlet；
12、17—液空出口管 Liquid air outlet；13—液氮回流管 Liquid nitrogen return pipe；14—筒体 Shell；
16—液面计及阻力计接管 Liqucd level gauge and resistance meter connection；
18—裙座 Skirt；20—液空吹除管 Liquid air blow off pipe；21—液面计接管 Liqud level gauge connection

· 799 ·

9.1.1.3 空冷塔 Air cooling tower

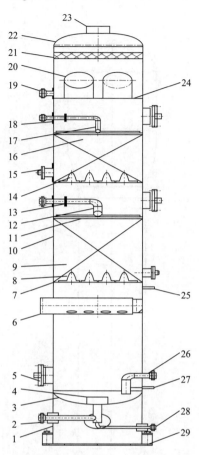

图 9.1.3 空冷塔结构图 Structure drawing of air cooling tower

1—裙座 Skirt；2—出水管 Water outlet；3—下封头 Bottom head；4—防涡流器 Vortex breaker；5—人孔 Manhole；6—进气管 Air inlet；7—支承环 Support ring；8、14—填料支承板 Packing support plate；9—散堆填料(阶梯环) Random Packing(Cascade ring)；10—筒体 Shell；11—填料限位格栅 Packing limit grille；12—冷却水分布器 Cooling water distributor；13—冷却水管 Cooling water pipe；15—手孔 Hand hole；16—散堆填料(共轭环) Random Packing(Conjugate ring)；17—冷冻水分布器 Chilled water distributor；18—冷冻水管 Chilled water pipe；19—溢流管 Overflow pipe；20—弯头 Elbow；21—丝网除沫器 Wire mesh demister；22—上封头 Upper head；23—空气出口 Air outlet pipe；24—圆板 Plate；25、27—液位计接管 Liquid level connection；26—回流管 Return pipe；28—排污管 Drain outlet；29—地脚螺栓 Anchor bolt

9 其他化工设备 Other chemical equipment

9.1.1.4 水冷塔 Water-cooling column

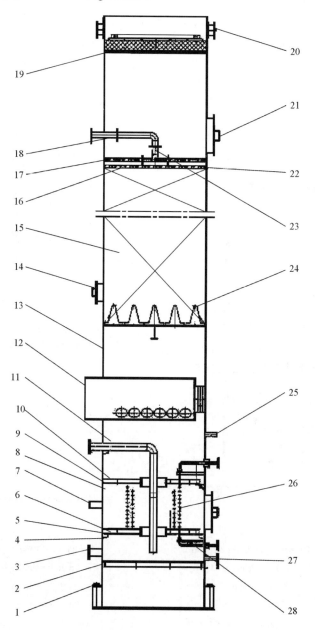

图9.1.4 水冷塔结构图 Structure drawing of water-cooling column

1—地脚螺栓 Anchor bolt；2—平封头 Flat head；3、27—接管 Connecting pipe；4—支板 Support plate；5、17—支承环 Support ring；6、10—支架 Support；7—铭牌 Nameplate；8、9、28—支承角钢 Angle steel；11—溢流管 Overflow pipe；12—进气管 Air inlet；13—筒体 Shell；14—手孔 Hand hole；15—散堆填料（共轭环）Random Packing(Conjugate ring)；16—填料限位格栅 Packing limit grille；18—冷却水管 Chilled water pipe；19—丝网除沫器 Wire mesh demister；20—吊耳 Lifting lug；21—人孔 Manhole；22—保温支承圈 Insulation branch palm circle；23—冷却水分布器 Chilled water distributor；24—填料支承板 Packing support plate；25—液位计接管 Liquid level measurement hole；26—换热管 Heat exchange tube

9.1.2 净化设备 Purification equipment

9.1.2.1 单层立式吸附器 Monolayer vertical adsorber
9.1.2.2 双层立式吸附器 Double layer vertical adsorber

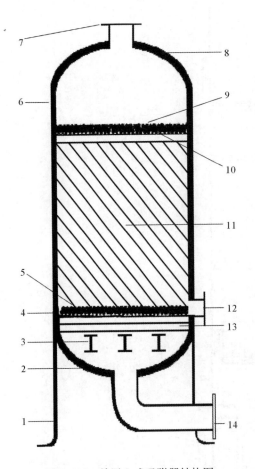
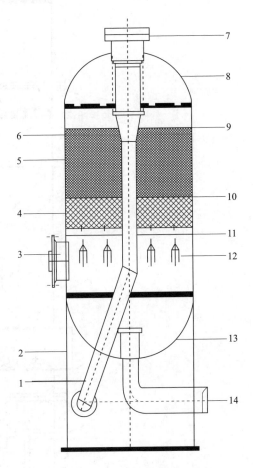

图 9.1.5 单层立式吸附器结构图
Structure drawing of monolayer vertical adsorber
1—裙座 Skirt；2—下封头 Bottom head；3—支承横梁 Support beam；4—支承网 Support network；5、9—隔离粗滤料 Isolation coarse filter material；6—筒体 Shell；7—进口管/装填口 Inlet pipe/ filling opening；8—上封头 Upper head；10—筛网 Screen cloth；11—分子筛吸附剂 Molecular sieve adsorbent；12—卸料口 Discharge outlet；13—支承栅 Support grid；14—出口管 Outlet pipe

图 9.1.6 双层立式吸附器结构图
Structure drawing of double layer vertical adsorber
1—空气出口/再生气入口 Air outlet/ Regeneration air inlet；2—裙座 Skirt；3—人孔 Manhole；4—活性氧化铝 Activated aluminium oxide；5—分子筛吸附剂 Molecular sieve adsorbent；6—筒体 Shell；7—加料口 Charging opening；8—上封头 Upper head；9—丝网 Screen mesh；10—中间隔网 Interval filter；11—支承栅 Support grid；12—支承横梁 Support beam；13—下封头 bottom head；14—空气入口/再生气出口 Air inlet/ Regeneration air outlet

9 其他化工设备 Other chemical equipment

9.1.2.3 单层卧式吸附器 Monolayer horizontal adsorber

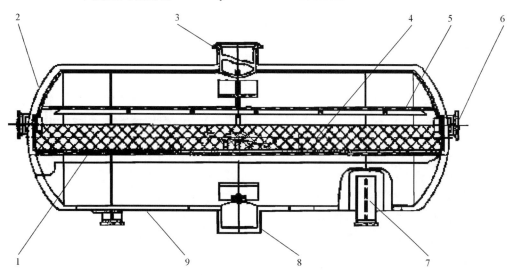

图 9.1.7 单层卧式吸附器结构图 Structure drawing of monolayer horizontal adsorber
1—支承栅 Support grid；2—封头 Head；3—空气出口/再生气进口 Air outlet/ Regeneration air inlet pipe；
4—分子筛吸附剂 Molecular sieve adsorbent；5—丝网 Screen mesh；6—人孔 Manhole；
7—支座 Support；8—空气进口/再生气出口 Air inlet/ Regeneration air outlet pipe；9—筒体 Shell

9.1.2.4 双层卧式吸附器 Double layer horizontal adsorber

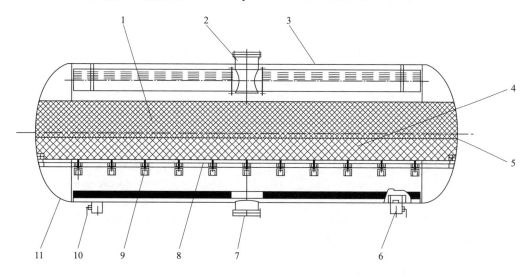

图 9.1.8 双层卧式吸附器结构图 Structure drawing of double layer horizontal adsorber
1—分子筛吸附剂 Molecular sieve adsorbent；
2—空气出口/再生气进口 Air outlet/ Regeneration air inlet pipe；
3—筒体 Shell；4—活性氧化铝 Activated aluminium oxide；5—中间隔网 Interval filter；
6—支座 Support；7—空气进口/再生气出口 Air inlet/ Regeneration air outlet pipe；
8—支承栅 Support grid；9—支承横梁 Support beam；
10—静电接地板 Electrostatic ground plate；11—封头 Head

9.1.3 空气过滤器 Air filter

9.1.3.1 自洁式空气过滤器 Self-cleaning air filter
9.1.3.2 袋式过滤器 Bag filter

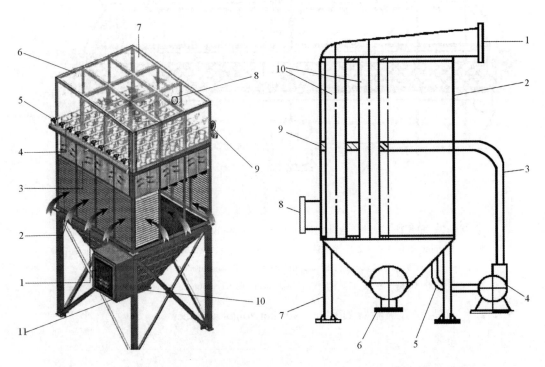

图9.1.9 自洁式空气过滤器外形图
Structure drawing of self-cleaning air filter
1—空气吸入口 Air suction inlet；
2—预过滤段 Pre filtering section；
3—滤筒 Filter cartridge；
4—文氏管 Venturi-tube；
5—储气筒 Air reservoir；
6—空气排出口 Air outlet；
7—净气室 Filtration chamber；
8—高精度微压差控制器 High precision micro differential pressure controller；
9—反吹电池阀 Back-flushing battery valve；
10—卸灰口 Discharge door；
11—控制器 Controller

图9.1.10 袋式过滤器结构图
Structure drawing of bag filter
1—入口管 Inlet pipe；
2—箱体 Tank；
3—风机出口管 Fan outlet pipe；
4—鼓风机 Air blower；
5—风机入口管 Fan inlet pipe；
6—放灰口 Dust discharging opening；
7—支腿 Support；
8—出口管 Outlet pipe；
9—吹刷环 Blow and brush ring；
10—过滤袋 Filter bag

9 其他化工设备 Other chemical equipment

9.1.4 加热器 The heater

9.1.4.1 空气加热器 Air heater

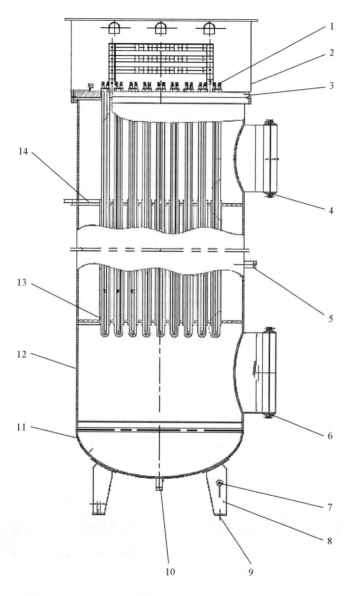

图 9.1.11 空气加热器结构图 Structure drawing of air heater
1—接线柱 Binding post；2—防雨罩 Rain cover；
3—螺栓 Bolt；4—再生气入口 Regeneration air inlet；
5—测温口 Temperature measuring port；
6—再生气出口 Regeneration air outlet；
7—接地板 Answer the floor；
8—支座 Support；9—地脚螺栓 Anchor bolt；
10—排污口 Drain outlet；11—封头 Head；
12—筒体 Shell；13—电热管束 Electric heating tube bundle；
14—支承板 Support plate

9.1.4.2 蒸汽加热器 Steam heater

1. 双管板式加热器 Double tube plate type heater

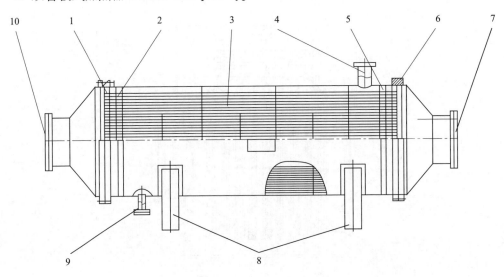

图 9.1.12 双管板式加热器结构图 Structure drawing of double tube plate type heater

1、6—外管板 Outer tube plate；2、5—内管板 Inner tube plate；
3—管束 Tube bundle；4—蒸汽入口管 Steam inlet pipe；
7—再生气出口 Regeneration air outlet；
8—支座 Support；
9—饱和水出口 Saturated water outlet；10—再生气入口 Gas inlet

2. U形管蒸汽加热器 U tube Steam heater

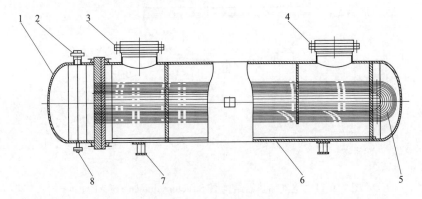

图 9.1.13 U形管蒸汽加热器结构图 Structure drawing of U tube Steam heater

1—管箱 Channel；2—蒸汽入口 Steam inlet；
3—再生气出口管 Regeneration air outlet pipe；
4—再生气进口管 Regeneration air inlet pipe；
5—管束 Tube bundle；6—筒体 Shell；7—支座 Support；
8—饱和水出口 Saturated water outlet

9.1.5 汽化器 Vaporizer

9.1.5.1 空浴式汽化器 Air bath type vaporizer

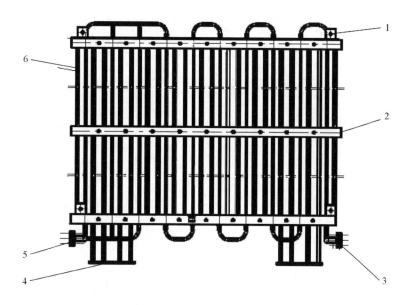

图 9.1.14　空浴式汽化器结构图
Structure drawing of air bath type vaporizer

1—吊耳 Lifting lug；
2—固定板 Dead plate；
3—出口 Outlet；
4—支座 Support；
5—入口 Inlet；
6—翅片管 Finned tube

9.1.5.2 水浴式汽化器 Water bath type vaporizer

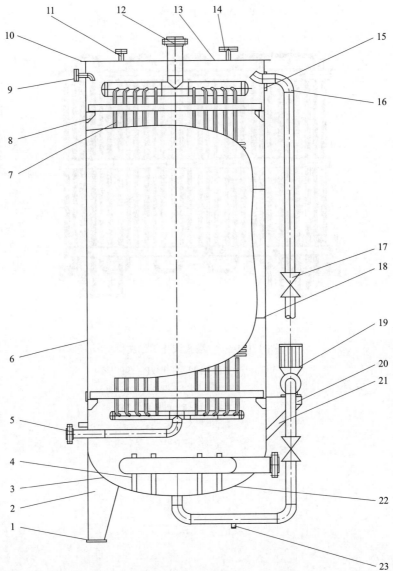

图 9.1.15 水浴式汽化器结构图 Structure drawing of water bath type vaporizer

1—地脚螺栓 Anchor bolt；2—支腿 Support；3、22—封头 Head；
4—U 形抱箍 U embracing hoop；5—进液集合管 Liquid feed collection tube；
6—筒体 Shell；7—盘管 Coil；8—支承板 Support plate；
9—溢流管 Overflow pipe；10—法兰环 Flange ring；
11—温度计接口 Thermometer connection；
12—出气集合管 Gas outlet collection tube；13—盖板 Cover plate；
14—放空管 Blow-down pipe；15—液位计接口 Liquid level gauge interface；
16—循环水管 Circulating pipe；17—阀 Valve；18—加强板 Reinforcing plate；
19—管道泵 Pipeline pump；20—泵支承 Pump support；21—支承角钢 Angle steel；
23—排污管 Drain outlet

9 其他化工设备 Other chemical equipment

9.1.6 储罐 Storage tanks

9.1.6.1 常压液体储罐 Ordinary pressure liquid storage tanks

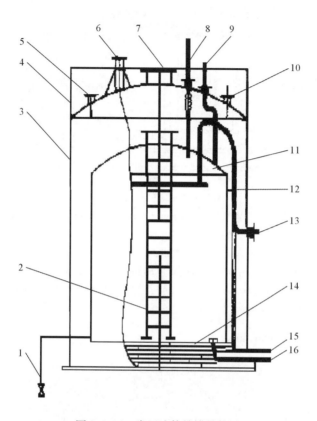

图 9.1.16 常压液体储罐结构图
Structure drawing of ordinary pressure liquid storage tanks

1—装卸车管线 Loading and unloading vehicle pipeline；
2—内槽检修梯子 Maintenance ladder；
3—外槽 Outside slot；4—围栏 Fence；
5、10—装填口 Loading port；
6—外槽安全阀接口 Outside slot safety valve pipe；
7—人孔 Manhole；
8—内槽安全阀口 Internal slot safety valve pipe；
9—内槽压力调节管 Internal slot pressure regulator tube；
11—液位计上接管 Level gauge upper tube；
12—内槽测温点 Internal slot temperature measuring point；
13—液体入口管 Liquid inlet pipe；
14—液位计下接管 Level gauge bottom tube；
15—液体回流管 Liquid return pipe；
16—液体外供管 Liquid outer supply pipe

9.1.6.2 真空液体储罐 Vacuum liquid storage tanks

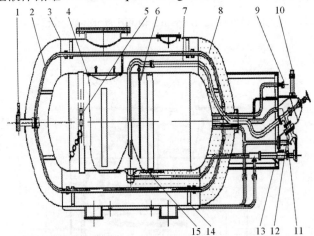

图 9.1.17 真空液体储罐结构图 Structure drawing of vacuum liquid storage tanks

1—真空阀 Vacuum valve；2—抽真空管 Vacuum tube group；3—外筒体 Outer container；4—内筒体 Inner container；5—悬吊带 Suspender equipment；6—进液管 Liquid inlet tube；7—排液管 Liquid inlet and outlet pipe；8—绝热材料 Heat insulating material；9—安全膜 Safety Film；10—安全阀 Safety valve；11—液体出口阀 Liquid outlet valve；12—气体吹除阀 Gas blowoff valve；13—增压阀 Pressure increasing valve；14—U 形蒸发器 U-type evaporator；15—吸附器 Adsorber

9.1.6.3 球罐 Spherical tank

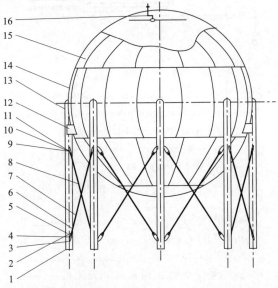

图 9.1.18 球罐结构图 Structure drawing of spherical tank

1—基础板 Foundation plate；2—支板 Support plate；3—下支耳 Bottom lifting lug；4—翼板 Wing panel；5—支柱 Supporting column；6—下拉杆 Lower link；7—松紧节 Elastic section；8—上拉杆 Upper link；9—上支耳 Upper lifting lug；10—销轴 Hinge pin；11—销 Pin；12—托板 Support plate；13—支柱垫板 Ball cover plate；14—赤道板 Ball plate；15—温带板 Ball plate；16—人孔 Manhole

9.1.6.4 立式气体储罐 Vertical gas storage tank

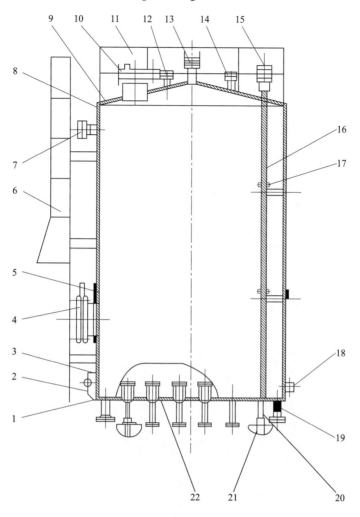

图 9.1.19 立式气体储罐结构图 Structure drawing of vertical gas storage tank

1—底板 Base plate；2—尾部吊耳 Rear lug；
3—罐壁 Tank shell；4—人孔 Manhole；
5—补强圈 Reinforcement ring；6—直梯 Vertical ladder；
7—出口管 Outlet pipe；8—包边角钢 Covered edge angle steel；
9—锥形罐顶 Conical tank roof；
10—罐顶人孔 Tank roof manhole；
11—护栏 Guardrail；12—备用管口 Reserve pipe orifice；
13—测试口 Test mouth；14—罐顶管口 Tank roof nozzle；
15—液位测量口 Level gauge tube；
16—插入管 Insert tube；17—导向管 Guide tube；
18—接地板 Grounding electrode；
19—排放口 Discharge outlet；20—支座 Support；
21—地脚螺栓 Anchor bolt；22—支承螺栓 Stay bolt

9.1.7 低温泵 Cryogenic pump

9.1.7.1 液氮泵 Liquid nitrogen pump

1. 离心式液氮泵 Centrifugal liquid nitrogen pump

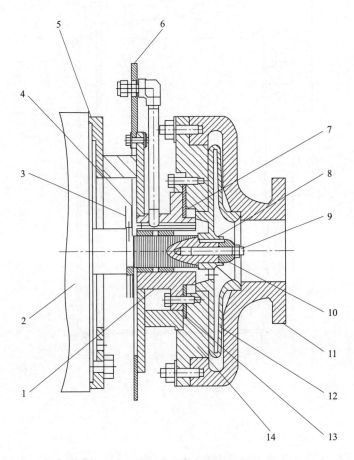

图 9.1.20 离心式液氮泵结构图 Structure drawing of centrifugal liquid nitrogen pump

1—迷宫密封 Labyrinth seal assembly；
2—电机 Motor；3—甩液环 Oil slinger；
4—O 形密封圈 O sealing ring；5—中间体 Intermediate part；
6—隔热板 Thermal baffle；
7—机械密封 Mechanical seal；
8—键 Key；
9—叶轮锁紧螺母 Impeller lock nut；
10—叶轮压盖 Impeller necking；
11—蜗壳 Volute；
12—叶轮 Wheel；
13—绝热板 Insulating plank；
14—泵盖 Pump cover

2. 活塞式液氮泵 Piston type liquid nitrogen pump

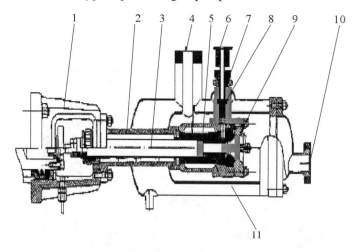

图 9.1.21　活塞式液氮泵结构图 Structure drawing of piston type liquid nitrogen pump

1—泵头支承 Pump head support；2—密封腔 Sealer；
3—活塞杆 Piston rod；4—排气口 Exhaust port；
5—缸套 Cylinder sleeve；6—排液口 Drainage interface；
7—泵体 Pump body；8—排出阀 Discharge valve；
9—吸入阀 Suction valve；10—进液口 Liquid inlet interface；
11—真空夹套 Vacuum jacket

9.1.7.2　液氧泵 Liquid oxygen pump

1. 活塞式液氧泵 Piston type liquid oxygen pump

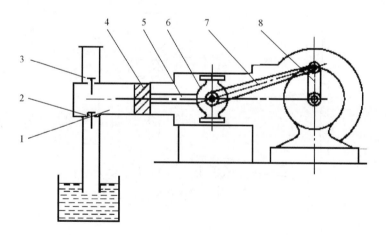

图 9.1.22　活塞式液氧泵外形图 Structure drawing of piston type liquid oxygen pump

1—泵缸 Working cylinder；2—吸入阀 Suction valve；
3—排出阀 Discharge valve；4—活塞 Piston；
5—活塞杆 Piston rod；6—十字头 Crosshead；
7—连杆 Connecting rod；8—曲轴 Crank shaft

2. 柱塞式液氧泵 Plunger type liquid oxygen pump

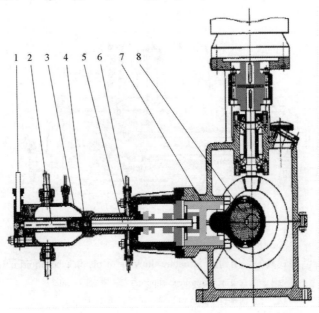

图 9.1.23 柱塞式液氧泵结构图 Structure drawing of plunger type liquid oxygen pump
1—出口阀 Outlet valve；2—前柱塞 Upper plunger；3—泵体 Pump body；
4—缸套 Cylinder sleeve；5—推杆 Push rod；6—密封腔 Sealer；7—十字头 Crosshead；8—凸轮 Cam

3. 离心式液氧泵 Centrifugal liquid oxygen pump

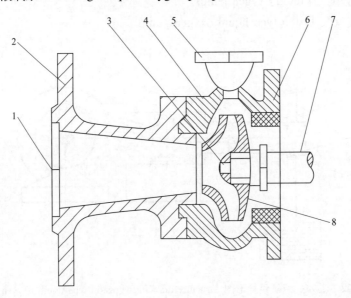

图 9.1.24 离心式液氧泵结构图 Structure drawing of centrifugal liquid oxygen pump
1—入口管 Inlet pipe；2—入口法兰 flange；3—叶轮锁紧螺母 Impeller nut；
4—泵壳 Pump case；5—出口管 Outlet pipe；6—泵盖 Rear cover；
7—主轴 Shaft；8—叶轮 Wheel

9 其他化工设备 Other chemical equipment

9.1.8 压缩机 Compressor

9.1.8.1 原料压缩机 Feed gas compressor

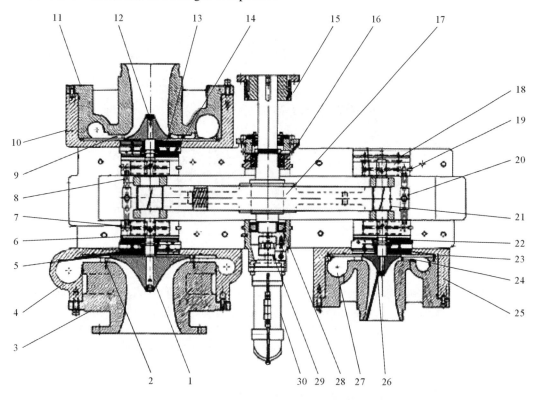

图 9.1.25 原料压缩机结构图 Structure drawing of feed gas compressor

1——级叶轮 First stage impeller；2——级扩压器 First stage diffuser；
3——级机壳 First stage machine shell；4——级蜗壳 First stage volute；
5、13、23—气封 Air seal；6、14、22—油封 Oil seal；7、18—滑动轴承 Pendulum block bearing；
8、21—小齿轮 small gear wheel；9—二级扩压器 Second stage diffuser；
10—二级机壳 Second stage machine shell；
11—二级蜗壳 Second stage volute；12—二级叶轮 Second stage impeller；
15—主轴联轴器 Main shaft coupling；16—止推轴承 Thrust bearing；
17—大齿轮 Big gear wheel；19—定位销 Locating pin；20—油路分配器 Oil-way distributor；
24—三级扩压器 Third stage diffuser；25—三级机壳 Third stage machine shell；
26—三级叶轮 Third stage impeller；27—三级蜗壳 Third stage volute；
28—主轴轴承 Main shaft bearing；
29—主油泵联轴器 Main oil pump coupling；
30—主油泵 Main oil pump

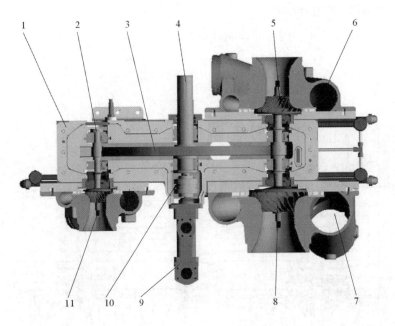

图 9.1.26 原料压缩机水平剖面结构图 Horizontal section structure drawing of feed gas compressor
1—轴承箱 Bearing box；2—三级高速轴 Third stage high speed shaft；3—大齿轮 Big gear wheel；4—低速轴 Low speed shaft；5—二级叶轮 Second stage impeller；6—二级蜗室 Second stage volute chamber；7——级蜗室 First stage volute chamber；8——级叶轮 First stage impeller；9—主油泵 Main oil pump；10—主油泵联轴器 Main oil pump coupling；11—三级叶轮 Third stage impeller

9.1.8.2 螺杆式压缩机 Screw-type compressor

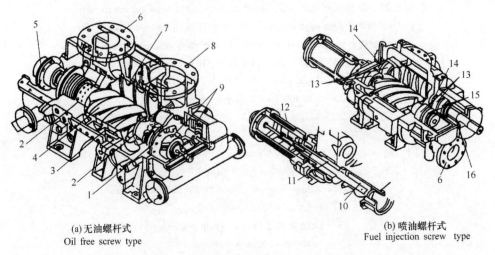

(a) 无油螺杆式 Oil free screw type

(b) 喷油螺杆式 Fuel injection screw type

图 9.1.27 螺杆式压缩机 Screw compressor
1—轴承 Bearing；2—轴封 Shaft seal；3—壳体 Casing；4—阳螺杆 Male screw；5—联轴器 Coupling；6—出口法兰 Out-let flange；7—阴螺杆 Female screw；8—进口法兰 Inlet flange；9—同步齿轮 Synchronous gear；10—滑阀 Slide valve；11—可变阀 Variable valve；12—油缸 Fuel tank；13—径向轴承 Radial bearing；14—平衡活塞 Balance piston；15—止推轴承 Thrust bearing；16—机械密封 Mechanical seal

9 其他化工设备 Other chemical equipment

9.1.8.3 往复式压缩机 Reciprocating compressor

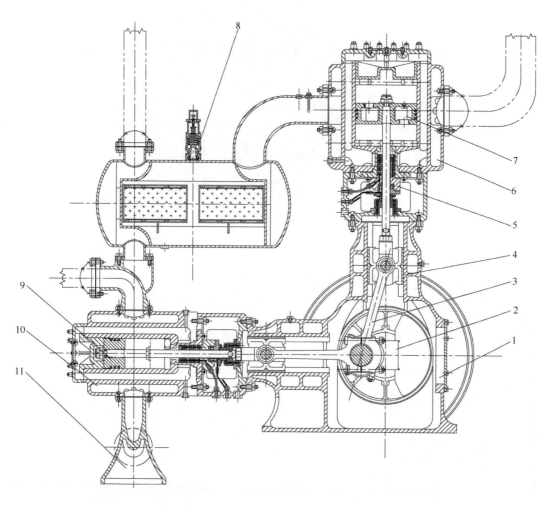

图 9.1.28 往复式压缩机结构图 Structure of reciprocating compressor

1—机身 Body；2—曲轴 Crankshaft components；3—连杆 Connecting rod parts；4—十字头 Crosshead；5—填料密封 Packing；6——级气缸 Primary cylinder；7——级活塞 Primary piston；8—安全阀 Safety valve；9—二级活塞 Secondary piston；10—二级气缸 Secondary cylinder；11—二级排气管 Secondary exhaust pipe

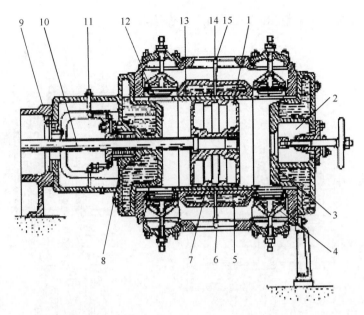

图 9.1.29　往复式压缩机气缸结构图 Structure drawing of reciprocating compressor cylinder

1—活塞 Piston；2—气缸端盖余隙腔 Cylinder end cover clearance；
3—气缸盖 Cylinder head；4—排气阀 Vent valve；5—气缸 Cylinder；
6—气体出口 Air outlet；7—活塞环 Piston ring；8—填料密封 Padding；
9—中间体 Connecting crank throw；10—活塞杆 Piston rod；11—吹扫口 Purging hole；
12—吸气阀 Inlet valve；13—气缸衬套 Cylinder bush；14—气体入口 Air inlet；15—活塞注油口 Piston pouring orifice

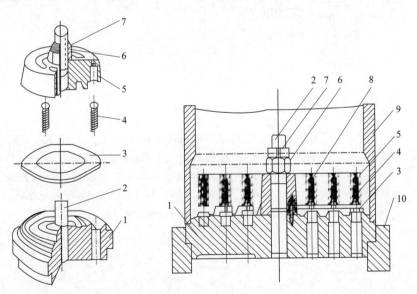

图 9.1.30　往复式压缩机气阀结构图 Structure drawing of reciprocating compressor air valve

1—阀座 Valve seat；2—连接螺栓 Cnnecting bolt；3—阀片 Valve block；
4—弹簧 Spring；5—升程限制器 Lift limiter；6—螺母 Nut；7—开口销 Cotter pin；
8—定位销 Locating pin；9—气阀罩 Pinch device；10—气缸 Cylinder

9 其他化工设备 Other chemical equipment

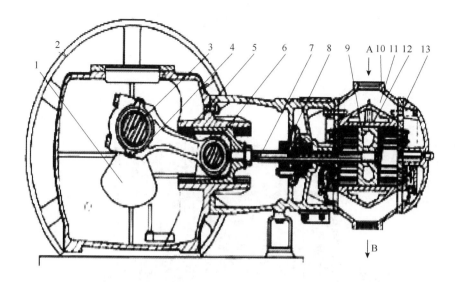

图 9.1.31 往复式压缩机曲轴箱结构图 Structure drawing of reciprocating compressor crankcase
1—平衡配重 Balance weight；2—飞轮 Flywheel；3—曲轴 Crank shaft；
4—连杆 Connecting rod；5—十字头滑道 Crosshead guide；
6—十字头 Crosshead；7—活塞杆 Piston rod；8—填料密封 Padding；
9—活塞 Piston；10—气缸 Air cylinder；11—空气通道 Air duct；
12—冷却水夹套 Cooling water jacket；13—气缸盖 Cylinder head；
A—气体进口 Gas inlet；B—气体出口 Gas outlet

9.1.9 冷水机组 Cold water unit

9.1.9.1 螺杆冷水机组 Screw cold water unit

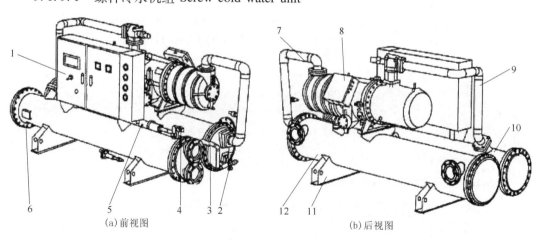

图 9.1.32 螺杆冷水机组外形图 Shape of screw chilled water unit
1—电控箱 Electric cabinet；2—热力膨胀阀 Thermostatic expansion valve；
3—视液镜 Level glass；4—电磁阀 Solenoid valve；5—干燥过滤器 Drying and filtering device；
6—易熔塞 Fusible plug；7—回气管 Muffler；8—压缩机 Compressor；9—排气管 Exhaust pipe；
10—冷凝器 Condensator；11—机组支座 Unit support；12—蒸发器 Evaporator

9.1.9.2 活塞冷水机组 Piston chilled water unit

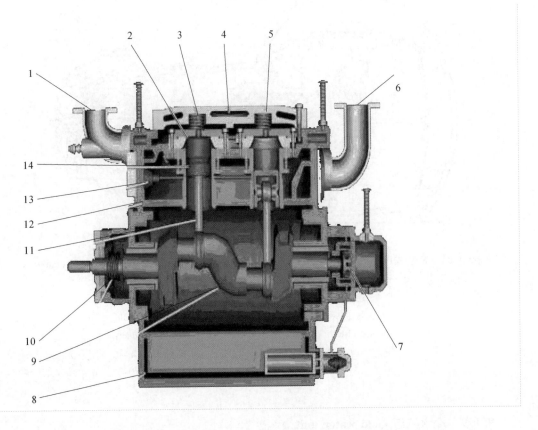

图 9.1.33 活塞冷水机组结构图 Structure drawing of piston chilled water unit
1—排气管 Exhaust pipe;
2—气缸套及进排气阀 Cylinder sleeve and intake and exhaust valve assembly;
3—缓冲弹簧 Buffer spring; 4—水套 Water jacket;
5—气缸盖 Cylinder head; 6—进气管 Intake-tube;
7—油泵 Oil pump; 8—油箱 Bearing housing;
9—曲轴 Crank shaft; 10—轴封 Shaft seal;
11—连杆 Connecting rod; 12—进气腔 Air chamber;
13—油压推杆机构 Hydraulic push rod mechanism;
14—活塞 Piston

9.1.10 膨胀机 Expansion machine

9.1.10.1 增压透平膨胀机 Supercharged turbo expander

9 其他化工设备 Other chemical equipment

1. 增压透平膨胀机外形 Shape of supercharged turbo expander

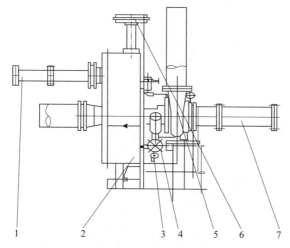

图9.1.34 增压透平膨胀机外形图 Shape of supercharged turbo expander

1—膨胀机过滤器 Expander filter；2—保冷箱 Cold insulation box；3—法兰 Flange；4—滤油器 Oil filter；5—增压透平膨胀机 Supercharged turbo expander；6—气动薄膜执行机构 Pneumatically operated diaphragm actuator；7—增压机过滤器 Supercharger filter

2. 增压透平膨胀机结构 Structure drawing of supercharged turbo expander

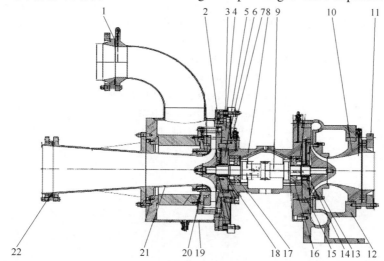

图9.1.35 增压透平膨胀机结构图 Structure drawing of supercharged turbo expander

1—入口管 Inlet pipe；2—可调喷嘴 Adjustable nozzle；3—垫板 Base plate；4—前隔板 Upper clapboard；5—前密封隔板 Front seal clapboard；6、7—保冷板 Cold insulation plate；8—机身 Frame；9—转子 Rotor；10—法兰 Flange；11—肩圈 Shoulder ring；12—增压机蜗壳 Supercharger volute；13—增压机叶轮密封盖 Supercharger seal cover；14—后密封套 Rear seal cartridge；15—后隔板 Rear clapboard；16—后轴承及后挡油环 Rear bearing and rear oil retainer；17—前轴承及前挡油环 Front bearing and front oil retainer；18—前密封套 Front seal cartridge；19—透平蜗壳 Turbine volute；20—工作轮密封盖 Running wheel cover gasket；21—扩压室 Diffusion chamber；22—出口管 Outlet pipe

· 821 ·

9.1.10.2 发电机制动膨胀机 Generator brake expander

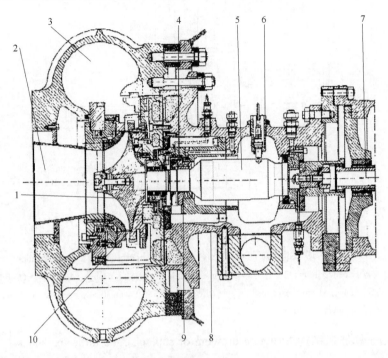

图 9.1.36 发电机制动膨胀机结构图 Structure drawing of generator brake expander

1—叶轮 Wheel；
2—膨胀端出口（扩压管）Expanding outlet(diffuser pipe)；
3—蜗室 Cochlear room；4—密封 Seal；
5—主轴 Shaft；6—转数探头 Revolutions probe；
7—发电机端 Generator end；8—中间体 Intermediate；
9—轴承座 Bearing seat；10—喷嘴 Nozzle

9.1.10.3 鼓风机制动膨胀机 Blower brake expander

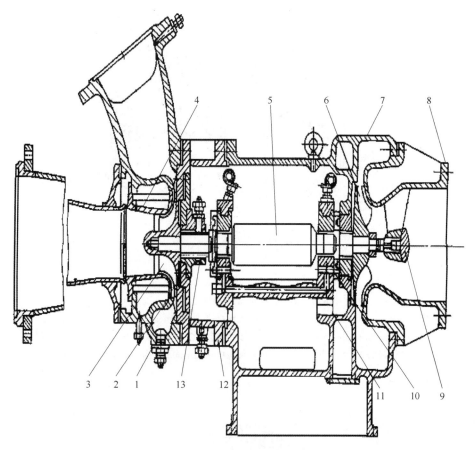

图 9.1.37 鼓风机制动膨胀机结构图
Structure drawing of blower brake expander

1—蜗壳 Volute；2—喷嘴 Nozzle；
3—工作轮 Running wheel；4—扩压器 Diffuser；
5—主轴 Shaft；6—风机叶轮 Fan impeller；
7—风机蜗壳 Fan volute；
8—风机端盖 Fan end cap；
9—转速计 Tachometer；
10—轴承座 Bearing seat；
11—机身 Frame；
12—中间体 Intermediate；
13—密封 Seal

9.1.10.4 油制动膨胀机 Oil brake expander

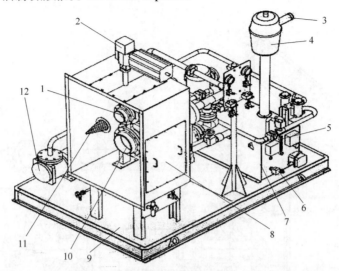

图9.1.38 油制动膨胀机外形图 Structure drawing of oil brake expander

1—膨胀机入口 Expander inlet；2—喷嘴调整执行机构 Nozzle adjustment actuators；
3—油雾器放空口 Atomized lubricator venting；4—除油雾过滤器 Demister filter；5—油箱液位计 Tank gauge；
6—油加热器 Oil heater；7—油箱 Oil tank；8—保冷箱 Cold insulation box；9—底座 Plate；
10—膨胀机出口 Expander outlet；11—膨胀机入口过滤网 Expander inlet filter screen；12—油冷却器 Oil cooler

9.1.10.5 活塞式膨胀机 Piston type expander

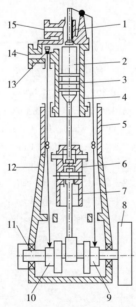

图9.1.39 活塞式膨胀机结构图 Structure drawing of piston type expander

1—排气阀 Vent valve；2—活塞 Piston；3—活塞环 Piston ring；4—气缸 Air cylinder；5—中间座 Intermediate seat；
6—十字头 Crosshead；7—连杆 Connecting rod；8—飞轮 Flywheel；9—排气凸轮 Exhaust cam；10—进气凸轮 Intake cam；
11—曲轴 Crankshaft；12—机身 Frame；13—进气阀 Inlet valve；14—进气口 Air inlet；15—排气口 Exhaust port

9 其他化工设备 Other chemical equipment

9.2 双氧水 Hydrogen peroxide

9.2.1 塔设备 Column equipment

9.2.1.1 萃取塔 Extraction column

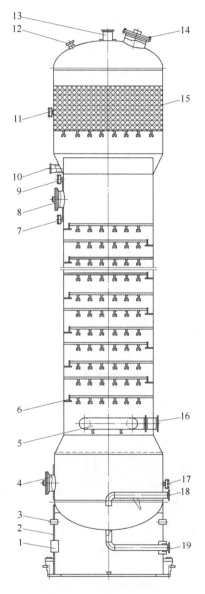

图 9.2.1. 萃取塔结构图 Structure drawing of extraction column

1—检修口 Access hole；2—裙座 Skrit；3—通气口 Air vent；4、8、14—人孔 Manhole；
5—环形管 Ring pipe；6—筛板 Sieve plate；7、9—液位计接口 Liquid level gauge interface；
10—排液口 Liquid outlet；11、17—测温口 Temperature measurement port；12—氮气入口 Nitrogen inlet；
13—放空口 Vent；15—填料 Padding；16—介质入口 Medium inlet；
18—双氧水出口 Hydrogen peroxide outlet；19—排放口 Discharge outlet

9.2.1.2 氧化塔 Oxidizing column

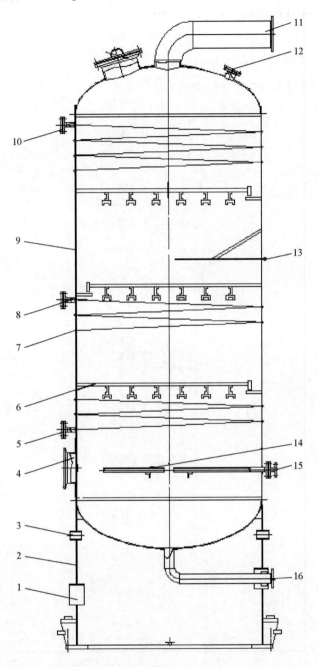

9.2.2 氧化塔结构图 Structure drawing of oxidizing column

1—检修口 Access hole；2—裙座 Skrit；3—通气口 Air vent；4—氢化液入口 Hydrogenated liquid inlet；
5—冷却水入口 Cooling water inlet；6—塔盘 Tray；7—伴热管 Heating pipe；
8、10—冷却水出口 Cooling water outlet；9—筒体 Barrel shell；11—气相出口 Gas phrase outlet；
12—安全阀接口 Safety valve interface；13、14—温度计套管 Thermometer nozzle；
15—分布器 Distributor；16—排放口 Discharge outlet

9 其他化工设备 Other chemical equipment

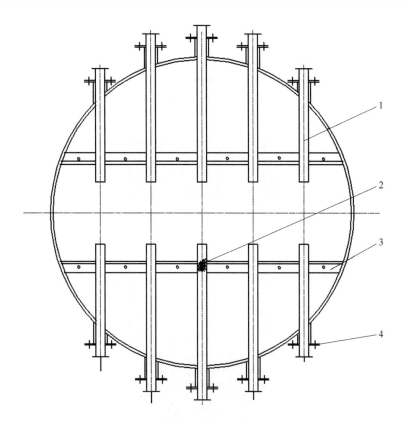

图 9.2.3 氧化塔分布器结构图 Structure drawing of oxidation tower distributor
1—空气分散管 Air scattered tube；2—钢丝网气孔 Wire mesh gas hole；
3—支承角钢 Angle steel；4—接管 Connecting pipe

9.2.1.3 碱塔 Alkali column

碱塔也称干燥塔，萃余液从塔底进入，被塔盘和填料分散后向塔顶漂浮，与塔中的碳酸钾溶液接触，以除去萃余液中的水分、中和酸和分解双氧水。

The alkali column is also known as the drying column. The raffinate passes from the bottom of the column, and floats to the top of the column when dispersed by the tray and the padding, then touches the potassium carbonate solution in the column in order to remove the water of the raffinate, neutralize the acid and decompose the hydrogen peroxide solution.

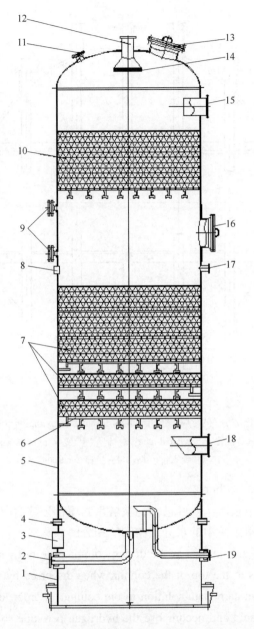

图 9.2.4 碱塔结构图 Structure drawing of alkali column
1—裙座 Skirt；2—排污口 Drain outlet；3—检修口 Access hole；4—通气口 Air vent；
5—筒体 Shell；6—塔盘 Tray；7、10—填料 Padding；8—视镜 Sight glass；
9—液位计接口 Liquid level gauge interface；
11—氢气入口 Hydrogen inlet；12—放空口 Vent；13、16—人孔 Manhole；
14—除沫器 Demister；15—工作液出口 Working solution outlet；
17—碳酸钾入口 Potassium carbonate inlet；18—工作液入口 Working solution inlet；
19—碳酸钾出口 Potassium carbonate outlet

9 其他化工设备 Other chemical equipment

9.2.1.4 氢化塔 Hydrogenation column

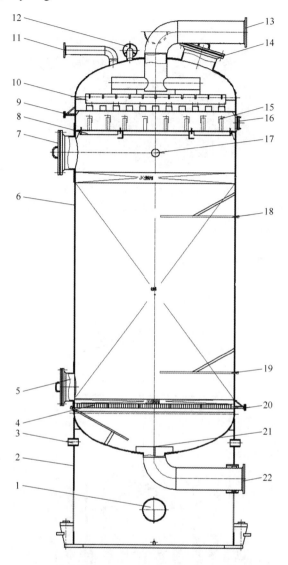

图 9.2.5 氢化塔结构图 Structure drawing of hydrogenation column

1—检修口 Access hole；2—裙座 Skrit；3—通气口 Air vent；
4—支承栅板 Support grid；5、7、14—人孔 Manhole；6—筒体 Shell；
8—支承 Support；9、20—压力表接口 Pressure gauge port；
10—液体分布器 Liquid distributor；11—接管 Connecting pipe；
12—放空口 Vent；13—气液入口 Gas and liquid inlet；
15—气液再分布器 Gas and liquid distributor；16、17—视镜 Sight glass；
18、19—温度计接口 Thermometer mouth；21—防涡器 Vortex breaker；
22—氢化液出口 Hydrogenated liquid outlet

· 829 ·

图 9.2.6 氢化塔液体分布器结构图
Structure drawing of hydrogenation column liquid distributor
1—顶盖 Top cover；2—加强筋 Stiffener；3—接管 Connecting pipe；4—筛板 Sieve plate

9.2.2 过滤设备 Filtration equipment

9.2.2.1 袋式过滤器 Bag filter

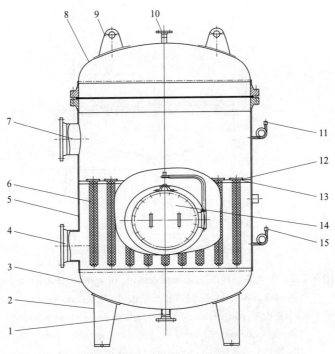

图 9.2.7 袋式过滤器结构图 Structure drawing of bag filter
1—排放口 Empting mouth；2—支座 Support；
3—下封头 Bottom head；4—物料出口 Feed outlet；
5—筒体 Shell；6—滤袋 Filter bag；7—物料入口 Material inlet；
8—上封头 Upper head；9—吊耳 Lifting lug；10—放空口 Vent；
11、15—压力计接口 Pressure gauge port；
12—芯筒 Core-tube；13—孔板 Tube nest；14—人孔 Manhole

9 其他化工设备 Other chemical equipment

9.2.2.2 金属烧结过滤器 Metal sintered filter

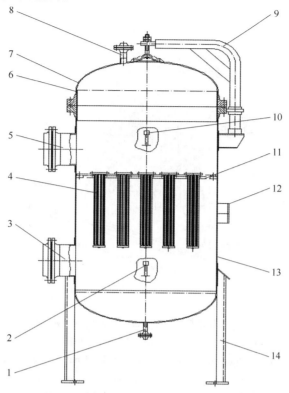

图 9.2.8 金属烧结过滤器结构图 Structure drawing of metal sintered filter

1—排污口 Drain outlet；2、10—压力计接口 Pressure gauge port；3—空气进口 Gas inlet；
4—烧结滤芯 Sinter filter element；5—空气出口 Gas outlet；6—顶盖 Top cover；7—封头 Head；
8—放空口 Vent；9—吊柱 Davit；11—孔板 Tube nest；12—铭牌 Nameplate；13—筒体 Shell；14—支座 Support

9.2.3 换热设备 Heat-exchange equipment

9.2.3.1 工作液预热器 Working liquid preheater

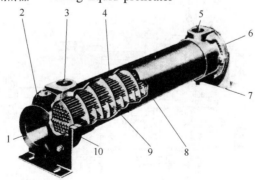

图 9.2.9 工作液预热器结构图 Structure drawing of working liquid preheater

1、5—流体出口 Fluid outlet；2—管箱 Channel；3、6—流体进口 Fluid inlet；4—换热管 Heat exchange tube；
7—支座 Support；8—筒体 Shell；
9—折流板 Baffle plate；10—管板 Tube sheet

· 831 ·

9.2.3.2 氢化冷却器 Hydrogenated cooler

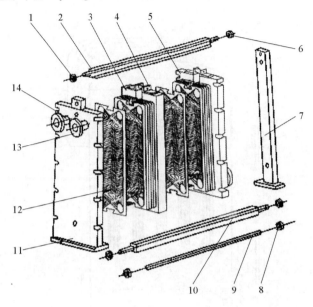

图 9.2.10 氢化冷却器结构图 Structure drawing of hydrogenated cooler

1、6、8—螺母 Nut；2—上导杆 Upper guide rod；3—中间隔板 Intermediate bulkhead；
4—滚动机构 Rolling mechanism；5—活动压紧板 Mobile compaction plate；
7—立柱 Column；9—夹柱螺柱 Clamping column stud；
10—下导杆 Lower guide rod；11—固定压紧板 Fixed pressure plate；
12—垫片 Gasket；13—法兰 Flange；14—接管 Connecting pipe

9.2.4 泵 Pump

9.2.4.1 氢化液泵 Hydrogenated liquid pump

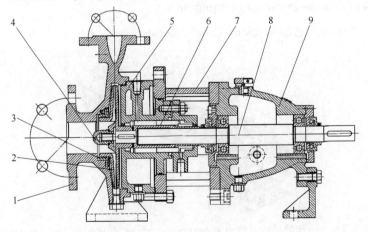

图 9.2.11 氢化液泵结构图 Structure drawing of hydrogenated liquid pump

1—泵壳 Pump case；2—叶轮 Wheel；3—密封环 Seal ring；4—叶轮锁紧螺母 Impeller locking nut；
5—泵盖 Pump cover；6—机械密封 Mechanical seal；7—中间体 Intermediate support；
8—泵轴 Pump shaft；9—轴承箱 Bearing box

9 其他化工设备 Other chemical equipment

9.2.4.2 磷酸计量泵 Phosphate measurement pump

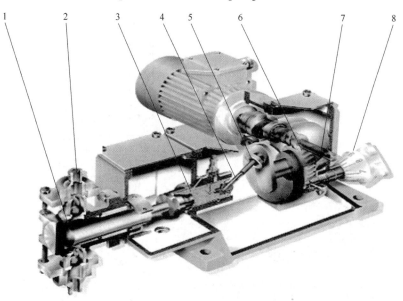

图 9.2.12　磷酸计量泵外形图 Shape of phosphate measurement pump

1—柱塞 Plunger；2—单向阀 Check valve；
3—滑块 Sliding block；4—连杆 Connecting rod；
5—曲柄 Crank；6—蜗轮 Worm wheel；
7—蜗杆 Worm rod；8—调节手柄 Regulating handle

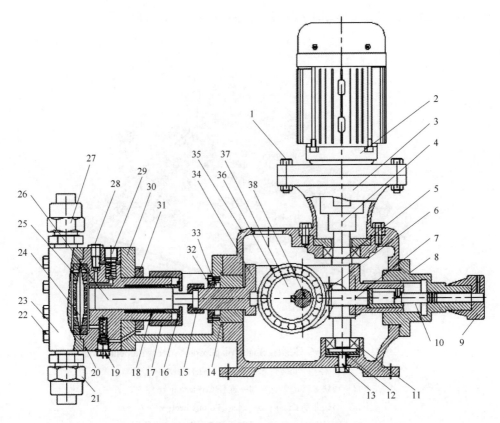

图 9.2.13 磷酸计量泵结构图 Structure drawing of phosphate measurement pump

1—螺栓 Bolt；2—电机 Motor；3—电机座 Motor base；

4—轴 Shaft；5—蜗杆 Worm rod；

6—蜗轮 Worm wheel；7—顶杆 Mandril；8—调节器 Regulator；

9—手轮 Hand wheel；10—调节杆 Regulating stem；

11—轴承 Bearing；12—调节底板 Adjusting baseplate；

13—调节螺栓 Adjusting screw；14—密封圈 Seal ring；

15—连接螺母 Coupling nut；16—填料压板 Padding clamping plate；

17—填料压紧螺母 Padding cap；18—填料 Padding；

19—补油阀 Oil supplementary valve；20—泵头底座 Pump head base；

21—进水阀 Inlet valve；22—泵头螺栓 Pump head bolt；

23—泵头 Pump head；24—膜片 Membrane；

25—膜片底座 Diaphragm base；26—柱塞 Plunger；

27—排水阀 Outlet valve；28—放气螺栓 Air bleed screw；

29—过载阀 Overload valve；30—泵头连接头 Pump head connector；

31—圆螺母 Round nuts；32—油封压板 Oil seal clamping plate；

33—油封 Oil seal；34—连杆 Connecting rod；

35—泵体 Pump body；36—滚轮 Roller；37—偏心轮 Eccentric wheel；

38—偏心轴 Eccentric axle

9 其他化工设备 Other chemical equipment

9.2.4.3 双氧水自吸泵 Hydrogen peroxide self-priming pump

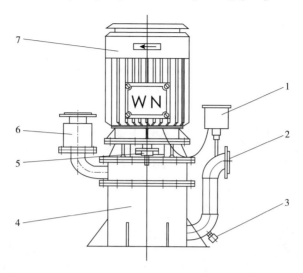

图 9.2.14 双氧水自吸泵结构图 Structure drawing of hydrogen peroxide self-priming pump
1—电磁通气阀 Electromagnetic breather valve;
2—泵入口 Pump inlet; 3—排放口 Discharge outlet; 4—底座 Base;
5—加液口 Filling hole; 6—泵出口 Pump outlet; 7—电机 Motor

9.2.4.4 潜水泵 Submersible pump

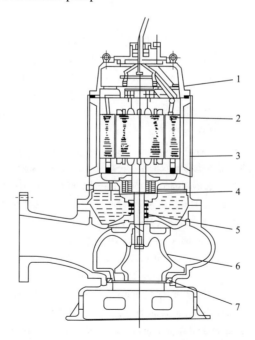

图 9.2.15 潜水泵结构图 Structure drawing of submersible pump
1—泵头护罩 Pump head shield; 2—电机转子 Motor rotor;
3—电机护罩 Motor shield; 4—主轴 Shaft; 5—机械密封 Mechanical seal;
6—叶轮 Wheel; 7—密封环 Seal ring

9.2.5 压缩机 Compressor

9.2.5.1 空气压缩机 Air compressor

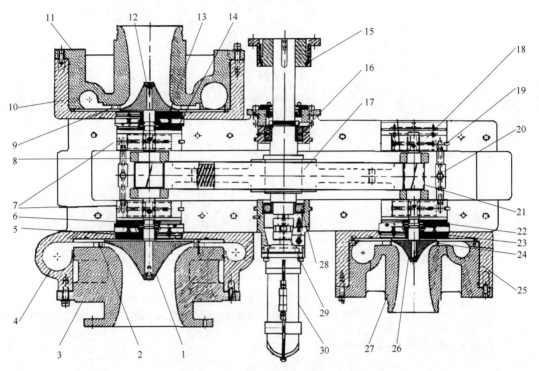

图 9.2.16 空气压缩机结构图 Structure drawing of air compressor

1——级叶轮 First stage impeller；2——级静叶扩压器 First stage static vane diffuser；
3——级机壳 First stage machine body；4——级蜗壳 First stage volute；
5——级气封 First stage air seal；6——级油封 First stage oil seal；
7—滑动轴承 Sliding bearing；8—小齿轮 Low speed pinion；
9—二级静叶扩压器 Second stage static vane diffuser；
10—二级蜗壳 Second stage volute；11—二级机壳 Second stage machine body；
12—二级叶轮 Second impeller；13—二级气封 Second stage air seal；
14—二级油封 Second stage oil seal；15—主轴联轴器 Main coupling；
16—主轴止推轴承 Main shaft thrust bearing；17—主轴 Shaft；
18—三级摆块轴承 Third stage swing block bearing；
19—半月型定位销 Half moon locating pin；20—油路分配器 Oil passage distributor；
21—小齿轮轴 Pinion shaft；22—三级油封 Third stage oil seal；
23—三级气封 Third stage air seal；
24—三级静叶扩压器 Third stage static vane diffuser；
25—三级机壳 Third stage machine body；26—三级叶轮 Third stage impeller；
27—三级蜗壳 Third stage volute；28—主轴轴承 Main shaft bearing；
29—主油泵联轴器 Main oil pump coupling；30—主油泵 Main oil pump

9.2.5.2 尾气膨胀机 Exhaust expansion machine

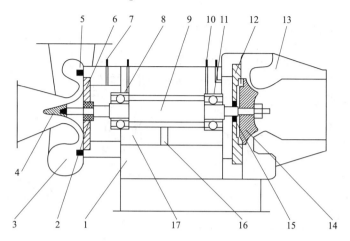

图 9.2.17 尾气膨胀机结构图 Structure drawing of exhaust expansion machine
1—机座 Machine base；2—蜗轮端衬套 Worm wheel end bushing；
3—蜗壳 Volute；4—蜗轮 Worm wheel；5—喷嘴环 Nozzle ring；
6—蜗轮隔板 Worm wheel clapboard；7—密封气进口 Seal gas inlet；
8、11—轴承 Bearing；9—轴 Shaft；10—润滑油进口 Lubricating oil inlet；
12—压气机隔板 Compressor clapboard；
13—压气机蜗壳 Compressor worm house；
14—压气机叶轮 Compressor impeller；
15—压气机端衬套 Compressor end bushing；
16—油气出口 Oil gas outlet；17—机体 Compressor body

9.2.6 尾气吸附设备 Oxidized exhaust gas adsorption equipment

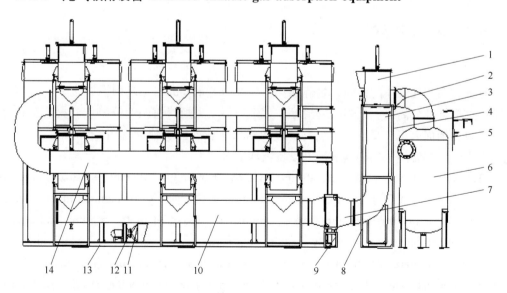

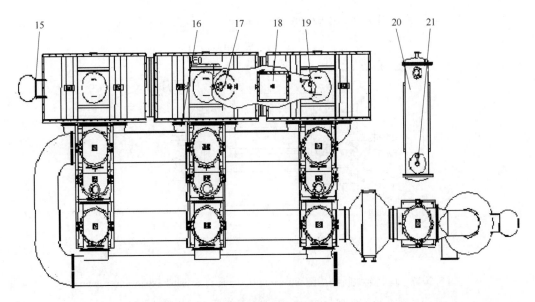

图 9.2.18 尾气吸附设备外形图 Shape of oxidation tail gas adsorption equipment
1—排气三通挡板阀 Exhaust tee-junction baffle valve；2—变径弯管 Variable diameter elbow；
3—连接管 Joint pipe；4—排气三通阀支架 Exhaust tee-junction valve support；
5—缓冲罐梯子 Buffer tank ladder；6—缓冲罐 Buffer tank；7—表冷器 Surface air cooler；
8—冷凝器支架 Condenser support；9—表冷器支架 Surface air cooler support；10、14—进气管 Intake tube；
11—干燥风机过滤器 Dry air blower filter；12—防爆风机 Blast blower；
13—主台架 Master frame；15—吸附塔人梯 Adsorption column ladder；
16—进气口垫 Air inlet gasket；17—储槽 Tank；18—分层槽 Layered slot；
19—螺旋板换热器 Spiral plate heat exchanger；20—碳罐 Carbon tank；21—冷凝器 Condenser

9.3 火炬 Flare

火炬系统是在炼油和石油化工装置中，安全、有效地处理事故或正常生产中排放的易燃、有毒、腐蚀性气体的设施，是炼油化工生产装置重要的安全环保设施之一。火炬系统通常由火炬气液分离罐、火炬气密封罐、火炬筒体、火炬管道四部分组成。火炬型式分为高空火炬和地面火炬。高空火炬由筒体、火炬头、长明灯、辅助燃料系统、高空点火器、地面点火器、感温热电偶和其他辅助设备组成。

Flare system is the facility of safe and effective handling of flammable, toxic, corrosive gas in the accident or in the normal production in the refinery or petrochemical plant, which is the important facility for the safety and environmental protection of the refining and chemical production plant. Flare system usually contains four parts, including the flare gas and liquid separation tank, the flare gas seal pot, the flare barrel and the flare pipeline. The type of the flare contains the aerial flare and the ground flare. The aerial flare contains the barrel, the flare tip, the ever-burning lamps, the auxiliary fuel system, the aerial ignition, the ground ignition, the thermal thermocouples and other auxiliary equipment.

9 其他化工设备 Other chemical equipment

9.3.1 火炬 Flare

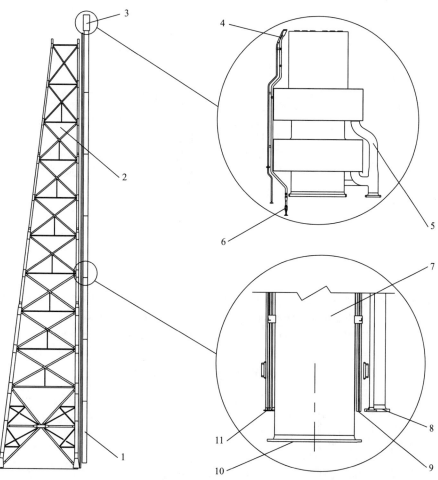

图 9.3.1 火炬结构图 Structure drawing of self-discharging bundled the flare
1—火炬筒体 Flare barrel；2—塔架 Tower；3—火炬头 Flare tip；
4—长明灯 Ever-burning lamps；5—消烟蒸汽管 Smoke-elimination steam；
6—高空电点火装置 Aerial electric ignition device；7—分段筒体 Subsection barrel；
8—蒸汽管 Steam pipeline；9—仪表电缆线 Instrument cable；
10—分段筒体连接法兰 Subsection barrel connecting flange；
11—长明灯燃料气管 Ever-burning lamps fuel gas pipeline

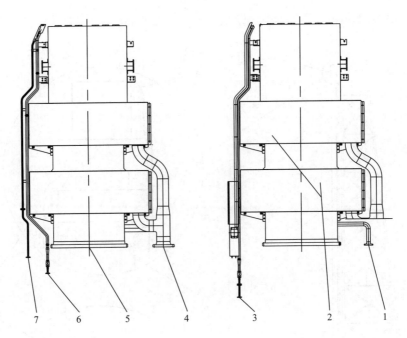

图 9.3.2 新型蒸汽消烟型火炬头结构图 Structure drawing of new type of the steam smoke-elimination flare tip
1—中心蒸汽 Center steam；2—高效盆式消音器 Efficient basin muffler；
3—点火枪燃料气管线 Firing gun fuel gas pipeline；4—引射蒸汽 Ejector steam；
5—火炬气通道 Flare gas channel；6—长明灯燃料气管线 Ever-burning lamps fuel gas pipeline；
7—内传焰点火器管线 Smuggled flame igniter pipeline

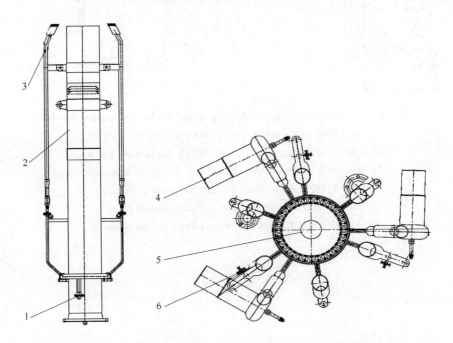

图 9.3.3 扩散燃烧型火炬头结构图 Structure drawing of diffusion combustion type of flare tip
1—伴烧器入口 Burning device inlet；2—火炬头 Flare tip；3、6—长明灯 Ever-burning lamps；
4—高能点火器 High-energy igniter；5—伴烧器 Tracing Burning device

9.3.2 火炬气分液罐 Flare gas knock-out drum

每根火炬排放气总管应设分液罐，用以分离气体夹带的液滴或可能发生的两相流中的液相。

Every collecting pipeline of the flare exhaust gas should set the knock-out drum, which separates the liquid droplets entrained by the gas or the liquid phase of the contingent two-phase flow.

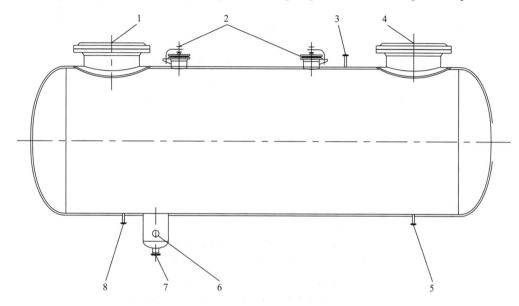

图 9.3.4 火炬气分液罐结构图 Structure drawing of flare gas knock-out drum

1—火炬气入口 Flare gas inlet; 2—人孔 Manhole; 3—放空口 Vent; 4—火炬气出口 Flare gas outlet; 5—伴热蒸汽出口 Tracing steam outlet; 6—凝液出口 Condensate outlet; 7—排污口 Sewage outlet; 8—伴热蒸汽入口 tracing steam inlet

9.3.3 火炬气水封罐 Flare gas water sealed drum

水封罐是火炬放空系统的安全设备，通过水封将火炬系统和上游装置有效隔离，确保上游装置的绝对安全。

The water sealed drum is the safe equipment of the flare flare system, by which could effectively isolate the flare system from the upstream device, and ensure the absolute safety of the upstream device.

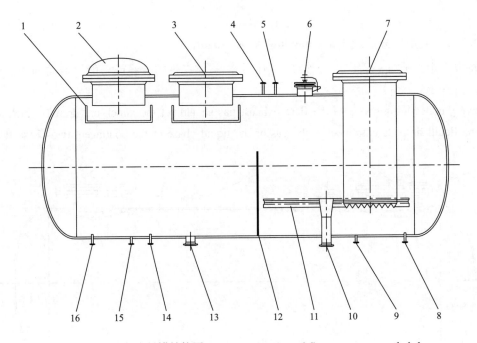

图 9.3.5 火炬气水封罐结构图 Structure drawing of flare gas water sealed drum
1—火炬气出口挡板 Flare gas outlet baffle；
2—火炬气备用出口 Flare gas alternate outlet；
3—火炬气出口 Flare gas outlet；4—放空口 Vent hole；
5—氮气入口 Nitrogen inlet；6—人孔 Manhole；
7—火炬气入口 flare gas inlet；8—补水口 Replenish water hole；
9、15—排污口 Sewage outlet；10—连通器出口 Link vessels outlet；
11—水封孔板 Water seal orifice；12—水封隔板 Water seal diaphragm；
13—连通器入口 Link vessels inlet；14—放水口 Dewatering outlet；
16—二次水封补水口 Secondary water sealed replenish water hole

9.4 装车设备 Truck-loading facility

装车设备分为铁路油罐车装车和公路油罐车装车两种，主要由鹤管、装卸台和相应的油气回收设施组成。

Truck-loading facilities contain the railway loading oil tank car and the highway loading oil tank car, which contain the crane tube, the platform and the corresponding oil and gas recovery facilities.

9 其他化工设备 Other chemical equipment

9.4.1 火车装车设施 Train loading facility

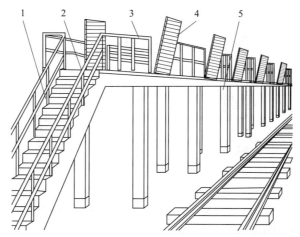

图 9.4.1 铁路栈桥示意图 Structure drawing of Railway bridge

1—斜梯 Inclined ladder；2—作业平台 Work platform；3—保护栏杆 Protective railings；
4—活动渡梯 Mobile ladder；5—立柱 Upright pillar

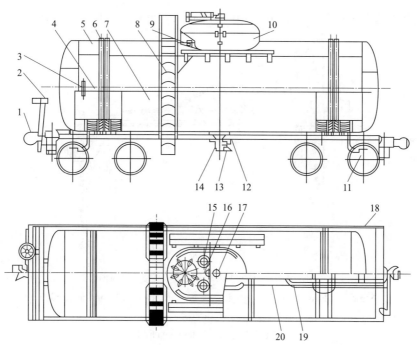

图 9.4.2 黏油罐车结构图 Structure drawing of sticky oil tank car

1—车钩 Car coupler；2—手制动装置 Hand brake device；3、15—出气阀 Gas outlet valve；
4—支架 Support；5—罐体 Tank；6—卡带 Cassette；7—加热套板 Heating jacket plate；
8、20—梯子 Ladder；9—内梯 Manhole ladder；10—罐颈 Tank neck；
11、13—排水阀 Necked-in；12、19—蒸汽管 Steam pipe；14—卸油器 Oil discharge device；
16—阀杆插入口 Valve insert；17—进气阀 Inlet valve；18—底架 Frame

· 843 ·

9.4.2 火车装车鹤管 Train loading crane tube

1 弹簧力矩平衡鹤管 Spring torque balance crane tube

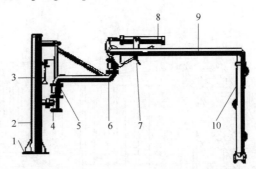

图 9.4.3 弹簧力矩平衡鹤管结构图 Structure drawing of train loading crane tube

1—安装底板 Mounting plate；2—立柱 Upright pillar；
3—内臂锁紧机构 Inner arm locking mechanism；4—法兰接口 Flange connection；
5—回转器 Gyrator；6—内臂 Inner arm；
7—外臂锁紧装置 Outer arm locking mechanism；8—平衡器 Balancer；
9—外臂 Outer arm；10—垂直管 Vertical tube

2 铁路油罐车密闭装油鹤管 Railway oil tank car airtight oil loading crane tube

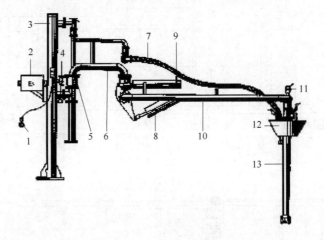

图 9.4.4 铁路油罐车密闭装油鹤管结构图
Structure drawing of railway oil tank car airtight oil loading crane tube

1—气源总阀 Air header valves；
2—液位控制箱 Liquid level control box；
3—立柱 Upright；4—内臂锁紧机构 Inner arm locking mechanism；
5—回转器 Gyrator；6—内臂 Inner arm；
7—气相管 Vapor pipe；8—气缸 Air cylinder；9—平衡器 Balancer；
10—外臂 Outer arm；11—滑管卷扬机构 Sliding pipe windlass mechanism；
12—密封盖 Sealed cap；13—垂直管 Vertical tube

9 其他化工设备 Other chemical equipment

9.4.3 汽车装车设施 Auto loading facility

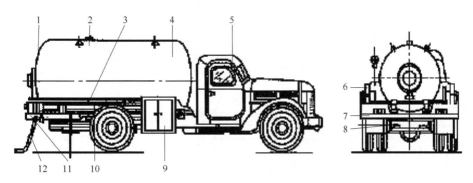

图 9.4.5 液化石油气汽车罐车结构图 Structure drawing of liquefied petroleum gas tank car
1—人孔、液面计 Manhole, liquid level meter；
2—安全阀 Safety valve；3—挡泥板 Mudapron；
4—罐体 Tank；5—驾驶室 Cab；6—干粉灭火器 Dry powder extinguisher；
7—后保险杠 Rear bumper；8—备用轮胎 Spare tire；9—阀门箱 Valve box；
10—走台 Walking board；11—尾灯 Taillight；12—接地带 Earth strip

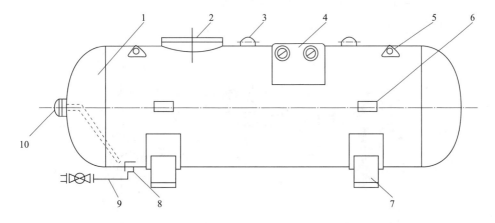

图 9.4.6 活动式槽车罐体结构图 Structure drawing of simple activities type of tank car tanks
1—封头和筒体 Head and cylinder；2—人孔 Manhole；
3—安全阀 Safety valve；4—仪表箱 Instrument box；
5—吊耳 Lifting lug；6—拉杆及固定装置 Tie rod and the fixed device；
7—支座 Support；8—紧急切断阀 Emergency cut-off valve；
9—装卸管 Loading and unloading pipeline；
10—液面计 Liquid level meter

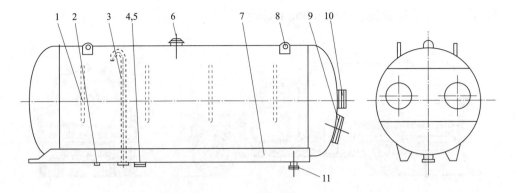

图 9.4.7　固定式汽车槽车罐体结构图 Structure drawing of stationary type of tank car tanks

1—防冲板 Impingement baffle；2—温度计接口 Thermometer connecting pipe；
3—气相管 Vapor pipe；4、5—紧急切断阀接口 Emergency cut-off valve tieing；
6—安全阀接口 Safety valve interface；7—V 行支座 V support；8—吊耳 Lifting lug；
9—人孔 Manhole；10—液位计接口 Liquid level gauge interface；
11—排污口 Sewage hole

9.4.4　汽车装车鹤管 Auto loading crane tube

1. 汽车顶部装车鹤管 Auto top loading crane tube

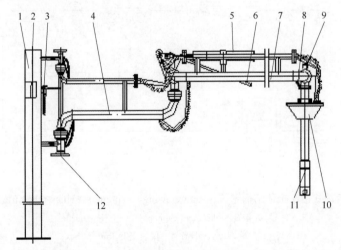

图 9.4.8　汽车顶部装车鹤管结构图 Structure drawing of auto top loading crane tube

1—立柱 Upright pillar；2—铭牌 Nameplate；
3—内臂锁紧器 Inner arm locking；4—内臂 Inner arm；
5—平衡器 Balancer；6—外臂锁紧器 Outer arm locking；
7—外臂 Outer arm；8—导静电带 Electrostatic conduction belt；
9—球阀 Ball valve；10—密封盖 Auto sealing cap；
11—垂直管 Vertical tube；12—入口法兰 Inlet flange

9 其他化工设备 Other chemical equipment

2. 夹套伴热鹤管 Jacket heat tracing crane tube

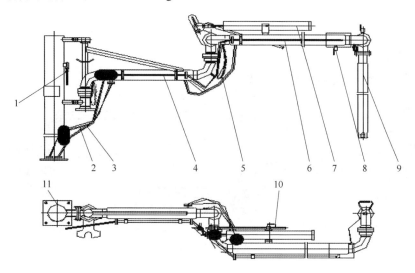

图 9.4.9 夹套伴热鹤管结构图 Structure drawing of jacket heat tracing crane tube

1—内臂锁紧器 Inner arm locking；2—蒸汽回管 Steam return pipe；3—蒸汽入口管 Steam inlet pipe；
4—伴热内臂 Heat tracing inner arm；5—导静电带 Electrostatic conduction belt；6—伴热外臂 Heat tracing outer arm；
7—平衡器 Balancer；8—蒸汽排放阀 Steam outlet valve；9—垂直管 Vertical tube；
10—锁紧杆 Manubrium；11—立柱 Upright pillar

3. 汽车装车鹤管 Auto loading crane tube

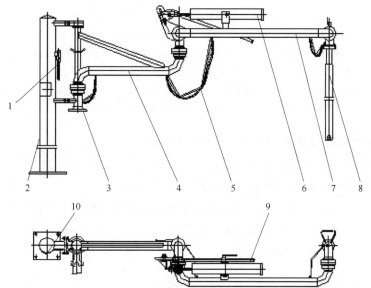

图 9.4.10 汽车装车鹤管结构图 Structure drawing of auto loading crane tube

1—内臂锁紧器 Inner arm locking；2、10—立柱 Upright pillar；3—入口法兰 Inlet flange；4—内臂 Inner arm；
5—导静电带 Electrostatic conduction belt；6—平衡器 Balancer；7—外臂 Outer arm；
8—伸缩垂管 Retractable verticae tube；9—锁紧杆 Locking lever

· 847 ·

4. 液化气底部装车鹤管 Liquefied petroleum gas bottom loading crane tube

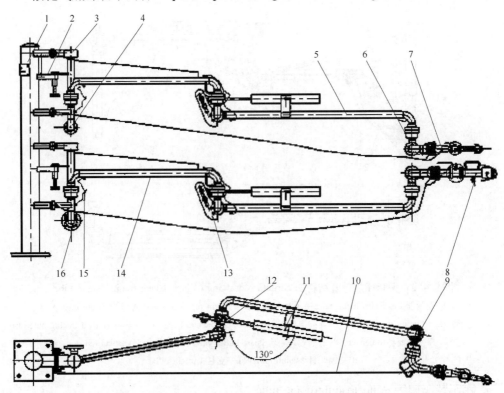

图 9.4.11　液化气底部装车鹤管结构图
Structure drawing of liquefied petroleum gas bottom loading crane tube
1—立柱 Upright pillar；2—内臂锁紧器 Inwall locking；3—轴承座 Bearing block；
4、16—接口法兰 Interface flange；5—外臂 Outer arm；6—回转器 Gyrator；7、8—垂管 Verticae tube；
9—钢把手 Steel handle；10—钢丝绳 Wirerope；11—平衡器 Balancer；12—连接管 Connecting pipe；
13—固位板 Retention plate；14—内臂 Inner arm；15—防静电带 Electrostatic conduction belt

9.4.5　油气回收装置 Vapor recovery equipment

油气回收装置用来收集挥发的汽油油气，使油气从气态变为液态，以解决油气对环境的污染和给安全生产带来的不利影响，并减少油品损失。

The vapor recovery equipment is used for collecting the volatile oil and gas, which makes the oil and gas change from the gaseous to liquid so that reducing the adverse effects on the environmental safety and reducing the oil loss.

9 其他化工设备 Other chemical equipment

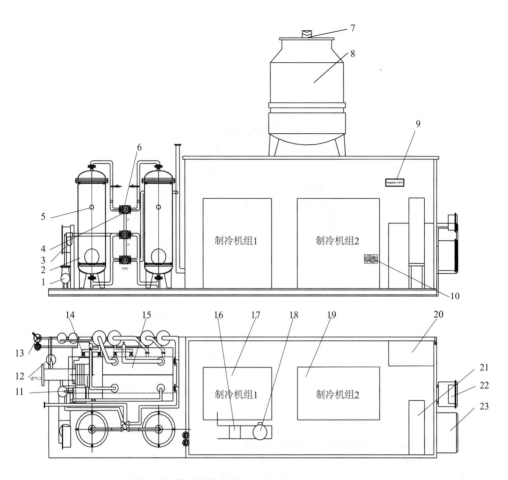

图9.4.12 油气回收装置结构图 Structure drawing of vapor recovery system

1—真空泵 Vacuum pump；
2—碳罐 Carbon container；3—过滤器 Filtrator；4—三通阀 Three-way valve；
5—温度传感器 Temperature sensor；6—电动执行器 Electric actuator；7—冷却塔风扇 Cooling tower fan；
8—冷却塔 Cooling tower；9—防爆标志牌 Explosion-proof sign；10—铭牌 Nameplate；
11—板式换热器 Plate heat exchanger；12—气液分离器 Gas and liquid separator；
13—油气分离器 Oil and gas separator；14—电磁阀 Electromagnetic valve；
15—冷箱 Refrigerated containers；16—离心水泵 Centrifugal water pump；
17、19—制冷机组 1 Refrigeration unit1；18—电子除垢仪 Electronic descaling instrument；
20—动力柜 Power supply cabinet；21—控制柜 Control cabinet；
22—防爆接电箱 Explosion-proof power box；
23—防爆控制柜 Explosion-proof hut control cabinet

9.5 循环水 Circulating water

9.5.1 冷却塔 Cooling column

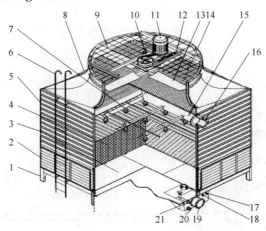

图 9.5.1 冷却塔结构图 Structure drawing of Cooling column

1—支腿 Support；2—进风窗 Air inlet window；3—消声毯 Silencing blanket；4—填料 Loading；5—塔体 Column body；6—扶梯 Staircase；7—喷嘴 Nozzle；8—风筒 Air duct；9—电机网罩 Motor guard；10—减速器 Reducer；11—电机 Motor；12—井架 Derrick；13—风机 Draught fan；14—收水器 Water cleaner；15—配水管 Distribution pipe；16—进水管 Water inlet；17—手动补水管 Manual filling water pipe；18—自动补水管 Automatic filling water pipe；19—出水管 Exhalent siphon；20—排污管 Drain outlet；21—溢流管 Overflow pipe

9.5.2 循环水泵 Circulating water pump

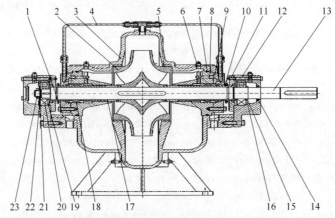

图 9.5.2 循环水泵结构图 Structure drawing of circulating water pump

1—泵体 Pump body；2—叶轮 Wheel；3—泵盖 Pump cover；4—水冲洗管 Water flushing tube；5—密封环 Sealing ring；6—管堵 Casing cap；7—O形密封圈 O sealing ring；8—密封函 Sealing body；9—密封压盖 Sealing gland；10—挡水圈 Water retaining ring；11—轴承压盖 Bearing gland；12—轴承箱 Bearing box；13—轴 Shaft；14—油封 Oil seal；15—弹性挡圈 Circlip；16—轴承挡圈 Bearing retaining ring；17—轴套 Sleeve；18—机械密封 Mechanical seal；19—轴承套 Bearing sleeve；20—轴承 Bearing；21—轴承垫圈 Bearing gasket；22—蝶形弹簧 Butterfly spring；23—锁紧螺母 Locknut

9.5.3 多功能水泵控制阀 Multifunctional water pump control valve

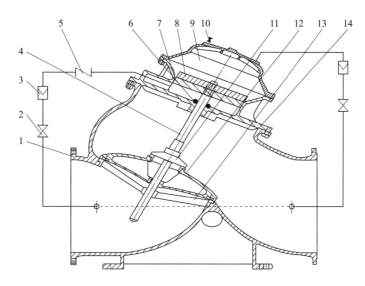

图 9.5.3 多功能水泵控制阀结构图 Structure drawing of multifunctional water pump control valve
1—阀体 Valve body；2—控制阀 Control valve；3—过滤器 Filter；4—阀杆 Valve rod；
5—止回阀 Wee-resistance buffer valve；6—O 形密封圈 O sealing ring；7—膜片 Membrane；
8—膜片压板 Diaphragm pressure plate；9—阀盖 Valve cover；10—排气阀 Vent valve；
11—缓闭阀板 Slowly closing valve plate；12—主阀板 Main valve plate；
13—主阀板座 Main valve plate seat；14—膜片座 Daphragm seat

9.5.4 监测换热器 Monitoring heat exchanger

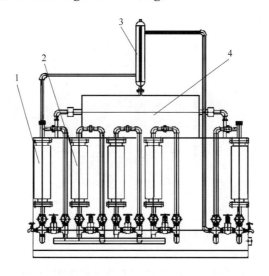

图 9.5.4(a) 监测换热器结构图 Structure drawing of monitoring heat exchanger
1—旁路挂片器 by‑pass hanging device；2—玻璃转子流量计 Glass rotameter；
3—冷凝器 Condenser；4—电加热箱 Electric heating box

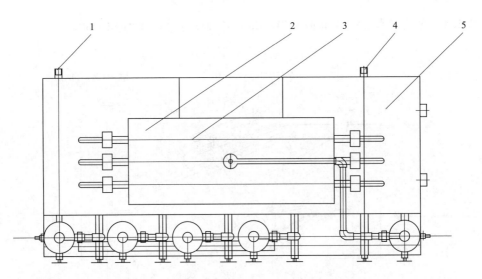

图 9.5.4(b) 监测换热器结构图 Structure drawing of monitoring heat exchanger

1—循环水入口 Circulating water inlet；2—循环水 Circulating water；
3—监测换热管束 Monitoring heat exchange tube bundle；
4—循环水出口 Circulating water outlet；5—机体 Compressor body

9.5.5 纤维球过滤器 Fiber ball filter

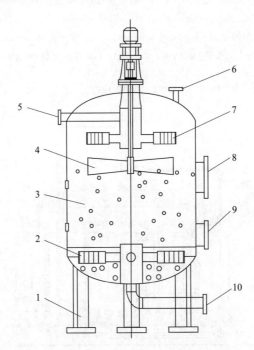

图 9.5.5 纤维球过滤器结构图 Structure drawing of fiber ball filter

1—支座 Support；2、5—配水管 distribution pipe system；3—纤维球 Fiber ball；
4—搅拌机 Agitator；6—滤液进口 Filter liquor inlet；7—搅拌桨叶 Stirring blade；8—人孔 Manhole；
9—出料口 Discharge port；10—滤后水出口 Filtration water outlet

9 其他化工设备 Other chemical equipment

9.5.6 纤维束过滤器 Fiber bundle filter

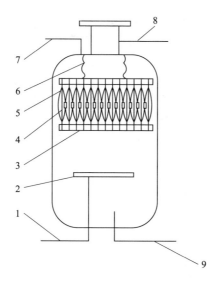

图 9.5.6 纤维束过滤器结构图 Structure drawing of fiber bundle filter
1—进气管 Intake-tube；2—气体分配器 Gas distributor；3—固定孔板 Fixed orifice；
4—滤料 Filter material；5—活动孔板 Mobile orifice；6—限位索 Limit line；
7—排气管 Exhaust pipe；8—原水进口 Raw water inlet；9—净水出口 Clear water outlet

9.6 污水处理 Sewage treatment

9.6.1 臭氧发生器 Ozonizer

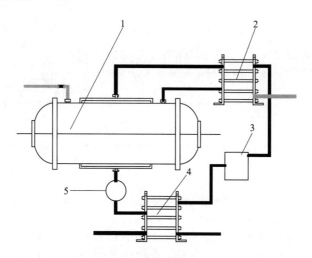

图 9.6.1 臭氧发生器结构图 Structure drawing of ozonizer
1—放电室 Discharge chamber；2、4—热交换器 Heat exchanger；3—调节水箱 Regulating tank；5—水泵 Water pump

9.6.2 气浮机 Flotation machine

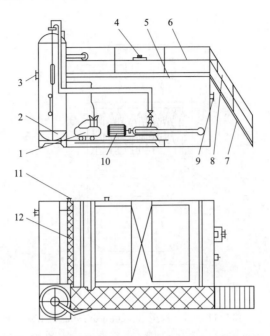

图 9.6.2 气浮机外形图 Shape of flotation machine

1—空压机 Air compressor；2—溶气罐 Dissolved air vessel；3—进水管 Water inlet；
4—刮渣机 Sediment scraper；5、12—操作平台 Operating platform；6、7、8—栏杆 Handrail；
9—出水管 Water outlet；10—溶气泵 Dissolved gas pump；
11—放空管 Blow-down pipe

9.6.3 污泥脱水机 Sludge centrifugal dewatering machine

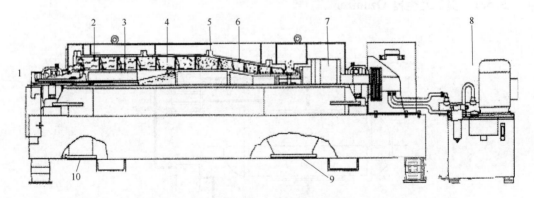

图 9.6.3 污泥脱水机外形图 Shape of sludge centrifugal dewatering machine

1—进料口 Feed inlet；2—澄清区 Settling section；3—螺旋 Spiral；
4—离心压缩区 Centrifugal compression zone；5—转鼓 Drum；
6—双向挤压区 Two-way extrusion area；7—液压马达 Hydraulic motor；
8—液压站 Hydraulic pressure station；9—固相出口 Solid phase outlet；
10—液相出口 Liquid phase outlet

9.6.4 刮泥机 Mud scraper

9.6.4.1 半桥周边传动刮泥机 Half bridge peripheral transmission mud scraper

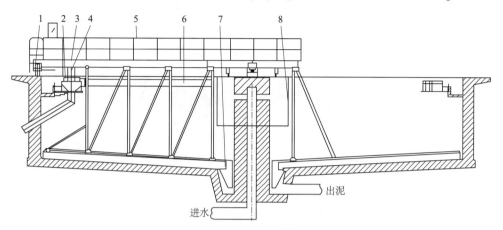

图 9.6.4 半桥周边传动刮泥机结构图
Structure drawing of half bridge peripheral transmission mud scraper

1—驱动装置 Transmission；2—浮渣漏斗 Scum funnel；
3—工作桥 Service bridge；4—浮渣耙板 Scum rake plate；
5—栏杆 Handrail；6—刮集装置 Scraping device；
7—小刮刀 Small scraper；8—稳流桶 Steady flow barrels

9.6.4.2 全桥周边传动刮泥机 Full bridge peripheral transmission mud scraper

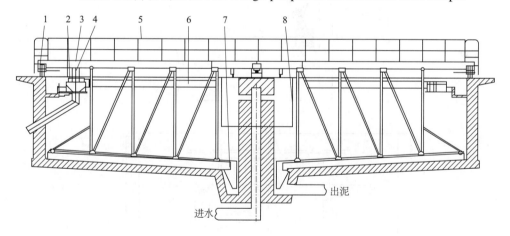

图 9.6.5 全桥周边传动刮泥机结构图
Structure drawing of full bridge peripheral transmission mud scraper

1—驱动装置 Transmission；2—浮渣漏斗 Scum funnel；
3—工作桥 Service bridge；4—浮渣耙板 Scum rake plate；
5—栏杆 Handrail；6—刮集装置 Scraping device；
7—小刮刀 Small scraper；8—稳流桶 Steady flow barrels

9.6.5 潜水搅拌器 Submersible mixer

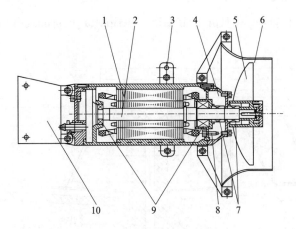

图 9.6.6 潜水搅拌器结构图
Structure drawing of submersible mixer

1—电机 Motor；2—轴 Shaft；3—吊钩 Lift hook；4—油室 Oil room；5—叶轮 Wheel；6—导流罩 Guide cover；
7—机械密封 Mechanical seal；8—漏水探头 Leak probe；9—轴承 Bearing；10—导轨 Guide rail

9.6.6 滗水器 Water decanter

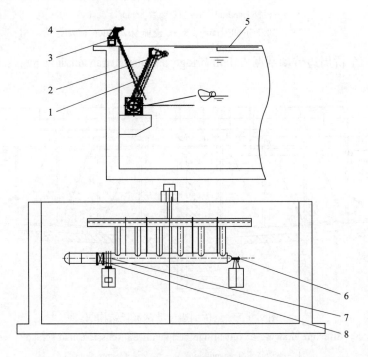

图 9.6.7 滗水器结构图
Structure drawing of water decanter

1—排水装置 Drainage device；2—撇渣浮筒 Skimming buoy；3—电动推杆支架 Electric drive pusher support；
4—电动推杆 Electric drive pusher；5—盖板 Cover plate；6—回转支承 Revolving support；
7—卡板轴承 Board bearing assembly；8—回转密封接头 Revolving hermetical seal

9.6.7 螺旋输送机 Shaftless screw conveyor

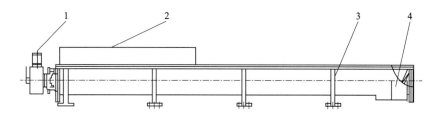

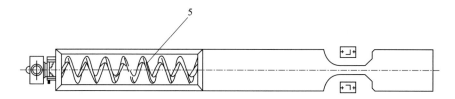

图 9.6.8 螺旋输送机外形图

Structure drawing of shaftless screw conveyor

1—驱动装置 Transmission；2—进料口 Feed inlet；3—支架 Support；
4—出料口 Feed outlet；5—无轴螺旋 Shaftless screw